COFFEE: BOTANY, BIOCHEMISTRY AND PRODUCTION OF BEANS AND BEVERAGE

COFFEE
Botany, Biochemistry and Production of Beans and Beverage

EDITED BY M.N. CLIFFORD AND K.C. WILLSON

CROOM HELM
London & Sydney

AMERICAN EDITION
Published by
THE AVI PUBLISHING COMPANY, INC.
Westport, Connecticut 1985

© 1985 M.N. Clifford and K.C. Willson
Softcover reprint of the hardcover 1st edition 1985

Croom Helm Ltd, Provident House, Burrell Row,
Beckenham, Kent BR3 1AT
Croom Helm Australia Pty Ltd, First Floor, 139 King Street,
Sydney, NSW 2001, Australia

British Library Cataloguing in Publication Data

Coffee: botany, biochemistry and production
   of beans and beverage.
   1. Coffee
   I. Clifford, M.N.   II. Willson, K.C.
   641.3'373   TX415

   ISBN-13: 978-1-4615-6659-5   e-ISBN-13: 978-1-4615-6657-1
   DOI: 10.1007/978-1-4615-6657-1

Published in the United States and dependencies, Canada,
Mexico, Central and South America 1985 by
The AVI Publishing Company, Inc.
250 Post Road East, P.O. Box 831
Westport, Connecticut 06881, USA

Library of Congress Cataloging in Publication Data
Main entry under title:

Coffee: botany, biochemistry, and production of beans
   and beverage.

   Bibliography: p.
   Includes index.
   1. Coffee. 2. Coffee — Processing. 3. Coffee trade.
I. Clifford, M.N. (Michael N.) II. Willson, K.C.
(Ken C.)
SB269.C587   1985   633.7'3   84-24324

CONTENTS

Acknowledgements
Foreword
Preface
List of Contributors

1	A History of Coffee *R. F. Smith*	1
2	Botanical Classification of Coffee *A. Charrier and J. Berthaud*	13
3	Coffee Selection and Breeding *H. A. M. van der Vossen*	48
4	Climate and Soil *K. C. Willson*	97
5	Physiology of the Coffee Crop *M.G.R. Cannell*	108
6	Mineral Nutrition and Fertiliser Needs *K. C. Willson*	135
7	Cultural Methods *K. C. Willson*	157
8	Pest Control *R. Bardner*	208
9	Control of Coffee Diseases *J. Waller*	219
10	Green Coffee Processing *R. J. Clarke*	230
11	World Coffee Trade *C. F. Marshall*	251
12	The Microscopic Structure of the Coffee Bean *Eliane Dentan*	284
13	Chemical and Physical Aspects of Green Coffee and Coffee Products *M. N. Clifford*	305

14	The Technology of Converting Green Coffee into the Beverage *R. J. Clarke*	375
15	The Physiological Effects of Coffee Consumption *K. Bättig*	394
Glossary		440
Index		450

ACKNOWLEDGEMENTS

We would like to thank our contributors for their efforts on our behalf. In this respect, Ron Clarke deserves special mention for producing Chapter 10 at very short notice, indeed at the eleventh hour, when another good friend was unable to deliver his manuscript because of circumstances beyond his control. Similarly, Herbert (Ham) van der Vossen deserves special recognition for his editorial assistance in perfecting the translation of Chapter 2. In addition, we would like to thank colleagues — secretarial, librarian and academic — at the University of Surrey and at the University of Liverpool for their help in many respects, but particularly with the production of our own chapters.

Last, but not least, we would like to thank our wives, Jane and Marjorie and Jane's junior assistants — Helen and Mary, for their support and encouragement and substantial editorial assistance.

M.N. Clifford
K.C. Willson

FOREWORD

We live in an era of constantly accelerating scientific and social change brought about by developments in education, technology and modern communication. This is a time of questioning and new perceptions affecting all facets of our daily lives. With increasing frequency issues are being raised which demand answers and new approaches. This increases the responsibility of those involved in determining the future shape of the world of coffee. The dependence of developing countries on income generated from trade in coffee, the emergence of new processing techniques, health implications and questions of quality of coffee in the cup are among the issues related to coffee. The knowledge required to form the basis to resolve these issues for the benefit of the multitudes of coffee drinkers will be generated only through the systematic build up of information and its subsequent evaluation. Science and modern technology provide essential tools for these endeavours.

This book should act as a stimulant to thought and creativity so the issues facing the industry may be fully analysed and a healthy future for coffee secured. It marks a step forward in laying the foundation for coffee's future.

Alexandre F. Beltrão
Executive Director
International Coffee Organisation
London

PREFACE

We have long been fascinated by coffee and on many occasions bemoaned the lack of a comprehensive text dealing with the varied scientific aspects. With the encouragement of Tim Hardwick of Croom Helm Ltd, we decided to pool our resources and produce just such a multi-author volume.

It has taken over two years and the consumption of some 2,500 litres of coffee brew to complete this task. We sincerely hope that the product will be of value to all scientists and technologists concerned with any stage in the supply of this important beverage, whether they be involved in immediate production or long-term research and development and irrespective of whether they are based in an exporting or an importing country.

However, a text such as this is never perfect — accordingly we will be pleased to have our errors corrected, to hear of other points of view and of new developments.

LIST OF CONTRIBUTORS

Richard Bardner BSc, MIBiol, is a Principal Scientific Officer in the Entomology Department, Rothamsted Experimental Station and the author of more than fifty publications in the field of agricultural entomology. His current research interests include the relationships between insect populations and their economic effects, and also the integrated control of agricultural pests. He spent two years on secondment to the Coffee Research Station, Ruiru, Kenya, where he was Senior Entomologist and has continued to advise on pest problems in Africa.

Karl Bättig was born in Switzerland in 1926. After studying in Zürich, Göttingen and Paris he received his MD degree in 1957. From 1957 to 1959 he was visiting scientist at the National Institute of Mental Health, Bethesda. Then he joined the Department of Hygiene and Physiology of Work at the Swiss Federal Institute of Technology in Zürich. His present position there is Professor of Behavioral and Comparative Physiology and Head of the Laboratory of Behavorial Biology.

Julien Berthaud has worked since 1970 in the Plant Breeding Department of ORSTOM (Office de la Recherche Scientifique et Technique Outre Mer) in the Ivory Coast. During this period he has been concerned with the genetics of wild African coffee trees and has taken part in several surveys and missions to collect specimens: Central African Republic, 1975; Kenya 1977; Tanzania 1982 and Ivory Coast 1976-1981. He is now studying the genetic diversity of wild coffee trees, their genetic organisation and their use in a coffee breeding programme.

Melvin G.R. Cannell obtained his BSc in Agricultural Botany at the University of Reading in 1966. From 1966 until 1971 he was a Research Officer at the Coffee Research Station, Ruiru, in charge of work on crop physiology. In 1971 he obtained his PhD from the University of Reading with a thesis entitled 'Effects of Season and Fruiting on Accumulation and Distribution of Dry Matter in Coffee Trees in Kenya East of the Rift'. From 1971-74 he worked at the Institute of Tree Biology, near Edinburgh, researching on shoot development in trees, and ecophysiology of trees. In 1974 he joined the Institute of Terrestrial Ecology, as a Principal Scientific Officer researching on tree crop physiology. In 1977/78 he worked as a visiting scientist on tree physiology and breeding with the Weyerhaeuser (Forestry) Company, USA. Consultancy interests include Overseas Development Administration (ODA) consultant to the Tea Research Institute of East Africa (1974, 1978) and consultant to the International Council for Research in Agroforestry (1980, 1982).

André Charrier obtained his PhD with a thesis concerned with the Wild Coffees of Madagascar and is now Head of the Genetics Department at ORSTOM and has wide experience in genetic resources and the breeding of tropical plants. He has worked in Africa since 1968, first on the genetics of the Malagasy Coffee gene pool, and now as Coordinator of the coffee breeding programme in the Ivory Coast.

Ronald J. Clarke MA (Oxon) CEng, FLChemE, FIFST, was a member of the Technical Research Division of General Foods Ltd, Banbury, Oxfordshire from 1957-83. He was chairman of the Food Engineering Panel, Society of Chemical Industry from 1972-4. In 1982 in recognition of his services he was made a medallist of the Association Scientifique Internationale du Café. He retired in January 1984, but remains active as a consultant.

Michael N. Clifford obtained his PhD at the University of Strathclyde in 1972 with a thesis entitled 'The Phenolic Compounds of Green and Roasted Coffee Beans'. Since then his research has centred upon the analysis of phenolic compounds in foods, particularly the chlorogenic acids. His current studies in the Department of Biochemistry at the University of Surrey are concerned with the behaviour of phenols during food processing and their influence upon food acceptability.

Eliane Dentan studied biology at the Universities of Lausanne and Brussels. From 1956-66 she worked at the Laboratory for Cytology and Experimental Cancerology at the Institut Jules Bordet in Brussels. In 1966 she joined the microscopy laboratory at Nestlé Research in Switzerland and since 1975 has been Head of Microscopy.

C.F. Marshall has been engaged in the green coffee trade since 1930 and is a former chairman of the UK Coffee Association, the Coffee Trade Federation and the Committee of European Coffee Associations. Until recently he was a member of the UK delegation to the International Agreement as a Trade Adviser, a committee member of the London Coffee Terminal Market Association and Consultant to the World Bank.

Reginald F. Smith BSc, CChem, FRIC, worked in the laboratories of J. Lyons & Co. Ltd for 44 years, the last 20 or so years being spent on the control, technology and fundamental research on coffee and tea and their products. He then spent 12 years as editorial assistant to Analytical Abstracts. He is now retired but continues to abstract for Analytical Abstracts and Fresenius' Zeitchrift für Lebensmittel-untersuchung und Forschung.

Herbert A.M. van der Vossen graduated with an MSc from the Agricultural University of Wageningen in 1964. From 1964-71 he was Research Officer in charge of the Oil Palm Research Centre at Kade, Ghana. He obtained his PhD in 1974 from the Agricultural University of Wageningen with a thesis on oil palm breeding. He was Head of the Coffee Breeding Unit at the Coffee Research Station, Ruiru from 1971-1981. Since January 1982 he has been

Breeding Manager for vegetable and flower crops at Sluis and Groot Research (Zaadunie Seed Company) at Enkhuizen, the Netherlands.

Jim Waller worked in Kenya from 1964-70, spending most of his time on coffee diseases, especially Coffee Berry disease. Subsequently he was at Rothamsted Experimental Station. He joined the Commonwealth Mycological Institute at the end of 1971 as the ODA Pathology Liaison Officer. This involves frequent short-term overseas visits to establish and monitor plant pathology projects in a wide range of tropical countries. He has undertaken many consultancy visits to Latin American countries to advise on coffee rust control and other coffee disease problems, both for ODA and FAO. These include Brazil, Bolivia, Colombia, Costa Rica, Nicaragua, Honduras and El Salvador. From 1976-78 he was part-time Consultant to the Coffee Research Station, Ruiru.

Ken C. Willson BSc, CChem, PhD, FRSC, MIInfSc, obtained his BSc in Chemistry at Swansea in 1944. From 1944-61 he was a Research Chemist and Plant Manager with ICI. In 1961 he joined Uganda Cement Industry as Works Chemist. From 1962 to 1968 he was Head of Chemistry Division at the Tea Research Institute of East Africa at Kericho, Kenya. Whilst working mainly on tea, some other crops were also studied. He was awarded his PhD in 1969 by the University of Reading for a thesis entitled 'The Mineral Nutrition of Tea with particular reference to Potassium'. From 1974-79 he was a Senior Information Officer with ICI, carrying out some consultancy work in Tropical Agriculture during this period. From 1974 to 1979 he was Senior Lecturer (Plantation and Perennial Crops) at the University of Papua New Guinea. Since 1979 he has been a Consultant in tropical agriculture specialising in perennial crops. In this capacity he has worked on coffee, tea, cocoa, coconuts and fruit crops in Fiji, Indonesia, Papua New Guinea and Sri Lanka for plantation companies, ICI, Sri Lanka Government and UNDP. He is now adviser on Fruit Tree Research in Senegal, a post funded by the UK Overseas Development Administration.

1 A HISTORY OF COFFEE

Reginald F. Smith

This first chapter is an attempt to blend historical fact with a number of apocryphal tales and beliefs on the origins and development of coffee to provide a light-hearted introduction to the serious text that follows. The sources from which most of the information is culled are listed in the Bibliography at the end of the chapter, and while I have integrated this material and added comments and drawn on my own experience with coffee, I am indebted to the authors of these books for simplifying my task. By virtue of their age and haphazard transmission, the apocrypha and legends are inevitably vague, but those about coffee blend with accepted facts. I trust that this chapter will provide the reader with an interesting and fascinating account of the story of coffee from the discovery of the plant up to the present day.

Over the ages numerous legends have accumulated about the discovery of coffee. Possibly the earliest references to the use of coffee are to be seen in the Old Testament e.g. Genesis **25**, 30; I Samuel **25**, 18 and II Samuel **17**, 28, this last saying, 'Brought beds, and basins, and earthen vessels, and wheat, and barley, and flour, and parched corn, and beans, and lentils, and parched pulse,'. Apparently, beans were distinct from pulse and lentils but whether or not they were coffee beans is open to conjecture.

Although cultivation may have begun as early as AD 575, the first written mention of coffee as such is by Razes, a 10th century Arabian physician. The oldest legend is that coffee was introduced by Mohammed, for when he lay ill and prayed to Allah, the angel Gabriel (as well as bringing the Koran) descended with a beverage 'as black as the Kaaba of Mecca' that gave him 'enough strength to unseat 40 men from their saddles and make love to the same number of women'.

The most well-known story of the discovery of the coffee plant is that concerning Kaldi, a goatherd tending his flock in the hills around a monastery on the banks of the Red Sea. He noticed that his goats, after chewing berries from bushes growing there, started to prance about excitedly. A monk from the monastery observed their behaviour, took some of the berries back to the monastery, roasted and brewed them and tried out the beverage on his brethren. As a result they were kept more alert during their long prayers at night.

Another story about the origin of coffee concerns a priest, Ali bin Omer who, charged with misconduct with a king's daughter, was banished into the mountains of the Yemen. There he discovered a tree with white flowers and drank and enjoyed the decoction of its beans. He took the beans with him on a pilgrimage to Mecca and there cured pilgrims of an epidemic of the itch. For this he was honoured on his return and made the patron saint of coffee growers, coffee-house keepers and coffee drinkers.

The word 'coffee' is derived from the arabic word 'quahweh' which originally some people say was a poetic term for wine. Since wine is forbidden to strict Muslims the name was transferred to coffee, and through its Turkish equivalent *kahweh* became *café* (French), *caffè* (Italian), *Kaffee* (German), *koffie* (Dutch) and coffee (English), and the Latin *Coffea* for the botanical genus. In Abyssinia coffee is called *bun* and the beverage *bunchung*; these words are the origin of the German *Bohn* and the English *bean*. Coffee is also called *Mocha*, a name taken from the port of Mocha on the Red Sea coast, from whence it was shipped.

The wild coffee plant (*Coffea arabica*) is indigenous to Ethiopia, where it was discovered in about AD 850, and was cultivated in the Arabian colony of Harar. It then spread to Mecca, from whence it was taken home by pilgrims to other parts of the Islamic world.

Two other species of coffee, *C. liberica* and *C. robusta*, a cross between *C. arabica* and *C. liberica*, were later discovered in Africa. *C. robusta* is noted for its resistance to disease, hence its specific epithet, and is grown on native smallholdings in West Africa. It was first imported into this country after the start of the Second World War to replace South American coffees that were no longer available on account of currency restrictions. I was very surprised when I first had samples of robusta coffee to analyse and found that the caffeine contents were over 2 per cent, compared with around 1 per cent for arabica coffees. Although robusta coffee has a flavour that is inferior to that of arabica, with a caffeine content more than double, it has in proportion a greater stimulating action, and also offered advantages to the manufacturers of instant coffee extracts, ensuring that they satisfied the wartime Statutory Instrument requirements regarding caffeine content. *C. robusta* also has a higher content of soluble extractives, and is therefore more economical in the manufacture of instant coffee.

In more recent times a cross of *C. arabica* and *C. robusta* (*C. arabusta*) has been developed in the Ivory Coast and is reported to have good flavour qualities. Attempts have been made to develop a coffee that is free of or deficient in caffeine, but with little success to date. Some wild coffees, e.g. the Mascarocoffea *C. vianneyi* from Madagascar, while virtually free from caffeine yield too bitter a beverage because of the presence of bitter diterpene glycosides such as mascaroside.

To return to my account of the spread of coffee from the Arabian continent to other parts of the world, there are many romantic stories. The raw beans were not allowed out of the country of origin without first being steeped in boiling water or heated to destroy their germinating power. Strangers were not allowed to visit the plantations, but it was Baba Budan, a pilgrim from India, who smuggled out the first seeds capable of germination. He is supposed to have smuggled out seven coffee seeds from Mecca to Mysore, strapped to his belly, at around AD 1600; but the present Mysore coffee is more likely to have been derived from that introduced later by the British. Many European travellers to the Levant had reported a 'strange black drink', and in 1615 Venetian merchants brought coffee beans from Mocha to Europe. This started a lucrative trade for the Arabians, which they guarded jealously for 100 years, during which time they were the sole providers of coffee. The first coffee plant brought to Europe was stolen by a Dutch trader in 1616; from it, plants were propagated in the Amsterdam Botanical Gardens and taken out to the East Indies to set up plantations.

The French had a more romantic way of introducing coffee into their colonies. A coffee plant was presented by the burgomaster of Amsterdam to King Louis XIV, from which plants were raised in the *Jardin des Plantes* in Paris. Also a young French naval officer, hearing about the plantations in Java, obtained a seedling by a clever intrigue and set sail with it in a glass box for Martinique. It withstood a fierce tempest, flooding and the unwelcome attention of another passenger and survived in Martinique under an armed guard and surrounded by thorny bushes. The plant provided for the rich estates of the West Indies and Latin America. Soon after, coffee was introduced by Spain into its West Indian colonies. Britain was the last country to cultivate coffee in its colonies, starting with Jamaica in 1730 and India in 1840. At the same time Brazil entered the field through a Brazilian officer who while on a vist to French Guiana in 1727, received a plant hidden in a bouquet of flowers, as a token of affection from the Governor's wife. This was the start of the coffee empire of Brazil, which now holds supremacy in world coffee production.

Who would believe that there is a connection between Brazilian coffee and aeronautics? Henriques, the father of Alberto Santos-Dumont, 'the first man in the world to fly', was called the 'Coffee King' because he had planted five million coffee trees in Brazil. Henriques had attached his wife's maiden name Santos to his own — hence the origin of the name for Santos coffee. Many more million trees were planted during Alberto's formative years, and 50 miles of railway track were laid. Among the boy's first memories was his being allowed to drive one of the imported English locomotives on the coffee plantation. When Alberto took up an interest in ballooning he received a generous allowance from his father, who had sold his estates for more than five million dollars.

At the present time the cultivation of coffee has spread to a belt of countries around the world in the area between the Tropics of Cancer and Capricorn. However, the crops in Ceylon, India and the East Indies were wiped out in the middle of the last century by the fungus disease *Hemileia vastatrix* (coffee rust), and in Ceylon (now Sri Lanka) has been replaced by tea, for which Sri Lanka is now so well-known. Coffee rust is still a threat to coffee crops in other producing countries, but is kept under control by precautions in cultivation, breeding of resistant strains, and use of copper sprays and modern fungicides.

In addition to Venetian merchants, invading Turkish armies brought coffee to Europe in the 17th century, but when they finally retreated after the siege of Vienna, 500 sacks of beans were left behind in a warehouse. In 1683 these were discovered by Franz Georg Kolschinski, interpreter to the Turkish army. He had learned how to roast the beans and prepare them for drinking, and sold cups of coffee from door to door. When he was rewarded with the gift of a house in Vienna by the Austrians for his wartime bravery, he turned it into a coffee house. This was the model for the Viennese coffee houses, which became, and still are, meeting places for journalists and politicians. The first coffee house in England was opened in 1650 by Jacob, a Jew, at the Angel in the High in Oxford; a plaque in the Cooperative grocery store between University College and the Examination Schools was erected to mark the spot where Jacob's coffee house once stood. The first coffee house in London was opened two years later in 1652 in St Michael's Alley, close to the Royal Exchange. A certain David Saunders brought home from Italy a native of Ragusa named Pasque Rosee, who used to prepare coffee for his master's breakfast each morning. The story goes that this exotic beverage attracted an inconvenient number of early callers at his house and that Saunders, with a good nose for business, set up his servant in an establishment for the sale of coffee. The venture was so successful that many competitors followed his example and coffee houses mushroomed all over the City of London. These had their detractors, for in 1757 James Farr, proprietor of the Rainbow Coffee House in Fleet Street, was prosecuted by the Wardmote for creating a public nuisance with the smell of his roasting coffee. (So there is nothing new in the concern for pollution of the atmosphere from the numerous coffee roasting plants in New York.)

In 1715 there were over 2,000 coffee houses in London, and these became centres for social, political, literary and commercial life. In due course they gave rise to insurance houses, merchant banks and the Stock Exchange in the City of London, and it was in such coffee houses that the *Spectator* and *Tatler* were born. Lloyd's Coffee House in Tower Street, the main thoroughfare between St Katharine Docks and Wapping, was a favourite haunt of seafaring men, including captains and ships' husbands, who met there to do their day-to-day business and to obtain shipping information. Lloyd moved in 1691 to a larger coffee house in a more central position at the junction of Lombard Street and Abchurch Lane, and eventually to the Royal Exchange. It was here that he started the business of insurance of ships and their cargoes, with the Foundation of Lloyd's of London, whose business is no longer confined to ships and shipping. To this day the transactions at Lloyd's are conducted across tables between pews in compartments like those in the original coffee houses. A more dubious legacy of the coffee houses was that of the 'tip' — a brass-bound box having the words TO INSURE PROMPTNESS — set up to receive contributions for the waiters.

During the second half of the 18th century a few coffee houses introduced reception rooms for ladies. This was quite an innovation and, although the ladies had to drink their coffee strictly apart from the men, their invasion was one of the first steps to emancipation. When the ladies were later permitted to enter the cafés, it is said that they showered their affections on the male cashiers in the reception rooms, just as the male customers were attracted by the female cashiers.

At various times in the past there has been a strong opposition to the sale and consumption of coffee, on social, religious, political, economic or health grounds. The accounts that follow may appear rather amusing and bizarre.

In 1511 Kair Bey, a young and enthusiastic Governor to the Sultan of Cairo in Mecca, on leaving a mosque after devotions, saw some people in a corner where they were planning to spend the night drinking coffee. This displeased Kair Bey, who said that it was against the teaching of Islam. The next day orders were given that all coffee houses should be closed, all coffee destroyed, and heavy fines were imposed. When the Sultan, a confirmed coffee drinker, heard of this he called Kair Bey and his associates for an explanation and repealed the order. Kair Bey and his associates are said to have had a tragic end. According to Burgin, W.H. Ukers in *All About Coffee* said at the end of his account of this incident, 'The most interesting fact of the coffee story is that revolution is encouraged by its introduction, and its function is always to make people think, and when people begin to think, the tyrant and opposition are in danger'.

In Italy the priests appealed to Pope Clement (1592-1605) to have the use of coffee forbidden among Christians as it was given by Satan to the infidel Muslims to replace wine in their religious ceremonies. It seems that the Pope liked the drink, and said that, 'As the Satan's drink was so delicious we shall cheat Satan by baptising it'. However, this did not stop the Council of Ten in Venice from trying to eradicate coffee, which they charged with immorality, vice and corruption.

In 1656 the coffee houses were charged with vice and corruption by the Ottoman Grand Vizir Koprilli. The first violation of this edict was punished with cudgelling, and for the second the offender was sewn up in a leather bag and thrown into the Bosphorus.

A *Woman's Petition Against Coffee* was published in 1674. They complained that at times of crisis they were deserted by their husbands, who frequented the coffee houses and were thereby rendered impotent by the drink. A reply was published at the same time by the men in their defence. In the following year King Charles II issued a Proclamation for the Suppression of Coffee Houses, because they 'were the resort of idle and disaffected persons who mis-spent their time, instead of being employed about their lawful callings, and published malicious and scandalous reports to the defamation of His Majesty'. Only eleven days later Charles published a second edict withdrawing the first, allowing the coffee houses to stay open out of 'Royal Compassion'. This quick turnround was a record for the most words eaten in the shortest time, unsurpassed until modern times.

In France opposition to coffee came from the wine merchants, who were afraid of the competition it introduced. At the wine merchants prompting a thesis was put forward by the physicians of Marseilles in 1679 that 'the vile and worthless foreign novelty ... the fruit of the tree discovered by goats and camels ... burned up the blood ... induced palsies, impotence and leaness ... hurtful to the greater part of the inhabitants of Marseilles'. This thesis was probably based upon a most unsettling accusation levelled against coffee in a tale by a seventeenth century German traveller. The story concerns a King of Persia who 'had become so habituated to the use of coffee that he took a dislike to women'. One day his Queen saw a stallion being emasculated; upon asking the reason why, she was told that the animal was too spirited, and was being gelded to tame it. Whereupon she suggested a simpler solution would be to feed it coffee every morning. This story which starkly contradicts the effects which coffee had upon Mohammed (make love to 40 women) was said to have virtually ruined the coffee trade for 50 years. In this connection it should be mentioned that a modern over-the-counter aphrodisiac is said to contain caffeine as the active principle along with ginseng powder and Vitamin E.

After the Boston Tea Party, when the citizens boarded the British ships in the harbour and threw their cargoes of tea overboard, coffee was crowned once and for all 'King of the American breakfast table'. It was thus that coffee became for ever linked by the Americans through the War of Independence with liberty and democracy. Since then the Americans have become one of the world's greatest coffee-drinking nations. However, this does not apply to Mormons, Christian Scientists and members of some 'fringe' religions, who abstain from drinking coffee, tea, alcohol and other stimulating beverages. This fact has been taken into account in various epidemiological surveys which have sought to detect effects that might be associated with the consumption of these items.

In 1773 the Landgrave Frederic of Hesse prohibited the use of coffee throughout his territories. 'If his subjects did not follow the regulations they were threatened with the punishment of hard labour on the roads. Persons of higher social level, however, whose means did not suffer from coffee consumption, should go unpunished.'

The banning of coffee consumption or state control of its roasting led to illegal roasting and to smuggling. Illegal roasting was detected by trained coffee 'sniffers' who roamed the streets in search of revealing smells coming from house windows. In Hamburg smugglers once loaded a coffin with roasted coffee and carried it past the customs post disguised as a funeral procession. One of the party went ahead calling 'Cholera, Cholera' to scare off the customs officials. Also, containers for coffee were strapped on to the backs of dogs, who were chased through the customs barrier. Tailors, haberdashers and shoemakers were specialists in creating clothing with secret pockets for smuggling purposes. It is surprising how many crippled persons passed through the customs post daily!

Frederick the Great of Prussia issued the following declaration in 1777: 'It is disgusting to note the increase in the quantity of coffee used by my subjects and the amount of money that goes out of the country in consequence. Everybody is using coffee. If possible it must be prevented. My people must drink beer. His Majesty was brought up on beer, and so were his officers. Many battles were fought and won by soldiers nourished on beer; and the King does not believe that coffee-drinking soldiers can be depended upon to endure hardships or to beat his enemies in case of the occurrence of another war. Only the upper classes will be allowed coffee.'

According to Rothfos the monks of Constantinople once presented the Mufti with a petition for the banning of coffee. Their reason is said to have been that roasted coffee resembled charcoal, and charcoal and similar substances were prohibited by the Koran. Eventually a new Mufti decided that coffee was definitely not charcoal and withdrew the ban. However discussion with Muslim students suggests that this story, if it ever had a basis of truth, is now corrupt. Firstly, it is unlikely that the Mufti would have heeded a petition from Christian monks, and secondly the Koran forbids alcohol, gambling and pork; there is no mention of charcoal.

In spite of all attempts to ban the sale and consumption of coffee, it has become a very popular beverage, and many claims have been made for its beneficial effects, right from the time of Kaldi's goats. In general it is considered that, in moderation, coffee is not harmful. The limit of moderation depends upon the individual since as Paracelsus (born 1493 at Einsiedlen, Switzerland) said:

Tout est poison	Everything is poisonous
Et rien n'est sans poison	And nothing is not poisonous
Seul la dose	Only the dose
Fait qu'une chose	Makes a thing
N'est pas un poison	Not poisonous.

In an English newspaper advertisement of 1657 coffee was described as 'having many excellent vertues, closes the Orifice of the Stomack, fortifies the heart within, helpeth Dijestion, quickeneth the Spirits, maketh the heart lightsome, is good against Eyesores, coughs and cold, Rhumes, Consumptions, Head-ache, Dropsie, Gout, Scurvey, King's Evil and many other'. Many of these virtues, but not all, are today confirmed by modern research, but the controversy as to its effects is still unresolved. At least we know that some of the old misconceptions, such as the promotion of impotence, barrenness and sterility have been exploded. A notable investigation to settle the controversy on the effects of coffee was made by King Gustav III of Sweden under the control of two physicians. The death sentences imposed for murder on two identical-twin brothers were commuted on condition that one was to take a daily dose of coffee and the other, as a control, a daily dose of tea. Eventually one of the physicians died, then the other. The judges who passed the death sentence passed away, then the King was murdered. The tea drinker died at the age of 83 and lastly the coffee drinker died. Is this the reason why the Swedes are to this day such heavy coffee drinkers?

When Sir William Harvey, who discovered the circulation of the blood, was on his death-bed he is said to have called his lawyer to his side and held up a coffee bean. 'This little fruit,' he whispered, 'is the source of happiness and wit'. Sir William bequeathed his entire supply of coffee, 56 pounds, to the London College of Physicians, so that they would commemorate the day of his death every month with a morning round of coffee. One wonders how long the supply of coffee lasted and in what condition it was at the end of its life?

The praises of coffee are sung in the Coffee Cantata 'Schweigt stille, plaudet nicht...' BMV 211, written (1732-35) by J.S. Bach who used the words of Picander (Christian Friedrich Henrici):

Ei! wie schmeckt der Coffee süsse, lieblicher als
tausend Küsse, milder als Muskatenwein
Coffee, Coffee musse ich haben: und wenn jemand
mich will laben, ach so schenkt mir Coffee ein!

which translates roughly into English as:

Oh! how sweet is the taste of coffee, choicer than
a thousand kisses, milder than muscatel wine
Coffee, Coffee must I have: and if anyone
wishes to comfort me, pour me out some coffee.

The cantata recounts the tale of an irascible, blustering Herr Schlerendrian and his daughter, Lieschen. She has a passion for coffee and also wants to marry, but her father forbids her to continue with coffee drinking, otherwise he will not allow the marriage. Bach introduces a happy ending: Lieschen swears that she will give up coffee if she is allowed to marry, but secretly a clause is written into the marriage contract such that her husband can allow her to drink coffee. The purpose of the cantata was to mock the physicians who were campaigning (in Germany) to discredit coffee.

The cantata concludes with the verse:

Die Katze Lässt das Mausen nicht, die Jungfern bleiben Coffeeschwestern. Die Mutter liebt den Coffeebrauch, die Grossmamma trank solchen auch wer will nun auf die Töchter lästern?

which roughly translates into English as:

The cat leaves the mouse alone, the spinsters continue to be coffee addicts. The mother loves the coffee habit, the grandmother also, and who will now forbid the daughters?

Prince Talleyrand (1754-1839) expressed a general feeling when he said that a cup of coffee 'detracts nothing from your intellect, your stomach is freed by it and no longer distresses the brain; it will not hamper your mind with troubles but give freedom to its working. Suave molecules of Mocha stir up your blood, without causing excessive heat: the organ of thought receives from it a feeling of sympathy; work becomes easier and you will sit down without distress to your principal repast, all of which will restore your body and afford a calm delicious night' — all of which is confirmed by modern thought.

Honoré de Balzac has been called the king of coffee drinkers. He used to go to bed at 6 pm and get up again at midnight in order to work for 12 hours, and is said to have consumed about 50,000 cups of coffee in his lifetime. It is said that Voltaire needed 72 cups of coffee per day. When he was told that coffee was a poison slowly affecting the body, he replied that he had known that for the last 80 years. Beethoven liked strong coffee, for which 60 beans per cup were counted off by him personally. However, the title of World Champion of Coffee Drinkers has more recently been given to Mr Gemsock of Cleveland, Ohio, when he consumed 85 large cups in 24 hours.

In more recent times a young woman (unnamed) wrote in the New York Evening Post that 'I know of the magic power of coffee and will brew good coffee for my husband. I want to do this for him every day: make his mornings even sweeter with coffee. Then even the most beautiful girl on earth will not be able to strike a flame in his heart.'

Isombard Kingdom Brunel, builder of the Great Western Railway, was very critical of the quality of the coffee that he drank and castigated the caterer, who contracted with the GWR for the catering at Swindon station, for the undrinkable coffee served there. Eventually the GWR had to buy out the contractor for £100,000.

Coffee has been shown to have definitely beneficial properties. During the Second World War, I was involved in research on the nicotinic acid content of coffee, as part of an investigation instigated by the late Professor Drummond on the nutritional value of our wartime diet. I was able to demonstrate how nicotinic acid is produced during the roasting of coffee by the decomposition of trigonelline (nicotinic acid N-methylbetaine). The nicotinic acid therefore contributes to the anti-pellagra action of coffee. This has since been confirmed in countries where the native population exists on a poor diet, significantly in coffee-producing countries, e.g. Angola, where daily consumption of cups of coffee has been shown to reduce the incidence of pellagra. Incidentally, some of the trigonelline is further decomposed, to the structurally related pyridine. This trace of pyridine contributes to the aroma, but in very large doses may have toxic effects; so 'what you gain on the swings you may lose on the roundabouts'.

Towards the end of the last century J. Lyons hit upon the idea of opening up teashops in London, the first in Piccadilly in 1895, thus providing places where respectable young women could call to partake of cups of tea and coffee and light refreshment, for at that time no respectable young woman would be seen alone in a bar or restaurant frequented by men. This was a sign of the emancipation of women, who were becoming more independent and leaving home to take up employment in offices in the city. To give Lyons teashops the appearance of respectability, the waitresses were dressed in black with white aprons and caps with ribbons, like those of maidservants in upper-class houses. In more recent times I can remember how the Jewish diamond merchants of Hatton Garden used to meet in the Lyons teashop there to carry out their business over cups of coffee. They would bring out their diamonds, wrapped in tissue paper, from their waistcoats and display them on the marble-topped tables, a reminder of the business carried out in the coffee houses in former times.

The introduction of soluble coffee powder, after the Second World War, was a major technological development. This was followed by an important sociological development — the appearance of coffee bars with exotic decor, appealing to the younger generation, and serving instant coffee. The coffee bars have only survived by serving meals, as are now seen in the chains of franchise restaurants serving hamburgers and barbecued chickens in addition to coffee. The social side of coffee is now rooted more in the home and restaurants, where a good cup of coffee provides a fitting finish to an excellent meal.

Since World War II, due partly to the introduction of soluble coffee powder by the Americans and the increase in foreign travel, there has been a considerable increase in the amount of coffee consumed in the United Kingdom, mainly at the expense of tea, which tends nowadays to be regarded as 'grandmother's drink'. At present the consumption of instant coffee represents about 85 per cent of the total import of coffee beans, and the consumption of whole roasted and ground coffee represents only about 10 or 12 per cent of the total, the remainder being liquid extract.

In former times coffee for domestic consumption was roasted and ground in the home. Later the beans were roasted in bulk and sold wholesale to the tea and coffee merchants, or they were roasted and ground in the grocer's shop and purchased for grinding and brewing in the home. The old-fashioned grocers used to scatter roasted beans on the floor of their shops so that the aroma was released when the beans were trodden underfoot. This practice has ceased but the attractive aroma is still provided by the coffee roasters to be seen in the windows of specialist shops, particularly in Europe where the sales of whole roast beans and roast and ground coffee are relatively high. In Great Britain it is more usual for ground coffee to be sold in hermetically sealed cans, either filled with inert gas (carbon dioxide or nitrogen) or vacuum-packed, which prevents staling by oxidation of coffee constituents on contact with atmospheric oxygen. In more recent times packing in sealed pouches, either gas-filled or vacuum-packed, has been introduced, and the coffee can remain in good condition for up to six months. Many methods may be used in the home for brewing coffee, and ground coffee may be purchased to suit the method of brewing.

In London, green coffee is offered for sale by merchants, dealers and brokers, the majority of whom are in or around Mark Lane in the City. There is also a Coffee Terminal Market operating in the Corn Exchange to deal with 'futures' contracts, i.e. buying and selling the crop in advance as a protection against price fluctuations.[1] Most coffee is bought through brokers, but trading for the cheaper robusta coffees is also carried out at the London Commodity Exchange. Arabicas and 'milds' are traditionally sold direct to the buyers.

The International Coffee Organisation of producing and consuming countries was formed in 1962 with headquarters in Berners Street, London, to stabilise the world coffee trade, by imposing export quotas and therefore controlling coffee production. But with conflicting interests on quotas, the agreement lapsed in 1973, and the producing countries took measures to maintain prices as high as possible. The coffee crop is sensitive to vagaries of weather, disease and natural disasters, such as the Black Frost that had a serious effect on the Brazilian crop during the night of July 17th, 1975, and also to political unrest in African countries. However, by careful control there is no longer the risk of a glut of coffee as in the 1930s, when surplus Brazilian coffee had to be destroyed by burning. At that time searches were even made for alternative uses of raw coffee.

For twenty years accounts of recent developments in coffee chemistry and technology have been reported in papers read by international experts at colloquia organised by the Association Scientifique Internationale du Café (ASIC). The first, which I attended was a more-or-less informal meeting of some 45 experts held in Paris in 1963. At the second in Paris in 1965 ASIC was founded. Since then meetings now with some 200 participants, have been held every two or two-and-a-half years, either in a coffee producing country, where the emphasis is mainly on agronomy, or in a European country, with an emphasis on the composition, technology and consumption of coffee. The papers presented at these meetings cover the fields of: agronomy, composition of raw and roasted coffee, technology including instant coffee, the beverage and physiological effects (beneficial or possibly harmful). These topics will be dealt with in the chapters of this book that follow. Following the inaugural meeting, Pierre Navellier, the Scientific Secretary of ASIC, knowing that I was keeping a bibliography on the chemistry of coffee as part of my involvement in coffee research, proposed that I should prepare a review of the literature for the next meeting. I have continued to do this at the subsequent colloquia and through these reviews have seen that there have been considerable advances in the quality of the green and roasted beans and hence the beverage, and in particular of soluble coffee powders, which has been considerably improved since their introduction at the end of the Second World War. Let us hope that this improvement will continue.

Note

1. I wish to thank A.W. Ayling of Sol Cafe Limited for this information on coffee buying, also on the consumption of coffee in the UK.

Bibliography

Bürgin, E.C. *Kaffee* (Sigloch Edition, Künzelsau, 1978)
Davids, K. *The Coffee Book* (Whittet Books, 1980)
Encyclopaedia Brittanica 14th edn. (Encyclopaedia Brittanica Inc. New York, 1973)
Flower, R. and Wynn-Jones, M. *Lloyds of London. An Illustrated History* (David and
    Charles, Newton Abbot, 1974)
Rosen, C. *Coffee*, (Penguin Books, London, 1981)
Rothfos, B. *Coffea Curiosa*, (Rothfos, Hamburg, 1968)

2 BOTANICAL CLASSIFICATION OF COFFEE

André Charrier and Julien Berthaud

Introduction

While the international coffee trade is concerned with only two coffee species — *Coffea arabica* and *C. canephora* — botanists regard as coffee trees all tropical plants of the *Rubiaceae* family, which produce seed resembling coffee beans. During botanical explorations of the tropical regions, from the sixteenth century onwards, wild coffees also attracted the attention of explorers and botanists. Their specimens are found in the herbaria and the names of the most famous explorers have been commemorated in both specific and generic epithets. Hundreds of species have been described, but the taxonomic classification of the genus *Coffea* has become very complex and rather confused.

Even the most authoritative classification system of Chevalier (1947) is now due for revision in view of the many new species discovered over the past 20 years on Madagascar (Leroy, 1961a, b, c, 1962, 1963, 1972a, b) and in East Africa (Bridson, 1982).

In addition the conventional methods of taxonomic classification, which are mainly based on morphological characteristics of specimens deposited in herbaria, are inadequate to give full justification to the tremendous variability encountered in the allogamous wild coffee populations.

In their efforts to develop a more exacting system of classification, coffee botanists and geneticists (Charrier, 1978; Berthaud, 1985) have followed the trend of modern taxonomy, which today makes use of a variety of scientific disciplines, to reflect the true genetic-historical relationship among plants and animals (Clausen, Keck and Hiesey, 1945; Dobshansky, 1970; Mayr, 1970; Harlan and de Wet, 1971).

This chapter presents a review of advances made in the taxonomy of coffee, starting with the standard classification based on herbaria specimens and living collections. This will then be followed by an analysis of data from studies of cytotaxonomy and cytogenetics, ecology and plant geography, as well as on biochemical and serological affinities, which have all contributed valuable knowledge needed for a better coffee classification.

Detailed botanical descriptions of cultivated coffee species have been presented earlier by various authors (Wellman, 1961; Haarer, 1962; Coste, 1968; Carvalho *et al.*, 1969) and therefore need not be repeated here.

14 *Botanical Classification of Coffee*

Taxonomy

The first botanical description of a coffee tree, under the name *Jasminum arabicanum*, was made in 1713 by A. de Jussieu, who studied a single plant originating from the botanic garden of Amsterdam. However, Linnaeus (1737) classified it as a separate genus *Coffea* with the then only one known species *C. arabica*.

Many new species of *Coffea* have been discovered during exploration of the tropical forests of Africa since the second half of the nineteenth century. Several botanists have tried to describe these species, but this led often to confusion and numerous epithets have proved to be synonymous. Special mention should be made of the extensive taxonomic work of Chevalier (1947) on the *Coffea* species of Africa and Madagascar and of Lebrun (1941), who paid particular attention to the coffees of Central Africa, especially those found in Zaire.

Of recent date are the detailed taxonomic studies of Leroy (1967, 1980) on the coffee species of Madagascar and of Bridson (1982) on the coffee species found in East Africa. Especially Leroy's (1980) efforts to indicate the relationship between species of the genus *Coffea* and those of *Psilanthus* and others are of particular importance to the understanding of the whole *Coffea* spectrum. His most important criteria for differentiating the

Figure 2.1: Coffee Type Placentation

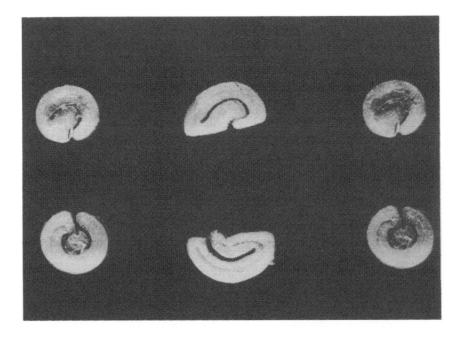

Table 2.1: Classification System for the Genera *Coffea* and *Psilanthus* According to Leroy (1980), with Indication of Geographical Distribution

Family:	*Rubiaceae,*	Subfamily:	*Cinchonoidea*
	Genus	SubGenus	Localisation
	Coffea L.	· *Coffea*	Africa, Madagascar
		· *Psilanthopsis* (Chev.)Leroy	Africa
		· *Baracoffea* (Leroy) Leroy	Africa, Madagascar
	Psilanthus (Hook.f.)	· *Paracoffea* (Miquel) Leroy	Africa, Asia New Guinea
		· *Psilanthus* (Hook. f.)	Africa

Table 2.2: The Two Criteria used by Leroy (1980) to Differentiate Genera and Subgenera

		Criterion 1	
		— Long corolla tube — Anthers not exserted — Short style	— Short corolla tube — Anthers exserted — Long style
		GENUS: *Psilanthus*	GENUS: *Coffea*
Criterion 2	· Axillary flowers · Monopodial development	P. Subgenus *Psilanthus*	C. Subgenus *Coffea*
	· Terminal flowers · Predominantly sympodial development	P. Subgenus *Paracoffea*	C. Subgenus *Baracoffea*

genus *Coffea* from all other genera within the family *Rubiaceae* are the type of gynaecium and placenta (Figure 2.1) The classification system proposed by Leroy (1980) for the genera *Coffea* and *Psilanthus*, with indication of geographical distribution of the sub-genera, is presented in Table 2.1. The differentiation between genera and sub-genera is based on two main criteria (Table 2.2):

(1) flower shape, more in particular the length of the corolla tube (long or short) associated respectively with exerted anthers and style (Figure 2.6-2) or inserted anthers and short style (Figure 2.6-1).

(2) growth habit and type of inflorescence: monopodial with axillary flowers or sympodial with terminal flowers.

However, the second criterion is not unconditional, as both growth habits can occasionally be found on the same plant in some species.

On the other hand, Chevalier (1947) tried to group the species within the genus *Coffea* into the following four sections:

Argocoffea, Paracoffea, Mascarocoffea, Eucoffea.

According to Leroy's (1967) classification *Argocoffea* should be excluded fom the genus *Coffea*, because the seeds do not resemble coffee beans, and the section *Paracoffea* should be considered as a sub-genus of *Psilanthus* (Table 2.1). The section *Eucoffea*, now more correctly named *Coffea*, and *Mascarocoffea* include most of the presently known coffee species.

Coffea

This is divided into five subsections according to very diverse criteria: tree height (*Nanocoffea*), leaf thickness (*Pachycoffea*), fruit colour (*Erythrocoffea, Melanocoffea*) and geographical distribution (*Mozambicoffea*). The species grouped under each subsection are shown in Table 2.3.

Lebrun (1941) proposed the following series for *Coffea* species found in Central Africa, using the increasingly complex structure of the inflorescence as main criterion:

Abyssinicae, Robustae, Libericae.

The *Libericae* series contain a single species, because all the species with thick leaves and large fruits recognised by Chevalier were placed in synonymy with *C. liberica* by Lebrun. Consequently, the subsection *Pachycoffea* of Chevalier becomes equivalent to the series *Libericae* of Lebrun and to *C. liberica* in a broad sense. Chevalier saw during his taxonomic studies only a few herbarium specimens of coffee species from East Africa, but nevertheless observed the originality of this material. Bridson (1982) carried out detailed studies on coffee species for the *Flora of Tropical East Africa* (not yet published). A list of coffee species and taxa found in this part of Africa, many of which are still incompletely known, is given in Table 2.4. Many of these coffee species are well adapted to drier climates by their xeromorphic characteristics and the short interval between flowering and mature seed (only three months versus 8-12 months for most other coffee species).

Mascarocoffea

The coffee species belonging to this section all have one characteristic in common: the absence of caffeine, which was first reported by Bertrand (1902). The tremendous variety in forms has hampered taxonomic classification. A first regrouping of species in series was made by Chevalier (1938), but in view of the more than 50 different forms (Portères, 1962; Leroy, 1961a, b, c, 1962, 1965, 1982) found within material collected during the early 1960s Leroy proposed a revision of the series within the *Mascarocoffea* section (Table 2.5). This classification takes into account

Table 2.3: The Grouping of Species in the Subsection *Eucoffea* According to Chevalier (1947)

Subsections	Species
Erythrocoffea	C. canephora
	C. arabica
	C. congensis
Pachycoffea	C. abeokutae
	C. liberica
	C. klainii
	C. oyemensis
	C. dewevrei
Melanocoffea	C. stenophylla
	C. carissoi
	C. mayombensis
Nanocoffea	C. humilis
	C. brevipes
	C. togoensis
Mozambicoffea	C. schumanniana
	C. eugenioïdes
	C. kivuensis
	C. mufindiensis
	C. zanguebariae
	C. racemosa
	C. ligustroïdes
	C. salvatrix

Table 2.4: List of Taxa of *Coffea* from East Africa Adapted from Bridson (1982)

C. fadenii	C. sp. I
C. mongensis	C. mufindiensis
C. kivuensis	— subsp. *mufindiensis*
C. salvatrix	— subsp. *lundaziensis*
C. pseudozanguebariae	— subsp. *autralis*
C. sp. A	C. pawekania
C. sp. B	C. ligustroïdes
C. sp. C	C. zanguebariae
C. sp. D	C. racemosa
C. sp. E	C. sp. J
C. sp. F	C. sp. K
C. sp. G	C. paolia
C. sp. H	

leaf, fruit and seed characteristics. Charrier (1978) gives in his synthesis on *Mascarocoffea* a concise botanical description of all known species with details of their geographical distribution on Madagascar.

Of considerable interest are also the efforts of Lobreau-Callen and Leroy (1980) and Chinnappa and Warner (1981) to establish a palyno-

18 Botanical Classification of Coffee

Table 2.5: List of the 'Series' proposed by Chevalier (1938) and Revised by Leroy for the *Mascarocoffea* of Madagascar

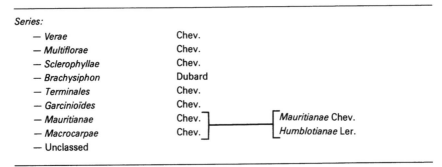

Series:
- *Verae* — Chev.
- *Multiflorae* — Chev.
- *Sclerophyllae* — Chev.
- *Brachysiphon* — Dubard
- *Terminales* — Chev.
- *Garcinioïdes* — Chev.
- *Mauritianae* — Chev. ⎤ ⎡ *Mauritianae* Chev.
- *Macrocarpae* — Chev. ⎦ ⎣ *Humblotianae* Ler.
- Unclassed

logical basis for coffee taxonomy. It enables differentiation at the generic level, as the 3-porate pollen of *Coffea* species is easily distinguished from the 4-5 porate pollen of *Psilanthus* and *Paracoffea* species. Unfortunately, the eight morphologically different types of pollen described by Chinnappa and Warner (1981) bear little relation to the subsections and series in the taxonomic classification of the genus *Coffea*.

The wealth of forms encountered in wild coffees is the result of the interaction of the genetic variability of coffee populations with an endless range of ecological 'microniches' (Forster, 1980) of natural habitats. Botanists have often failed to take this variation into account by restricting themselves mostly to morphological characters, when they tried to establish clear-cut distinctions between species in which genetic differentiation is still occurring.

Souces of Botanical Information

Information on coffee species in relation to habitats, taxonomic characteristics, genetic diversity and geographical distribution can be found in herbaria, travel reports of botanical explorers and in the living collections of research stations.

Major Herbaria

Because of the historical ties with Africa and South East Asia most European botanists of earlier times sent their collections to herbaria in their home countries, with the result that these herbaria are characterised by a high degree of geographical specialisation. This applies also to the herbarium specimens of coffee, which are therefore important sources of information for botanists when preparing new missions to collect and preserve coffee genetic resources in centres of genetic diversity in Africa. Below follows a brief description of the herbaria most relevant for coffee:

(1) The Royal Botanic Garden at Kew and the Natural History Department of the British Museum in London, UK: comprehensive collections from the Sudan, Uganda, Kenya, Tanzania and other eastern and southern African countries, as well as from English speaking West Africa and Angola.
(2) Jardin Botanique National de Belgique at Meise, Belgium: with emphasis on Zaire.
(3) Museum National d'Histoire Naturelle at Paris, France: with comprehensive collections from Guinea, Ivory Coast, Cameroon, Gabon, Congo, the Central African Republic and the Malagasy Republic.
(4) Botanische Garten und Botanische Museum at Berlin-Dahlen, Germany: with collections from Togo, Cameroon and Tanzania, which were however almost completely destroyed in 1943.
(5) Herbarium Vadense of the Department of Plant Taxonomy and Geography of the Agricultural University of Wageningen and the Rijksherbarium of the University of Leiden, The Netherlands: with coffee specimens from respectively Africa (especially Cameroon) and Indonesia.
(6) Erbario Tropicale di Firenze, Florence, Italy: collections from Ethiopia and Somalia.
(7) Botanical Institute of the University of Coimbra and Centro de Botanica da Junta de Investigaçoes Cientificas do Ultramar at Lisbon, Portugal: with collections from Angola and Mozambique.

These European herbaria may already give a very good impression of the overall distribution of the wild coffees. Nevertheless, the importance of national herbaria in Africa should not be underestimated, as they often represent part of the duplicate herbarium specimens sent to Europe and some have also extended their collections through their own explorations.

A comprehensive list covering all the coffee specimens present in the herbaria of Europe and Africa is lacking. It would be very difficult to realise, because determinations made are sometimes inaccurate or only approximate descriptions. Many of the collected plants have still to be determined taxonomically and to be classified.

Natural Habitats of the Wild Coffees

Wild coffee trees are components of the understorey of tropical forests in Africa. The observations made by collectors include precise descriptions of habitats, which are quite distinct for *C. arabica* and other coffee species.

C. arabica. All botanists, who have explored the forests on the southwestern highlands of Ethiopia, agree in their observation that this is the centre of diversity of *C. arabica*, but that it is very difficult to find truly wild populations (Sylvain, 1955; Von Strenge, 1956; Meyer, 1965; Friis, 1979).

This species is indeed very common to the understorey of the forest, but most of the trees are regularly harvested by the local people, who usually carry out some form of maintenance by clearing the bush to facilitate picking of the ripe cherries. Guillaumet and Hallé (1978) distinguished the following stages from practically wild to truly cultivated coffee plots: (a) natural populations maintained *in situ*; (b) populations improved by introduction of young coffee trees from elsewhere; (c) farmers' plots established with plant material derived from wild coffee trees.

On the other hand, Thomas (1942) found in secondary forest on the Boma plateau in south-eastern Sudan truly wild populations of *C. arabica*, which apparently grew there without human interference.

Semi-wild populations of *C. arabica* can also be found at an altitude of about 1500 m in the upland forest on Mt Marsabit in north Kenya (Berthaud, Guillaumet, Le Pierrès and Lourd, 1980). It is not clear, whether these trees are truly wild or in earlier times brought from Ethiopia by man.

The Other Coffee Species. Apart from some subspontaneous populations of *C. liberica* and *C. canephora* found in Ivory Coast and Central Africa, most other wild coffee species appear to grow in their natural habitat, which consists of the understorey of tropical forests. However, these forests are an extremely complex and organised environment. All the collectors have noticed a really specific adaptation of coffee trees to elevation, altitude, rainfall and soil types.

In East Africa rainfall is largely controlled by altitude and mountains stand out as wet areas covered in dense forest (Lind and Morrison, 1974). Typical examples are found in Kenya: Mt Marsabit with *C. arabica* and the Taita Hills with *C. fadenii*. In contrast, where variations in altitude are much less pronounced as in West Africa, the well-drained hill tops are the driest zones. For example, in the Ira forest of Ivory Coast *C. stenophylla* grows on the top of the hill, while *C. canephora* and *C. liberica* occupy the lower humid zones (Figure 2.2). This location of *C. stenophylla* populations is common in the western part of the Ivory Coast and is an indication of its specific adaptation to dry conditions. In the eastern part of that country *C. stenophylla* is found in the dry lowlands with a semi-deciduous type of forest.

C. humilis represents a good example of the importance of the biotope: whatever the altitude, this species grows in the small talwegs and at the border of swamps of south-western Ivory Coast. This species is able to grow in very wet environments, which would explain its limited area of distribution. Thomas (1944) reported that in Uganda *C. eugenioides* is always found in the drier zones of the forest and *C. canephora* in more humid environments.

In the Central African Republic Berthaud and Guillaumet (1978)

Figure 2.2: Spatial Distribution of the Coffee Species According to Topography in the Ira Forest (Ivory Coast)

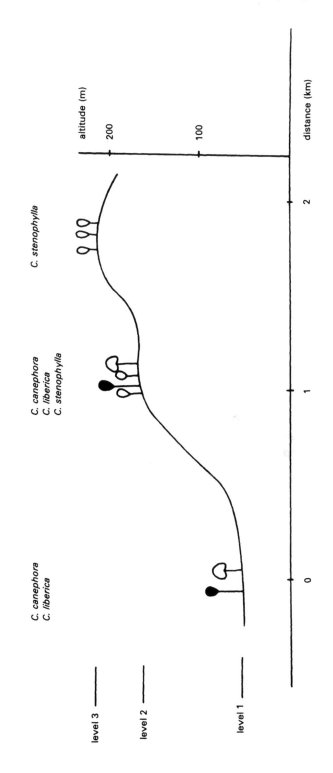

observed also the presence of *C. dewevrei* in the periphery of gallery forests bordering savannah land, where the soil is well-drained (Figure 2.3-1). In the same country *C. congensis* was found on the eroded banks of the Oubangui river and on sandy islands, but always in seasonally flooded areas. When the river is high, the coffee roots hold the sandy soil in place, with the result that each coffee tree stands on a little hillock (Figure 2.3-2).

On Madagascar, within the same geographical area where *C. resinosa* and *C. richardii* are found together, the distribution of the two species is closely associated with pedological characteristics: spodosols for *C. resinosa* and oxysols for *C. richardii.*

The habitats of the various coffee species correspond closely with specific biotopes and these should, therefore, be very well known in order to be able to discover wild coffee populations. Such ecological data are also essential for selecting appropriate growing conditions for living collections and testing of the agronomic qualities for cultivation.

Figure 2.3: Cross Section of Wild Coffee Habitats. 1. *C. dewevrei* (in dark) at the periphery of a gallery forest (Lihou — Central African Republic).
2. *C. congensis* (in dark) on sandy banks and small hillocks (Louma island — Central African Republic). (From Berthaud and Guillaumet, 1978.)

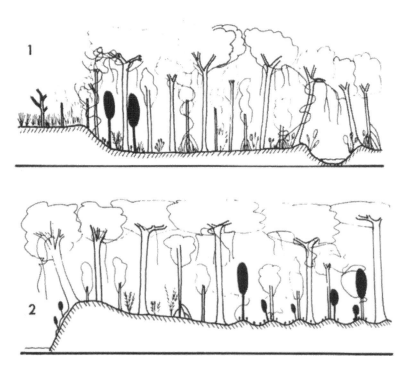

Botanical Classification of Coffee

Collecting Missions for Wild Coffee Species

As already mentioned, exploration for wild coffee species started together with that of other tropical plants in the sixteenth century and was particularly intense in Africa towards the end of the last century and during the first part of the twentieth century. Evidence for that era is mostly found in herbaria and very little in the existing living collections.

Of course, all coffee research centres maintain collections of the cultivated coffee species. These working collections are used for the improve-

Table 2.6: Coffee Collecting Missions Since 1960

Years	Countries explored	Organisations	Collectors' names	Coffee species	Germplasm maintenance countries
1964	Ethiopia	FAO	Meyer Monaco Narasimhaswamy Fernie Greathead	C. arabica	Ethiopia India Tanzania Costa Rica
1966	Ethiopia	ORSTOM	Guillaumet Hallé	C. arabica	Ethiopia Madagascar Ivory Coast Cameroon
1960 to 1974	Madagascar Mauritius Reunion Island Comoro Islands	Museum IFCC ORSTOM	Leroy Portères Vianney-Liand Guillaumet Charrier	Mascarocoffea (up to 50 taxa)	Madagascar
1975	Central Africa Republic	ORSTOM IFCC	Guillaumet Berthaud	C. congensis C. dewevrei C. canephora Caféier de la Nana	Ivory Coast Central African Republic
1975 to 1980	Ivory Coast	ORSTOM	Berthaud	C. liberica C. stenophylla C. canephora C. humilis Paracoffea sp. Psilanthus sp.	Ivory Coast
1977	Kenya	ORSTOM IFCC	Berthaud Guillaumet Lourd	C. arabica C. eugenioïdes C. zanguebariae C. fadenii	Kenya Ivory coast
1982	Tanzania	ORSTOM IFCC	Berthaud Anthony Lourd	C. zanguebariae C. mufindiensis C. sp.	Tanzania Ivory Coast
1983	Cameroon	ORSTOM IFCC	Anthony Couturon de Namur	C. canephora C. liberica C. congensis C. brevipes C. sp. Psilanthus sp.	Cameroon Ivory Coast

ment of *C. arabica* and *C. canephora*. Such collections have been extended regularly by introductions of more or less selected plant material from other coffee stations, botanic gardens or local plantations. The other coffee species are usually scarcely represented.

Awareness of the lack of variability in the existing coffee collections made the FAO and French organisations (ORSTOM, IRCC, the Museum at Paris) intensify their efforts to collect coffee germplasm during the last twenty years. The most important collecting missions are listed in Table 2.6, with indication of the countries covered, names of the collectors, the species collected and the countries where the material is maintained (Meyer *et al.*, 1968; Charrier, 1980; Berthaud, Guillaumet, Le Pierrès and Lourd, 1977; Leroy, 1961a, b, c, 1962, 1963, 1982). Emphasis in the collection of coffee germplasm was particularly on *C. arabica* because of its economic importance, but a number of non-cultivated species have also been collected, such as those of the section *Mascarocoffea* (caffeine free), the subsection *Pachycoffea*, *C. congensis* (progenitor of the Congusta hybrids), *C. eugenioides* (presumed progenitor of the allotetraploid *C. arabica*) and of the related genus *Psilanthus*.

Living Collections

A comprehensive botanical study should also include the World living collections of coffee. The FAO prepared an inventory of the existing coffee collections in 1960 (Krug, 1965) and this was updated in 1978-79 (unpublished). However, these reports do not fully reflect the actual importance and extent of the genetic diversity of the living collections. The location of living collections is depicted in Figure 2.4.

Important collections of *C. arabica* with material of the Ethiopian centre of genetic diversity are present at Jimma (Ethiopia), Turrialba (Costa Rica), Campinas (Brazil), Chinchina (Colombia), Lyamungu (Tanzania), Ruiru (Kenya), Foumbot (Cameroon), Man (Ivory Coast) and Ilaka-Est (Madagascar). There is a unique collection of species of the section *Mascarocoffea* at Kianjavato (Madagascar).

The main African coffee species are kept in the living collections at Divo and Man in Ivory Coast: more than 10 species of the section *Coffea* are represented by hundreds of thousands of genotypes collected from natural populations in different countries since 1965 (Table 2.6).

At the same time the working collections of *C. arabica* and *C. canephora* maintained in India, Cameroon, Togo, Angola and other countries also contain very valuable material.

In tropical Africa one can find, at least for the time being, wild coffee trees in undisturbed forests and subspontaneous coffee in traditional agricultural areas. However, it has become a matter of urgency to continue with the collection of coffee germplasm, where the natural habitats are being threatened by human activities.

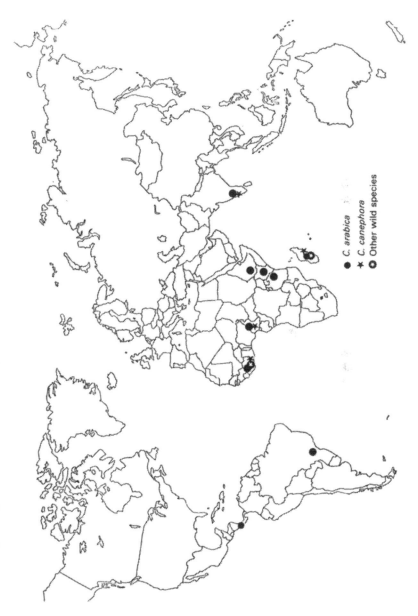

Figure 2.4: Existing Living Collections of Coffee Species in the World

26   *Botanical Classification of Coffee*

Figure 2.5 presents a map of Africa, indicating the areas of origin or high genetic diversity for coffee species of the sections *Coffea* and *Mascarocoffea*. This map is based on accumulated information from herbaria, collecting missions and living collections.

Cytotaxonomy and Reproductive Systems

Chromosome Number

Results of studies on chromosome numbers in coffee carried out since the 1930s have been reviewed by Sybenga (1960). The basic genome of the genus, $x = 11$ chromosomes, is typical for most of the genera of the family *Rubiaceae*. Chromosome counts were made for most species of the genus *Coffea* and for some representatives of the genus *Psilanthus*.

In the section *Coffea* all species are diploid with $2n = 22$ chromosomes, except for the tetraploid *C. arabica* which has $2n = 4x = 44$ chromosomes.

In the section *Mascarocoffea* the chromosome number of more than 20 species has been determined (Portères, 1962; Leroy and Plu, 1966; Friedman, 1970; Louarn, 1972). Species belonging to this section are all diploid ($2n = 22$) and the very large variability encountered in this section cannot therefore be attributed to variation in chromosome number.

In the genus *Psilanthus* the chromosome number $2n = 22$ has been confirmed for *P. humbertii* (Leroy and Plu, 1966), *P. horsfieldiana* (Bouharmont, 1959), *P. bengalensis* (Fagerlind, 1937) and *P. mannii* (Couturon, personal communication).

Deviations from the normal chromosome number do occur in exceptional cases. For example, in *C. arabica* a whole series of polyploids have been found (Sybenga, 1960): triploids ($3n = 33$), pentaploids ($5n = 55$), hexaploids ($6n = 66$) and octoploids ($8n = 88$). Haploids, or more correctly called di-haploids with $n = 2x = 22$ chromosomes, occur in low frequencies in seedling offspring as weak plants with narrow leaves and have been called *monosperma* (Mendes and Bacchi, 1940).

Diploid *C. canephora* with sectorial tetraploid chimaeras have been found and induction of artificial autotetraploidy by colchicine treatment of germinating seed (Mendes, 1939) or shoot apices (Berthou, 1975; Noirot, 1978) is possible. Methods of recovering the naturally occurring haploid plants in *C. canephora* have been developed recently (Couturon and Berthaud, 1982). Doubling with colchicine results in homozygous plants of *C. canephora*, which are of great interest to coffee breeding (see Chapter 3).

The morphology of coffee chromosomes was studied by Mendes (1938) and Bouharmont (1959, 1963). The latter author described some 10 species and prepared an average idiogram for the 11 chromosomes of the basic genome for the genus *Coffea*. Coffee chromosomes are relatively

Botanical Classification of Coffee 27

Figure 2.5: Natural Distribution of Coffee Species in Africa and Madagascar

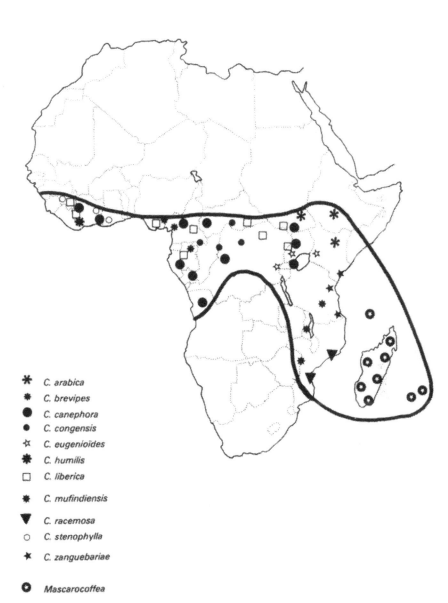

small (1 − 3μm), but modern methods of staining chromosomes (banding) open up possibilities of studying the chromosome morphology in more detail.

Reproductive Systems

Psilanthus. In this genus flowers have a short style and long corolla tube (Figure 2.6-1). System of incompatibility appears to be absent and most species, e.g. *P. bengalensis*, are autogamous. However, occasional cross pollination and consequently genetic exchange may occur, as heterozygous plants have been found after analysis by electrophoresis.

Coffea arabica. Flowers of this species are typical of the genus *Coffea*: short corolla tube, long style and exerted stamens (Figure 2.6-2). Such morphology would permit natural cross pollination, but nevertheless *C. arabica* is largely autogamous. Fruit set after self pollination is 60 per cent or higher (Carvalho *et al.*, 1969). In Ivory Coast, Le Pierrès (personal communication) effected self pollination on 32 trees of F_1 progenies of crosses between various accessions of *C. arabica* from Ethiopia. Of the 8,400 flowers 5,400 set fruit after self pollination. This means a success rate of 65 fruits to 100 flowers. However, this percentage varied from 3 to 91 per cent between trees, which could indicate variation in the degree of self fertility in wild populations of *C. arabica*. Meyer (1965) reported 40-60 per cent cross pollination in wild populations of *C. arabica* in Ethiopia.

Most studies on the degree of natural cross pollination were carried out on cultivars of *C. arabica*, which underwent many cycles of selection. By using the recessive marker genes *Cera* (yellow endosperm) and *Purpurascens* (purple leaves) Carvalho and Krug (1949) in Brazil and Van der Vossen (1974) in Kenya found percentages of natural out-pollination ranging from 7 to 15 per cent.

Figure 2.6: 1. Flowers of *Psilanthus*, on the Right: Flower with Removed Corolla; 2. Flowers of *Coffea*

Diploid Species of the Genus Coffea. Most diploid coffee species have proved to be highly self incompatible including 24 tested species of the section *Mascarocoffea.* Devreux, Vallayes, Pochet and Gilles (1959) describe how, after self pollination, pollen tube growth on the stigma of *C. canephora* becomes distorted and further penetration into the style is blocked. Berthaud (1980) produced further evidence for a gametophytic system of incompatibility in *C. canephora*, which is controlled by one gene with multiple alleles. A similar mechanism appears to operate also in Congusta coffee (a hybrid between *C. congensis* and *C. canephora*). On the other hand, observations on *C. liberica* indicate that the incompatibility reaction can be delayed until the day of anthesis. Penetration of the pollen tubes in the stigma and style could take place with 'bud pollination' (Hamon, personal communication).

One notable exception appears to be an accession in the living coffee collections in Ivory Coast of unknown origin but resembling *C. brevipes* and self-compatible (Le Pierrès and Louarn, personnal communication). We observed that some pollen tubes grow through the style while others were blocked at the papillary zone of the stigma, as in incompatible combinations. The offspring of such trees was very homogeneous. Could this mean mutation of an S allele making it inoperative?

Because of the high degree of self-incompatability in most diploid species, a high level of heterozygosity will be maintained in populations and this has great consequences to breeding (see Chapter 3).

Wild Populations of Coffea

The variability existing in natural coffee populations has been studied to a limited extent so far. Portères (1937) was one of the first to describe the variability of populations of *C. canephora, C. liberica* and *C. stenophylla* and their offspring in Ivory Coast. Considerable work was also carried out on Java on numerous progenies of introduced coffee species (Cramer, 1957).

Botanists and geneticists have in their recent efforts to explore and preserve coffee genetic resources applied various methods to describe the variability present in wild coffee populations, including: (a) morphological observations and numerical taxonomy; (b) analysis of electrophoretic variants; (c) studies of the frequency distribution of incompatibility alleles within and between populations; and (d) genetic analyses with progenies of controlled crosses.

A few examples of natural coffee populations may serve as an illustration.

Coffea arabica

The great variability in natural populations of *C. arabica* has been apparent to most botanists and geneticists, who visited and explored the southwestern highlands of Ethiopia.

An electrophoretic analysis of six enzyme systems produced similar homogeneous patterns, both for accessions from Ethiopia and from Mt Marsabit in Kenya (Berthou and Trouslot, 1977, and unpublished). This made it possible to describe the electrophoretic type for *C. arabica*. It is of interest to note here, that this analysis generally pointed to an expression of two alleles at each locus, which would support earlier conclusions from classical genetic analyses (Carvalho *et al.*, 1969) that the allotetraploid *C. arabica* is a functional diploid for most gene expressions.

This uniformity of the species is lost if the morphological characteristics of germplasm collections (origin differences) or their progenies (family differences) are examined. Actually, hierarchical variance analysis carried out by Reynier, Pernes and Chaume (1978) and Louarn (1978) showed that differences between origins and between families are both significant, but with the interorigin variance component being several times higher.

The caffeine content of *C. arabica* germplasm from Ethiopia varies from 0.8 to 1.9 per cent (of dry matter), a variation not correlated with geographical distribution.

It is difficult to differentiate 'varieties' within this (semi-)wild material because of the considerable variation existing between trees of the same subpopulation and such trees produce also heterogeneous progenies. The term variety (or cultivar) should therefore be reserved for the cultivated forms of this species, which are indeed homogeneous due to several generations of line selection.

C. canephora: the 'Nana' Population

This population is located in a gallery forest of the savanna zone in the Central African Republic and it represents the northern limits for the distribution of *C. canephora*. This material is well adapted to the climatic and pedological conditions of that area and it has been cultivated locally since 1923. The original population was rediscovered in 1975 (Berthaud and Guillaumet, 1978) and about 100 plants raised from seed and cuttings were added to the coffee collection in Ivory Coast.

Observations on morphological characteristics and floral biology indicate that there is considerable variation, but no sub-populations could be detected. Some characters like colouration of mature fruits appear to be fixed, while others are heterozygous. This was also confirmed by electrophoretic analysis. An analysis of incompatibility points to a considerable number of S alleles being present in that population.

Taxonomically, there is now sufficient evidence to consider the Nana population as part of the species *C. canephora*. The diversity found within this material appears similar to that encountered in other populations of *C. canephora*.

C. stenophylla

Wild populations of this species exist in west and east Ivory Coast. Phenotypically these populations are easily distinguishable as is shown in Table 2.7. Electrophoretic analysis indicates that these populations are very homogeneous but the alleles of the western populations differ from those fixed in the eastern ones. In a population of 1000 trees of western origin, the total number of S alleles was less than 10, which would suggest that the population originates from five parent trees at the most. Notwithstanding the expected relatively low genetic diversity, caffeine content determinations gave a range from 0.9 to 1.9 per cent (on dry matter), a variation similar to *C. arabica*.

The geographical isolation apparently produced considerable genetic diversity between populations, as interpopulation crosses gave distinct hybrid vigour.

C. zanguebariae

When examining herbarium specimens one realises that the specific name applies to two taxa, one without and one with fruit stalks. These forms are indicated here as A and B, which is synonymous with the taxa *Coffea* sp. A and *C. pseudozanguebariae* described by Bridson (1982). Various populations, collected in Kenya by Berthaud *et al.* (1980), are now under observation in Ivory Coast. The main characters of these two taxa are compared in Table 2.8. Analysis by electrophoresis has also indicated differences between the two taxa. Three of the studied populations were found to belong to form B, while one included both forms as well as intermediary types. Plants of the A form have all flowers open on the same day, while trees of form B flower over three consecutive days (Hamon *et al.*, 1984)

Hybridisation between the two forms is possible, although the chances of natural cross pollination between trees of the B form only should be higher.

Table 2.7: Morphological Characteristics of the Western and Eastern Forms of *C. stenophylla* in Ivory Coast

Observed character	Eastern form	Western form
Secondary branching pattern	very numerous branches	numerous branches
Leaves	very small	small
Flowers per fascicle	1	2
Flower shape	globulous	oblong
Fruit colour	black	black

Table 2.8: Morphological Characteristics of the A and B Forms of *C. zanguebariae*. (After Hamon and Anthony, to be published.)

	A form	B form
Leaf-thickness	thick	thin
Stipule length	long	short
Domatia	small	big
Day range between shower and first flowering	6	6
Number of flowering days	1	3
Corolla lobes number	5	6-7
Fruit stalk length	very short	long
Fruit colour before ripening	green	brown

However, Anthony (personal communication) and Louarn (1982) carried out crosses between trees of the A and B forms as well as between the A form and several other species. Crosses between A and B forms gave a considerably lower fruit set than crosses between the A form and for instance *C. eugenioides* and *C. racemosa*. It appears that a genetic barrier has evolved between the two forms but not due to geographic separation.

Such a genetic barrier had not evolved, however, between the earlier mentioned subpopulations of *C. stenophylla*, regardless of the considerable morphological differences.

C. liberica

This species also shows distinctly different forms: *C. liberica* var. *liberica* from West Africa (Guinea, Liberia, Ivory Coast) and *C. liberica* var. *dewevrei* from Central Africa (usually called Excelsa coffee). Forms of *C. liberica* found in Cameroon appear to be intermediate between these two extremes, but this still requires verification.

Mascarocoffea

A large number of different species (or taxa) have been distinguished already (Charrier, 1978). However, the relatively high success of interspecific hybridisation within this section indicates that only weak genetic barriers have evolved. Figure 2.7 presents a hierarchical classification, mainly based on morphological characteristics.

Multispecific Populations

The existence of multispecific wild coffee populations is supported by observations made during various collecting missions carried out in the past 20 years.

Apart from the coexistence of the A and B forms of *C. zanguebariae* in one population in Kenya, the following associations have been encountered in Ivory Coast and in the Central African Republic:

Figure 2.7: Differentiation of the Taxa of *Mascarocoffea* by a Hierarchical Classification

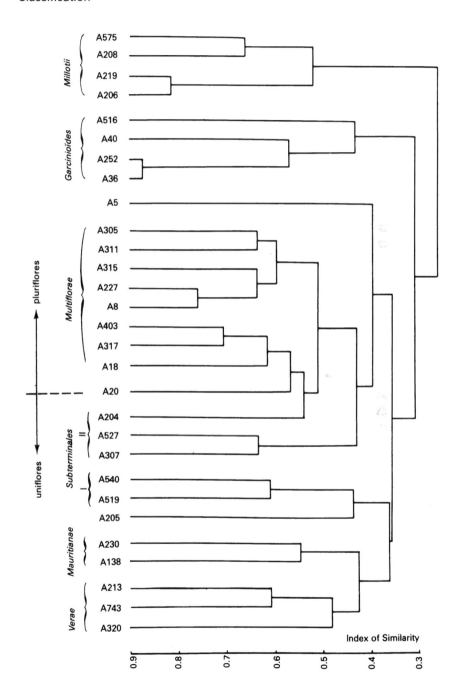

- *C. liberica* + *C. canephora* (Gbapleu, Ivory Coast)
- *C. liberica* + *C. humilis* (Tai, Ivory Coast)
- *C. liberica* + *C. stenophylla* + *C. canephora* (Ira, Ivory Coast)
- *C. liberica* + *C. congensis* + *C. canephora* (Bangui, Central African Rep.)

Multispecific populations have also been found on Madagascar with species of the section *Mascarocoffea* (Charrier, 1978), while in Uganda associations of *C. canephora* with *C. liberica* or *C. eugenioides* have been reported by Thomas (1944).

These observations made on wild coffee populations strongly suggest that:

(1) hybridisation between geographically adjoining species in wild populations occurs only sporadically. On Madagascar hybrids between two species were found on two occasions, while in Ivory Coast occasional interspecific hybrids were noticed in the seed progeny of wild coffee trees;
(2) interspecific hybrids represent a transitory state, which may explain why they are rarely encountered in natural conditions.

Various factors appear to restrict chances of gene exchange between species in multispecific populations:

(1) species coexist more side by side within a limited area rather than in a real mixture: their separation can be due to microclimatic (different sides of a hill), pedologic (different soils) or topographic differences;
(2) flowering patterns differ between species: *C. canephora* usually flowers early in the year, while *C. liberica* and *C. stenophylla* tend to flower later: the time interval between a flowering inducing rainshower and anthesis is also often one day longer for *C. canephora* than for *C. liberica*.

Clearly, studies of wild coffee populations have contributed considerably to a better understanding of the relation between species of multispecific associations. Genetic exchange between coffee species through interspecific hybridisation should be taken into account when formulating theories on the evolution of coffee species.

Evolution of the Coffee Gene Pool

Most studies on the genetic relationships of species in the genus *Coffea* before 1960 were almost entirely restricted to the subsection *Erythro-*

Table 2.9: Review of the Cytogenetic Behaviour of Interspecific Crosses Between Diploid *Coffea* Species

Diploid F₁ hybrids	Chromosome associations			PMC with 11 bivalents %	Pollen fertility	References
	Univalents	Bivalents				
C. canephora × *C. congensis*	0.04-0.74	10.63-10.98			high	Leliveld (1940)
	0.20-0.52	10.74-10.90		74-90	89-93	Charrier (1978)
C. liberica × *C. dewevrei*	Normal meiose (Rhoades)				high	Carvalho and Monaco (1968)
C. canephora × *C. dewevrei*	1.44	10.28			middle	Leliveld (1940)
C. canephora × *C. liberica*	0.30-1.40	9.93-10.66			39	Chinnappa (1970)
C. canephora × *C. neo-arnoldiana*	1.16-1.20	10.40-10.42		50-58	64	Louarn (unpublished)
C. canephora × *C. eugenioides*	1.30-2.22	9.89-10.35		23-44	43	Louarn (1976)
Liberio-excelsoides × *C. eugenioides*			about 2 I			Vishveshwara (1975)
C. canephora × *C. kapakata*	1.28-1.64	10.18-10.36		42-48	35	Louarn (unpublished)
	1.50	10.25			middle	Leliveld (1940)
C. canephora × *C. lancifolia*	3.20	9.40		8	8	
C. canephora × *C. resinosa*	4.40-6.40	7.80-8.80		0-2	4-7	Charrier (1978)
C. canephora × *C.* sp A311	5.04	8.48		4	6	

coffea. More recently, however, the crossability and chromosomal homology among other coffee species have received much attention, particularly in Brazil (Carvalho and Monaco, 1968), India (Vishveshwara, 1975), Madagascar (Charrier, 1978) and in Ivory Coast (Capot, 1972; Louarn, 1982). Coffee taxonomists have also started to establish the biochemical and serological affinities of coffee taxa through research on enzyme polymorphism, cytoplasmic DNA and serological reactions.

Considerable progress has thus been made in the understanding of the genetic-historical relationship among coffee species.

The Diploid Coffee Species

From the extensive cytogenetic studies on interspecific hybrids — a summary of the most relevant data together with major references is presented in Table 2.9 — the following main conclusions can be drawn as regards the genetic differentiation of the diploid coffee species

(1) Absolute crossing barriers appear to be absent within the genus *Coffea*, although considerable variation in the degree of successful hybridisation exists between species of different taxa.

(2) Exact quantitative information on the genetic relation of species is difficult to obtain due to the influence of genotype, crossing techniques and environment on the rate of success of interspecific hybridisation.

(3) Maternal effects can be very significant in crosses between species, which differ in the time interval between flowering and ripe fruits: the best results are usually obtained when the species with the quickest formation of the albumen is used as female parent.

(4) In general, crosses between species within one group present the least difficulty: for instance *C. canephora* and *C. congensis*, which both belong to the subsection *Erythrocoffea*, or species of the section *Mascarocoffea*. However, considerable success can sometimes be obtained as well with taxa of different taxonomic groups: for instance *C. liberica* and *C. eugenioides* or *C. dewevrei* and *C. stenophylla*. The species *Psilanthopsis kapakata* (erroneously classified in another genus by Chevalier) can easily be crossed with *C. canephora* and *C. eugenioides*.

(5) *C. eugenioides* performs in general much better in crosses with all other diploid taxa than *C. canephora*. This difference is particularly evident in crosses with species of the section *Mascarocoffea* (Figure 2.8)

(6) Intergeneric hybridisation between *Coffea* and *Psilanthus* has been unsuccessful so far.

All diploid species of the genus *Coffea* have maintained a similar chromosomal structure, which would arise from the same basic A genome (monophyletic origin). Charrier (1978) observed a variable rate of PMCs

Botanical Classification of Coffee 37

Figure 2.8: Genetic Relationships Between Species of *Coffea* (according to number of hybrids per 100 flowers). From Louarn, 1982 (1,2,3); Charrier, 1978 (4); Carvalho and Monaco, 1968 (5)

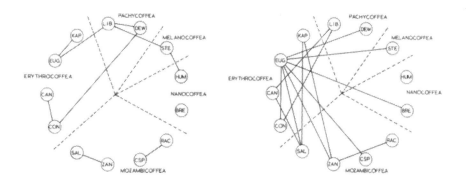

(1) More than 19 hybrids per 100 flowers

(2) 6-18 hybrids per 100 flowers

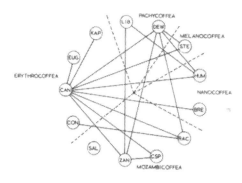

(3) Less than 5 hybrids per 100 flowers

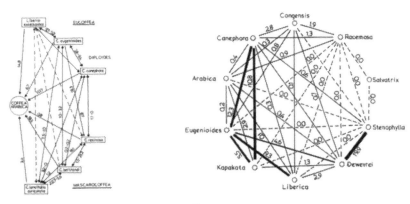

(4)

(5)

(pollen mother cells) with 11 bivalents in F_1 hybrids between diploid species and found this to be highly correlated with fertility. It appears that all diploid coffee species have preserved the identity of their common origin during the evolutionary history notwithstanding the geographical isolation. This lack of chromosomal differentiation was also observed by Bouharmont (1959; 1963) in the karyotypes of ten African coffee species. An exception is the chromosomal aberrations found in hybrids between *C. kianjavatensis* and *C. canephora* (Lanaud, 1979). The genetic diversification of the diploid coffee species would therefore result essentially from genetic differentiation, which causes various levels of interspecific incongruity.

A good example of species which are morphologically and ecologically clearly distinct, but show almost normal chromosome pairing and good fertility in hybrids, are *C. canephora* (distribution in West and Central Africa) and *C. congensis* (limited to the banks and flooded islands of the Congo and Oubangui rivers). The latter species occupies a specific ecotype well isolated from the former by a very strict ecological barrier (Berthaud and Guillaumet, 1978). In the subsection *Pachycoffea* the exact position of *C. liberica sensu stricto* (berries with thick mesocarp) in relation to *C. dewevrei* (berries with thin mesocarp) is not yet clear. On the other hand, the abundant number of taxa in the section *Mascarocoffea* described by Leroy often correspond to allopatric populations with distinct morphological characteristics but nevertheless without strongly developed crossing barriers. Apparently, effective geographical isolation has produced genetically divergent populations through genetic drift and natural selection pressure. A similar situation could also be found in East African coffee species (Leroy, 1982; Hamon, Anthony and Le Pierrès, 1984). From a taxonomic point of view some taxa found in East Africa should not be classified as different species.

On the other hand, genetic divergence is in species like *C. canephora*, *C. liberica* and *C. eugenioides* exhibited as partial incongruity. For example, Louarn (1982) obtained fairly fertile F_1 hybrids between these three species, but their meiotic behaviour was characterised by a reduced pairing of the chromosomes (only 40-50 per cent of the PMCs had 11 bivalents). In addition there is a genetic effect on the chromosome pairing and the hybrid fertility within the interspecific combination *C. liberica* × *C. canephora* (Louarn, 1980). Similar genetic divergence appears to exist also between some series of *Mascarocoffea*.

There is a particularly sharp differentiation between *C. canephora* and the section *Mascarocoffea*: the rare F_1 hybrids obtained are weak and almost completely sterile. This genetic divergence is less pronounced for *C. eugenioides*, since crosses between this species and *Mascarocoffea* produces a higher proportion of fairly fertile F_1 hybrids.

It is in fact astonishing that a geographic isolation, started in the Upper

Cretaceous, when Madagascar was separated from the main African continent, did not have a more profound effect on the chromosomal differentiation between coffee species. In this respect it is interesting to note the affinity of the Malagasy coffee species with East African coffee species (Charrier, 1978), in particular the similarity between *C. grevei* and *C. rhamnifolia* (ex *C. paolia*) (Leroy, 1980).

Little work has so far been done on the intergeneric relationship between *Coffea* and *Psilanthus*.

Cytogenetic Evidence for the Origin of Coffea arabica

C. arabica, the only tetraploid species in the genus *Coffea*, is indigenous to the highlands of south-western Ethiopia and south-eastern Sudan. The diploid meiotic behaviour and the fact that its centre of genetic diversity is situated outside the area of distribution of the diploid coffee species, indicate an allotetraploid origin (Carvalho, 1952). According to Grassias and Kammacher (1975), *C. arabica* has to be considered as a segmental allotetraploid.

C. eugenioides and *C. canephora* (or *C. liberica* or *C. congensis*) have often been assumed to be the ancestral parents of *C. arabica* (Carvalho, 1952; Cramer, 1957; Narasimhaswamy, 1962). However, meiotic pairing of chromosomes of the two genomes in interspecific hybrids of *C. eugenioides* and *C. canephora* was observed to be better (Louarn, 1976) than in dihaploid plants of *C. arabica* (Mendes and Bacchi, 1940; Berthaud, 1976). Besides, duplication of such interspecific hybrids (allodiploids) produces tetraploids with autotetraploid meiotic behaviour, such as the formation of quadrivalents. Louarn (personal communication) has observed this in different interspecific combinations.

Triploid hybrids, as a result of crosses between *C. arabica* and diploid species, show vigorous growth, but they are almost completely sterile as would be expected. Table 2.10 summarises the observations on meiotic behaviour of interspecific hybrids between *C. arabica* and different diploid coffee species made by various coffee geneticists. The number of bivalents plus trivalents formed during meiosis is, with few exceptions, close to 11. This would suggest that one genome of *C. arabica* has close affinity to the genome present in the diploid species, therefore that in the genus *Coffea* all species share the same basic genome and have a monophyletic origin.

C. arabica could have arisen from natural hybridisation between two ancestral diploid coffee species followed by unreduced gamete formation (see Demarly, 1975, for a description of the rare events that can lead to the occurrence of unreduced gametes). The degree of homology of the two genomes could have been high as a consequence of the monophyletic origin of the participating species, but various mechanisms (preferential pairing, genetic regulation of the synapsis) are likely to have played an

Table 2.10: Review of the Cytogenetic Behaviour of Interspecific Crosses Between *C. arabica* and Diploid *Coffea* Species

Triploid F₁ hybrids	Chromosome associations				PMC with 11 bivalents %	Pollen fertility	References
	I	II	III	Others			
C. arabica × *C. canephora*	14.4	5.4	2.60	—	—		Krug and Mendes (1940)
	7.80	9.75	1.61	0.21	89		Kammacher and Capot (1972)
	7.98	9.55	1.93	0.04			Chinnappa (1968)
	9.87	9.57	1.33	—		4	Louarn (unpublished)
C. arabica × *Liberio-excelsoides*	9.28	9.64	1.44	0.03	90	6	Charrier (1978)
C. arabica × *C. eugenioides*	9.70	9.64	1.34	—	77	7	Louarn (unpublished)
C. arabica × *C. kapakata*	10.07	9.45	1.33	—	92		Monaco and Medina (1965)
C. racemosa × *C. arabica*	11.3	9.7	0.80	—	62		Medina, 1963
C. arabica × *C. betrandi*	11.9	8.6	1.3	—	49	1	
C. arabica × *C. perrieri*	12.4	8.2	1.4	—	24	2	Charrier (1978)
C. arabica × *C. pervilleana*	14.3	8.6	0.5	—	21	1	
C. arabica × *C.* sp. A311	17.1	7.7	0.2	—	10	1	
					2		

important role in the progressive diploidisation from the archetype tetraploid to the present amphidiploid *C. arabica*.

Lobreau-Callen and Leroy (1980) observed in their palynological studies that *C. arabica* produces two types of pollen: one type related to *C. canephora* and one closely resembling pollen of *C. rhamnifolia*, a xerophytic species indigenous to the coastal regions of Somalia and Kenya. This would not necessarily contradict the foregoing cytogenetic evidence, but a definite conclusion on the origin of *C. arabica* will have to await further studies, including the intergeneric level between *Coffea* and *Psilanthus*.

Biochemical and Serological Affinities in Coffea

Enzyme Polymorphism. Studies on enzyme polymorphism in *Coffea* started with Payne and Fair-Brothers (1976), who compared total crude proteins and malate dehydrogenase in seeds of *C. arabica* and *C. canephora*, and Guedes and Rodrigues (1974), who studied the phenoloxidase variability after polyacrylamide electrophoresis of extracts taken from genotypes of *C. arabica*, used as differentials for the identification of physiological races of leaf rust (*Hemileia vastatrix*).

Methods of analysing enzyme polymorphism have been further developed to study the genetic affinities between various coffee populations by Berthou and Trouslot (1977) and Berthou *et al.* (1980) in Ivory Coast. The horizontal starch-gel electrophoresis technique was applied initially to three and later to eight enzyme systems and this led to the identification of a number of loci. Allozymatic frequencies were estimated for natural coffee populations. Estimates of genetic distances between species, whereby an index developed by Nei (1972) for 3-8 enzyme systems was used, gave the following information:

(1) *C. canephora* and *C. congensis* have the same allozymes with different frequencies; the genetic distance between the species is larger than between populations of the same species: the Nana taxon is therefore clearly to be considered as part of *C. canephora*.

(2) The genetic distance between *C. liberica* from Ivory Coast and *C. dewevrei* from the Central African Republic is of the same order as that between *C. liberica* and *C. humilis* from Ivory Coast. This evidence justifies a clear distinction between *C. liberica* and *C. dewevrei*, which both belong to the subsection *Pachycoffea*.

(3) The genetic distances between the species *C. canephora, C. liberica* and *C. eugenioides* are considerable.

(4) The enzymic affinities of *C. arabica* with all different diploid coffee species are similar. According to Berthou and Trouslot (1977) *C. arabica* would result from the complementary electrophoretic bands of acid phosphatases and esterases of *C. eugenioides* and *C. canephora* or *C. congensis*.

(5) Lower enzymic affinities have been found between species of the genus *Coffea* and the genus *Psilanthus* such as *Paracoffea ebracteolata*.

Cytoplasmic DNA. Berthou et al. (1980, 1983) have applied methods of identifying DNA of cytoplasmic organelles (chloroplasts and mitochondria) to coffee. Such methods include separation by electrophoresis on agarose slab gels of fragments of DNA obtained by bacterial restriction enzymes.

Electrophoresis of fragments of chloroplast DNA, obtained by the Hpa II enzyme (from *Haemophilus parainfluenzae*) suggest that the following species have a similar origin: *C. arabica* and *C. eugenioides*; *C. canephora*, the Nana taxon and *C. congensis*

Electrophoresis of fragments of mitochondrial DNA, obtained by action of the Sal I enzyme (from *Streptomyces albus*), suggested the following affinities: (a) great similarity between *C. canephora* and the Nana taxon; (b) great similarity of *C. arabica* with *C. eugenioides* and *C. congensis*; (c) considerable divergence between *C. canephora* and *C. arabica* or *C. eugenioides*; (d) wide genetic divergence between *C. dewevrei* and *C. liberica* and *Paracoffea ebracteolata*

Evidently, electrophoresis of mitochondrial DNA confirms the distinction between the subsections *Abyssinicae* (*C. arabica*, *C. eugenioides* and *C. congensis*) and *Robustae* (*(C. canephora*) earlier proposed by Lebrun (1941). The affinity of *C. canephora* with *C. congensis*, as indicated by the similarity of chloroplast DNA is clear, but these two species show differences in their mitochondrial DNA. This type of analysis has also indicated that the Nana taxon should be considered as an ecotype of *C. canephora*.

Serological Affinities. Antisera obtained for *C. arabica* and *C. canephora* were used in serological reactions with soluble antigens from seed of various coffee species (Hofling and Oliveira, 1981). It was possible to establish the following relationship: *C. arabica* has more affinity with *C. congensis and C. eugenioides* than with *C. canephora*.

Other Chemical Affinities. Chromatographic analysis of the flavonoid component (Longo, 1975) indicated that *Psilanthopsis kapakata* belongs to the genus *Coffea* rather than *Psilanthus* as earlier proposed by Chevalier.

Additional data, based on simple chemical tests applied by Ram, Sreenivason and Vishveshwara (1982), suggest that *C. eugenioides*, *C. racemosa* and *P. kapakata* are more closely related to each other than to *C. salvatrix*. Nevertheless, all four species are rightly classified under the subsection *Mozambicoffea* (see also Chapter 13: 334).

Conclusions

From this review of advances in the taxonomy and botany of coffee the following main conclusions can be drawn.

(1) A general revision of coffee taxonomy should take into account the new insights into the relationships of coffee species, acquired by the extensive exploration of natural habitats and centres of high genetic diversity, as well as by the application of modern concepts of the biological series and of biosystematic classification systems.

(2) During the evolutionary organisation of the coffee gene pool differentiation into species was not accompanied by the development of strong crossing barriers. This offers considerable prospects for introgressing desirable characters into cultivated species or the development of new cultivars by interspecific hybridisation. Of particular interest are the taxonomic affinity of *C. congensis* and *C. canephora* (Congusta hybrids) and the crossability of *C. eugenioides* with many other diploid species. There are also indications that *C. eugenioides, C. congensis, C. canephora* and the allotetraploid *C. arabica* have originated from common ancestral forms.

(3) There is a great urgency for intensifying studies of the still existent wild coffee populations, especially in those geographic areas where such populations are most threatened by extinction. The preservation of this coffee germplasm should be secured, either by conservation of the natural forest habitats or by establishing living collections. The following regions require special attention in this respect:

— The highlands of south-western Ethiopia and south-eastern Sudan (Boma Plateau): the centre of high genetic diversity for *C. arabica*;
— Uganda and Central Africa (Gabon, Congo, Zaire): the major areas of distribution for many diploid coffee species;
— The Malagasy Republic: the centre of genetic diversity for species of the section *Mascarocoffea.*

References

Berthaud, J. (1976) 'Etude cytogénétique d'un haploide de *Coffea arabica* L' *Café Cacao Thé, 20*, 91-6

Berthaud, J. (1980) 'L'incompatibilité chez *Coffea canephora*: méthode de test et déterminisme génétique', *Café Cacao Thé, 24*, 267-74

Berthaud, J. (1985) 'Les Ressources Génétiques Chez les Caféiers Africains: Populations Sylvestres, Échanges Génétiques et Mise en Culture' Doctoral Thesis, University of Paris XI, Orsay (in press).

Berthaud, J., Guillaumet, J.L , Le Pierrès, D. and Lourd, M. (1977) 'Les prospections des caféiers sauvages et leur mise en collection' in *8th International Colloquium on the Chemistry of Coffee*, ASIC, Paris, pp. 365-72

Berthaud, J. and Guillaumet, J.L., (1978) 'Les caféiers sauvages en Centrafrique' *Café Cacao Thé, 22,* 171-86

Berthaud, J., Guillaumet, J.L., Le Pierrès, D. and Lourd, M. (1980) 'Les caféiers sauvages du Kenya: prospection et mise en culture', *Café Cacao Thé, 24,* 101-12

Berthou, F. (1975) 'Méthode d'obtention de polyploides dans le genre *Coffea* par traitements localisés de bourgeons à la colchicine', *Café Cacao Thé, 19,* 197-202

Berthou, F. and Trouslot, P. (1977) 'L'analyse du polymorphisme enzymatique dans le genre *Coffea*: adaptation d'une méthode d'électrophorèse en série: premiers résultats', in *8th International Colloquium on the Chemistry of Coffee,* ASIC, Paris, pp. 373-83.

Berthou, F., Trouslot, P., Hamon, S., Vedel, F. and Quetier, F. (1980) 'Analyse en électrophorèse du polymorphisme biochimique des caféiers: variation enzymatique dans dix-huit populations sauvages: *C. canephora, C. eugenioides* et *C. arabica*' *Café Cacao Thé, 24,* 313-26

Berthou, F., Mathieu, C. and Vedel, F. (1983) 'Chloroplast and mitochondrial DNA variation as indicator of phylogenetic relationships in the genus *Coffea* L' *Theoretical and Applied Genetics, 65,* 77-84

Bertrand, G. (1902) 'Recherche et dosage de la caféine dans plusieurs espèces de Café' *Bulletin de la Société de Pharmacie, 5,* 283-5

Bouharmont, J. (1959) 'Recherches sur les affinités chromosomiques dans le genre *Coffea*', *Publication INEAC,* Série Scientifique, *77*

Bouharmont, J. (1963) 'Somatic chromosomes of some *Coffea* species', *Euphytica, 12,* 254-7

Bridson, D. (1982) 'Studies in *Coffea* and *Psilanthus* (*Rubiaceae* sub fam. *Cinchonoideae*) for part 2 of the 'Flora of Tropical East Africa: *Rubiaceae*', *Kew Bulletin* (Royal Botanic Garden), *36,* 817-59

Capot, J. (1972) 'L'amélioration du Caféier en Côte d'Ivoire. Les hybrides 'arabusta', *Café Cacao Thé, 16,* 3-18

Carvalho, A. (1952) 'Taxonomia de *Coffea arabica* L. Caractères morphologicos dos haploides', *Bragantia, 12,* 201-12

Carvalho, A. and Krug, C.A. (1949) 'Genetica de Coffea XII. Hereditariedade da côr amarela da semente', *Bragantia, 9,* 193-202

Carvalho, A. and Monaco, L.C. (1968) 'Relaciones geneticas de especies selectionadas de *Coffea'*, *Café, 4,* 3-19

Carvalho, A., Ferwerda, F.P., Frahm-Leliveld, J.A., Medina, D.M., Mendes, A.J.T. and Monaco, L.C. (1969) 'Coffee (*Coffea arabica* L. and *C. canephora* Pierre ex Froehner)' in F.P. Ferwerda and F. Wit (eds). *Outlines of Perennial Crop Breeding in the Tropics,* Veenman & Zonen, Wageningen, pp. 189-241

Charrier, A. (1978) 'La structure génétique des caféiers spontanés de la région Malgache (*Mascarocoffea*). Leurs relations avec les caféiers d'origine africaine (*Eucoffea*)', *Mèmoires ORSTOM (Paris), no. 87*

Charrier, A. (1980) 'La conservation des ressources génétiques du genre *Coffea*', *Café Cacao Thé, 24,* 249-57

Charrier, A. (1982) 'L'amélioration génétique des cafés', *La Recherche, 13* (136), 1006-16

Chevalier, A. (1938) 'Essai d'un groupement systématique des caféiers sauvages de Madagascar et des îles Mascareignes', *Revue Botanique Appliquée et d'Agriculture Tropicale, 18,* 825-43

Chevalier, A. (1947) 'Les caféiers du globe III. Systématique des caféiers et faux caféiers. Maladies et insectes nuisibles', *Encyclopedie biologique 28,* Fascicule III, P. Lechevalier, Paris

Chinnappa, C.C. (1968) 'Interspecific hybrids of *Coffea canephora* and *C. arabica*', *Current Science, 37,* 676-7

Chinnappa, C.C. (1970) 'Interspecific hybrids of *C. canephora* and *C. liberica*', *Genetica, 41,* 141-6

Chinnappa, C.C. and Warner, B.G. (1981) 'Pollen morphology in the genus *Coffea* (*Rubiaceae*) and its taxonomic significance', *Botanical Journal of the Linnean Society, 83,* 221-36

Clausen, D., Keck, D. and Hiesey, W.M. (1945) 'Experimental studies on the nature of species II. Plant evolution through amphiploidy and autoploidy, with examples from *Maidinae*', *Carnegie Institute,* Washington, *Publ. no. 564*

Coste, R. (1968) *Le Caféier*, G.P. Maisonneuve & Larose, Paris
Couturon, E. and Berthaud, J. (1982) 'Presentation d'une méthode de récupération d'haploïdes spontanès et d'obtention de plantes diploides homozygotes chez *C. canephora*', in *10th International Colloquium on the Chemistry of Coffee*, ASIC, Paris, pp. 385-91
Cramer, P.J.S. (1957) in F.L. Wellman, *Review of Literature of Coffee Research in Indonesia*. SIC Editorial International American Institute of Agricultural Sciences, Turrialba, Costa Rica
Demarly, Y. (1975) 'Amélioration du caféier liée aux progrès génétique', in *7th International Colloquium on the Chemistry of Coffee*, ASIC, Paris, pp. 423-35
Devreux, M., Vallayes, G., Pochet, P. and Gilles, A. (1959) 'Recherches sur l'autostérilité du caféier robusta (*C. canephora* Pierre)', *Publication INEAC Série Scientifique, 78*
Dobshansky, T. (1970) *Genetics of Evolutionary Process*, Colombia University Press, New York
Fagerlind, F.O. (1937) 'Embryologische, zytologische und bestaubungsexperimentelle Studien in der Familie *Rubiaceae* nebst Bemerkungen uber einige Polyploiditat problem. *Acta Horticulturae, 11*, 195-470
Forster, R.B. (1980) 'Heterogeneity and disturbance in tropical vegetation' in M.E. Soule and B.A. Wilcox (eds.), *Conservation Biology*, Sinauer Associates Inc., Sunderland, pp. 75-92
Friedman, F. (1970) Etude biogéographique de *Coffea buxifolia* Chev.', *Café Cacao Thé, 14*, 3-12
Friis, I.B. (1979) 'The wild populations of *Coffea arabica* L., and cultivated coffee', in G. Kunkel (ed.). *Taxonomic Aspects of African Economic Botany*, Les Palmas de Gran Canaria. Ayuntamiento de las Palmas. pp. 63-8
Grassias, M. and Kammacher, P. (1975) 'Observations sur la conjugaison chromosomique de *Coffea arabica* L.', *Café Cacao Thé, 19*, 177-90
Guedes, M.E.M. and Rodrigues, C.J. Jr. (1974) Disc electrophoretic patterns of phenoloxidase from leaves of coffee cultivars', *Portugaliae Acta Biologica, 13*, 169-78
Guillaumet, J.L. and Hallé, F. (1978) 'Echantillonnage du materiel *Coffea arabica* recolté en Ethiopie' in A. Charrier (ed.) *Etude de la Structure et de la Variabilité Génétique des Caféiers, IFCC (Paris) Bulletin no 14*, pp. 13-18
Haarer, A.E. (1962) *Modern Coffee Production*. Leonard Hill, London
Hamon, S., Anthony, F. and Le Pierrès, D. (1984) 'Etude de la variabilité génétique des caféiers spontanés de la section des *Mozambicoffea Coffea zanguebariae* Bridson et *C. sp. novo* Bridson', *Adamsonia*, in press
Harlan, J.R. and de Wet M.J. (1971) 'Towards a rational classification of cultivated plants', *Taxon, 20*, 509-17
Hofling, J.F. and Oliveira, A.R. (1981) 'A serological investigation of some *Coffea* species with emphasis on the origin of *C. arabica*', *Ciencia e Cultura, 33*, 66-72
Kammacher, P. and Capot, J. (1972) 'Sur les relations caryologique entre *Coffea arabica* and *C. canephora*', *Café Cacao Thé, 16*, 289-94
Krug, C.A. (1965) 'Ensayo mundial del cafe' *FAO, Rome*
Krug, C.A. and Mendes, A.J.T. (1940) 'Cytological observations in *Coffea*', *Journal of Genetics, 39*, 189-203
Krug, C.A. and Carvalho, A. (1951) 'The genetics of coffee', *Advances in Genetics, 4*, 127-68
Lanaud, C. (1979) 'Etude de problemes génétiques posés chez le caféier par l'introgression de caractères d'une espèce sauvage (*C. kianjavatensis: Mascarocoffea*) dans l'espèce cultivée *C. canephora: Eucoffea*', *Café Cacao Thé, 23*, 3-28
Lebrun, J. (1941) Recherches morphologiques et systematiques sur les caféiers du Congo, *Memoires TX1 (3), Institute Royal Colonial Belge, Bruxelles*
Leliveld, J.A. (1940). 'Nieuwe gezichtspunten voor het selectie onderzoek bij soortskruisingen naar aanleiding van cytologische gegevens', *Bergcultures, 14*, 370-86
Leroy, J.F. (1961a) '*Coffea novae madagascariensis*', *Journal d'Agriculture Tropicale et de Botanique Appliquée, 18*, (1, 2, 3), 1-20
Leroy, J.F. (1961b) 'Sur les trois caféiers endemiques de l'archipel des Mascareignes', *Journal d'Agriculture Tropicale et de Botanique Appliquée, 8*, (1, 2, 3), 21-9
Leroy, J.F. (1961c) 'Sur deux caféiers remarquables de la forêt sèche du Sud-Ouest de

Madagascar (*C. humbertii* J.F. Ler, et *C. capuronii* J.F. Ler.)', *Comptes Rendus de l'Academie des Sciences, Paris, 252,* 2285-7

Leroy, J.F. (1962) 'Coffeae novae madagascariensis et mauritianae', *Journal d'Agriculture Tropicale et de Botanique Appliquée, 9* (11, 12), 525-30

Leroy, J.F. (1963) 'Sur les caféiers sauvages des îles Mascareignes', *Comptes Rendus de l'Academie des Sciences, Paris, 256,* 2897-9

Leroy, J.F. (1967) 'Recherches sur les caféiers. Sur la classification biologique des caféiers et sur l'origine et l'aire du genre *Coffea*', *Comptes Rendus de l'Academie des Sciences, Paris, 265,* 1043-5

Leroy, J.F. (1968) *Notice sur les titres et travaux scientifiques de J.F. Leroy* Monnoyer, Le Mans fascicule II

Leroy, J.F. (1972a) 'Prospection des caféiers sauvages de Madagascar: deux espèces remarquables (*Coffea tsirananae* n. sp., *C. kianjavatensis* n. sp.)', *Adamsonia, 2* (12), 317-28

Leroy, J.F. (1972b) 'Prospection des caféiers sauvages de Madagascar: sur deux espèces sympatriques du Nord', *Adamsonia, 2* (12), 345-58

Leroy, J.F. (1980) 'Evolution et taxogenèse chez les caféiers. Hypothese sur leur origine', *Comptes Rendus de l'Academie des Sciences, Paris, 291,* 593-6

Leroy, J.F. (1982) 'L'origine kenyane du genre *Coffea* et la radiation des espéces a Madagascar', in *10th International Colloquium on the Chemistry of Coffee,* ASIC, Paris, pp. 413-20

Leroy, J.F. and Plu, A. (1966) 'Sur les nombres chromosomiques des *Coffea* malgaches', *Café Cacao Thé, 10 (3),* 216-18

Lind, E.M. and Morrison, M.E.S. (1974) *East African Vegetation.* Longman, London

Linnaeus, C. (1737) *Hortus Cliffortianus.* Amsterdam

Lobreau-Callen, D. and Leroy, J.F. (1980) 'Quelques données palynologiques sur le genre *Coffea* et autres genres du cercle des caféiers', in *9th International Colloquium on the Chemistry of Coffee,* ASIC, Paris, pp. 479-506

Longo, C.R.L. (1975) 'The use of phenolic pigments in phylogenetic studies on *Coffea*', *Baletin de Divulgação,* Escola Superior de Agricultura "Luis de Queiroz", Sao Paulo, *20,* 171-4, 392-5.

Louarn, J. (1972) 'Introduction à l'étude génétique des *Máscarocoffea* Nouvelles déterminations de leurs nombres chromosomiques', *Café Cacao Thé, 16,* 312-15

Louarn, J. (1976) 'Hybrides interspecifiques entre *Coffea canephora* Pierre et *C. eugenioides* Moore', *Café Cacao Thé, 20,* 433-52

Louarn, J. (1978) 'Diversité comparée des descendances de *C. arabica* obtenues en autofécondation et en fécondation libre au Tonkoui', in A. Charrier (ed.), *Etude de la Structure et de la Variabilité Génétique des Caféiers. IFCC Bulletin no 14,* pp. 75-8

Louarn, J. (1980) 'Hybrides interspécifiques entre *Coffea canephora* et *C. liberica.* Résultats préliminaires sur les hybrides F_1', *Café Cacao Thé, 24,* 297-304

Louarn, J. (1982) Bilan des hybridations interspécifiques entre caféiers africains diploides en collection en Côte d'Ivoire', in *10th International Colloquium on the Chemistry of Coffee,* ASIC. Paris, pp. 375-84

Mayr, E. (1970) *Populations, Species and Evolution.* Belknap Press of Harvard University Press, Cambridge, Mass

Medina, D.M. (1963) 'Microsporogenesé em um hibrido triploide de *C. racemosa* × *C. arabica*', *Bragantia, 22,* 299-318

Mendes, A.J.T. (1938) 'Morfologia dos cromosomios de *Coffea excelsa*', *Instituto Agronomico de Campinas, Boletin Tecnico, 56*

Mendes, A.J.T. (1939) 'Duplicaçao do numero de cromosomios em cafe, algodao e fumo, pela açao da colchicina', *Instituto Agronomico de Campinas, Boletin Tecnico, 57*

Mendes, A.J.T. and Bacchi, O. (1940) 'Observaçoes citologicas em *Coffea* V. Uma variedade haploide (di-haploide) de *C. arabica* L.' *Instituto Agronomico de Campinas, Boletin Tecnico, 77*

Meyer, F.G. (1965) 'Notes on wild *Coffea* arabica from southwestern Ethiopia, with some historical considerations', *Economic Botany, 19,* 136-51

Meyer, F.G., Fernie, L.M., Narasimhaswamy, R.L., Monaco, L.C. and Greathead, D.J. (1968) *FAO Coffee Mission to Ethiopia 1964-65.* FAO, Rome

Monaco, L.C. and Medina, D.M. (1965) 'Hibridaçoes entre *Coffea arabica* e *C. kapakata*. Analise citologica de um hibrido triploide', *Bragantia*, 24, 191-201

Narasimhaswamy, R.L. (1962) 'Some thoughts on the origin of *C. arabica* L. *Coffee* (Turrialba)', 4, 1-5

Nei, M. (1972) Genetic distance between populations. *American Naturalist*, 106, 283-92

Noirot, M. (1978) Polyploidisation de caféiers par la colchicine. Adaptation de la technique sur bourgeons axillaires aux conditions de Madagascar. Mise en évidence de chimères périclines stables. *Café, Cacao, Thé*, 22, 187-94

Payne, R.C. and Fair-Brothers, D.E. (1976) 'Disc electrophoresic evidence for heterozygosity and phenotypic plasticity in selected lines of *Coffea arabica* L.' *Botanical Gazette*, 137, 1-6

Portères, R. (1937) 'Etude sur les caféiers spontanés de la section *Eucoffea*', *Annales agricoles de l'Afrique occidentale et Etrangères*. 1ère partie: Répartition, habitat, 1 (1), 68-91: 2ème partie: Espèces, varietés, formes, 1 (2), 219-63

Portères, R. (1962) 'Sur quelques caféiers sauvages de Madagascar', *Journal d'Agriculture Tropicale et de la Botanique Appliquée*, 9 (3-6), 201-10

Ram, A.S., Sreenivasan, M.S. and Vishveshwara, S. (1982) 'Chemotaxonomy of Mozambicoffea', *Journal of Coffee Research* (Chikmagalur), 12, 42-6

Reynier, J.F., Pernes, J. and Chaume, R. (1978) 'Diversité observée sur les descendances issues de pollinisation libre au Tonkoui', in A. Charrier (ed.). Etude de la structure et de la variabilité génétique des caféiers. *IFCC Bulletin no. 14*, pp. 69-74

Sybenga, J. (1960) 'Genetics and cytology of coffee: a literature review', *Bibliographia Genetica*, 19, 217-316

Sylvain, P.G. (1955). 'Some observations on *Coffea arabica* L., in Ethiopia', *Turrialba* (Costa Rica), 5, 37-53

Thomas, A.S. (1942) 'The wild Arabica coffee on the Boma plateau, Anglo-Egyptian Sudan', *Empire Journal of Experimental Agriculture*, 10, 207-12

Thomas, A.S. (1944) 'The wild coffees of Uganda', *Empire Journal of Experimental Agriculture*, 12, 1-12

Van der Vossen, H.A.M. (1974) 'Plant Breeding', *Coffee Research Foundation of Kenya, Annual Report 1973-74*, pp. 40-51

Vishveshwara, S. (1975) 'Coffee: commercial qualities of some newer selections', *Indian Coffee*, 39, 363-74

Von Strenge, H. (1956) 'Wild coffee in Kaifa province of Ethiopia', *Tropical Agriculture*, 38, 297-301

Wellman, F.L. (1961) *Coffee: Botany, Cultivation and Utilization*. Leonard Hill Books, London

3 COFFEE SELECTION AND BREEDING

Herbert A.M. van der Vossen

Introduction

In spite of the tremendous genetic potential revealed by recent botanical and (cyto)genetic research into the genus *Coffea* (see Chapter 2), coffee breeding is still largely restricted to the two species, *Coffea arabica* and *C. canephora*, that dominate world coffee production.

C. liberica and *C. excelsa* have lost most of their earlier significance, although they are still grown to some extent in a few countries (Krug and de Poerck, 1968). However, *C. liberica* and other species like *C. congensis* have contributed useful characters to the gene pools of respectively *C. arabica* and *C. canephora* through natural and artificial interspecific hybridisation (Cramer, 1957; Carvalho *et al.*, 1969).

Arabica coffee is preferred over all other species because of its superior quality and it would certainly have continued to be the exclusive producer of all coffee in the World, as it had been for more than 150 years until the end of the 19th century, if it had not been so vulnerable to diseases, particularly coffee leaf rust (*Hemileia vastatrix* Berk. & Br.), when grown at lower altitudes in the tropical zones. At the present time it still contributes about 75 per cent of the World coffee exports, mainly due to the fact that more than two-thirds of all coffee is produced by the Central- and South-American countries, where arabica coffee was free from *Hemileia* epidemics until 1970.

Although most of the arabica coffee cultivated there is of the same narrow genetic origin and therefore as susceptible as the coffee wiped out by *Hemileia* in Sri Lanka and Indonesia a century ago, it is unlikely that the past will repeat itself for three reasons.

First, almost all coffee is cultivated in subtropical areas (Brazil) or at altitudes above 1,000 m in the tropical zones, where climatic conditions favour the host rather than the pathogen.

Secondly, the availability of modern fungicides makes effective control possible, though at considerable cost and therefore decreased profitability of coffee production.

Thirdly, breeding programmes to develop disease-resistant varieties, which had started in countries like Brazil (Monaco, 1977) and Colombia (Castillor and Moreno, 1980) already in the 1960s in anticipation of possible *Hemileia* epidemics, have made considerable progress and the first rust-resistant varieties are already being released in some countries. However, in a perennial crop like coffee a change to new varieties is expensive and a long-term process. The economic impact of the breeding efforts on coffee production in these countries may, therefore, not be noticed for another decade.

At the time of the *Hemileia* epidemics in South-East Asia (1870-1900) fungicides had not yet been invented and little was known about the genetics of host and pathogen. While in Sri Lanka coffee cultivation was almost completely replaced by tea, the coffee industry of Indonesia was saved by the introduction of the robusta form of *Coffea canephora*, which combined vigorous growth and production with a high level of resistance to coffee leaf rust. However, the price to be paid was the loss of good quality.

From the pioneering work on coffee biology and selection carried out on East Java during the period 1900-1942 (Ferwerda, 1948; Cramer, 1957) evolved the systematic breeding designs, which became exemplary to all subsequent breeding programmes of robusta coffee in India and Africa.

In arabica coffee actual breeding was given serious attention at a much later date and most varieties grown commercially at present in Latin America as well as in Africa and Asia have arisen from simple systems of line selection within genetically homogeneous parent populations carried out in the period 1920-1940. Emphasis was mostly on good adaptation to local ecological conditions, yield and quality. Disease resistance was given low priority, either because of the absence of serious diseases as in Latin America, or the relative ease of control by chemical means as with *Hemileia* in the highlands of East Africa. An exception are the rust-resistant arabica varieties released in India after 1946 (Narasimhaswamy, 1961), but these are not grown much commercially outside Asia mainly because of their inferior bean and liquor quality.

This situation has completely changed during the last two decades. Breeding efforts have been greatly intensified in Latin America in the face of coffee leaf rust (Monaco, 1977), while the grave threats of coffee berry disease (*Colletotrichum coffeanum* Noack *sensu* Hindorf) to arabica coffee in the highlands of Eastern and Central Africa have prompted a number of entirely new programmes, particularly in Kenya (Van der Vossen and Walyaro, 1980, 1981) and in Ethiopia (Van der Graaff, 1978, 1981). Considerable progress has been made in a relatively short period, not in the least due to the application of advanced breeding and selection methods. However, results would have been far less spectacular without the basic information on coffee genetics produced by the eminent coffee scientists of earlier days (reviewed by Sybenga, 1960; Carvalho *et al.*, 1969) and the fundamental work on *Hemileia* carried out by the Coffee Rust Research Centre in Portugal (Rodrigues, Bettencourt and Rijo, 1975). Of similar importance are the tireless efforts of botanists and geneticists to collect germplasm of arabica coffee from its centres of high genetic diversity in Ethiopia (Meyer *et al.*, 1968; Guillaumet and Hallé, 1967, 1978).

Successful hybridisation of the two species *Coffea arabica* and *C.*

canephora, originally achieved in Brazil around 1950 (Monaco, 1977), has paved the way for an entirely new approach in robusta coffee breeding. The arabusta hybrids developed in Ivory Coast (Capot, 1972) appear to have the potential, on account of their better liquor quality and lower caffeine content, of eventually replacing much of the traditional robusta coffee of the humid tropical regions of lowland Africa and Asia.

Breeding Populations

Coffea arabica

The large number of named varieties and selections of arabica coffee encountered in the variety collections at coffee research centres and also grown commercially (Wellman, 1961; Haarer, 1962) belies the actually very narrow genetic diversity of the base populations from which they were selected during the first half of this century. Historical evidence (see Chapter 1) indicates that these base populations all descended from the few trees that survived various efforts to introduce arabica coffee from Southern Arabia, now Yemen, into Asia and Latin America during the early part of the eighteenth century and into East Africa towards the end of the nineteenth century. But not until some 50 years ago did coffee breeders start to realise that even the coffee in Yemen represented just a small fraction of the potential genetic variability found in the real centre of origin, or at least the centre of high genetic diversity, for arabica coffee in the southwestern highlands of Ethiopia (Sylvain, 1955; Carvalho, 1959; Monaco, 1968).

The coffee from Yemen gave rise to two distinct types: *C. arabica* var. *arabica*, usually called *typica*, which was the earliest grown coffee in Asia and Latin America and *C. arabica* var. *bourbon* which came to South America through the island of La Réunion, formerly called Bourbon (Carvalho *et al.*, 1969). The *bourbon* type has a more compact and upright growth habit than the *typica* coffee and is generally higher yielding and produces better quality coffee.

Similar *bourbon* types of coffee were brought to East Africa, either from La Réunion or directly from Aden by missionaries. The name 'French Mission' given to the coffee which was used to establish most of the coffee estates in Kenya until 1940 bears testimony to these early-day agricultural pioneers.

The genetic uniformity within these populations is further enhanced by the predominantly self-pollinating nature of *C. arabica*. The subsequently encountered variation, which gave rise to so many cultivars, is generally believed to be more the result of spontaneous mutations of major genes conditioning plant, fruit and seed characters than of residual heterozygosity (Carvalho *et al.*, 1969). However, where *typica* and *bourbon* were planted in close proximity as on research stations, hybridisation between the two types due to natural cross-pollination occasionally gave rise to very vigorous and productive cultivars, such as Mundo Novo in Brazil (Carvalho *et al.*, 1969). The considerable success of the early programmes of line selection for yield, quality and drought resistance in the French Mission and *bourbon* material in East Africa (Fernie, 1970) would suggest a somewhat larger degree of genetic variability of these base populations.

The very detailed studies undertaken in Brazil (Krug and Carvalho, 1951) on more than 40 mutants found in arabica coffee contributed to a very much better understanding of the genetics of arabica coffee and clearly demonstrated the diploid mode of inheritance of all characters in this allotetraploid species. The main characteristics of a few mutants with practical significance to breeding are presented in Table 3.1.

The compact growth character of Caturra (Figure 3.1) plays an essential role in almost all breeding programmes, as it offers a tremendous potential for intensification of coffee production, i.e. increased productivity by high-density planting. *Purpurascens* and *Cera* have been used as marker genes

Table 3.1: Mutants in *C. arabica* Which are of Practical Significance to Breeding

	Gene Notation	Dominance relation	Original population	Main characteristics
Caturra	Ct	almost complete dominance	bourbon, Brazil (1935)	compact growth due to short internodes, small beans
Purpurascens	pr	recessive	found several times	purple leaves
Cera	ce	recessive	Brazil (1935)	yellow endosperm
Erecta	Er	incomplete dominance	Brazil, Indonesia, Kenya in typica and bourbon	orthotropic branch growth
Maragogipe	Mg	complete dominance	typica, Brazil (1870)	large leaves and seeds, long internodes, vigorous growth, low yield
Laurina	lr	recessive	bourbon, Brazil	narrow leaves, low productivity, seed narrow and pointed at one end, low caffeine content
Mokka	mo, lr	double recessive	very old introduction from Yemen (?)	very small leaves and beans, very compact plant
Sao Bernardo	SB	almost complete dominance	typica, Brazil	short internodes and compact growth like Caturra
San Ramon	SR	complete dominance	typica, Brazil	extremely short internodes

Figure 3.1: *Coffea arabica*. Young coffee plants, one year after field planting at Ruiru, Kenya; on the left cultivar SL28 about 1.60 m high, on the right a plant of the new compact growing hybrid variety resistant to Coffee Berry Disease and Leaf Rust, about 1.10 m high.

to determine the percentage natural cross-pollination in arabica coffee. Erecta enhances the possibilities of planting at still closer spacings. Maragogipe has been grown commercially to a limited extent in a few Latin American countries (e.g. Colombia), as its large beans and special flavour were of interest to some European consumers. Laurina has been used in breeding because of its low caffeine content. Coffee breeders have from time to time shown interest in other dwarf type coffees such as Mokka, Sao Bernardo and San Ramon, but their low productivity and small bean size have proven to be a major handicap. For detailed descriptions of most mutants and types found in arabica coffee reference is made to review papers by Cramer (1957) and Sybenga (1960).

Of the numerous selections made within arabica coffee populations in various countries outside the centre of genetic diversity, the following should be mentioned because of their importance as cultivars and/or progenitors in coffee breeding:

— Kent: a variety developed from a single tree selection in Mysore, India, around 1920 (Narasimhaswamy, 1950), probably a hybrid between *typica* and an unknown other type; it is high yielding and resistant to race II of *Hemileia vastatrix*; selections within Kent like the

cv. K7 showed also some resistance to coffee berry disease (CBD) in Kenya;
— S288, S333, S795 and BA selections of the Balehonnur Coffee Research Station in India: they are resistant to many rust races and are believed to have originated from a natural hybrid between *C. arabica* and *C. liberica* (Narasimhaswamy, 1960);
— Blue Mountain: a variety originating from *typica* coffee in Jamaica and found to have some resistance to CBD in Kenya;
— Hibrido de Timor: a variety believed to be the result of natural hybridisation between *C. arabica* and *C. canephora* (Rodrigues *et al.*, 1975) and grown widely on the island of Timor. It is the most important progenitor for resistance to coffee leaf rust and has also good resistance to CBD (Van der Vossen and Walyaro, 1980).

Systematic collection of germplasm of arabica coffee in its centre of high genetic diversity started with the FAO Coffee Mission to Ethiopia in 1964-65 (Meyer *et al.*, 1968). However, on various earlier occasions material had been collected in Ethiopia by individual workers. The Abyssinian coffee grown on Java arose from seed collected by Cramer (1957) during his visit to Ethiopia in 1928. Varieties like Geisha, Amfillo and Harar, which had been sent to various coffee research centres before 1940, were supplemented with varieties like Dalle, Dilla, Gimma Mbuni and others, collected by officers of the British East African Forces in Ethiopia during 1941-42 (Jones, 1956). The expedition of Thomas (1942) to the Boma Plateau in south-eastern Sudan resulted in varieties like Rume Sudan, Barbuk Sudan and Boma Plateau, of which the first one would eventually prove to be one of the best progenitors for CBD resistance (Van der Vossen and Walyaro, 1980).

The botanical and genetic studies on natural arabica populations in Ethiopia by Sylvain (1955), Lejeune (1958) and others, as well as on a small population of Ethiopian origin in Brazil by Carvalho (1959) prepared the ground for the subsequent efforts by the FAO (Meyer *et al.*, 1968), the ORSTOM (Office de la Recherche Scientifique et Technique Outre-Mer) (Guillaumet and Hallé, 1967, 1978) and the Ethiopian Institute of Agricultural Research (Fernie, 1971) to collect and preserve valuable germplasm of arabica coffee before it becomes permanently lost. Countries where germplasm of arabica coffee is presently being maintained have been listed in Table 2.6. The characteristics of this material and its significance to breeding will be discussed later.

Coffea canephora (Figure 3.2)

This diploid coffee species is indigenous to the African equatorial lowland forest zone from Guinea to Uganda. It is characterised by its great variation in forms or ecotypes, many of them originally described under different

54 Coffee Selection and Breeding

Figure 3.2: *Coffea canephora.* a. A recently rejuvenated coffee tree of the Nganda type at Kawanda, Uganda, bearing a heavy crop. b. Upper branch of a Robusta and Lower branch of a Nganda coffee tree taken at Kawanda, Uganda. Note the larger leaves and bolder berries of the Robusta coffee (measure at centre is 10 cm).

a.

b.

botanical names (Cramer, 1957; Wellman, 1961). Central Africa and especially the Congo basin are particularly rich centres of genetic diversity for *C. canephora*, as well as for many other diploid coffee species. It was from the borders of the Lomani, a tributary of the Congo river, that seeds were collected by Gallain around 1890 (Chevalier, 1937) to establish a coffee plot of some 500 trees. This would constitute the base population of robusta coffee, by far the most important cultivated form of *C. canephora*.

The history of introduction of robusta coffee from Zaire via a nursery in Brussels to Java around 1900 is related in detail by Cramer (1957). Early selection work at the Bangelan research station and other institutes on Java was very successful and improved seed from Java was even used to establish robusta plantations in other countries like India, Uganda, Ivory Coast and also Zaire, from where robusta coffee originated. However, the African countries within the geographic distribution of *C. canephora* would eventually follow the initiative of the INEAC, which started robusta coffee breeding in Zaire around 1930, to exploit the locally available natural populations.

Robusta coffee is characterised by a more erect growth habit, larger leaves and berries than the local forms of *C. canephora*, such as the Petit Indenié of Ivory Coast (Krug and de Poerck, 1968) and the Nganda coffee of Uganda (Leakey, 1970). 'Kouilou' coffee originates from Central Africa and derives its name from the Kouillou river in Congo or the Kwilu river in Zaire (Berthaud, 1983). It much resembles the Petit Indenié and Nganda types and 'Kouilou' coffee is often synonymous with all *C. canephora* forms with a spreading type of bush, smaller leaves and fruits than robusta coffee. Susceptibility to the disease tracheomycosis and low productivity have accelerated the replacement of 'Kouilou' by robusta coffee in Ivory Coast, but it still produces part of the local coffee in Malagasy and Uganda. Kouilou coffee was introduced into Indonesia in 1901 from Congo, but it never attained economic importance. However, Kouilou coffee accounts for 50 per cent of the coffee production in the state of Esperito Santo in Brazil and this coffee has been the subject of an important study of the inheritance of partial (race-non-specific) resistance to coffee leaf rust (Eskes, 1983).

Recent studies on germplasm of *C. canephora* — existing breeding populations of the IRCC in Ivory Coast and accessions collected by ORSTOM missions in Central and West Africa in 1975-1983 (see also Table 2.6 and Chapter 2) — have confirmed the existence of genetically diverse subpopulations. Berthaud (1983, 1985) emphasises a distinction of two main groups: (a) *C. canephora* of Central African ('Congolese') origin, which includes robusta coffee, and (b) *C. canephora* of West African ('Guinean') origin. In the analysis of enzyme polymorphisms these two groups were found to produce different zymogrammes for seven enzyme systems. The highest yielding clones of the IRCC breeding programme also turned out to be derived from crosses between trees of different groups (Berthaud, 1983). It appears that herewith the importance of interpopulation crosses to breeding for increased productivity is now also established for *C. canephora* and breeding strategies should be revised accordingly.

Selection Criteria

Basically, coffee breeding is concerned with the development of cultivars, which under the circumstances of climate, soil, cultural practices, as well as disease and pest incidence of a specific region, give the maximum economic return to coffee growers.

Yield

Increasing the genetic yield potential has for obvious reasons been the prime objective in arabica as well as in robusta coffee. In the latter species this is still largely achieved by increased productivity per tree, since all selection is carried out at conventional (1,100-1,400 trees/ha) plant densities (Leakey, 1970; Capot, 1977). On the other hand, in arabica coffee breeding emphasis has been shifted from increased production per tree to developing cultivars which are well-adapted to intensive coffee production (Van der Vossen and Walyaro, 1981), as close-spacing systems appear to offer the best prospects of dramatically increasing productivity (Mitchell, 1975).

Yield is usually expressed on a kg or tonnes per ha per year basis of clean coffee, i.e. beans dried to 10-12 per cent moisture content. It may vary from 200 kg in traditional smallholder plots to 2 tonnes/ha for arabica (Fernie, 1970) and 3.5 tonnes/ha for robusta coffee (Capot, 1977) in experimental plots at normal (1,350 trees/ha) plant densities. Yields of 5 tonnes/ha have been obtained in high density planting systems of arabica coffee in Colombia (Uribe and Mestre, 1980) and in Kenya (Walyaro, 1983).

Yield Stability

This means high productivity under a wide range of environmental conditions and the ability to overcome biennial bearing.

Growth Characters and Yield Components

Recent biometrical genetic studies in arabica coffee (Walyaro and Van der Vossen, 1979; Walyaro, 1983) have shown that the selection efficiency for higher yield is increased considerably by taking into account various growth parameters and components of yield, such as stem girth, canopy radius, percentage of bearing primaries, percentage of bearing nodes, number of berries per node. Internode length (short internodes as with Caturra) and angle of primaries with the main stem (erect growth) are important characters for adaptation to close-spacing systems of planting.

De Reffye (1982, 1983) developed a mathematical model to simulate the growth of a robusta coffee tree.

Outturn

This is the fraction of clean coffee over freshly harvested cherry and may vary from 0.12 to 0.22 in arabica and robusta coffee.

Quality

There exists a great variation in systems for quality assessment from grading based on size and percentage defective beans in unwashed robusta to the comprehensive methods of bean and cup quality determination evolved for the arabica coffees referred to as 'Colombian Milds' and applied in the breeding programme in Kenya (Van der Vossen and Walyaro, 1977). The following criteria have been used to measure quality:

(1) bean size, shape and density by mechanical and pneumatic separation:
   PB the fraction of round beans retained by a piano wire screen with 4.43 mm spaces
   AA fraction of heavy beans retained by a no. 18 (7.15 mm) screen
   AB fraction of heavy beans retained by a no. 15 (5.95 mm) screen
   TT fraction of light beans separated from AA and AB
   C   fraction of beans retained by a piano wire screen with 2.90 mm spaces

The remaining fraction are E (elephant beans) retained by a no. 21 (8.33 mm) screen and T (very small beans and broken bits).

(2) bean colour, shape of the centre cut and general appearance by visual scoring;
(3) roast beans: general appearance by visual scoring;
(4) cup quality by organoleptic means: samples are submitted to professional coffee liquorers who assess body, acidity, flavour and general standard, while also a great number of off-flavours are distinguished.

Beans of commercial robusta coffee are generally smaller than of arabicas: average 100-bean weight 12-15g against 18-22g (Coste, 1968). However, clones have been selected within robusta coffee with 100-bean weights of more than 20g (Cramer, 1957). The colour of robusta and unwashed arabica coffees are brownish, while the top quality washed arabicas have a bluish-green appearance. The cup quality depends very much on the method of processing, but even the ideally prepared robusta coffee will at the most have a neutral cup without the flavour and pointed acidity characteristic of well-prepared arabica coffee.

Caffeine Content

So far, this has been an important selection criterion only in robusta coffee in which the average caffeine content (2-3 per cent on dry weight basis) is almost twice as high compared to arabica (1-1.3 per cent). Variation in caffeine content between and within *Coffea* species was studied by Charrier and Berthaud (1975) (see Chapter 13).

Resistance to Diseases

Breeding for disease resistance is mainly restricted to the two major diseases of arabica coffee:

(1) *Hemileia vastatrix*, coffee leaf rust, now endemic in all countries except Colombia and Papua New Guinea. (e.g. Rodrigues *et al.*, 1975).
(2) *Colletotrichum coffeanum*, coffee berry disease, which is restricted to East and Central Africa (e.g. Van der Vossen, Cook and Murakaru, 1976).

Other diseases of local importance in arabica coffee, for which variation in levels of resistance have been found and selection programmes have started are:

(3) *Fusarium stilboides*, fusarium bark disease, in Southern Africa (Siddiqi, 1980);
(4) *Cercospora coffeicola*, brown eye spot and berry blotch, in some countries of Latin America and East Africa (e.g. Soto and Campos, 1971; Van der Vossen and Cook, 1975);
(5) *Gibberella xylarioides*, tracheomycosis, in Ethiopia (Pieters and Van der Graaff, 1980);
(6) *Hemileia coffeicola*, occurring in Cameroon (Muller, 1980) and other Central and West African countries.

Except for the red blister disease in Uganda, also caused by *Cercospora coffeicola* (Leakey, 1970), robusta coffee is remarkably tolerant of most earlier mentioned diseases (see also Chapter 9).

Resistance to Pests

Very little progress has been made with breeding for resistance to any of the insect or nematode pests of arabica or robusta coffee, although different levels of tolerance/resistance between and within species have been observed against nematodes (Cramer, 1957; Fazuoli, Monaco, Carvalho and Reis, 1977) and against leaf miner, *Leucoptera* spp., in arabica coffee (Wanjala, 1980). In robusta coffee variation in tolerance has been found against the berry borer, *stephanoderes hampeii* (Cramer, 1957) and the branch borer, *Xyleborus morstatti* (Coste, 1968) (see also Chapter 8).

Drought Resistance

Arabica with its deeper root system is generally more drought resistant than robusta coffee, but there are also marked varietal differences within the two species in the ability to overcome periods of water stress.

Suitability for Mechanical Harvesting

A new element of selection is the adaptation of the coffee tree to mechanical harvesting, where this is possible due to a concentrated harvesting season as for instance in Brazil (Watson, 1980). Compact growth, uniform flowering and berry ripening, as well as a good response to induced berry abscission will be among the relevant selection criteria.

Reproductive Systems and Controlled Pollination

Floral Biology

The biology of flowering of arabica and robusta coffee has been described many times (e.g. Sybenga, 1960; Wellman, 1961; Coste, 1968) and only the aspects most relevant for coffee breeding will be mentioned here.

(1) Flowers are borne on first year lateral branches in clusters of 15-20 for arabica and up to 80 flowers per node for robusta coffee. The morphology of coffee flowers — a corolla of five petals to which the stamens are fused (epipetalous) — facilitates emasculation work.

(2) Flowering in both species is remarkably periodical. Major flowerings normally coincide with the onset of a rainy season. Only when preceded by an extended period of water stress will a shower (of at least 20 mm) break the dormancy of flower buds initiated 6-7 months earlier (Wormer and Gituanja, 1970). Such flowerings will occur once or twice a year depending on uni- or bimodal rainfall distribution. Coffee was reported to be an intermediate to short-day plant (Franco, 1940; Piringer and Borthwick, 1955; Went, 1955), but later work (Cannell, 1972; Monaco *et al*, 1978) showed that the effect is minimal, so that it is unlikely that day length plays much part in regulating the cycle of growth (see page 110). In coffee areas further from the equator, such as the major coffee belt in Brazil, the concentration of flowering can be explained by other factors. On the other hand, even in equatorial regions with a more evenly distributed rainfall, periodicity in flowering is usually maintained although less pronounced. (See also Chapter 5).

(3) The time between breaking of the dormancy and anthesis may vary from 4 to 10 days depending on temperature and atmospheric humidity. The total period of flowering is normally not more than three days with the majority of flowers opening on the first and second day. Pollen shedding starts very soon after opening of the flowers early in the morn-

ing and the stigma is then also receptive. Flowers wither in one or two days after pollination. Flowers of robusta coffee produce somewhat larger quantities of pollen than arabica flowers.

(4) Arabica coffee is autogamous with a low (10 per cent) degree of natural cross-pollination (Carvalho and Monaco, 1962; Van der Vossen, 1974) in contrast to robusta coffee, which is strictly allogamous with a gametophytic system of self-incompatibility (Berthaud, 1980). The styles of robusta flowers are somewhat longer compared to those of arabica flowers, which may also facilitate cross-pollination. On the other hand, in arabica coffee self-pollination shortly before opening of the flower buds is not uncommon (Stoffels, 1936).

Artificial Hybridisation

Techniques of artificial cross-pollination applied in most breeding programmes of arabica coffee are very similar to those described by Carvalho and Monaco (1969). Emasculation, necessary to prevent selfing, is effected by removing the whole corolla with a pair of specially adapted scissors one or two days before flowering (Figure 3.3a). The section of a branch with emasculated flowers is then enclosed in a paper bag. According to experiments carried out in Kenya (Walyaro and Van der Vossen, 1977) emasculation and bagging could be safely done until the evening before the first day of anthesis (Figure 3.3b). The stigmas of unfertilised flowers were found to remain receptive for at least nine days (Figure 3.4). This means that isolation bags should not be removed until two weeks after anthesis to prevent out-pollination. On the other hand, in large crossing programmes pollination can be spread over some days without affecting fruit set.

Flowering can also be induced by applying overhead irrigation to a coffee field, provided the preceding period of water stress has been adequate, in practice only towards the end of a dry season (Browning, 1975). This offers the opportunity to stagger the flowering time and so to increase the opportunity of completing large programmes of hybridisation for breeding or hybrid seed production in one season.

In the self-incompatible robusta coffee emasculation should, strictly speaking, not be necessary. However, a small amount of selfing occurs occasionally and techniques of artificial cross-pollination, similar to those described for arabica coffee, are usually applied in robusta coffee breeding (Ferwerda, 1969).

Pollen Storage

Efficient pollen storage is of practical importance in coffee breeding, since crosses may have to be made between parents which do not flower simultaneously, or which grow far apart. It is also of great advantage in the arabica × robusta interspecific hybridisation programmes, such as those carried out in Ivory Coast (Capot, 1972) and in Kenya (Owuor and Van

Figure 3.3: Artifical Cross-pollination in *Coffea arabica*. (a) Emasculation of the flowers in the candle stage, with the aid of an adapted pair of scissors, two days before anthesis. (b) Application of pollen to the receptive stigmas of emasculated flowers on day of anthesis.

(a)

(b)

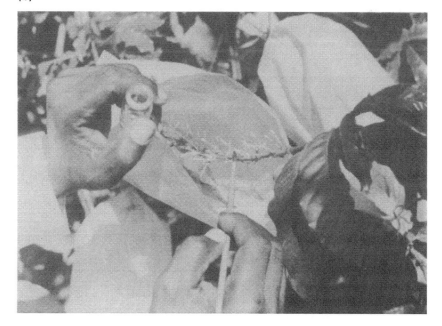

Figure 3.4: Percentage Fruit Set in arabica Coffee when Pollination is Delayed 0-12 Days after Anthesis

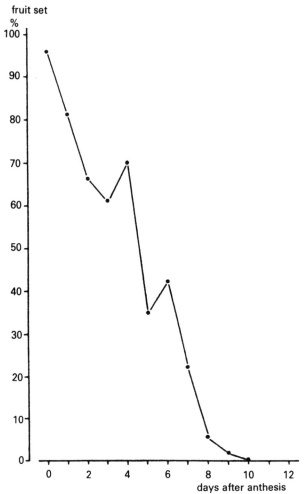

der Vossen, 1981), where large-scale testing for fertility by pollen germination *in vitro* necessitates a time interval of several months between pollen collection and actual viability testing. Walyaro and Van der Vossen (1977) demonstrated for arabica pollen, that a high pollen viability can be maintained for more than two years by storing it under vacuum at sub-freezing ($-18°C$) temperature. In fact, coffee pollen stored in this manner for about six years at the Coffee Research Station in Kenya still gave 75 per cent *in vitro* germination (Van der Vossen and Walyaro, 1981).

Fruit set with pollen stored under such conditions for about one year was found to be as good as with fresh pollen, but effects of longer periods of pollen storage on fruit set have not yet been investigated. For breeding purposes one year is generally adequate, as it is unlikely that fresh pollen could not be collected within that period. However, Charrier (1980) suggests the possibility of long storage of coffee germplasm by pollen storage, especially since long storage of coffee seed (beyond 2-3 years) has not been realised so far.

Seed Development and Storage

The length of time from pollination to mature berries is for arabica 8-9 months and for robusta coffee 9-11 months. There are normally two seeds per berry. Abortion of one ovule in the early stage of fruit development will give rise to one round seed, called peaberry. The normal percentage peaberries is in arabica 10-15 per cent and 10-25 per cent in robusta coffee. Higher percentages are found in crosses between divergent genotypes of arabica coffee and in interspecific hybrids, such as the arabica × robusta crosses, where it is an indication of poor female fertility (Grassias, 1977). In robusta coffee it is usually caused by cross-incompatibility or inadequate pollination.

The anatomical and physiological aspects of the growth of the coffee berry have been reviewed by Leliveld (1938), Leon and Fournier (1962) and Wormer (1964, 1966).

Coffee seeds are always stored with the parchment retained. The 100-seed weight (at 18 per cent moisture content) is 45-50 g for arabica and 36-40 g for robusta coffee (Coste, 1968).

There is no seed dormancy, but coffee seeds are relatively short-lived and their viability decreases rapidly after two (robusta) to six (arabica) months when stored at ambient temperatures. Ultée (1930) and de Fluiter (1939) extended this period for robusta seed to seven months and Bouharmont (1971) to one year for arabica coffee by storing wet parchment seed (40-45 per cent water content) in moist charcoal in a cool place. Recent investigations by Van der Vossen (1979) in Kenya and Couturon (1980) in Ivory Coast have demonstrated that a high viability (80 per cent germination) can be maintained for some 30 months by storing wet (40-41 per cent moisture content) parchment seed in air-tight polythene bags at a constant temperature of 15-19°C. This was also the best method for seed of robusta coffee and *C. stenophylla* (Couturon, 1980), although in these cases viability could not be extended beyond 15 months (Table 3.2). Chilling to temperatures below 10°C and seed moisture contents of 13-35 per cent resulted in a rapid loss of viability. A high germination capacity could also be maintained for about two years in arabica coffee when seed with a low moisture content of 10-11 per cent was stored at 15°C. However, the germinative energy was considerably better preserved in seed with a moisture content of about 41 per cent (Van der Vossen, 1979). These improved methods of seed storage have been very useful in the arabica breeding programme of Kenya, where seeds of crosses have sometimes to be stored for more than one year before they can be sown out. Similarly,

Table 3.2: The Viability of Coffee Seed in Storage under Optimum Conditions

Species	Temp. °C	\multicolumn{6}{c}{Months storage}	Reference
		0	8	15	24	30	36	
C. arabica	15	95	95	90	91	80	—	Van der Vossen (1979)
C. arabica	19	95	94	95	90	80	60	Couturon (1980)
C. canephora	19	95	91	64	0	0	0	Couturon (1980
C. stenophylla	19	90	80	50	0	0	0	Couturon (1980)

Note: seed stored at 40—41% moisture content in air-tight polythene bags

progeny testing for resistance to coffee berry disease (see page 68) of thousands of trees from the selection fields requires collection of seed several months before the screening test on germinated seed can be applied. However, the length of storability is still insufficient to be of use in the preservation of genetic resources of coffee.

Vegetative Propagation

Since most commercially grown arabica cultivars are sufficiently true-breeding, multiplication by seed has always been standard practice. In contrast, the great variation found in seedling populations of the strictly allogamous robusta coffee made breeders soon realise the great advantages of clonal multiplication to establishing uniform plantations.

In developing practical methods of vegetative propagation for robusta coffee grafting, mainly topworking by cleft grafting on new suckers of mature trees or side grafting on seedlings, became the standard method of clonal multiplication in Indonesia (Ferwerda, 1934, 1948; Feilden, 1940; Cramer, 1957), while rooting of softwood cuttings was preferred in Africa (Capot, 1968; Boudrand, 1974). Similar methods of grafting (Gillet, 1935; Feilden, 1940) and cuttings (Fernie, 1940; Gaie, 1957) have been applied in arabica coffee, but only for breeding purposes and for the establishment of seed gardens. Grafting is also particularly useful in maintaining collections of coffee germplasm outside their normal environment (Berthaud, Guillaumet, Le Pierrès and Lourd, 1977).

More recently, large-scale clonal propagation has also been considered in arabica coffee because of the hybrid nature of the disease-resistant genotypes developed by breeding programmes in Brazil and East Africa. Grafting on one-year old seedlings (Van der Vossen and Walyaro, 1981) was found to have several advantages over rooted cuttings (Van der Vossen and Op de Laak, 1976) or grafting on mature trees (Van der Vossen, Op de Laak and Waweru, 1977).

Nevertheless, all conventional methods of *in vivo* vegetative propagation permit only slow increases in clonal plants because of the restricted amount of orthotropic shoots initially available for scions or cuttings. The production of suckers (plagiotropic branches will not give normal upright growing plants) can be stimulated by horticultural means (Boudrand, 1974) or by hormones (Stemmer, van Adrichem and Roordas, 1982, but newly developed *in vitro* techniques of clonal multiplication appear to offer much better prospects.

In vitro enhancement of axillary shoot formation (Custers, 1980; Dublin, 1980) is preferred over somatic embryo formation from callus (Staritsky, 1970; Söndahl and Sharp, 1977, 1979; Dublin, 1981) as it will ensure genetically uniform new plants without the risk of mutation, as may be expected in the latter method (Murashige, 1974).

Resistance to Coffee Leaf Rust

Breeding efforts to obtain durable resistance to coffee leaf rust in arabica coffee have been complicated by the great variability in pathogenicity of *Hemileia vastatrix*, a common feature of most biotrophic pathogens such as rust fungi. The first physiological races started to appear in India after the introduction of resistant cultivars, such as Kent and S333 — 795 and BA series, for large-scale planting by the Central Coffee Research Institute of Ballehonnur (Mayne, 1935; Narasimhaswamy, 1960). Concern about the potential threat of coffee leaf rust to arabica coffee in the rest of the world led to the establishment of the Coffee Rust Research Centre (Centro de Investigaçao des Ferrugens do Cafeeiro, CIFC) at Oeiras, Portugal, in 1955. The first director was Dr B. d'Oliveira, who had already been conducting research work on screening coffee for rust resistance and on differentiating rust races for some years (D'Oliveira, 1958). Here research on coffee rust could be carried out under international cooperation and without the danger of introducing new races of the pathogen into coffee producing countries (D'Oliveira and Rodrigues, 1961).

More than 30 physiological races have been identified among several hundreds of isolates collected in some 33 countries. This was done with the help of a set of host differentials consisting of clones of arabica varieties, of interspecific hybrids of *C. arabica* with *C. canephora* and *C. liberica* and of other *Coffea* (*congensis, canephora, excelsa, racemosa*) species (Rodrigues *et al.*, 1975). Inheritance studies revealed the presence of five major genes within arabica coffee, $S_H1 - S_H5$, and a sixth gene, S_H6, in Hibrido de Timor and produced evidence for at least three other major genes in Hibrido de Timor and in *C. canephora* (Noronha-Wagner and Bettencourt, 1967; Bettencourt and Noronha-Wagner, 1971; Bettencourt, Lopes and Godinho, 1980).

The resistance factor S_H5, which is present in *typica* and *bourbon* coffee, gives no protection against race II, the most widespread of all and also the first one identified when coffee leaf rust arrived in Brazil around 1970.

Resistance based on genes S_H2 (present in Kent's types), S_H1 and S_H4 (both from Ethiopian origin) is easily matched by respectively, the races I, III and XV, as was experienced in Brazil within four years of the first occurrence of coffee leaf rust (Ribeiro, Sugimori, Moraes and Monaco, 1975). All these new races are believed to have originated from race II by mutation (Eskes, 1983). So far, the resistance conferred by factor S_H3, which must have originated from *C. liberica* and is present in the Indian selections such as S795 (Rodrigues *et al.*, 1975) has not been matched by virulent races outside India. Nevertheless, the concept of multilines, as applied in the Brazilian variety Iarana (Carvalho, Fazuoli and Monaco, 1975) and based on resistance genes $S_H1 - S_H5$, is apparently not an effective answer to the coffee leaf rust problem. Multilines could actually facilitate the build up of complex races by stepwise increases in virulence, as was indicated by the quick appearance of new races in the selection fields of the Instituto Agronomico at Campinas (Eskes, 1983).

Most Latin American coffee breeders have placed their hope in the near complete resistance to all known races of the Hibrido de Timor genotypes CIFC 832/1 and CIFC 1343/269. The new cultivar Catimor, which is derived from crosses between Caturra and Hibrido de Timor and homozygous for the S_H6 and possibly also for some of the other, unidentified resistance genes, as well as for the Ct gene for compact growth, is now being tested for resistance and general agronomic value in several countries in collaboration with the CIFC.

Advanced backcross generations of an interspecific cross between *C. arabica* and *C. canephora*, called Icatu in Brazil (Monaco, 1977), have similar resistance genes to Catimor, but this material is still very heterogeneous and inferior to Catimor especially as regards quality.

However, the chances of the pathogen overcoming even this resistance should not be underestimated. A combination of several resistance genes in one cultivar is therefore recommended as a better guarantee for durable resistance. There are indications that *Hemileia vastatrix* is not always able to combine certain virulence factors and in cases where the rust is still able to overcome resistance based on multiple genes, the resulting virulence is expected to be low (Eskes, 1983).

Polygenically inherited partial or incomplete resistance appears to confer more durable protection to crop pathogens, such as rust fungi, than the hypersensitivity type of resistance by major genes (Parlevliet, 1979). Investigations carried out at the Instituto Agronomico at Campinas indicated the presence of incomplete resistance in arabica breeding populations including germplasm from Ethiopia and in *C. canephora* var. Kouilou (Eskes, 1983). A leaf disc inoculation method (Eskes, 1983) proved to be suitable for assessing complete (major gene) and incomplete resistance and has also been successfully applied in Kenya (Van Dongen, 1979; Owuor, 1980). However, incomplete resistance to coffee leaf rust can also be race-specific, as was shown to exist in the Icatu population, while the S_H4 gene may also give an incomplete resistance reaction. Selection for incomplete resistance, therefore, requires proof of the polygenic nature of resistance (Eskes, 1983). In view of these difficulties and of the practical aspects of accumulating polygenes in one genotype, breeding for durable resistance to coffee leaf rust based on incomplete host-pathogen reaction types is likely to be a very long-term process.

Resistance to Coffee Berry Disease (CBD)

CBD is an anthracnose of the green and ripening berries caused by *Colletotrichum coffeanum* Noack *sensu* Hindorf and was first detected in 1922 on newly established arabica coffee plantations south of Mt Elgon in western Kenya (McDonald, 1926). It is now generally accepted that this disease originated from *C. eugenioides*, which is indigenous to the mountain forests in western Kenya and Uganda (Mogk, 1975; Robinson, 1976). By 1939 the CBD pathogen had spread to arabica coffee east of the Rift Valley in Kenya (Rayner, 1952) and to the Kivu province in Zaire (Hendrickx, 1939). Subsequently, its incidence was reported in Angola and Cameroon (Muller, 1964), Uganda (Butt and Butters, 1966), on Mt Kilimanjaro in Tanzania (Fernie, 1970) and finally in Ethiopia (Mulinge, 1973). Apparently, the free movement of coffee plant material from CBD infected areas has been the main factor in distribution of this disease throughout all important arabica coffee growing areas in Africa.

Latin American countries have been free from CBD, but climatic conditions in some high-altitude areas are comparable to coffee regions of East Africa, where CBD infection is usually very severe. Clearly, only strict plant quarantine regulations can prevent unintentional introduction of the CBD pathogen from the African continent.

CBD may cause crop losses of 50-80 per cent in years favourable to a severe disease epidemic (prolonged wet and cool weather), if not controlled by an intensive programme of fungicide sprays (up to 12-14 rounds) aimed at continuously protecting the crop (Griffiths, Gibbs and Waller, 1971). The frequency of fungicide sprays necessary for effective control of CBD can, according to Muller (1980), be reduced by inducing flowering some 6-8 weeks before the onset of the rainy season, since the most susceptible stages in the development of the berries (when they are rapidly expanding) will in that way not coincide with wet weather. However, this method can only be applied in areas with a uni-modal rainfall distribution, resulting in a single flowering season like in the highlands of Cameroon. It has been found ineffective in Kenya and other countries of East Africa, where overlapping crops of many different flowerings make frequent sprays indispensable.

The cost of CBD control by fungicides may account for one third of the total production costs at plantation management level (Njagi, 1981) and even then crop losses of 20-30 per cent cannot always be prevented at higher altitudes during excessively wet weather conditions, as happened in Kenya in the years 1977-79. Smallholders, who in most countries of East Africa produce the majority of the arabica coffee, are usually unable to carry out the recommended complete spray programme. They may end up with particularly severe crop losses due to CBD, especially in wet years, since occasional fungicide sprays on susceptible arabica coffee induce higher levels of CBD infection than would occur in the total absence of fungicide sprays (Griffiths, 1972).

It was realised that only cultivars with durable resistance to CBD could under these circumstances guarantee a stable cash income for the coffee growers and preserve an agricultural industry, which is vital to the national economies of Kenya, Tanzania, Ethiopia and other arabica coffee producing countries in Africa.

Basically, two different strategies have been followed. In Kenya a breeding programme was started in 1971 aimed at introgressing CBD resistance from exotic arabica coffee germplasm into local cultivars, thereby taking care to maintain the productivity and the high quality of the typical Kenyan coffee (Van der Vossen and Walyaro, 1981). On the other hand, the work in Ethiopia concentrated on selecting and multiplying CBD-resistant genotypes from the semi-natural coffee populations right in the centre of genetic diversity of arabica coffee (Robinson, 1974; Van der Graaff, 1981), whereas productivity and quality were only of secondary importance.

The tremendous variation in field infection to CBD caused by macro- (Cook, 1975) and micro-climatic (Waller, 1971) factors seriously biases selection for true resistance. However, the development of a reliable preselection test, based on the inoculation of hypocotyl stems of 6-week-old seedlings with a pure spore suspension of the pathogen, has greatly accelerated selection progress (Van der Vossen et al., 1976). In contrast to other laboratory tests, such as inoculation of detached berries or young leaves, the results of the inoculation test on hypocotyl stems were closely correlated ($r = 0.73-0.80$) with mature plant resistance.

The preselection test on 6-week-old seedlings requires exact conditions of temperature and humidity during infection and incubation periods (Van der Vossen and Waweru, 1977). Especially high ($> 20°C$) temperatures during the two weeks incubation period can cause inaccurate results as was experienced in Ethiopia (Van der Graaff, 1981).

This preselection test has been applied successfully in Kenya for the selection of resistant genotypes within segregated breeding populations and to confirm by progeny testing the resistance of thousands of mature trees in the selection fields.

The presence of fungistatic substances in the cuticular wax of resistant varieties (Lampard and Carter, 1973; Steiner, 1972) could play a role in slowing down initial infection. Histological evidence produced by Masaba and Van der Vossen (1982) suggests, however, that the resistance mechanism is to an important extent based on the formation of cork barriers in the pericarp of berries (still attached to the tree) or the cortex of hypocotyl stems in response to infection. In resistant genotypes this effectively seals off the invading pathogen from healthy host tissue. Such barriers were absent or incompletely formed in susceptible genotypes. The almost identical response to infection observed in berries and hypocotyl stems explains the earlier found high correlation between the preselection test and mature plant resistance and gives further proof of the reliability of the hypocotyl inoculation test as an efficient method of selection for CBD resistance.

Such a resistance mechanism is usually race-non-specific and together with the apparent absence of differential pathogenicity of CBD isolates in Kenya (Masaba, 1980) and in Ethiopia (Van der Graaff, 1981) there is good reason to assume that the CBD resistance is of a durable nature.

Nevertheless, genetic studies carried out in Kenya (Van der Vossen and Walyaro, 1980) have indicated that the resistance to CBD is controlled by maximally two major genes in the variety Rume Sudan (a dominant R and recessive K gene) and by one gene, different from the genes in Rume Sudan, in Hibrido de Timor (the T gene with incomplete dominance action). The moderate resistance of varieties like K7 and Blue Mountain was based on one recessive gene, identical to the K gene of Rume Sudan.

It may be interesting to note here that the resistance genes were found in host populations separated from the area where the pathogen is indigenous. Such resistance types tend to be more durable (Abdallah and Hermsen, 1971).

After the identification of CBD resistance genes in the main progenitors, the breeding programme in Kenya was adjusted to ensure that all three genes will be present in one genotype as a means of enhancing the stability of the resistance (Parlevliet and Zadoks, 1977).

The frequency of resistant genotypes in arabica coffee germplasm of Ethiopian origin is very low (Van der Graaff, 1981). Claims of polygenically inherited resistance in such populations by Robinson (1974) and Van der Graaff (1978, 1981) — inspired by the apparently continuous variation in resistance to CBD noticed under field conditions — could not be confirmed in Kenya. On the contrary, Walyaro and Van der Vossen (1985) could detect only one or at the most two major genes in resistant genotypes originating from the Ethiopian gene centre. It is not yet clear, whether these genes were different from the earlier detected CBD resistance genes present in Rume Sudan.

The presence of resistance in robusta coffee was confirmed from studies on populations of arabica × robusta interspecific hybrids (Carvalho, Monaco and Van der Vossen, 1976). It is likely that the resistance gene

70 *Coffee Selection and Breeding*

found in Hibrido de Timor, which derives from natural hybridisation between arabica and robusta coffee, has a robusta origin (Van der Vossen and Walyaro, 1980).

Phytotonic Effects of Fungicides

Long before regular spray programmes became necessary to control coffee leaf rust and CBD, coffee growers in Kenya and Tanzania had discovered, that a few fungicide applications could increase yields by 50-100 per cent, independent of disease control (Gillet, 1942; Rayner, 1957). The initial effect of such 'tonic' sprays of fungicide is a 2-3 months delay in leaf fall and this results in increased yields, because the number of flowers is directly related to the number of leaves present on the tree during periods of most intensive flower initiation (G. Browning, unpublished) — see Figure 3.5.

Two factors were found to contribute to accelerated senescence in coffee leaves: the activity of the phylloplane microflora, and water stress.

Figure 3.5: The Phytotonic Effect of Fungicides. This causes the leaves to remain attached much longer (upper branch) than in the absence of fungicide sprays (lower branch) under the climatic conditions of Kenya. (Measure at bottom is 10cm)

Figure 3.6: Leaf Abscission Induced by CEPA Sprays (15 August 1975) at Concentrations of 250, 500, 1,000 and 2,000 ppm on Trees Receiving Tonic Sprays with Fungicides (●_____●) and Trees Never Sprayed with Fungicides (○— —○) in Expt. B3b. The vertical bars are SEs of each treatment mean.

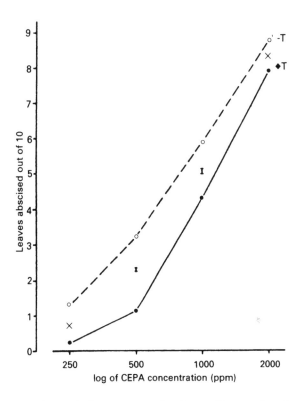

Both would act through the same mechanism of increasing the endogenous ethylene levels in the leaves (Van der Vossen and Browning, 1978). Broad-spectrum fungicides (e.g. copper oxychloride and captafol) effect a marked reduction of the microflora on coffee leaves (Achwanya, 1980) and thus remove one of the factors, which induces premature leaf fall. Fungicide sprayed trees showed indeed a significantly lower leaf abscission response to CEPA (2-chloroethanephosphonic acid), which releases ethylene *in vivo*, than unsprayed leaves (Figure 3.6)

A screening of all available germplasm of arabica coffee showed a high variation in thresholds for leaf abscission, but the locally selected cultivars, such as SL28, were among the most resistant to exogenously applied ethylene. This could indicate that selection pressure for an ability of coffee trees not to shed leaves readily under conditions of water stress — firstly, in Yemen and subsequently in East Africa — produced coffee genotypes, which were able to withstand much more severe water stress conditions than exist in the natural environment of the Ethiopian forests. There are apparently no genotypes of arabica coffee that would exhibit a better leaf retention capacity under water stress conditions than the East African cultivars.

Since even a cultivar like SL28 gave clear yield responses to fungicide applications, it was concluded that the new disease-resistant cultivars would also require tonic sprays of fungicide for increased yields.

Incomplete spray programmes are known to aggravate the incidence of coffee leaf rust and CBD (Gibbs, 1972; Mulinge and Griffiths, 1974). However, field experiments with twice-yearly fungicide sprays applied to a full range of highly susceptible to completely resistant genotypes (Van der Vossen, 1982) showed that this was only the case with disease-susceptible genotypes (Figure 3.7).

Yields over four years were almost doubled in the tonic-sprayed subplots of disease-resistant progenies, while the dramatic increase of CBD and leaf rust incidence in the sprayed subplots of susceptible varieties or progenies caused such heavy crop loss, that the yield was the same or even lower than in the non-sprayed plots.

In Kenya it is now common practice to apply a few fungicide sprays to selection fields as a means of differentiating more efficiently susceptible from resistant genotypes.

On the other hand, occasional applications of broad-spectrum fungicides will be an effective means of obtaining economically attractive yield increases at low additional costs with arabica cultivars resistant to CBD and coffee leaf rust.

Selection for Yield and Quality

Once effective methods of selection for disease resistance are available, the duration of breeding programmes aimed at developing disease resistant varieties will largely depend on the efficiency of selection for high yield and good quality. This becomes even more critical in countries where traditionally coffees of the highest quality have been produced.

The response to selection for yield in coffee was found to be considerably lower than for bean size (Stoffels, 1941; Ferwerda, 1948; Castillo, 1957; Cramer, 1957; Carvalho and Monaco, 1969). Significant phenotypic correlations observed between the first 2-3 years' yields and production accumulated over 5-6 years and between growth parameters and yield for arabica as well as robusta coffee (Machado, 1952; Dublin, 1967; Dhaliwal, 1968; Srinivasan, 1969, 1980) indicated possibilities of early selection for yield potential.

Otherwise, information on the inheritance of these quantitative characters which is required to decide on methods of selection which should

Coffee Selection and Breeding 73

Figure 3.7: The Effect of Tonic Sprays of Fungicide on Yield, CBD and Leaf Rust Infection of Susceptible and Resistant Genotypes of *Coffea arabica*

●———●   ■   ▨ — + tonic sprays
○- - -○   □   □ — – tonic sprays
×—·—× — complete spray programme

P1 — progeny of cross (Erecta × Padang)
P2 — progeny of cross (Pad. × K7) (Maragogipe × Pad.)
P3 — progeny of cross (Rume Sudan × SL 28)(Bourbon × Hibrido de Timor)
P4 — progeny of cross (SL 34 × RS) × HT
LSD ($p = 0.05$) between tonic and non-tonic subtreatments within cultivar or progeny:

	1975	1976	1977	1978
Cum. clean coffee (t ha^{-1})	NS	1.2	1.6	2.1
CBD (%)	5.7	4.1	7.3	4.1
Leaf rust (0-10)	—	0.2	1.0	1.0

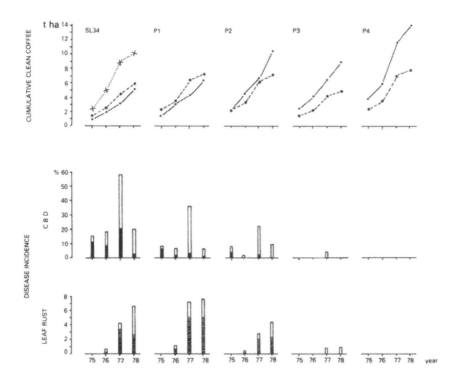

74 Coffee Selection and Breeding

lead to rapid genetic improvement, has been generally lacking.

However, recently published data from biometrical studies carried out in Kenya have contributed to a better understanding of the genetic basis of variation and covariation of growth, yield and quality characters in arabica coffee (Walyaro and Van der Vossen, 1979; Walyaro, 1983). The most relevant conclusions can be summarised as follows.

(1) *Heritability of Growth, Yield and Quality Characters*

Heritability is the ratio of genetic variance to the total observed, or phenotypic variance. This concept was developed to enable the relative importance of genetic and environmental factors to be determined and to show whether selection progress is relatively easy or difficult to obtain. Estimates of heritability for a number of characters from the analysis of a diallele cross among 11 varieties of arabica coffee are summarised in Table 3.3.

Table 3.3: Heritability Estimates of Growth, Yield and Quality Characters in Arabica Coffee from a Diallele Cross Between 11 Varieties (Walyaro, 1983)

Character	Heritability (h_w^2) estimates based on	
	Means of 4 trees	Individual tree
Growth		
Girth of main stem	0.64	0.31
Height of tree	0.70	0.37
Canopy radius	0.65	0.31
Internode length	0.74	0.42
Angle of primary with main stem	0.60	0.28
Yield components		
% bearing primaries	0.39	0.14
% bearing nodes	0.23	0.07
Flowers per node	0.10	0.03
Berries per node	0.22	0.06
Yield (kg/tree)		
First year	0.29	0.10
Total of two years	0.72[†]	
Total of 10 years	0.81[†]	
Bean size*		
100—bean weight	0.74	0.41
% PB	0.67	0.35
% AA	0.70	0.36
% AB	0.64	0.31
% TT	0.25	0.08
% C	0.73	0.40
Cup quality[†]		
Acidity	0.15	
Body	0.08	
Flavour	0.23	
Overall standard	0.48	

* see under selection criteria (page 57) for a description of bean size and cup quality characters

[†] from Walyaro and Van der Vossen (1979)

Most growth and bean size characters (except % TT) had a high heritability, particularly when estimated from means of four trees. Single measurements of these characters on young trees (two years after field planting) gave already a reliable indication of the actual genotypic values, so that selection progress should be high. Growth characters like erect branches and short internodes could be selected effectively on 1-year-old nursery seedlings. Yield and its components showed a low heritability, as would have been expected from earlier observations. Only yield taken over several years and based on plot means of at least four trees gave reasonable heritability values. The fairly high heritability observed for the overall standard of cup quality (range 0-7) would indicate that good selection progress for this character is possible with the assistance of an experienced coffee liquorer. The separate components of cup quality — acidity, body and flavour — showed low heritabilities, possibly because of the small range of the scoring system (0-4).

(2) *Early Determination of the Genetic Yield Potential*

High genotypic correlations existing between growth characters and 10-year yield totals, as well as between 2-3 years yield records and 10-year totals, suggested the possibility of early selection. The expected genetic advance based on an index comprising a few growth parameters (girth, canopy radius, percentage of bearing primaries) and the first 2-3 years yield data of individual trees was shown to be as large as that obtained by straight selection based on yield totals of plot means over several years. In that way, the breeding cycle in arabica coffee need not be longer than five years for maximum selection progress for improved productivity.

(3) *Hybrid Vigour*

There was evidence for considerable hybrid vigour, due to the effects of complementary epistatic genes, for yield, particularly in F_1 crosses between varieties of diverse origin (Table 3.4). Cases of hybrid vigour have also been reported by Srinivasan and Vishveshwara (1978) for arabica coffee in India. Hybrid vigour can be best exploited in F_1 hybrid varieties. Variation in bean size and cup quality characters was mainly due to additive genetic effects. The performance of progenies can for these characters therefore be predicted from midparent values.

(4) *Stability of Yield and Quality Characters*

There were pronounced genotype-environment interaction effects on yield, but it should be possible to select for high yielding genotypes with a high linear response to environments. This means that the highest yielders remain so under different environmental conditions (e.g. plant densities and climate). It was also shown that yield stability could be combined with compact growth, as both characters are inherited independently. It was also

Table 3.4: Cumulative Yield of Clean Coffee for the First Three Years of Production of Varieties and F_1 Hybrids of Arabica Coffee at Two Plant Densities. Data taken from a diallele cross (Walyaro, 1983)

Variety or F_1 Hybrid	Yield (tonnes/ha)			
	3,333 trees/ha (1.5 × 2m)		6667 trees/ha (1 × 1.5m)	
Padang	11.7		13.1	
SL34	11.6		10.7	
SL28	10.9		9.3	
Caturra	8.6		10.1	
Hibrido de Timor	8.5		7.9	
Erecta	7.3		9.5	
Rume Sudan	2.5		2.9	
Laurina	2.7		5.1	
Padang × SL34	13.5	(11.7)	14.3	(11.9)
Padang × Caturra	12.4	(10.2)	12.4	(11.6)
Padang × Erecta	10.7	(9.5)	15.6	(11.3)
Padang × Rume Sudan	11.4	(7.1)	14.0	(8.0)
Padang × Laurina	12.7	(7.2)	14.9	(9.1)
SL28 × Hibrido de Timor	12.8	(9.7)	12.5	(6.1)
SL28 × Rume Sudan	10.9	(7.1)	14.2	(6.8)
Laurina × Hibrido de Timor	12.3	(5.6)	13.3	(6.5)

Notes — in parentheses midpart values assuming additive genetic variance only
— a complete fungicide spray programme was applied to control CBD and coffee leaf rust

higher in F_1 hybrids than in homogeneous varieties. Quality characters, such as bean size and cup quality were less influenced by the effects of genotype-environment interaction.

It should be realised that estimates of genetic parameters are not only a property of the particular character, but also of the population from which the parameters have been derived and of the environmental circumstances to which the population has been subjected (Falconer, 1981). Strictly speaking, above conclusions apply therefore only to similarly structured breeding populations and under comparable environmental conditions. Nevertheless, a number of them should be relevant to other arabica breeding programmes and even to robusta coffee breeding.

As an illustration of the variation of quality characters encountered in varieties and crosses of arabica coffee, some examples taken from the breeding programme in Kenya are presented in Table 3.5. The cultivar SL28 has a large bean (46 per cent AA) and excellent cup quality. The small bean size and lower cup quality of varieties like Caturra and Rume Sudan are apparent. Hibrido de Timor has fairly large beans, but the cup quality is poor. The percentage peaberries in the varieties Rume Sudan and Hibrido de Timor and in all crosses is about twice as high compared to varieties like Padang, Caturra and SL28, and is a sign of lower female fer-

Table 3.5: Examples of Bean Size and Cup Quality Characters for Varieties, Single and Multiple Crosses and Backcrosses in Arabica Coffee (Van der Vossen and Walyaro, 1981)

Variety or cross	Bean grading (%)			Cup quality			
	PB	AA	AB	Acidity	Body	Flavour	Standard
SL28	14	46	18	1.0	1.0	2.0	2.0
Padang	15	37	20	1.8	1.8	3.6	3.6
Caturra	11	15	25	2.0	1.5	3.7	3.8
Rume Sudan (RS)	29	3	27	2.5	2.0	3.3	3.3
Hibrido de Timor (H de T)	23	39	21	2.0	1.6	4.1	4.0
SL28 × Caturra	23	24	21	1.5	1.5	3.0	3.3
SL28 × Rume Sudan	35	15	27	1.5	1.0	2.5	3.0
SL28 × Hibrido de Timor	26	35	18	2.0	2.0	3.8	3.6
Caturra × Hibrido de Timor	31	29	19	3.0	2.5	4.5	3.8
a. (RS × SL28)(Bourbon × H deT)	17	15	40	1.0	1.0	3.0	3.0
b. (RS × K7)(H de T × SL34)	28	18	44	2.0	2.0	3.3	3.5
c. (SL34 × RS) × H de T	30	10	46	2.5	2.5	4.0	4.0
SL28 × (a)	28	31	17	1.0	1.0	1.5	2.0
SL34 × (b)	29	42	17	0.5	0.5	1.5	2.0
SL28 × (c)	35	38	12	1.5	1.5	3.3	3.0
Caturra × (a)	26	21	21	1.5	2.0	3.3	3.3
Caturra × (b)	40	18	20	2.0	1.5	4.0	3.8
Caturra × (c)	23	14	31	1.5	2.0	3.5	3.3
Catimor ex Colombia F_3 progeny 1	19	48	13	1.5	1.7	3.5	3.2
Catimor ex Colombia F_3 progeny 2	23	47	15	1.5	2.0	4.0	3.5
Catimor ex Colombia F_3 progeny 3	19	33	28	2.0	1.7	3.8	3.3

Notes — see under section on selection criteria (page 57) for a description of bean and cup quality characters
— scores for acidity, body and flavour 0-4, for standard 0-7; 0 means very good, 4 intermediate, 7 is very poor
— all coffee is prepared according to the wet processing method, sun dried to 11% moisture content and the cup quality is assessed by the same professional coffee liquorer

tility possibly caused by genetic imbalance.

The bean size and cup quality in the single and 4-way crosses are mediocre and vary according to the progenitors included in the crosses. However, almost complete restoration to the required standards of Kenyan coffees could be attained by backcrossing to SL28.

While the bean size of the cross between Caturra and Hibrido de Timor was fairly small, some of the Catimor selections from Colombia — F_3 progenies of the cross Caturra × Hibrido de Timor — had a very satisfactory bean size (47-48 per cent AA). It is proof of the good selection progress possible for this character by strict selection. The cup quality is inferior to SL28, but appreciably better than that of Hibrido de Timor.

Arabica germplasm of Ethiopian origin is characterised by low productivity, very small and light beans and generally inferior cup quality, when compared to Kenyan cultivars. The examples given in Table 3.6 are

Table 3.6: Variation in Yield and Quality of Arabica Germplasm of Ethiopian Origin (FAO Mission, 1964)

Introduction number		Place	Province	Yield (3 years totals)		100-bean weight (g)	Out turn	Bean grading %				Cup quality standard
				Cherry (kg/tree)	clean coffee (t./ha)			TT	PB	AA	AB	C	
F4	(E237)	Yergalem	Sidamo	8.1	3.7	16	0.17	25	6	0	55	9	3.6
F5	(E238)	Aleta	Sidamo	11.5	4.9	18	0.16	28	4	1	31	26	3.8
F19	(E252)	Wush-Wush	Kaffa	6.4	2.4	22	0.14	38	6	1	44	10	3.3
F24	(E257)	Core-Alle	Illubabor	8.5	3.6	14	0.16	35	9	1	28	24	4.5
F26	(E259)	Core-Matta	Illubabor	8.4	3.5	19	0.16	37	7	3	35	14	4.0
F54	—	Bonga	Kaffa	10.9	5.2	19	0.18	19	17	16	39	8	4.5
F59	(E126)	Geisha-Gore	Kaffa	9.3	4.0	15	0.16	45	4	0	14	33	4.3
F68	(E123)	Ainamba	Kaffa	9.2	3.7	16	0.15	27	5	1	19	44	4.3
F96	(E87)	Ghera-Teka	Kaffa	7.0	3.2	16	0.17	26	10	0	30	29	5.0
F99	(E114)	Mizan-Tafari	Kaffa	7.8	3.1	14	0.15	35	2	0	25	33	4.5
F125	(E193)	Teppi	Illubabor	9.4	3.5	18	0.14	21	8	12	42	12	3.8
F130	(E224)	Teppi	Illubabor	8.8	3.5	18	0.15	17	6	1	34	29	5.5
F146	(E441)	Teppi Forest	Illubabor	9.9	5.0	19	0.19	18	10	11	48	11	4.5
F184	(E552)	Ainamba	Kaffa	7.5	2.6	17	0.13	14	10	0	37	35	5.0
Mean over 173 Introductions				6.4 ± 2.0	2.9 ± 0.9			8	16	42	29	3	2.0
Standard cultivar SL28				12.1 ± 4.1	6.2 ± 2.1	21	0.19						

Notes — in parentheses the FAO introduction number (Meyer et al., 1968)
— the field was never sprayed with fungicides (no tonic effect and disease control) and yields are therefore low
— plant density: 2,665 trees/ha (1.37 × 2.74 m)
— see under selection criteria (page 57) for a description of bean size and cup quality characters and footnote Table 3.5

Table 3.7: Examples of Bean Size and Liquor Quality of F_1 Progenies of Crosses Between Accessions of the Ethiopian Collection (FAO, 1964) and Arabica Variety SL28

Progeny of cross	PB	Bean grading % AA	AB	Cup quality Standard
SL28 × F5	25	23	25	4.0
SL28 × F14	32	22	25	3.3
SL28 × F24	31	20	30	4.0
SL28 × F45	33	24	22	4.0
SL28 × F70	35	30	23	3.3

Notes: — first year quality samples taken from a selection field planted in 1977 (CRF Annual Report, Section Plant Breeding, 1980)
— see further footnotes to Table 3.5

accessions already selected for higher yield. The overall average yield of 173 accessions was almost half that of SL28. Crossing with SL28 gave considerable improvement in bean size and a slightly better cup quality and yield (Table 3.7). This kind of germplasm may become of significance to arabica coffee breeding (only in the long term).

In robusta coffee priority is given to selection for higher yield and regular bearing. Considerable progress is to be expected from exploiting the hybrid vigour of interpopulation crosses (Berthaud, 1983, 1985). Bean size is usually a character of secondary importance and as regards cup quality, body and neutral flavour are emphasised. Availability of compact growing plants, such as the Nana form of *C. canephora* found in Central Africa (Berthaud and Guillaumet, 1978), offers the possibility of breeding for plants suitable for high density planting systems analogous to the Caturra type in arabica coffee.

Interspecific Hybridisation

Of the many natural and artificially induced interspecific hybrids studied in coffee (Cramer, 1957; Sybenga, 1960; Monaco and Carvalho, 1964; Ferwerda, 1969; Chinnappa, 1970; Capot, 1972; Charrier, 1978; Louarn, 1976, 1980; Berthaud, 1978) only a few have attained economic importance so far.

A spontaneous hybrid between *C. canephora* var. *ugandae* and *C. congensis*, discovered on Java at the beginning of this century and called congusta, proved to be of considerable commercial value (Ferwerda, 1958). The vigour and productivity, adaptation to cooler climates, tolerance to temporary waterlogging and good bean size of congusta have attracted the attention of plant breeders outside Indonesia because of its potential to replace robusta coffee. The IFCC started a selection programme within congusta populations on Madagascar and more recently such a programme of crosses has been initiated also in Ivory Coast (Charrier, 1982) between robusta coffee and germplasm of *C. congensis* collected in central Africa (Berthaud and Guillaumet, 1978).

However, breeders in Brazil, Ivory Coast and Kenya have focussed their attention on interspecific hybrids between arabica and robusta coffee for

Figure 3.8: Arabica × Robusta Interspecific Hybridization, a. Tetraploid Robusta plant at Kawanda, Uganda; note the large leathery leaves. b. A comparison of infructescence and leaf size and shape of tetraploid Robusta (upper left), Arabica (upper right), Arabusta (lower left) and first backcross of Arabusta to Arabica (lower right). Pieces of bearing branches have been added to give an impression of berry size; measure at bottom is 10 cm.

a

two main reasons: (1) to improve the quality of robusta coffee and (2) to introgress the vigour and disease resistance of robusta into arabica coffee.

Crosses between the allotetraploid *C. arabica* and diploid *C. canephora* produce vigorously growing and profusely flowering trees, which are nevertheless almost completely sterile due to their triploid nature. Doubling the chromosome number by colchicine treatment (Mendes, 1947; Berthou, 1975) produces hexaploid and completely homozygous plants with normal fertility (Krug and Carvalho, 1952; Orozco and Cassalett, 1975; Berthaud, 1978). With two-thirds of the chromosomes from *C. arabica*, they closely resemble this species not only in bean size, cup quality and caffeine content, but also in their poor adaptation to lowland tropical climates. Breeders of arabica coffee appear to have lost interest in such hexaploid hybrids, mainly because introgression of disease resistance from *C. robusta* can be achieved much easier through Hibrido de Timor — a natural arabica × robusta hybrid very much resembling arabica coffee — or the artificially made arabusta hybrid, both with 44 chromosomes. Possibilities of adapting hexaploid arabica × robusta hybrids to the humid tropical lowlands, through selection and recombination of different genotypes, are presently under investigation in Ivory Coast (Le Pierrès and Anthony, 1980).

Capot (1972) applied the name arabusta to fertile interspecific F_1 hybrids from crosses between arabica and induced autotetraploid robusta coffee. Such hybrids had been produced already around 1950 in Brazil (Monaco, 1977) after Mendes (1939, 1947) had developed methods of producing tetraploid plants of robusta coffee by colchicine treatment of germinating seed or shoot tips (Figure 3.8a). The encouraging results of the first arabusta hybrids, especially as regards vigour, good adaptation to tropical lowlands and improved cup quality, made the IFCC in Ivory Coast to embark on a large programme of clonal selection within the highly variable F_1 populations with emphasis on improved fertility and productivity (Grassias, 1977; Capot, 1975; Duceau, 1980). Yields of 1.5 t/ha clean coffee have been obtained after a first cycle of selection and a development programme with a target of 15,000 ha arabusta coffee is presently in progress (IFCC — SATMACI, 1977; IFCC, 1981) — see Figure 3.8b.

The Icatu populations of the Instituto Agronomico at Campinas in Brazil are the result of repeated backcrossing of the interspecific robusta × arabica hybrids to arabica cultivars Mundo Novo and Caturra (Monaco, Carvalho and Fazuoli, 1974). This material is still very variable in most agronomic aspects, but the presence of genotypes with complete resistance to coffee leaf rust makes this programme very important for the Brazilian coffee industry (Monaco, 1977).

Advanced backcrosses of arabusta to arabica coffee offer also perspectives for coffee production outside the main arabica coffee zones in Kenya because of their better adaptation to lower altitudes (Owuor and Van der Vossen, 1981).

A comparison of the performance of arabusta and backcrosses (BC_1 and BC_2) to arabica in relation to the arabica parent, as regards vigour, fertility, yield and quality is presented in Table 3.8. The data are derived from a study carried out in Kenya (Van der Vossen and Owuor, 1981). Of particular interest is the hybrid vigour of the F_1 hybrids, as is reflected in the growth parameters. The low number of secondary lateral branches and high number of flowers per node of the F_1 hybrids and characteristic of robusta coffee, was not found in the backcrosses. Fertility, as is indicated by pollen viability, fruit set and percentage peaberries, was progressively restored in the backcrosses. This applied also to the degree of self-compatibility (fruit set after selfing). Bean size and cup quality were considerably inferior to the arabica parent SL28, even in the BC_2 generation. This slow improvement in quality is in fact the major deficiency of the interspecific hybridisation programme for arabica coffee improvement.

The ORSTOM has started a comprehensive programme of interspecific hybridisation between diploid coffee species, including *C. canephora, C. congensis, C. eugenioides, C. liberica* and species of the *Mascarocoffea* group (Charrier, 1978; Louarn, 1976, 1980; Lanaud and Zickler, 1980) for genetic and cytological purposes in the first place. In the long-term these programmes are also expected to contribute useful characteristics, such as low caffeine content, disease and pest resistance, drought resistance and other characters of agronomic value, to the improvement of the robusta coffees of the lowland tropical zones in particular.

Table 3.8: A Comparison of Vigour, Fertility, Bean Size and Cup Quality of the F_1, BC_1 and BC_2 Generations of Interspecific arabica × robusta Crosses with the arabica Parent

		Growth and flowering				In vitro pollen viability (%)	Fruit set		Bean grading (%)			Cup quality standard
	Height (cm)	Girth of stem (cm)	Radius canopy (cm)	Second. laterals (no.)	Flowers per node (no.)		on open pollination (%)	after selfing (%)	PB	AA	AB	
C. arabica cv SL28	163	16	54	13	14	72	63	60	15	42	16	2.0
F_1 Arabusta (mean of 6 progenies)	196	18	73	6	23	8	19	0	60	11	11	4.3
BC_1 (mean of 2 progenies)	156	14	60	12	13	22	36	12	26	8	27	4.4
BC_2 (mean of 5 progenies)	167	15	58	12	13	46	52	36	18	9	39	4.2

Notes: — see under section on selection criteria (page 61) for a description of growth, flowering, bean grading and cup quality characters
— secondary laterals on primary branches

Breeding Strategies: Seed or Clonal Varieties

Arabica coffee

C. arabica is one of the very few examples of a perennial plant to which breeding methods common to self-pollinated diploid crops have been applied successfully. The resultant varieties were sufficiently homogeneous to permit propagation by seed and practically all arabica grown in the world today has been established from seedlings.

The new demands for disease resistance, higher yield and improved quality, as well as a changed tree architecture (e.g. compact growth), have initiated breeding programmes in Africa and Latin America which involve recombination of many characters by complex programmes of crossing and backcrossing. From the review presented in the preceding paragraphs of this chapter it can be seen that considerable advances have been made in developing efficient selection methods to ensure rapid selection progress.

All the same, even the most advanced breeding material should, with few exceptions, still be heterozygous for many factors, as it has passed through maximally 3-4 cycles of breeding so far.

An additional 4-5 generations, or 20-25 years, of inbreeding and selection, necessary to arrive at true-breeding varieties, can however be evaded by resorting to clonal propagation of the superior genotypes. The *in vitro* method of vegetative propagation by nodal cultures (Custers, 1980; Dublin, 1980) has the great advantage of rapid multiplication over conventional methods. It produces also plants with a root system similar to seedlings, i.e. a clear tap root, while rooted cuttings tend to have a more superficial root system (Ferwerda, 1969).

In their concern for possible failure of introducing clonal varieties requiring sophisticated technology yet untried in insufficiently advanced agricultural systems, Van der Vossen and Walyaro (1981) worked out a scheme for F_1 hybrid seed production in Kenya. It offers the possibility of quick realisation of plant material homogeneous for all required characters of resistance to CBD and coffee leaf rust, yield, quality and compact growth. One factor essential to that breeding plan was the successful selection of female lines homozygous for compact growth, for the S_H6 gene for resistance to coffee leaf rust and for one major gene for CBD resistance, from Catimor families.

The benefits of F_1 hybrid seed — heterosis for yield, better yield stability and standard procedures of raising seedlings — far outweigh the problems of organising large-scale hand pollination in the seed gardens.

The essential features of alternative breeding plans for arabica coffee leading to either clonal or seed varieties, are presented diagrammatically in Figure 3.9. For a critical evaluation of breeding plans II to IV reference is made to Walyaro (1983). Breeding plan 1 applies to the selection programme carried out in Ethiopia (Van der Graaff, 1981), where sufficient

Coffee Selection and Breeding 85

Figure 3.9: Outlines of Breeding Schemes for the Self-pollinating arabica Coffee

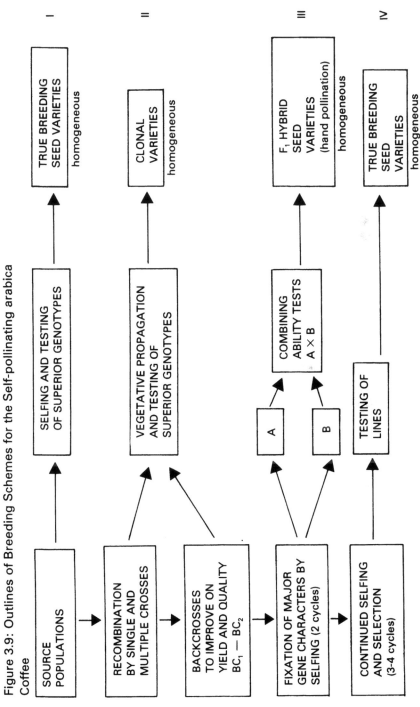

genetic variability is already present in the natural source population for effective pure-line selection. Each selected genotype will be homozygous enough for most characters, because of the mating system (self-pollination) of arabica coffee, to give rise to true-breeding offspring within one or two generations of inbreeding and selection. Such a breeding scheme can lead to rapid results when the objectives are relatively simple and within the range of the existing population.

Robusta Coffee

In robusta coffee, as in most other perennial outbreeders (Simmonds, 1979), all trees are highly heterozygous and outstanding yielders probably represent favourable heterotic combinations. Strict self-incompatibility precludes inbreeding, but instant fixation of such genotypes is possible by vegetative propagation.

Although robusta coffee orchards planted with cross-compatible clones are known to outyield even the best seedling plots by 40-50 per cent (Dublin, 1967; Capot, 1977), the advantages of multiplication by seed, particularly in socio-economically less developed agricultural systems, made robusta coffee breeders in Asia and Africa develop breeding plans leading to polycross seed (Ferwerda, 1948, 1969; Capot, 1977).

Breeding strategies in *C. canephora* should also exploit the potential of increased yields exhibited by interpopulation crosses (Berthaud, 1983, 1985). Berthaud's general distinction in populations of 'Congolese' and 'Guinean' origins could be extended to other, still to be identified, sets of genetically highly variable and divergent subpopulations of *C. canephora*

Simplified outlines of breeding plans, which apply the principle of interpopulation crosses, are presented in Figure 3.10. The production potential of F_1 hybrid varieties should be further enhanced by establishing seed gardens with clones of homozygous genotypes, generated by diploidisation of haploid plants.

In vitro production of haploid plants in coffee (e.g. anther culturing) has been unsuccessful so far (Charrier, 1982), but refined techniques of isolating and raising naturally occurring (as twin embryos in seed) haploid plants in robusta coffee (Couturon and Berthaud, 1979; Couturon, 1982) have brought this possibility much closer to reality. However, thousands of homozygous plants will have to be raised and tested for combining ability as well as cross-compatibility, before large-scale production of F_1 hybrid seed in well-isolated bi-clonal gardens could start (plan IIa in Figure 3.5). A mixture of F_1 hybrid seed from different bi-clonal gardens will still be necessary to ensure good fruit set (cross-compatibility) in the commercial plantations.

Arabica × Robusta Interspecific Hybrids

The potential of such hybrids for the improvement of arabica and robusta

Figure 3.10: Outlines of Breeding Schemes for the Cross-pollinating (Self-incompatible) robusta Coffee

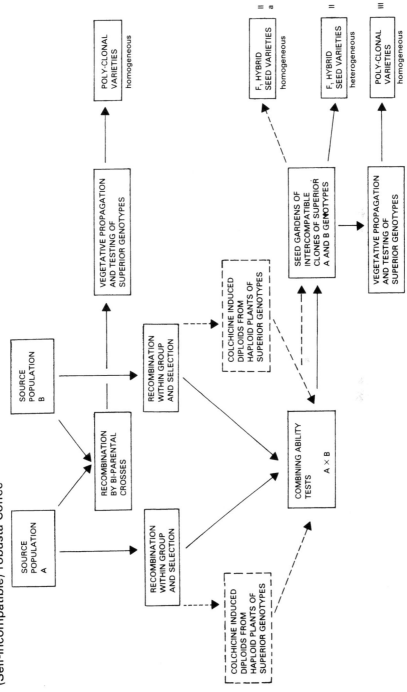

Figure 3.11: Outlines of Breeding Schemes Starting from arabica × robusta Interspecific Hybridisation

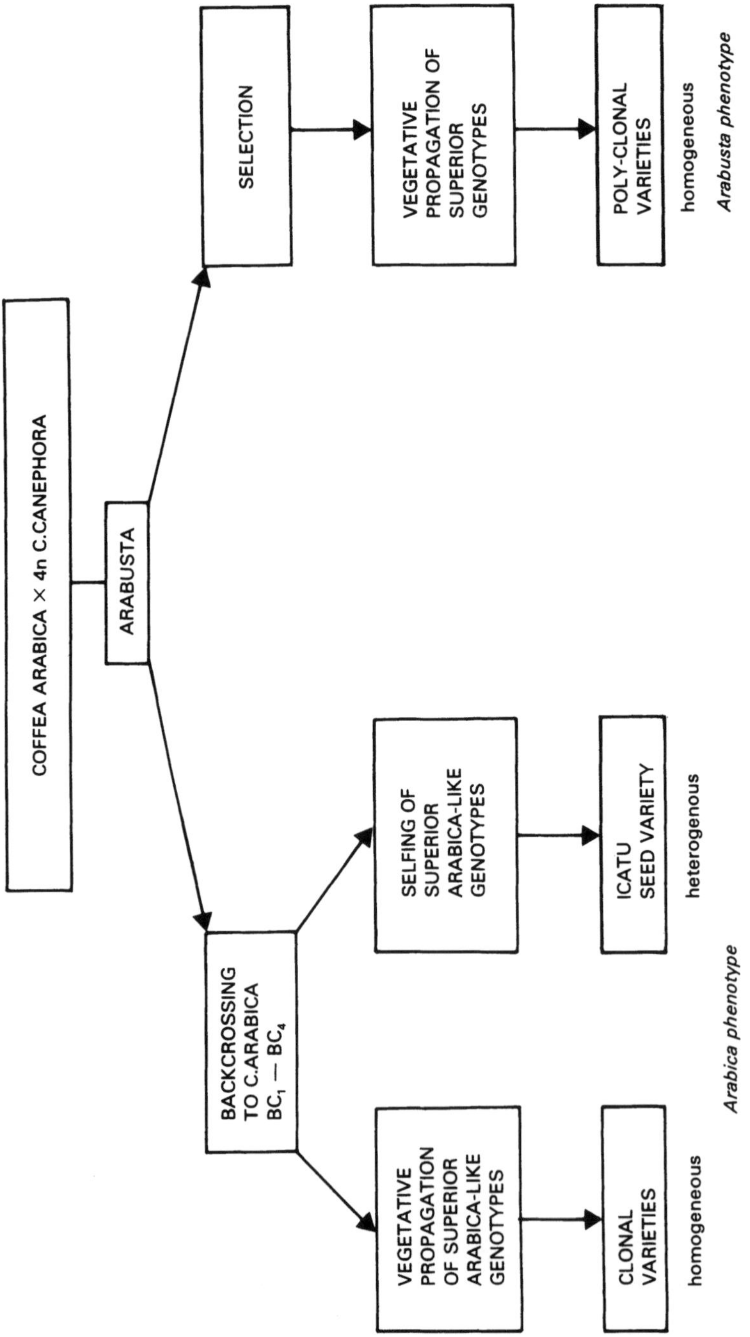

```
                    COFFEA ARABICA × 4n C.CANEPHORA
                                   │
                               ARABUSTA
                              ╱         ╲
                             ╱           ╲
                            ╱             ╲
                   BACKCROSSING          SELECTION
                   TO C.ARABICA              │
                   BC₁ — BC₄                 ▼
                    ╱      ╲          VEGETATIVE
                   ╱        ╲         PROPAGATION OF
                  ╱          ╲        SUPERIOR
         VEGETATIVE        SELFING OF  GENOTYPES
         PROPAGATION       SUPERIOR        │
         OF SUPERIOR       ARABICA-LIKE    ▼
         ARABICA-LIKE      GENOTYPES   POLY-CLONAL
         GENOTYPES             │       VARIETIES
             │                 ▼       homogeneous
             ▼              ICATU       Arabusta phenotype
         CLONAL             SEED VARIETY
         VARIETIES          heterogenous
         homogeneous
                  Arabica phenotype
```

coffee have been discussed earlier (page 79). General outlines of the two main breeding schemes are presented in Figure 3.11. In the case of arabusta programmes, where breeding and selection is focused on the F_1 hybrids, there appears to be no alternative to clonal propagation. Doubling of haploids will in this case not produce the desired level of homozygosity, since arabusta hybrids are segmental autotetraploids (Grassias, 1977).

Clonal propagation will also remain preferable in the programmes aimed at introgressing robusta characters into arabica coffee, because in interspecific hybridisation segregation of unfavourable characteristics can still occur after several generations of backcrossing and selfing. It is therefore unlikely that Icatu populations have attained a sufficient degree of uniformity required for seed propagated arabica cultivars. Modern methods of clonal propagation will have the dual advantage of instant fixation of superior genotypes and exploiting hybrid vigour still present in early breeding generations.

Concluding Remarks

Considering the fact that coffee has been subjected to relatively few generations of selection and that breeders have, so to speak, only just started to exploit the genetic resources of the genus *Coffea*, the expectations of further improvement of productivity, disease and pest resistance, as well as many other characteristics, by conventional and advanced breeding methods, are very good.

On the other hand, continuous world overproduction of coffee and the urgent demand for diversification of fertile land to food crop production, because of high population growth, does not allow further expansion of coffee production in most countries.

The spectacular potential of close spacing for intensive coffee production indicated by agronomic research — e.g. triple to fourfold yields compared to conventional plant densities in arabica coffee — can soon be realised thanks to the advances made in breeding coffee types well adapted to high density planting. The otherwise unsurmountable management problems of disease control and harvesting are solved to a great extent by the combination of resistance to the major diseases, coffee leaf rust and/or CBD, with compact growth, productivity and good quality.

Assuming a gradual stabilisation of export quota through the International Coffee Organisation and a conservative estimate of doubling yield by close spacing, it would mean that the allotted quota could in future be produced from less than half the land presently under coffee. In Kenya for instance, an estimated 50,000 ha of high-potential land would thus become available for food crop production at a fraction of the costs otherwise spent on land reclamation projects.

Genetic improvement of robusta and arabusta coffee could in a similar manner lead to crop intensification and diversification in the lowland tropics (Coste, 1981).

The change to intensive coffee production systems will no doubt raise many still unforeseen problems. These can only be solved by a multidisciplinary approach, in which plant breeding should continue to play a central role.

References

Abdallah, M.M. and Hermsen, J.G. Th. (1971) 'The concept of breeding for uniform and differential resistance and their integration', *Euphytica*, 20, 351-61

Achwanya, O.S. (1980) 'Studies on the Mycoflora of Coffee Leaf Surfaces', MSc Thesis, University of Nairobi

Berthaud, J. (1978) 'L'hybridation interspécifique entre *Coffea arabica* et *C. canephora*. Obtention et comparaison des hybrides triploides, arabusta et hexaploides. *Café Cacao Thé*, 22, 1-11, 87-109

Berthaud, J. (1980) L'incompatibilité chez *Coffea canephora*. Méthode de test et déterminisme génétique. *Café Cacao Thé*, 24, 267-73

Berthaud, J. (1983) 'Etude de la diversité génétique des caféiers: les résultats d'analyse des zymogrammes de *C. canephora*, leur apport dans l'interpretation des résultats de croisements intra-canephora. Propositions pour une strategie d'amélioration 'raisonnee' des *C. canephora*', *ORSTOM internal report*, mimeo

Berthaud, J. (1985) Les Ressources Génétiques chez les Caféiers Africains: Populations Sylvestres, Echanges Génétiques et Mise en Culture, Doctoral Thesis, *University of Paris XI*, Orsay (in preparation)

Berthaud, J., Guillaumet, J.L. Le Pierrès D. and Lourd, M. (1977) Les prospections des caféiers sauvages et leur mise en collection, in *8th International Colloquium on the Chemistry of Coffee, ASIC*, Paris, pp. 365-72

Berthaud, J. and Guillaumet, J.L. (1978) Les caféiers sauvages en Centrafrique. *Café Cacao Thé*, 22, 171-86

Berthou, F. (1975) 'Méthode d'obtention de polyploides dans le genre *Coffea* par traitements localisés de bourgeons à la colchicine', *Café Cacao Thé*, 19, 197-202

Bettencourt, A.J. and Noronha-Wagner, M. (1971) 'Genetic factors conditioning resistance of *Coffea arabica* to *Hemileia vastatrix*', Agronomia Lusitiana, 31, 285-92

Bettencourt, A.J., Lopes, J. and Godinho, I.L. (1980) 'Genetic improvement of coffee. Transfer of factors for resistance to *Hemileia vastatrix* into high yielding cultivars of *Coffea arabica*' in *9th International Colloquium on the Chemistry of Coffee*, ASIC, Paris, pp. 647-58

Boudrand, J.N. (1974) 'Le bouturage du caféier canephora a Madagascar', *Café Cacao Thé*, 18, 31-48

Bouharmont, P. (1971) 'La conservation des graines de caféier destinées à la multiplication du Cameroun', *Café Cacao Thé*, 15, 202-10

Browning, G. (1975) 'Perspectives in the hormone physiology of tropical crops — responses to water stress', *Acta Horticulurae*, 49, 113-23

Butt, D.J. and Butters, B. (1966) 'The control of coffee berry disease in Uganda', *Proceedings of 1st Specialist Meeting on Coffee Research, East African Common Services Organisation, Nairobi, Kenya*, pp. 103-14

Cannell, M.G.R. (1972) Photoperiodic response of mature trees of Arabica coffee. *Turrialba*, 22, 198-206

Capot, J. (1968) 'La production de boutures de clones sélectionnées de caféier canephora', *Café Cacao Thé*, 10, 114-26

Capot, J. (1972) 'L'amélioration du caféier en Côte d'Ivoire. Les hybrides 'arabusta'', *Café Cacao Thé*, 16, 3-16

Capot, J. (1975) 'Obtention et perspectives d'un nouvel hybride de caféier en Côte d'Ivoire: l'Arabusta', in *7th International Colloquium on the Chemistry of Coffee*, ASIC, Paris, pp. 449-59

Capot, J. (1977) 'L'amélioration du caféier robusta en Côte d'Ivoire', *Café Cacao Thé, 21*, 233-42

Carvalho, A. (1959) 'Genetica de *Coffea* 24 — Mutantes de *Coffea arabica* procedentes da Etiopia', *Bragantia, 18*, 353-71

Carvalho, A., Fazuoli, L.C. and Monaco, L.C. (1975) 'Characteristicas do cultivar Iarana de *Coffea arabica*', *Bragantia, 34*, 263-72

Carvalho, A., Ferwerda, F.P., Frahm-Leliveld, J.A., Medina, P.M., Mendes, A.J.T. and Monaco, L.C. (1969) 'Coffee', in F.P. Ferwerda and F. Wit (eds.) *Outlines of Perennial Crop Breeding in the Tropics*, Veenman & Zonen NV, Wageningen, pp. 189-241

Carvalho, A. and Monaco, L.C. (1962) 'Natural cross-pollination in *Coffea arabica*', *Proceedings of 16th International Horticultural Congress (Brussels), 4*, 447-9

Carvalho, A. and Monaco, L.C. (1969) 'The breeding of Arabica coffee', in F.P. Ferwerda and F. Wit (eds.) *Outlines of Perennial Crop Breeding in the Tropics*, Veenman & Zonen NV, Wageningen, pp. 198-216

Carvalho, A., Monaco, L.C. and van der Vossen, H.A.M. (1976) 'Cafe Icatu como fonte de resistencia a *Colletotrichum coffeanum*', *Bragantia, 35*, 343-7

Castillo-Z., J. (1957) 'Mejoramento de café arabica var. typica: utilizaciòn de registro de producciòn de arboles madres', *Cenicafe (Chinchina), 8*, 117-44

Castillo-Z, J. and Moreno-R, G. (1980) 'Seleccion de cruziamientos derivados del Hibrido de Timor en la obtenciòn de variedades mejoradas de café para Colombia', in *9th International Colloquium on the Chemistry of Coffee*, ASIC, Paris, pp. 731-45

Charrier, A. (1978) 'La structure génétique des caféiers spontanés de la région Malgache (*Mascarocoffea*). Leur relations avec les caféiers d'origine africaine (*Eucoffea*)', *Memoires ORSTOM (Paris) no. 87*

Charrier, A. (1980) 'La conservation des ressources génétiques du genre *Coffea*', *Café Cacao Thé, 24*, 249-57

Charrier, A. (1982) 'L'amélioration génétique des cafés', *La Recherche, 13 (136)*, 1006-16

Charrier, A. and Berthaud, J. (1975) 'Variation de la teneur en caféine dans le genre *Coffea*', *Café Cacao Thé, 19*, 251-64

Chevalier, A. (1937) 'Les plantations de cafe au Congo Belge; leur histoire (1881-1935), leur importance actuelle' (Review of a book by E. Leplae), *Revue Botanique Appliquée et d'Agriculture Tropicale, 17*, 66-8

Chinnappa, C.C. (1970) 'Interspecific hybrids of *Coffea canephora* and *C. liberica*', *Genetica, The Hague, 41*, 141-6

Cook, R.T.A. (1975) 'The effect of weather conditions on infection by coffee berry disease', *Kenya Coffee, 40*, 190-7

Coste, R. (1968) *Le Caféier*, G.P. Maisonneuve & Larose, Paris

Coste, R. (1981) 'Caféiers et cafés: propos d'un agronome', *Café Cacao Thé, 25*, 203-7

Couturon, E. (1980) 'Le maintien de la viabilité des graines de caféiers par la controle de leur teneur en eau et la temperature de stockage', *Café Cacao Thé, 24*, 27-31

Couturon, E. (1982) 'Obtention d'haploides spontanés de *Coffea canephora* par utilization du greffage d'embryons', *Café Cacao Thé, 26*, 155-60

Couturon, E. and Berthaud, J. (1979) 'Le greffage d'embryons de caféiers; mise au point technique', *Café Cacao Thé, 23*, 267-70

Cramer, P.J.S. (1957) in F.L. Wellman (ed.) *Review of Literature of Coffee Research in Indonesia* SIC Editorial International American Institute of Agricultural Sciences, Turriabla, Costa Rica

Custers, J.B.M. (1980) 'Clonal propagation of *Coffea arabica* by nodal culture', in *9th International Colloquium on the Chemistry of Coffee*, ASIC, Paris, pp. 589-96

Dhaliwal Singh, T. (1968) 'Correlations between yield and morphological characters in Puerto Rican and Columnaris varieties of *Coffea arabica*', *Journal of the Agricultural University of Puerto Rico, 52*, 29-37

D'Oliveira, B. (1958) 'Selection of coffee types resistant to the *Hemileia* leaf rust', *Coffee and Tea Industry Flavor Field, 8*, 112-20

D'Oliveira, B. and Rodrigues, C.J. Jr. (1961) 'O problema das ferrugens do cafeeiro'. *Revista de Cafe Portugues, 8*, 5-50

Dublin, P. (1967) 'L'amélioration du caféier robusta en République Centrafricaine. Dix annees de selection clonale', *Café Cacao Thé, 11,* 101-36

Dublin, P. (1980) 'Inductions de bourgeons neoformes et embryogenèse somatique. Deux voies de multiplication végétative *in vitro* des caféiers cultivés', *Café Cacao Thé, 24,* 121-8

Dublin, P. (1981) 'Embryogenèse somatique directe sur fragments de feuilles de caféier arabusta', *Café Cacao Thé, 25,* 237-41

Duceau, P. (1980) 'Critères de selection pour l'amélioration des hybrides arabusta en Côte d'Ivoire', *Café Cacao Thé, 24,* 275-9

Eskes, A.B. (1983) Incomplete Resistance to Coffee Leaf Rust (*Hemileia vastatrix*), PhD Thesis, Agricultura University of Wageningen

Falconer, D.S. (1981) *Introduction to Quantitative Genetics,* 2nd Edn, Oliver & Boyd, Edinburgh and London

Fazuoli, L.C., Monaco, L.C., Carvalho, A. and Reis, A.J. (1977) 'Resistencia do cafeeiro a nematodes I: Testes em progenies e hibridos para *Meloidogyne exigua*', *Bragantia, 36,* 297-307

Feilden, G. St. Clair (1940) 'Vegetative propagation of tropical and subtropical plantation crops', *Technical Communication Bureau of Horticulture, East Malling, No. 13.* Maidstone, Kent, UK

Fernie, L.M. (1940) 'The rooting of softwood cuttings of *Coffea arabica*', *East African Agricultural Journal, 5,* 323-9

Fernie, L.M. (1970) 'The improvement of Arabica coffee in East Africa', in C.L.A. Leakey (ed.) *Crop Improvement in East Africa,* Technical Communication no. 19 of the Commonwealth Bureau of Plant Breeding and Genetics. CAB, Farnham Royal, pp. 231-49

Fernie, L.M. (1971) 'A review of coffee production methods in Ethiopia', *Institute of Agricultural Research, Jimma Research Station Annual Report 1970-71,* Mimeo, pp. 8-16

Ferwerda, F.P. (1934) 'The vegetative propagation of coffee', *Empire Journal of Experimental Agriculture, 2,* 189-99

Ferwerda, F.P. (1948) 'Coffee breeding in Java', *Economic Botany, 2,* 258-72

Ferwerda, F.P. (1958) 'Advances in coffee production technology. The supply of better planting material II. Canephoras (Robustas)', *Coffee Tea Industries (New York), 81,* 58-63

Ferwerda, F.P. (1969) 'Breeding of Canephora coffee', in F.P. Ferwerda and F. Wit (eds.) *Outlines of Perennial Crop Breeding in the Tropics.* Veenman & Zonen NV, Wageningen, pp. 216-41

Fluiter, H.J. de (1939) 'Waarnemingen betreffende het bewaren van zaadkoffie' (Observations on storing of coffee seed), *Bergcultures, 14,* 1506-12

Franco, C.M. (1940) 'Fotoperiodismo em Cafeeiro (*C. arabica* L.). Instituto de Cafe do Estado de Sao Paulo, Brazil', *Revista, 27,* 1586-92

Gaie, W. (1957) 'Bouturage du Caféier d'Arabie', *Bulletin Information de l'INEAC, 6,* 175-96

Gibbs, J.N. (1972) 'Effects of fungicides on the population of *Colletotrichum* and other fungi in the bark of coffee', *Annals of Applied Biology, 70,* 35-47

Gillet, S. (1935) 'Vegetative propagation in coffee', *East African Agricultural Journal, 1,* 76-83

Gillet, S. (1942) 'Results and observations of spraying trials using bordeaux mixture on coffee at the Scott Agricultural Laboratories', *Monthly Bulletin of the Coffee Board of Kenya, 1,* 30-1

Grassias, M. (1977) 'Analyse des differentes méthodes d'estimation de la fertilité chez les hybrides 'arabusta' et la rélation avec leur comportement meiotique', in *8th International Colloquium on the Chemistry of Coffee,* ASIC, Paris, pp. 553-60

Griffiths, E. (1972) 'Negative effects of fungicides in coffee', *Tropical Science, 14,* 79-89

Griffiths, E., Gibbs, J.N. and Waller, J.M. (1971) 'Control of coffee berry disease', *Annals of Applied Biology, 67,* 45-74

Guillaumet, J.L. and Hallé, F. (1967) 'Etude de la variabilité du *Coffea arabica* dans son aire d'origine', *ORSTOM Mimeo, Report*

Guillamet, J.L. and Hallé, F. (1978) 'Echantillonnage du material *Coffea arabica* récolté en

Ethiopie', in A. Charrier (ed.) *Etude de la Structure et de la Variabilité Génétique des Caféiers*, IFCC, Paris, *Bulletin* no. 14, pp. 13-18

Haarer, A.E. (1962) *Modern Coffee Production*, Leonard Hill, London

Hendrickx, F.L. (1939) 'Observations phytopathologiques à la station de Mulungu en 1938', *Rapport Annuel pour l'exercise 1938 (2me partie) Publication INEAC*, pp. 117-28

IFCC (1976-1980) *Rapports d'Activités*

IFCC – SATMACI (1977) 'Le caféier arabusta produit de la recherche et son utilization pour la dévelopment de la Côte d'Ivoire', in *8th International Colloquium on the Chemistry of Coffee*, ASIC, Paris, pp. 417-24

Jones, P.A. (1956) 'Notes on the varieties of *Coffea arabica* in Kenya', *Monthly Bulletin of the Coffee Board of Kenya*, 21, 305-9

Krug, C.A. and Carvalho, A. (1951) 'The genetics of *Coffea*', *Advances in Genetics*, 4, 127-58

Krug, C.A. and Carvalho, A. (1952) 'Melhoramento do Cafeeiro V. Melhoramento par hibridacao', *Bragantia*, 12, 141-52

Krug, C.A. and de Poerck, R.A. (1968) *World Coffee Survey*. Agricultural Studies FAO, Rome

Lampard, J.F. and Carter, G.A. (1973) 'Chemical investigations on resistance to coffee berry disease in *Coffea arabica*', An antifungal compound in coffee cuticular wax', *Annals of Applied Biology*, 73, 31-7

Lanaud, C. and Zickler, D. (1980) 'Premières informations sur la fertilité des hybrides pentaploides et hexaploides entre *Coffea arabica* (Eucoffea) et *C. racemosa* (Mascarocoffea)', *Café Cacao Thé*, 24, 169-75

Leakey, C.L.A. (1970) 'The improvement of Robusta coffee in East Africa', in C.L.A. Leakey (ed.) *Crop Improvement in East Africa*. Technical Communication no. 19 of the Commonwealth Bureau of Plant Breeding and Genetics, Commonwealth Agricultural Bureau, Farnham Royal, pp. 250-77

Le Pierrès, D. and Anthony, F. (1980) 'Les hybrides interspecifiques hexaploides *Coffea arabica* × *C. canephora*: influence du milieu et de la structure génétique sur les potentialities agronomiques', *Café Cacao Thé*, 24, 291-6

Lejeune, J.B.H. (1958) 'Rapport au Gouvernement Imperial d'Ethiopie sur la production caféiers', *FAO Rapport PEAT 797*, Rome

Leliveld, J.A. (1938) 'Vruchtzetting bij koffie' (Fruit set in coffee), *Archieven Koffiecultuur Nederlands Indië*, 12 127-61

Leon, J. and Fournier, L.A. (1962) 'Crecimiento y desarollo del fruto de *Coffea arabica*', *Turrialba*, 12, 65-72

Louarn, J. (1976) 'Hybrides interspécifiques entre *Coffea canephora* et *C. eugenioides*', *Café Cacao Thé*, 20, 33-52

Louarn, J. (1980) 'Hybrides interspécifiques entre *Coffea canephora* et *C. liberica*: résultats preliminaires sur les hybrides F1', *Café Cacao Thé*, 24, 297-303

Machado, S.A. (1952) 'El uso de la correlacion y de la regression en los sistemas de la investigacion', *Cenicafe (Chincina, Colombia)*, 3, 24-44

Masaba, D.M. (1980) 'Differential pathogenicity of isolates of the CBD pathogen', *Coffee Research Foundation Kenya, Annual Report 1978-79*, pp. 171-2

Masaba, D.M. and Van der Vossen, H.A.M. (1982) 'Evidence of cork barrier formation as a resistance mechanism to berry disease (*Colletotrichum coffeanum*) in arabica coffee', *Netherlands Journal of Plant Pathology*, 88, 19-32

Mayne, W.W. (1935) 'Annual reports of the coffee scientific officer 1930-35', *Mysore Coffee Experimental Station Bulletin no. 13*

McDonald, J. (1926) 'A preliminary account of a disease of green coffee berries in Kenya Colony', *Transcriptions of the British Mycological Society*, 2, 145-54

Mendes, A.J.T. (1939) 'Induced polyploidy by treatment with colchicine', *Nature (London)*, 143, 299

Mendes, A.J.T. (1947) 'Observaçoes citologicas em *Coffea* XI. Metodos de tratamenta pela colchicina', *Bragantia*, 7, 221-30

Meyer, F.G., Fernie, L.M., Narasimhaswamy, R.L., Monaco, L.C. and Greathead, D.J. (1968) *FAO Coffee Mission to Ethiopia 1964-65*. FAO, Rome

Mitchell, H.W. (1975) 'Research on close-spacing systems for intensive coffee production in

Kenya', *Coffee Research Foundation Kenya, Annual Report 1974-75*, pp. 13-58
Mogk, M. (1975) 'Investigations on the origin of leaf rust and coffee berry disease in the provinces of western Kenya', *Mimeo. Report CRF, Kenya*
Monaco, L.C. (1968) 'Considerations on the genetic variability of *Coffea arabica* populations in Ethiopia' in F.G. Meyer *et al.* (eds.) *FAO Coffee Mission to Ethiopia, Rome*, pp. 49-69
Monaco, L.C. (1977) 'Consequences of the introduction of coffee rust into Brazil', *Annals of the New York Academy of Sciences*, 287, 57-71
Monaco, L.C. and Carvalho, A. (1964) 'Genetica de Coffea. Hibridos entre especies', *Ciencia e Cultura (Sao Paulo)*, 16, 144
Monaco, L.C., Carvalho, A. and Fazuoli, L.C. (1974) 'Melhoramento do cafeeiro: germoplasma do cafe Icatu e seu potencial no melhoramento', in *Resumos Segundo Congresso Brasilieiro de Pesquisas Cafeeiras, Poços de Caldas (M.G., Brazil) 1974*, p. 103
Monaco, L.C., Filho, H.P.M., Sondahl, M.R. and Alves, M.M. (1978) 'Efecto de dias longos no crescimento e florescimento de cultivares de cafe', *Bragantia*, 37, 25-31
Mulinge, S.K. (1973) 'Coffee berry disease in Ethiopia', *FAO Plant Protection Bulletin*, 21, 85-6
Mulinge, S.K. and Griffiths, E. (1974) 'Effects of fungicides on leaf rust, berry disease, foliation and yield in coffee', *Transcriptions of the British Mycological Society*, 62, 495-507
Muller, R.A. (1964) 'L'anthracnose des baies du caféier d'Arabie (*Coffea arabica*) due à *Colletotrichum coffeanum* au Cameroun', *IFCC Bulletin*, 6, 38
Muller, R.A. (1980) 'Contribution à la connaissance de la phytomycenose constitué par *Coffea arabica, Colletotrichum coffeanum, Hemileia vastatrix* et *Hemileia coffeicola*', *IFCC, Paris, Bulletin 15*
Murashige, T. (1974) 'Plant propagation through tissue cultures', *Annual Revue of Plant Physiology*, 25, 135-66
Narasimhaswamy, R.L. (1950) 'A brief history of coffee breeding in South India', *Indian Coffee Board Monthly Bulletin*, 14, 83-6, 112-3
Narasimhaswamy, R.L. (1960) 'Arabica selection S795. Its origin and performance', *Indian Coffee*, 24, 197-204
Narasimhaswamy, R.L. (1961) 'Coffee leaf disease (*Hemileia*) in India', *Coffee (Turrialba)*, 3, 33-9
Njagi, S.B.C. (1981) 'Current costs in coffee farming and farmers profit', *Kenya Coffee*, 46, 49-57
Noronha-Wagner, M. and Bettencourt, A.J. (1967) 'Genetic study of the resistance of *Coffea* spp. to leaf rust I. Identification and behaviour of four factors conditioning disease reaction in *Coffea arabica* to 12 physiological races of *Hemileia vastatrix*', *Canadian Journal of Botany*, 45, 2021-31
Orozco, F.J. and Cassalett, C. (1975) 'La fertilidad y el diametro de los granos de pollen en un hibrido interspecifico de cafe', *Cenicafe (Chinchina)*, 26, 38-48
Owuor, J.B.O. (1980) 'Selection for resistance to coffee rust', *Coffee Research Foundation, Kenya, Annual Report 1979-80*
Owuor, J.B.O. and Van der Vossen H.A.M. (1981) 'Interspecific hybridization between *Coffea arabica* and tetraploid *C. canephora* I. Fertility in F1 hybrids and backcrosses to *C. arabica*', *Euphytica*, 30, 861-6
Parlevliet, J.E. (1979) 'Components of resistance that reduce the rate of epidemic development', *Annual Review of Phytopathology*, 17, 203-22
Parlevliet, J.E. and Zadoks, J.C. (1977) 'The integrated concept of disease resistance: a new view including horizontal and vertical resistance in plants', *Euphytica*, 26, 5-12
Pieters, R. and Van der Graaff, N.A. (1980) 'Resistance to *Gibberella xylarioides* in *Coffea arabica*: evaluation of screening methods and evidence for the horizontal nature of the resistance', *Netherlands Journal of Plant Pathology*, 86, 37-43
Piringer, A.A. and Borthwick, H.A. (1955) 'Photoperiodic responses in coffee', *Turrialba*, 5, 72-7
Rayner, (1952) 'Coffee berry disease, a survey of investigations carried out up to 1950', *East African Agricultural Journal*, 18, 130-58
Rayner, R.W. (1957) 'Tonic spraying in coffee', *Proceedings of the Nairobi Scientific and Philosophical Society*, 9, 12-16
Reffye, Ph. de (1982) 'Modèle mathématique aléatoire et simulation de la croissance et de

l'architecture du caféier Robusta, 3me partie: Etude de la ramification sylleptique des rameaux primaires, et de la ramification proleptique des rameaux secondaires', *Café Cacao Thé, 26*, 77-96

Reffye, Ph. de (1983) 'Modèle mathématique aléatoire et simulation de la croissance et de l'architecture du caféier Robusta, 4me partie: Programmation sur micro-ordinateur de tracé en trois dimensions de l'architecture d'un arbre; application au caféier', *Café Cacao Thé, 27*, 3-19

Ribeiro, I.J.A., Sugimori, M.H., Moraes, S.A. and Monaco, L.C. (1975) Raças fisiologicas de *Hemileia vastatrix* no Estado de Sao Paulo. *Summa Phytopathologica, 1*, 19-22

Robinson, R.A. (1974) 'Terminal report of the FAO coffee pathologist to the Government of Ethiopia', *FAO Rome, AGO/74/443*

Robinson, R.A. (1976) *Plant Pathosystems*, Springer, Berlin, Heidelberg, New York

Rodrigues, C.J. Jr, Bettencourt, A.J. and Rijo, L. (1975) 'Races of the pathogen and resistance to coffee rust', *Annual Review of Phytopathology, 13*, 49-70

Siddiqi, M.A. (1980) 'The selection of Arabica coffee for Fusarium Bark Disease resistance at Bvumbwe', *Kenya Coffee, 45*, 55-9

Simmonds, N.W. (1979) *Principles of Crop Improvement*. Longman, London and New York

Söndahl, M.R. and Sharp W.R. (1977) 'High frequency induction of somatic embryos in cultured leaf explants of *Coffea arabica*', *Zeitschrift fur Pflanzenphysiologie, 81*, 395-408

Söndahl, M.R. and Sharp, W.R. (1979) 'Research in *Coffea* spp. and application of tissue culture methods' in W.R. Sharp *et al.* (eds.) *Plant Cell and Tissue Culture — Principles and Applications*, Ohio State University Press, Columbus, pp. 527-84

Soto Salazar, C.A. and Campos Gonzales, C. (1971) 'Nuevos fungicidas en el control de *Cercospora coffeicola* en almacigos de cafe', *Ministerio de Agricultura y Ganaderia San Jose (Costa Rica) Boletin Tecnico 57*

Srinivasan, C.S. (1969) 'Correlation studies in coffee I. Preliminary studies on correlation between stem girth and ripe cherry yield in some coffee selections', *Indian Coffee, 33*, 318-20

Srinivasan, C.S. (1980) 'Association of some vegetative characters with initial yield in coffee (*Coffea arabica*)', *Journal of Coffee Research (Mysore), 10*, 21-7

Srinivasan, C.S. and Vishveshwara, S. (1978) 'Heterosis and stability for yield in arabica coffee', *Indian Journal of Genetics, 38*, 416-20

Staritsky, G. (1970) 'Embryoid formation in callus tissues of coffee', *Acta Botanica Neerlandica, 19*, 509-14

Steiner, K.G. (1972) 'The influence of surface wax obtained from green berries of six selections of *Coffea arabica* on germination of conidia of *Colletotrichum coffeanum*', *Kenya Coffee, 37*, 179

Stemmer, W.P.C., Van Adrichem,J.C.J. and Roorda, F.A. (1982) 'Inducing orthotropic shoots in coffee with the morphactin chloroflurenol-methylester', *Experimental Agriculture, 18*, 29-35

Stoffels, E. (1936) 'La sélection du caféier arabica à la station de Mulungu', *Primaires Communications INEAC, Série Scientifique no 11*, pp. 1-41

Stoffels, E.H.J. (1941) 'La sélection de caféier arabica à la station de Mulungu (2me Comm.)', *INEAC, Série Scientifique no. 25*

Sybenga, J. (1960) 'Genetics and cytology of coffee: a literature review', *Bibliographica Genetica, 19*, 217-316

Sylvain, P.G. (1955) 'Some observations on *Coffea arbica* in Ethiopia', *Turrialba, 5*, 37-53

Thomas, A.S. (1942) 'The wild Arabica coffee on the Boma Plateau of Anglo-Egyptian Sudan. *Empire Journal of Experimental Agriculture, 10*, 207-12

Ultée, A.J. (1930) 'Over bewaren van zaadkoffie' (Storing coffee seed), *Bergcultures, 4*, 647-8

Uribe-Henao, A and Mestre-Mestre, A. (1980) 'Efecto de la densidad de poblacion y su sistema de manejo sobre la producciòn de cafè', *Cenicafe (Chinchina, Colombia), 31*, 29-51

Van der Graaff, N.A. (1978) 'Selection for resistance to coffee berry disease in arabica coffee in Ethiopia: evaluation of selection methods', *Netherlands Journal of Plant Pathology, 84*, 205-15

Van der Graaff, N.A. (1981) *Selection of Arabica Coffee Types Resistant to Coffee Berry Disease in Ethiopia* PhD Thesis (Communications no. 81-11), Agricultural University

of Wageningen
Van der Vossen, H.A.M. (1974) *Plant Breeding. Coffee Research Foundation, Kenya, Annual Report 1973-74*, pp. 40-51
Van der Vossen, H.A.M. (1979) 'Methods of preserving the viability of coffee in storage', *Seed Science and Technology*, 7, 65-74
Van der Vossen, H.A.M. (1982) 'Consequences of phytotonic effects of fungicide to breeding for disease resistance, yield and quality in *Coffea arabica*', *Journal of Horticultural Science*, 57, 321-9
Van der Vossen, H.A.M. and Browning, G. (1978) 'Prospects of selecting genotypes of *Coffea arabica* which do not require tonic sprays of fungicide for increased leaf retention and yield', *Journal of Horticultural Science*, 53, 225-33
Van der Vossen, H.A.M. and Cook, R.T.A. (1975) 'Incidence and control of Berry Blotch caused by *Cercospora coffeicola* on Arabica coffee in Kenya', *Kenya Coffee*, 40, 58-61
Van der Vossen, H.A.M., Cook, R.T.A. and Murakaru, G.N.W. (1976) 'Breeding for resistance to coffee berry disease caused by *Colletotrichum coffeanum* in *Coffea arabica* I. Methods of preselection for resistance', *Euphytica*, 25, 733-45
Van der Vossen, H.A.M. and Op de Laak, J. (1976) 'Large-scale rooting of softwood cuttings of *Coffea arabica* in Kenya I. Type of propagator, choice of rooting medium and type of cuttings', *Kenya Coffee*, 41, 385-99
Van der Vossen, H.A.M., Op de Laak, J and Waweru, J.M. (1977) 'Defining optimum conditions and techniques for successful topworking by cleft grafting of *Coffea arabica* in Kenya', *Kenya Coffee*, 42, 207-18
Van der Vossen, H.A.M. and Owuor, J.B.O. (1981) 'A programme of interspecific hybridization between Arabica and Robusta coffee in Kenya', *Kenya Coffee*, 46, 131-7
Van der Vossen, H.A.M. and Walyaro, D.J. (1977) 'Plant Breeding', *Coffee Research Foundation, Kenya, Annual Report 1976-77*, pp. 76-95
Van der Vossen, H.A.M. and Walyaro, D.J. (1980) 'Breeding for resistance to coffee berry disease in *Coffea arabica* II. Inheritance of the resistance', *Euphytica*, 29, 777-91
Van der Vossen, H.A.M. and Walyaro, D.J. (1981) 'The coffee breeding programme in Kenya: a review of progress made since 1971 and plan of action for the coming years', *Kenya coffee*, 46, 113-30
Van der Vossen, H.A.M. and Waweru, J.M. (1976) 'A temperature controlled inoculation room to increase to efficiency of preselection for resistance to coffee berry disease', *Kenya Coffee*, 41, 164-7
Van Dongen, L.L. (1979) 'Selection for horizontal resistance to coffee rust', *Coffee Research Foundation, Kenya, Annual Report 1978-79*, pp. 98-101
Waller, J.M. (1971) 'The incidence of climatic conditions favourable to coffee berry disease in Kenya', *Experimental Agriculture*, 7, 303-14
Walyaro, D.J. (1983) *Considerations in Breeding for Improved Yield and Quality in Arabica Coffee (Coffea arabica)*, PhD Thesis, Agricultural University, Wageningen
Walyaro, D.J. and Van der Vossen, H.A.M. (1977) '(Pollen longevity and artificial cross-pollination in *Coffea arabica*', *Euphytica*, 26, 225-31
Walyaro, D.J. and Van der Vossen, H.A.M. (1979) 'Early determination of yield potential in Arabica coffee by applying index selection', *Euphytica*, 28, 465-72
Walyaro, D.J. and Van der Vossen, H.A.M. (1985) 'Breeding for resistance to coffee berry disease in *Coffea arabica* III. Resistance in germ plasm of Ethiopian origin', *In Press*
Wanjala, F.M.E. (1980) 'A survey of the leaf miner (*Leucoptera meyricki*) populations in different coffee varieties in Kenya', *Kenya Journal of Science and Technology*, 1, 5-9
Watson, A.G. (1980) 'The mechanisation of coffee harvesting', in *9th International Colloquium on the Chemistry of Coffee*, ASIC, Paris pp. 681-6
Wellman, F.L. (1961) *Coffee: Botany, Cultivation and Utilization*, Leonard Hill Books, London
Went, F.W. (1955) 'IRI coffee studies at California Inst. of Technology I. Report on coffee research in Earhart carried out for IBEC Research Institute, October 1954
Wormer, T.M. (1964) 'The growth of the coffee berry', *Annals of Botany*, 28, 47-55
Wormer, T.M. (1966) 'Shape of bean in *Coffea arabica* in Kenya', *Turrialba*, 16, 221-36
Wormer, T.M. and Gituanja J. (1970) 'Floral initiation and flowering of *Coffea arabica*', *Experimental Agriculture*, 6, 157-70

4 CLIMATE AND SOIL

K.C. Willson

Rainfall

Cannell discusses in Chapter 5 the physiological effects of moisture status on the coffee plant. The effects of the changes from wet to dry seasons and vice versa are vital for flower initiation, flower dormancy breaking and vegetative growth induction. It follows that the distribution of rainfall is as important as the annual rainfall.

All coffee species are evergreen so transpiration is continuous. However, the natural habitat of coffee is the understorey of rainforest and therefore there has not been an incentive to develop a mechanism to reduce water loss in times of stress. Accordingly the plants will lose water continuously and the rate of loss will be dependent entirely on the meteorological conditions.

The rate of loss of water by evaporation from an open water surface is dependent on air temperature and relative humidity, wind speed and the amount of radiant solar energy arriving at the evaporating surface. Penman (1948) derived an equation by which evaporation from an open-water surface (open-pan evaporation, E_o) can be calculated from solar radiation, mean air temperature, mean temperature of dew point and mean run of wind. If the instrumentation necessary to measure solar radiation directly is not available, hours of bright sunshine can be used in its place. Ripley (personal communication, quoted by McCulloch, 1965) found that some factors in the equation were markedly dependent on altitude. McCulloch (1965) published a series of tables from which E_o can be calculated. These tables are intended for tropical conditions and include the adjustment for altitude.

The rate of loss of water from plants (evapotranspiration, E_t) is always less than E_o. The actual evapotranspiration for any crop can be determined by measurement of the water-balance using a lysimeter (Pereira and McCulloch, 1962). The results are normally expressed as the annual E_t/E_o ratio. McCulloch (1965) quotes some results. Evergreen montane rain forest has E_t/E_o equal to 0.95, mature pines 0.86, pasture 0.80 and mature bamboo forest 0.75. The ratio is usually between 0.75 and 0.95 for a complete vegetative cover. For a full discussion of this subject see Pereira (1973).

The water requirement of arabica coffee was studied in Kenya by Wallis (1963) and Blore (1966). The ratio E_t/E_o varied from 0.5 in dry months to 0.8 in wet months. The lower values in dry months arose because coffee was planted at the then standard spacing which left gaps between the tree canopies. The inter-rows would be either bare ground or mulched; these surfaces lose little moisture when dried out in the dry season. The average annual requirement of water over twelve years was found to be 951 mm. The monthly evapotranspiration varied between 60 and 115 mm. Achtnich (1958) calculated the water requirement for coffee in relation to its latitude. Where the dry season (or seasons) is not too long and the soil has a high water retention capacity, arabica coffee could, theoretically, be grown satisfactorily without irrigation where the average total rainfall is about 1,100 mm. In practice, imperfect rainfall distribution and adverse soil and weather conditions create a considerable risk of failure where average annual rainfall is less than about 1,300 mm. As conditions become more adverse a higher rainfall will be necessary but soil water capacity will become limiting when the dry season is prolonged. The length of the dry season may be critical; for example Robinson (1964: 10) recommended 'Not more than a maximum of four months dry weather should occur at any one stretch, unless a fair proportion of this period is cloudy and/or dull and cool.' Cultural methods, e.g. mulching to minimise water loss from the soil, become more important as conditions become more adverse.

Similar considerations apply to robusta coffee, but as it is more suited to lower altitudes where temperatures are higher, the total water requirement will be higher. Forestier (1969: 21) quotes a minimum annual rainfall of 1,250 mm and states that rainfall within the range 1,550 mm to 2,000 mm is preferable.

The relationship between coffee crop and rainfall has been studied in several countries. Sylvain (1959) reviewed the relationship between coffee and water. Physiological effects are discussed. A relationship between transpiration and light intensity for Brazil reported by de Franco (1947) is quoted, also a comparison of monthly transpiration with rainfall in Brazil, reported by de Franco and Inforzato (1950). The water available in various soils and the effects thereon of mulch and shade are quoted (see Chapter 7). Sylvain (1959) reported work by McFarlane in a very wet region (annual rainfall over 3000 mm) in Costa Rica which gave a negative correlation between crop and rainfall in the previous year. In Hawaii, Dean (1939) found a positive relationship between crop and rainfall in the previous year. De Castro (1960) reported from El Salvador a positive relationship between coffee yield and rainfall in January, February and March of the same year, i.e. the last three months of the dry season during which period the cherries mature, the bean swells and the crop is harvested. Clearly there is no universal relationship betwen rainfall and coffee crop.

Because of the effects of the changes of season, the pattern of distribution of rainfall controls the cropping pattern. Where the rainfall distribution is unimodal, there is one major flowering and therefore one period, usually lasting from three to four months, during which the crop matures and can be harvested. With a bi-modal rainfall there are two flowerings and

Table 4.1: Mean Monthly Rainfall, Banz, Western Highlands Province, Papua New Guinea; Mean of Recordings 1972 to 1983 Inclusive

	January	February	March	April	May	June	July	August	September	October	November	December
Mean rainfall (mm)	247	247	258	168	123	85	92	114	119	156	177	232

Source: Author's records (unpublished)

two major harvest periods. Cannell (Chapter 5, Figure 5.1) shows how unimodal rainfall is typical of the sub-tropical areas. The rainfall periods, which are in different months north and south of the equator, approach and overlap in equatorial regions, in which, therefore, there are typically two flowerings. However, in some high rainfall equatorial areas, there is only one period when rainfall is sufficiently infrequent to produce the necessary changes in water tension in the plants. In such areas there is only one major flowering and harvest period. The Western Highlands Province of Papua New Guinea is one such location; Table 4.1 gives the monthly average rainfall.

Soil conditions interact with rainfall in two ways. The ability of the plants to withstand a long dry season without adverse effects depends on the quantity of water held in the soil which is available to the plants. Two parameters control this; the depth of the soil explored by coffee roots and the capacity of the soil to hold water. In Kenya, for example, a three metre depth of soil is essential; Blore (1966) shows that the soil becomes very close to wilting-point (i.e. all available water has been extracted) over a three-metre depth at the end of a long dry season. At the other extreme coffee is grown in the Western Highlands of Papua New Guinea on very shallow soils. Here, the topsoil is in some places only 15 cm deep over a very heavy clay which is not penetrated by coffee roots. The amount of water available to the plants in the soil is therefore very small. The usual short dry season and the reduced transpiration due to periods with cloud cover ensure that unirrigated coffee only suffers when there is an unusually lengthy period without rain.

Where coffee is grown on such very shallow soils, excessive rainfall can have an adverse effect. The author found, in the Western Highlands of Papua New Guinea, that there was a reduction of yield when rainfall from November to July exceeded 1,500 mm. The greater the amount by which rainfall exceeded 1,500 mm, the greater was the reduction in yield. This effect is presumably a consequence of waterlogging which will reduce root efficiency, although low light levels may have some effect by reducing photosynthesis, and heavy rainfall at the commencement of the wet season may reduce pollination. Kumar (1982) reported that wet conditions affected hormone production; flowering abnormalities ensued which reduced crop yield.

The effects of rainfall can be modified by irrigation, which can be used to change the yield pattern in addition to supplementing rainfall. This is discussed in Chapter 7.

Temperature

The effects of temperature on the physiological activity of the coffee tree are discussed by Cannell in Chapter 5.

The optimum temperature ranges for the various species will be similar to those of their natural habitat. For arabica the range is 15 to 24°C. Above 25°C the photosynthetic rate is reduced and leaves are damaged by continuous exposure to high temperatures (over 30°C). Leaves exposed continually to high irradiance develop chlorotic symptoms (Huxley, 1967). Low temperatures produce a white or yellow discolouration of the leaves. The discoloured areas are not uniform, although the initial effect is often limited to the leaf margins. Severely affected leaves are reduced in size, often distorted and mottled and may eventually scorch and fall. The symptoms are more severe when a high temperature, as in bright sunlight, is followed by a low temperature. De Franco (1956) induced these symptoms by cooling seedlings to 3°C in a refrigerator. In severe cases excessive branching to secondary and tertiary stems occurs and shoot tips blacken, distort and shrivel. Because of the effect of wide diurnal temperature variations the condition is known as 'hot-and-cold disease'. Frost destroys arabica leaves and fruit.

For robusta coffee the temperatures are higher; the optimum range is 24 to 30°C. The tree will withstand an occasional temperature as low as 7°C but long periods of 15°C are harmful. Fruit and leaves are destroyed at 5°C.

Humidity and Cloud Cover

Atmospheric humidity is an important factor in determining the loss of moisture by evapotranspiration. High humidity will reduce water loss, whilst low humidity will increase it. The humidity level during the dry season is therefore important; high humidity reduces the stress on the plants and extends the rainless period through which the plants will survive without damage.

Cloud cover commonly results in increased humidity, in addition to lowering temperature, and can therefore be advantageous during a long dry season. The low assimilation rate of coffee leaves (see Chapter 5) means that the reduction of light by cloud will not have a significant effect. In extreme cases of severe reduction in light intensity lower leaves will not receive sufficient light to achieve their normal assimilation rate; this may be partly compensated by the upper leaves which will not reach temperatures at which their activity is reduced.

Cloud at ground level, i.e. mist, often adds to the water in the soil from condensation on the trees. In some marginal areas this water can be of critical importance. Parsons (1960) recorded about 250 mm of drip water from a pine tree near San Francisco.

Jaramillo-Robledo and Valencia-Aristizabal (1980) found that, in the absence of moisture deficits, growth was dependent on factors related to insolation.

Wind

Excessively strong winds can cause physical damage to the trees. At wind speeds below those necessary to break stems, wind increases water loss by evapotranspiration and therefore moisture stress in the trees. If the wind is cold the effects of low temperature are accentuated so that 'hot-and-cold disease' is more severe. If the wind is hot, exposed leaves may wilt and even die. In either case vegetative growth and crop yield are reduced. Good windbreaks are essential in exposed situations.

Soil; Physical Characteristics

The most important requirement for coffee growing is good drainage. Waterlogging will reduce yield by a substantial amount and kill trees if it is prolonged. The soil structure should therefore be such that the water from peak precipitation rates will drain away quickly from the root zone. Roots will be able to grow freely in such a soil. Heavy clay soils are undesirable because of slow drainage; root growth may be restricted and some very heavy clays will not be penetrated at all.

A high water capacity is necessary so that sufficient available water can be stored to maintain evapotranspiration throughout the dry season. Rocks in a soil of otherwise suitable texture reduce water capacity and the scope for root growth. Abdoellah (1982) calculated the water requirement of coffee on two estates in Indonesia, using experimental data for soil moisture contents at field capacity and wilting-point, together with weight of soil explored by coffee roots. Available moisture varied between 8.67 and 19.40 mm on one estate with a light soil, 30.84 and 40.90 mm on another estate with a heavier soil. He published a diagram giving the available water content of a range of soils.

A deep soil provides a greater volume for root proliferation which brings more water and nutrients within range of the tree. In areas of lower rainfall and a long dry season a deep soil is esssential; 3 metres depth is ideal. Nutman (1933) reported that roots explored the profile extensively down to 3.05 metres. Wallis (1963) reported that, in Kenya, on some occasions the soil was at its wilting-point (i.e. no more water could be extracted by plants) down to 3.05 metres at the end of the dry season. At the other extreme coffee is grown successfully in the Western Highlands of Papua New Guinea where rainfall is high, the dry season short and cloud cover frequent, in clay soils only 15 to 20 cm deep over heavy clay which is not penetrated by coffee roots. In such conditions, crop yields can be markedly reduced in years with excessive rainfall or an unusually long dry season.

Soil; Chemical Characteristics

Soil Chemistry and Analysis

When rock breaks down to the particles which form soil, the rupture of the crystal structure leaves free electric charges on the surface of the particles. Negatively charged ions (anions) are absorbed where there are positive charges, positively charged ions (cations) at negative charges. Each charge must be neutralised by an ion with the opposite charge. Ions can exchange one with another which allows them to move within the solution surrounding the particles so they can be absorbed by plants or lost by leaching in drainage water. The plant nutrients which are bases, for example potassium, calcium and magnesium, form positive ions (cations). The potential of a soil to absorb these bases is known as the Cation Exchange Capacity (CEC); this is a measure of the number of negative charges at the soil particle surfaces.

When an exchange site is not occupied by a base it will be occupied by hydrogen. The higher the proportion of sites occupied by hydrogen, the more acid will be the reaction of the soil/water system.

One measure of the fertility of a soil is the quantity of nutrients available to plants. This depends on two factors: the number of exchange sites, and the proportion of exchange sites which are occupied by elements of nutritive value to the plant. The number of exchange sites depends on the particles of which the soil is formed. The crystal structure of the parent rocks determines the number of sites on inorganic particles. Organic material usually has a high capacity for adsorption of ions, which will vary with the source of the material. The proportion of exchange sites which are occupied will be increased by the elements which have entered the system, by degradation of rock, decomposition of organic litter, fertiliser application and other means. The proportion will be decreased by nutrients absorbed by plants, those lost by leaching and those made unavailable (fixed) by chemical reaction to form insoluble compounds.

The aim of analysis of soil in relation to fertility assessment is to measure the amount of each nutrient which would be available to plants. Elements which are combined in rock particles will not be available until the rock has weathered.

For analysis, the absorptive activity of plant roots is simulated by the use of an extractant solution, of which a number are in use.

Methods for determination of the amount of each nutrient are well documented. Their accuracy and reproducibility are well known. The unknown in soil analysis is the degree to which the extractant represents extraction by roots. Different extractants usually give different results on the same sample. Nevertheless valid comparisons can be made of different soils when analysed by the same method. Comparisons of analyses made using different extractants can be valid at qualitative level.

Coffee Soils

Coffee is growing, on various sites, on soils varying from extremely acid (pH below 4.0) to slightly alkaline (pH up to 8.0). Neither of these extremes is suitable for economic high output production. A slightly acid soil is preferred; various writers suggest various ranges of pH. For example, Robinson (1964: 9) pH 5.2 to 6.2 for arabica, Acland (1971: 64) pH 5.3 to 6.0 for arabica, Forestier (1969: 26) higher than pH 4.5 for robusta. Robusta is more tolerant of neutral and slightly alkaline soils; Deuss (1969) reported that after clearance of forest and burning, the soil pH rose to 8.0, compared to 5.7 without burning. Robusta coffee planted in the high pH soil gave a crop yield, averaged over the first five harvests, which was 47.2 per cent greater than the yield from the coffee planted in the soil where the forest trash was not burnt.

Acidity within the correct range is not the only requirement. The necessary nutrients must be available at a reasonable level and in appropriate relative proportions. The more the nutrient levels depart from the optimum, the more costly will be the fertiliser applications necessary to provide the correct nutrition for maximum yields.

Table 4.2 lists a number of analyses of soils on which coffee is grown. The Kenya soil is reasonably fertile; with a good deep root system coffee should give a good yield with only nitrogen fertiliser. The first Papua New Guinea arabica samples show how soils deteriorate with time under coffee. The second sample is poor compared with the Kenya soil; potash fertiliser is essential here. The robusta soils from the Central African Republic show the effects of different clearing methods. The Papua New Guinea and Fiji robusta soils show extremes of potassium availability. One would expect coffee to need more potash fertiliser in Papua New Guinea.

The ability of coffee to grow in a wide range of soils means that fertiliser application levels will vary over a wide range. The optimum for any particular site will have to be determined on that site with reference to local conditions. The requirements of the main species are similar, so that it is unlikely that hybrids will be greatly different or pose special problems.

Soils with gross imbalances such as saline soils and those derived from, for example, pure limestone, will pose special problems which may be insoluble at an economic cost.

Day-length

Cannell shows in Chapter 5 that coffee is not sensitive to day-length. There will therefore be no problems arising from the variation in day-length throughout the tropics and sub-tropics where the other parameters are suitable for coffee.

Table 4.2: Analyses of Soils on Which Coffee is Growing

Country	Coffee Species	Age (years)	Depth	pH	P (µg/ml)	K (meq %)	Ca (meq %)	Mg (meq %)	CEC	K (% CEC)	Ca (%CEC)	Mg (% CEC)	N	Organic Matter	Source
Kenya	Arabica	5	0-10 cm	5.4		1.5	7.0	3.8					1.39	16.77	1
			10-20 cm	5.4		1.5	7.0	3.0							
			20-30 cm	5.3		1.2	5.2	2.5							
Papua New Guinea	Arabica	30	0-9 cm	4.3	69	0.46	0.5	0.41	20	2.3	3	2.1			
			9-27 cm	4.4	12	0.33	0.3	0.20	15	2.2	2	1.3			
			27-60 cm	4.9	2	0.47	2.3	1.39	10	4.5	22	13.3			2
			60-100cm	5.4	1	0.43	5.2	3.32	12	3.5	42	26.8			
	Arabica	2	0-15 cm	5.2	26	0.37	5.1	1.11	21	1.8	25	5.4			
Central African Republic	Robusta	a		8.0		0.83	9.50	1.18							
		b		5.7		0.41	4.40	0.73							3
Papua New Guinea	Robusta	3		6.0	10	0.47	17.4	4.72	30	1.6	59	16.0			
Fiji	Robusta	3		6.5	20	1.80	15.0	6.40	28	6.5	53	23.0			2

Notes: a. Ready for planting. Forest trash burned. b. Ready for planting. No burning.
Sources: 1. Robinson and Hosegood (1965). 2. Author's personal records, unpublished. 3. Deuss (1969).

Altitude

Altitude relates to temperature. Therefore in equatorial areas arabica is a highland crop, 1,000 metres to 2,000 metres, with robusta growing from sea level to 700 metres. As the distance from the equator increases, so robusta will become uneconomic, whereas arabica will continue to be a valuable crop at decreasing altitudes until restricted by frequent or lengthy frosts.

Site Aspect and Topography

These factors must be considered together in some degree. Steep slopes raise problems of erosion control and access. In the sub-tropics north- or south-facing slopes, depending on the hemisphere, will receive more sun, which may be important where conditions are marginal. In areas with a risk of frost, valley bottoms should be avoided, whilst planting on slopes will assist the movement of cold air away from the coffee. Wind is an important factor; exposed sites will need windbreaks. Sites close to the sea may suffer from the effects of salt spray. Areas with a fair risk of flooding at periods of heavy rainfall should be avoided.

References

Abdoellah, S. (1982) 'Keperluan air tanaman kopi di Kebun Gambar dan di Kebun Bangelan', *Menara Perkebunan*, 50, 73-6

Achtnich, W. (1958) *Sprinkling in Coffee Cultivation*, Perrot Regnerbau Gmbh. & Co., Calw/Wurtemberg, Federal Republic of Germany

Acland, J.D. (1971) *East African Crops*, FAO/Longmans, Rome

Blore, T.W.D. (1966) 'Further studies of water use by irrigated and unirrigated arabica coffee in Kenya', *Journal of Agricultural Science (Cambridge)*, 67, 145-54

de Castro, F.S. (1960) 'Relationships between rainfall and coffee production', *Coffee (Turrialba)*, 2, 85-9

Dean, L.A. (1939) 'Relationships between rainfall and coffee yields in the Kona district, Hawaii', *Journal of Agricultural Research*, 59, 217-22

Deuss, J. (1969) 'Influence du mode d'ouverture de plantation, avec ou sans brulis, sur la fertilite du sol et la productivite des cafeiers Robusta en zone forestiere centrafricaine', *Café, Cacao, Thé, XIII*, 283-9

Forestier, J. (1969) *Culture du Cafeier Robusta en Afrique Centrale*, Institut Francais du Cafe et du Cacao, Paris

de Franco, C.M. (1947) 'Pesquisas sobre a fisiologica do cafeeiro', *Boletim de Agricultura*, 48, 335-48

de Franco, C.M. and Inforzato, R. (1950) 'Quantidade de aqua transpirado pelo cafeeiro cultivado ao sol', *Bragantia*, 10, 247-57

de Franco, C.M. (1956) 'Descoloracao em folhas de cafeeiro causado calo frio', *Bragantia*, 15, 131-5

Huxley, P.A. (1967) 'The effects of artificial shading on some growth characteristics of arabica and robusta coffee seedlings. 1. The effects of shading on dry weight, leaf area and derived growth data', *Journal of Applied Ecology*, 4, 291-308

Jaramillo-Robledo, A. and Valencia-Aristizabal, G. (1980) 'Los elementos climaticos y el desarrolo de *Coffea arabica* L., en Chinchina, Colombia', *Cenicafé, 31*, 127-43

Kumar, D. (1982) 'Primary investigations into some flowering abnormalities of coffee in Kenya', *Kenya Coffee, 47*, 16-24

McCulloch, J.S.G. (1965) 'Tables for the rapid computation of the Penman estimate of evaporation', *East African Agriculture and Forestry Journal, 20*, 286-95

Nutman, F.J. (1933) 'The root system of *Coffea arabica*. Part 1. Root systems in tropical soils in British East Africa', *Empire Journal of Experimental Agriculture, 1*, 271-84

Parsons, J.J. (1960) 'Fog drip from coastal stratus', *Weather, 15*, 58

Penman, H.L. (1948) 'Natural evaporation from open water, bare soil and grass', *Proceedings of the Royal Society, Series A, 193*, 120-45

Pereira, H.C. (1973) *Land Use and Water Resources in Temperate and Tropical Climates*, Cambridge University Press, Cambridge

Pereira, H.C. and McCulloch, J.S.G. (1962) in E.W. Russell, (ed.) *Handbook of Natural Resources in East Africa*, Hawkins, Nairobi, Kenya

Ripley, L.A. Private communication, quoted by McCulloch (1965)

Robinson, J.B.D. (1964) *A Handbook on Arabica Coffee in Tanganyika*, Tanganyika Coffee Board, Moshi

Robinson, J.B.D. and Hosegood, P.A. (1965) 'Effects of organic mulch on fertility of a latosolic coffee soil in Kenya', *Experimental Agriculture, 1*, 67-80

Sylvain, P.G. (1959) 'El cafeto en relacion el aqua', *Materiales de ensenanza de Cafe y Cacao. 11*, 1-46

Wallis, J.A.N. (1963) 'Water use by irrigated coffee in Kenya', *Agricultural Science, 60*, 381-8

5 PHYSIOLOGY OF THE COFFEE CROP
M.G.R. Cannell

Introduction

The purpose of this review is to summarise existing information on the physiology of the coffee crop, with emphasis on whole-plant physiology and on those characteristics that influence the yield of beans. Information has been drawn from work in Kenya, which is well known to the author, from published reviews on coffee crop physiology (Huxley, 1970; Cannell, 1975), water relations (Nunes, 1976), eco-physiology (Maestri and Barros, 1977) and flowering (Alvim, 1973; Browning, 1975b; Barros, Maestri and Coons, 1978), and from the more recent literature. Most of the statements made here refer specifically to arabica coffee (*Coffea arabica* L.) but many will also be true for robusta coffee.

To provide a framework, this review is divided into sections dealing with (a) the seasonal cycle of growth, (b) determinants of bean yield, (c) net carbon fixation, (d) flower bud development, and (e) fruit growth and effects of fruiting on the vegetative structure. It should be borne in mind that arabica coffee was brought into cultivation relatively recently — most trees in cultivation are no more than three or four generations from the wild. Consequently, many features of coffee's physiology seem to reflect its adaptation to its native habitat, namely the cool, shady environment in the understorey of forests in the Ethiopian highlands, where there is one 'winter' dry season. As a conclusion to this review, some of the salient and peculiar features of coffee physiology are discussed with regard to the likely selection pressures in this kind of habitat.

Seasonal Cycle of Growth

Phenology

In non-Equatorial coffee-growing regions such as southern India, Ethiopia, Hawaii, Central America, south-central Brazil and Zimbabwe, coffee follows a single annual cycle of growth and fruiting. This is so even though there may be no marked dry season (e.g. in Costa Rica, Cannell, 1972a; Maestri and Barros, 1977). In general, rates of floral initiation are fastest, and rates of shoot growth are slowest, during the dry and/or cool 'winter' months, and blossoming and rapid shoot growth are triggered by 'spring' rains (Figure 5.1). The experience of low internal water potentials during the dry season seems to increase the unusually low hydraulic conductivity

Physiology of the Coffee Crop 109

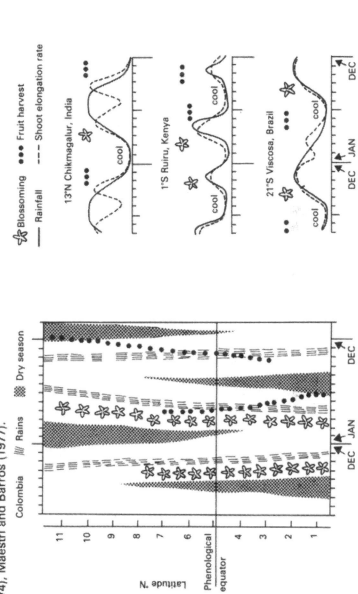

Figure 5 1: Seasonal Phenology of Arabica Coffee. Left: Times of blossoming and fruit harvesting at different latitudes in Colombia. Data from Trojer (1968). Right: Generalised seasonal changes in rainfall, shoot elongation rates, and times of flowering and fruit harvest, in coffee-growing regions of India, Kenya and Brazil. Data for India from Vasudeva and Ramaiah (1979), for Kenya from Cannell (1971c, 1971d), and for Brazil from Barros and Maestri (1974), Maestri and Barros (1977).

of coffee roots, and so predispose the trees to rapid rehydration following the first rains (Browning and Fisher, 1975). This rapid rehydration triggers both blossoming (see below) and 'flushes' of new shoot growth (Browning 1975a; Browning and Fisher, 1975). The fruits develop during the hot, 'summer' rainy season, and ripen at the start of the next dry and/or cool season (Figure 5.1). Further shoot growth during the rains provides more branch nodes at which flower buds initiate for the next season (see Figure 5.3). However, shoot growth is not continuous: workers in Brazil and India have both observed a 'mid-summer' depression in shoot elongation. This may be due to (a) inherent shoot growth periodicity, (b) supra-optimum tissue temperatures (see below) and daytime water stress owing to high evaporative demand, (c) fruit development, which diverts assimilates from the shoots and inhibits lateral bud growth (Cannell and Huxley, 1969; Cannell, 1971a, b; Clowes and Wilson, 1977) and/or (d) nutrient leaching. There is little evidence that periodic shoot growth in coffee is controlled directly by feedback mechanisms operating via old leaves (involving abscisic acid) or the roots (involving cytokinins).

In parts of Equatorial Kenya and Colombia the inter-tropical convergence zone traverses twice per year, giving two wet and dry seasons. In those regions coffee has two periods of rapid shoot growth, blossoming and fruiting (Figure 5.1). In Colombia it is possible to define a 'phenological equator' for coffee (Figure 5.1) although in Equatorial regions there are marked year-to-year differences in the relative importance of the two rainy seasons and the two annual growth cycles (Cannell, 1971c; Maestri and Barros, 1977).

Role of Daylength

Studies in the 1930s and 1940s suggested that coffee seedlings behaved like short day plants; floral initiation was promoted in short days (optimum 8h, critical value 13h) and shoot elongation was fastest in long days (Franco, 1940; Piringer and Borthwick, 1955). It has since been widely assumed that the seasonal cycle of growth and flowering of mature trees at latitudes of 10 to 25° N or S is regulated partly by daylength changes, because the dry and/or cool 'winters' coincide with periods of short days. But studies on mature trees growing near the equator showed that it took over 6 months for 15h photoperiods even partially to inhibit floral initiation and enhance shoot elongation (Cannell, 1972a), and Monaco *et al.* (1978) found that 18h photoperiods actually increased floral initiation on seedlings, compared with 12h photoperiods. Given that the shortest and longest days at latitude 25° N or S are 10.5h and 13.7h, it seems unlikely that daylength plays a major role in regulating the seasonal cycle of growth in the field. Seasonal changes in temperature and tree water status — the latter playing an important role in blossom timing (see below) — together with the fruiting cycle itself, are sufficient to explain most of the phenological observations reported in the literature (Trojer, 1968; Cannell, 1972a; Maestri and Barros, 1977; Gomez, 1977; Barros et al., 1978; Vasudeva and Ramaiah, 1979; Clowes, 1980; Robledo and Aristizabel, 1980, and many others).

Effects of Cultural Practices

The overall seasonal cycle of coffee growth cannot normally be altered by cultural practices, but the cycle can be amplified and small changes can be made. For instance, irrigation can advance blossoming and shoot 'flushing'; selective pruning can alter the balance of 'early' and 'late' crops in equatorial regions (Wormer and Gituanja, 1970); spraying with gibberellic acid can delay floral initiation (Cannell, 1971d); and spraying with ethrel (2-chloroethyl phosphonic acid) can hasten and thereby spread or synchronise fruit ripening (Browning and Cannell, 1970; Opile and Browning, 1975; Oyebade, 1976; Gopal, 1976; Snoeck, 1977; Clowes, 1977a; Rao, Venkataramanan and Rao, 1977).

Dry Matter Partitioning

In Kenya, seasonal changes in the rate of shoot growth were shown to be associated with marked seasonal changes in dry matter distribution within the trees (Figure 5.2; Cannell, 1971a, b). During the period of rapid shoot growth at the start of the 'long rains', large 'thin' leaves were produced (i.e. with large specific leaf areas), taking over 60 per cent of the total dry weight increment including roots. By the middle of the rains, when conditions were favourable for photosynthesis, the leaf area and leaf area ratio (leaf area/total tree weight) per tree had doubled. The thin feeder roots ($<$ 3 mm diameter) continued to grow almost unchecked during this period of rapid shoot growth, drawing assimilates from mature and old leaves, and using carbohydrate reserves in the thicker roots (Cannell, 1972b). During the ensuing cool, dry season, few new leaves were produced, but the old ones 'thickened'. Meanwhile, the thin roots took about 20 per cent of the total dry weight increment, compared with about 80 per cent during the 'long rains' (Figure 5.2). This pattern was repeated during the 'short rains' and following hot, dry season. Because the total dry weight increment was greater during the rains than during the dry seasons, the roots grew more continuously than the shoots, which was confirmed in root observation trenches (Huxley and Turk, 1975). Indeed, the fact that the shoots grew in 'flushes', faster than the roots, may have produced temporary imbalances in shoot/root functional relationships, which Borchert (1973) claimed could account for the inherent tendency for periodic shoot growth in many tropical trees.

112   *Physiology of the Coffee Crop*

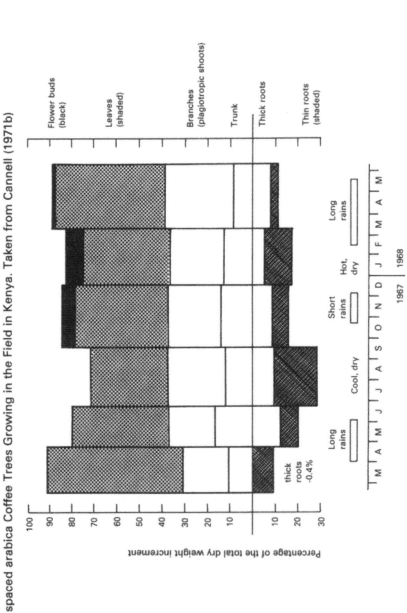

Figure 5.2: Seasonal Differences in the Distribution of Dry Matter Increment in 4-5 year-old De-blossomed, Widely spaced arabica Coffee Trees Growing in the Field in Kenya. Taken from Cannell (1971b)

Figure 5.3: Shoot Morphology of Arabica Coffee

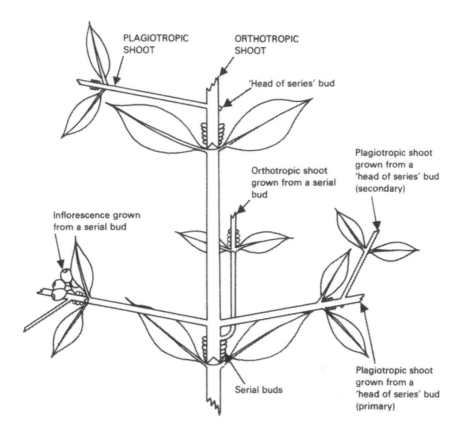

Determinants of Bean Yield

Shoot Structure

Coffee shoots have two distinctive structural features, namely: (a) the axil of each opposite and decussate leaf contains not one, but a series, of buds, and (b) branching is dimorphic (Figure 5.3). The orthotropic (vertical) shoots produce plagiotropic (horizontal) branches from the topmost 'head of series' bud, while the lower buds remain dormant or produce more orthotropic shoots (or, occasionally, inflorescences) — their dormancy can be broken using auxin-transport-inhibiting morphactins (Stemmer, Adrichen and Roorda, 1982). The inflorescences develop from the buds at each node on the plagiotropic branches. These buds can also develop into more plagiotropic branches, but cannot develop into orthotropic shoots, although there is one unconfirmed report that this also can be achieved using morphactins (Kumar, 1979). These simple rules governing shoot development have been stated in mathematical models, which, given the growth rates of orthotropic and plagiotropic shoots, can be used to simulate clonal differences in node production and hence potential yield (Reffye and Snoeck, 1976; Reffye, 1981).

Each plagiotropic branch node usually produces flowers only once, so pruning has to be done to ensure a continued supply of flowering nodes. This can be done by selectively pruning the plagiotropic branches growing on one permanent vertical shoot (single-stem pruning) or by allowing several vertical shoots to grow and replacing them with new ones (suckers) from the stumps every few years (multiple-stem pruning). See also Figure 7.1.

Components of Yield

The components of yield in coffee can be partitioned as follows.

(1) Yield per hectare = No. trees/hectare × yield/tree
(2) Yield per tree = No. fruits/tree × wt. of beans/fruit
where, no. fruits/tree = no. fruiting nodes/tree × no. fruits/node and wt. of beans/fruit = wt./fruit × bean/fruit weight ratio.

Yield per Hectare

Most of the world's coffee has been planted with fewer than 2000 trees per hectare, with one (single stem) or 2-3 (multiple-stem) orthotropic shoots per tree, pruned to 2 m height or less. There is then less than 50 per cent ground cover (Cannell, 1972b). There is ample evidence from many countries that closer planting increases yields (Mitchell, 1976) without increased water loss (Fisher and Browning, 1979), although with increased problems of pest and disease control. The greatest yield per hectare is obtained with 4,000-10,000 trees/ha (depending on the climate, cultivar, etc.) even though the yield per tree usually decreases with over 2,000 trees/ha, owing to a decrease in numbers of fruits produced (Figure 5.4). The mean weight per bean remains fairly constant. The decrease in fruit numbers can probably be attributed to an effect of mutual shading on floral initiation, which is critically dependent on solar radiation receipt at the potential flowering nodes. Reduced flowering is a well-known effect of overstorey shade (Ostendorf, 1962; Willey, 1975) and Castillo and Lopez (1966) demonstrated that 50 per cent shading of 18-month-old trees in Colombia reduced (i) the percentage of nodes with flowers, (ii) the number of inflorescences per node and (iii) the number of flowers per inflorescence, by about 30, 40 and 50 per cent, respectively (see Cannell, 1972b). Kumar (1978) associated these effects with increased levels of endogenous gibberellins. In general, the old practice of providing overstorey shade in coffee plantations is justified only deliberately to keep yields low so they are in balance with low nutrient supplies — unless of course, the overstorey trees fix N_2 or yield some other product such as fuelwood (Willey, 1975).

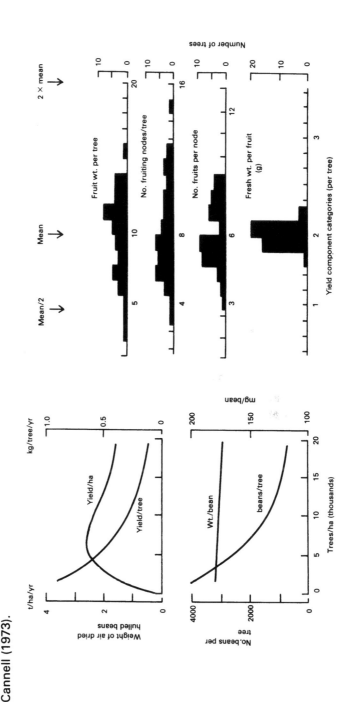

Figure 5.4: The Components of Yield in Arabica Coffee. Left: Effect of tree population per hectare on yield and its components. Data were derived from Browning and Fisher (1976) and Kuguru et al. (1977) for 4-year-old coffee grown in Kenya in experiments with fan-shaped systematic designs. Bean yields differed greatly between sites, but the trends with tree population were the same. Right: Variation in the yield and its components among 48 mature widely spaced trees in Kenya receiving various agronomic treatments (irrigation, mulch, fertilisers etc.). Reproduced from Cannell (1973).

Yield per Tree

The yield of individual trees at conventional spacings is highly dependent upon the number of potential flowering nodes produced the previous year (e.g. Beaumont and Fukunaga, 1958; Montoya, Sylvain and Umana, 1961; Gebre-Egziabher, 1978). The number of fruiting nodes per tree is the most variable component of yield among trees, and it is the component that is increased most by treatments such as irrigation, mulching and N-fertilisation (Figure 5.4). In Kenya, those treatments have little effect on either the seasonal fruiting pattern or the average bean size (Cannell, 1974).

Fruit Set

The number of fruits per node stands out as an important yield component when comparing cropping seasons, because fruit set differs between flowerings, in the range 20-80 per cent in arabica coffee (Nogueira, Carvalho and Antunes, 1959; Reddy and Srinivasan, 1979). Poor fruit set can be caused by (a) the development of atrophied flowers, attributed to prolonged drought or excessive rainfall during critical stages of flower bud development (Huxley and Ismail, 1969; Kumar, 1982), and/or (b) incomplete pollination or fertilisation owing to heavy rain, low temperatures or a shortage of pollinators at blossom times. In Kenya, the stigmas of arabica coffee are receptive for only 48 h; the pollen, which is shed predominantly after flower opening, remains viable for only 24-36 h; and the pollen tubes must grow rapidly enough to reach the ovaries within 2-3 days (Figure 5.5). Carvalho *et al.*, (1969) quoted reports of even shorter periods of pollen tube growth. Conditions favouring wind and insect pollination are especially important in the self-incompatible robusta coffees, but even in self-fertile arabica coffee there is considerable cross-pollination (4 to 93 per cent, Carvalho *et al.*, 1969); Taschdijan, (1932), and Raw and Free (1977) showed that final fruit sets on arabica trees caged with and without bee pollinators were 45 per cent and 32 per cent respectively. (See also Chapter 3.)

Fruit Drop and Overbearing

In favourable conditions 12-20 fruits can be set per node, each of which carries a maximum of two 30-40 cm^2 leaves. Studies on arabica coffee indicate that about 20 cm^2 of leaf is needed to support each fruit without severely checking vegetative growth, leading to bi-seasonal or biennial bearing (Cannell, 1974; Vasudeva and Ratageri, 1981). Consequently, even counting leaves at non-fruiting nodes, coffee is able to set more fruits than it can sustain. Some fruit shedding does occur, principally during the period of rapid fruit swelling (Figures 5.5 and 5.9); this shedding is exacerbated by drought, nutrient deficiency, defoliation (and ethrel sprays), and can be decreased using auxin sprays (Browning and Cannell, 1970; Gopal,

Physiology of the Coffee Crop 117

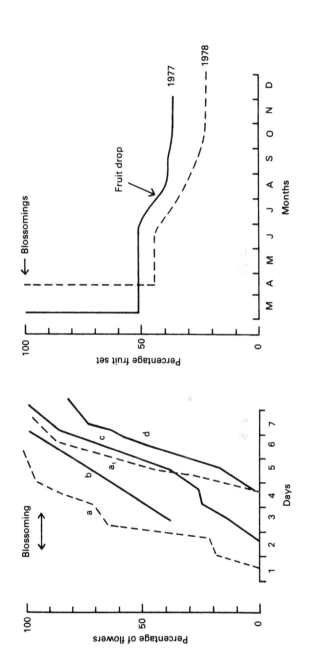

Figure 5.5: Pollination and Fruit Set in Arabica Coffee. Left: a = flowers open, a_1 = flower corollas half dead, b = germinating pollen present on the stigmas, c = tips of stigmas dead and no longer receptive, d = pollen tubes entering the ovaries. Unpublished data for coffee in Kenya (Cannell and Biddington). Right: Percentage fruit set in two years in India, taking the means of 16 varieties and assuming all losses prior to June occurred owing to poor pollination and/or fertilisation. Data taken from Reddy and Srinivasan (1979).

1971; Suryakantha, Srinivasan and Vishveshwara, 1974; Clowes, 1977b). But very often, even after a drought, the remaining fruiting load is very large and the trees are said to 'overbear'. In extreme cases the vegetative shoots 'die back' and very few are produced for the next season's flowering. This problem is especially serious on young trees which often need to be deblossomed.

Net Carbon Fixation

Leaf Photosynthesis

Coffee species exhibit all the characteristics of plants with predominantly C_3 photosynthetic pathways: $^{14}CO_2$ is incorporated first into C-3 phosphorylated compounds (Nunes, Brumby and Davies, 1973), the leaves produce a burst of photorespired CO_2 on being placed in darkness (Decker and Tio, 1959), their CO_2 compensation points lie in the range 50-100 vpm at 20-25°C (Heath and Orchard, 1957; Jones and Mansfield, 1970), and photosynthetic rates are depressed by increased O_2 partial pressures (Sondahl, Crocomo and Sodek, 1976). Sondahl *et al* (1976) estimated the photorespiration rate of coffee to be about 2 µmol $m^{-2}s^{-1}$ 2.8 times the dark respiration rate estimated by Nutman (1937) and the same order of magnitude as in other trees.

There are four notable features concerning the photosynthetic rate of leaves of arabica coffee, all of which seem to reflect its recent evolutionary history as a shade-adapted species.

(1) The maximum net photosynthetic rates of *sun* leaves are low: around 7 µmolCO_2 m^{-2} s^{-1} at 20°C, (Figure 5.6) which is in the lower part of the range recorded for trees (Altmann and Dittmer, 1968), and is much below the 15-25 µmolCO_2 m^{-2} s^{-1} commonly recorded for leaves of C_3 field crops. However, Kumar and Tieszen (1980a) recorded rates of up to 14 µmolCO_2 m^{-2} s^{-1} for *shade* leaves (Figure 5.6), which contain more chlorophyll per cm^2 than sun leaves (Hollies, 1967), possibly indicating the presence of larger and more efficient light-gathering units.

(2) The saturating irradiances for coffee leaf photosynthesis are low: 500-600 µE m^{-2} s^{-1} of photosynthetically active radiation for sun leaves and only about 300 µE m^{-2} s^{-1} for shade leaves (Yamaguchi and Friend 1979; Kumar and Tieszen, 1980a). Irradiances at mid-day on sunny days in the tropics are often around 2500 µE m^{-2} s^{-1}.

(3) The net photosynthetic rates of coffee leaves decrease markedly with increase in leaf temperature above 20-25°C (Figure 5.6). Kumar and Tieszen (1980a) showed that this decrease can be attributed to increased mesophyll resistance, but in the field high temperatures will often be associated with low leaf water potentials, which cause mid-day stomatal closure and an increase in internal CO_2 concentrations (Nutman, 1937; Heath and

Physiology of the Coffee Crop 119

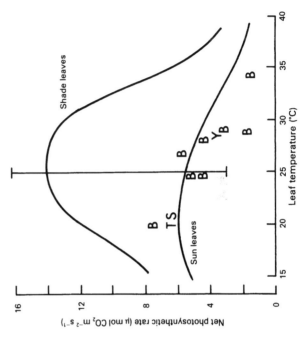

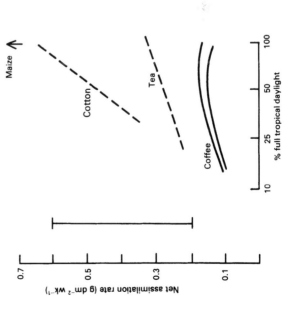

Figure 5.6: Net Carbon Fixation Characteristics of Coffee. Left: Net assimilation rates of coffee seedlings under various levels of shading, compared with cotton, tea and maize, all growing outside in the tropics. The vertical bar shows the range of values for broad-leaved tree species in temperate regions (Jarvis and Jarvis, 1964). The curvilinear data for coffee, and the data for cotton, are from identical experiments conducted by Huxley (1967) in Uganda. The tea data are from Barua (1959, Nigeria). Right: The net photosynthetic rates of coffee leaves at light saturation. The lines are redrawn from Kumar and Tieszen (1980a) for shade and sun leaves. The letters refer to sun leaves: B = Nunes et al. (1968, 1969) and Bierhuizen et al. (1969); T = Tio (1962); Y = Yamaguchi and Friend (1979); S = Sondahl et al. (1976), mean of 4 cultivars. The vertical bar shows the range of values for tree species according to Altmann and Dittmer (1968). 1 μmol $CO_2 m^{-2} s^{-1}$ ≃ 1 mg $dm^{-2} h^{-1}$.

Orchard, 1957; Nunes, Bierhuizen and Ploegman, 1968; Kumar and Tieszen, 1980b). In any case the net photosynthetic rate of exposed coffee leaves will normally be low during sunny days in the tropics, because the leaves can then reach temperatures of 35-40°C, as much as 10-15°C above ambient (Cannell, 1971c; Gomez and Robledo, 1974; Butler, 1977; Castano and Robledo, 1978).

(4) Even with the light-adapted sun leaves of coffee, the photosynthetic apparatus seems to be physically damaged by continued exposure to high irradiances, perhaps by disruption of Photosystem II (Bjorkman, 1968). Several workers have noticed that exposed leaves of coffee exhibit chlorotic symptoms (reviewed by Huxley, 1967).

Primary Production per Tree and per Hectare

From what has been said, one would expect that coffee seedlings, in which there is little mutual shading of leaves, will grow fastest in partial shade. This is indeed the case. Uncertainty over the need for overstorey shade in coffee plantations led to at least 15 different shading experiments on coffee seedlings, and the most critical of them all showed that maximum dry weights were attained at moderate levels of shade, attributed almost entirely to increased net assimilation rates, Ea (Figure 5.6; Huxley, 1967). Those same experiments showed that the Ea values of even shaded coffee seedlings (with mostly *sun* leaves) were smaller than in other crops (Figure 5.6).

The primary production rates of mature trees benefit from mutual shading of leaves, although the canopies of individual pruned trees can be poorly structured. Thus Robledo and Santos (1980) showed that up to 90 per cent of the total solar radiation can be intercepted by the top 'layer' of leaves (Figure 5.7). Nevertheless, the radiation and heat load is sufficiently spread to enable 4-5-year-old trees in Kenya to attain Ea values equal to those of seedlings (0.13 to 0.16 g dm^{-2} wk^{-1}; Cannell, 1971b) whereas one would expect them to be much lower owing to a several-fold decrease with age in leaf area ratio.

The potential primary production rates of closed canopies of unshaded coffee seem to be 20-30 t ha^{-1} yr^{-1}, equivalent to those of tea, oil palm and many tropical forest plantations (Cannell, 1972b), because, at close-spacings, unpruned coffee trees can produce large, well-structured canopies (Figure 5.7). Coffee has an inherently large 'leaf to total growth' ratio, at all ages, compared with other trees. Trees 4-5 years old in Kenya invested 40-45 per cent of their net annual dry matter increment in leaves (Figure 5.2), compared with, for instance, 27 per cent in mature tea in Kenya (Magambo and Cannell, 1981) and 22 per cent in 2-year-old apple in England (Maggs, 1961). Furthermore, coffee leaves can be retained to produce large leaf area indices (L), especially when sprayed with 'tonic' fungicides which prolong leaf life owing apparently to the removal of leaf

Figure 5.7: Vertical Profiles of Leaf Area Index and Transmitted Total Solar Radiation in Arabica Coffee Canopies. Left: A Continuous closed canopy of unpruned 3-year-old trees with erect upper branches, growing in Kenya at 0.9 × 1.2m spacing. Total leaf area index = 5.3. Taken from Cannell (1972b). Right: Measurements close to the centre of a mature tree (planted 4.0 × 3.5m) growing in Brazil (cv. Bourbon). The tree must have had an umbrella of foliage blocking lateral light. Total leaf area index = 2.2. Taken from Robledo and Santos (1980).

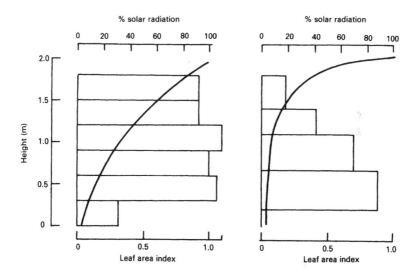

surface microbes that stimulate endogenous ethylene production (Mulinge and Griffiths, 1974; Vossen and Browning, 1978). In Kenya, young unpruned, closely planted, arabica coffee produced canopies with L = 5.3, with much better light transmission characteristics than isolated pruned trees (Figure 5.7). This may be because the upper branches were quite erect, there was 40 cm of height per unit L, and the individual trees had conical crowns (Jahnke and Lawrence, 1965). Huerta and Alvim (1962) produced canopies of closely planted seedlings with L = 7.4, and recorded L = 11 for 4-year-old trees planted in hedgerows; Boyer (1969) recorded L = 7.5 for closely planted robusta coffee; and Valencia (1973) found that the optimum L for yield in arabica cv. caturra was 8, obtainable in 3 years with 10,000 trees/ha or 4 years with 5000 trees/ha.

In practice, coffee canopies have to be discontinuous to allow access, and the optimum L for sustained yield over many years will be less than the optimum for primary production. The best manageable canopy might be produced by north-south, thin hedgerows, no taller than their distance apart (Huxley and Cannell, 1969; Jackson and Palmer, 1972).

Fruit Photosynthesis and the Sink Effect

Developing coffee fruits are green, have 30-60 stomata per mm^2 of surface (compared with 240-260/mm^2 on the undersurfaces of mature sun leaves), and being small, they have a large ratio of surface to volume. Also they develop about 75 per cent of their final surface area before the period of bean 'filling' (see page 126) and they can represent 20-30 per cent of the total photosynthetic surface on heavily bearing trees. Consequently, the fruits contribute a great deal to net carbon fixation, both directly by CO_2 uptake and probably, like other fruits, by re-assimilation of respiratory CO_2. They also seem to enhance leaf photosynthetic rates by providing 'strong' assimilate-accepting sinks.

Experiments using ring-barked attached branches on mature trees in Kenya showed that individual branch Ea values (calculated per unit leaf area over 88 days) of fruiting branches were twice as great as those of deblossomed branches (Figure 5.8; Cannell, 1971a). When the fruits were kept in the dark (Figure 5.8, treatment 2) the dry weight increment of the branches was decreased by an amount equivalent to 31 per cent of the dry weight increment of the fruits. That is, photosynthesis by the fruits accounted for nearly a third of their own dry weight gain. But those branches, with darkened fruits, and those without darkened fruits, had greater Ea values per unit leaf area, and per unit leaf plus fruit surface area, than deblossomed branches. That is, leaf photosynthesis was enhanced by the presence of fruits as assimilate-accepting sinks. Furthermore, the Ea values of the fruiting branches increased significantly with increase in leaf/fruit ratio (Figure 5.8).

Growth analysis studies on whole 4-5 year-old trees supported these findings. The Ea of deblossomed trees over successive 2-3 monthly periods was in the range 0.09 to 0.14 g dm^{-2} wk^{-1}, whereas the range for fruiting trees was 0.13 to 0.19 g dm^{-2} wk^{-1}, and when Ea values were calculated per unit leaf plus fruit surface area, the values for fruiting trees were still 9-23 per cent greater than those for deblossomed trees (Cannell, 1971b). Moreover, during those months when the fruits represented 20-27 per cent of the total photosynthetic surface, the Ea values of individual trees were negatively related to their leaf/fruit ratios, as found on the ring-barked branches (Cannell, 1971a). This increase in carbon fixation efficiency resulting from the sink-effect of fruits, more than compensated for the smaller leaf area produced as a result of fruiting, because fruiting trees increased 10-15 per cent more in total dry weight than deblossomed trees. This is not the case, for instance, on apple trees, where there is also sink-enhanced photosynthesis (Avery, 1970).

Figure 5.8: Coffee Fruit Photosynthesis and Sink Effect. The net assimilation rate (Ea) of attached branches, ring-barked at the base, as a function of their leaf/fruit ratio. 1. Ea of fruiting branches calculated as g dm^{-2} leaf area wk^{-1}. 2. Ea calculated as g dm^{-2} leaf area week^{-1} of fruiting branches on which fruits were kept in the dark, in opaque, cooled, ventilated, cloth sleeves. 3. As for 1, except Ea is calculated as g dm^{-2} of leaf plus fruit surface area week^{-1}. 4. Ea of branches with fruits removed. There were 60 replicate branches per treatment, and Ea was recorded over an 88-day period. Taken from Cannell (1971a).

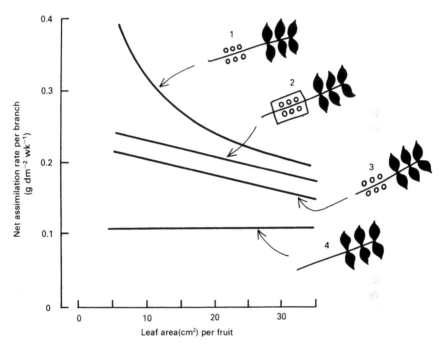

Flower Bud Development

The opening of coffee flower buds that are initiated at different times is synchronised in 'gregarious' blossomings by mechanisms that have been well-researched, and recently reviewed by Alvim (1973), Browning (1975b) and Barros et al. (1978). From the current state of knowledge it is possible to define four successive stages of coffee flower bud development from initiation to blossoming.

Dormancy Development

After initiation the flower buds develop over several months to an average length of 4-6 mm, and then they stop growing. The buds are then at their maximum state of dormancy, which is thought to be a true dormancy, associated with high levels of endogenous abscisic acid (ABA) (Browning, 1973a), and not simply an environmentally imposed quiescence (Gopal and Vasudeva, 1973). The microspores are fully developed, but no meiotic divisions have yet occurred (Mes, 1957; Dublin, 1960), and the xylem connections between the buds and the parent shoots are poorly developed (Mes, 1957).

Dormancy Breakage

Flower bud dormancy is broken, or rather progressively decreased, during the period of no visible growth (usually 1-4 months) by continued experience of low water potentials (water stress). This is analogous to the effect of chilling on the winter dormancy of buds of temperate trees. Water stress can occur in the buds even though the roots are well-watered, and the longer the period of water stress the easier it becomes to stimulate the buds to regrow (Porteres, 1946; Alvim, 1960; Capot, 1964; Browning, 1975b). Water stress appears to decrease endogenous ABA levels in the buds but it may have the following two effects: (a) a decrease in cell permeability to water (a known effect of ABA; Browning, 1973a) both in the buds and the tree roots, which, as mentioned above, increase in hydraulic conductivity during periods of drought (Browning and Fisher, 1975), and (b) a build-up of conjugated or bound gibberellic acid (GA) in the flower buds. Water stress appears to be mandatory for normal flower development: if the buds do not experience it they develop abnormally or not at all.

Stimulation of Regrowth

After several weeks of water stress the flower buds are no longer fully dormant, but they need a special stimulus before they will regrow. This stimulus has been defined as a sudden relief of water stress (or rehydration), in the buds themselves, and/or a sudden drop in temperature, of about 4°C per hour (Barros *et al.*, 1978). In the field, both of these signals are often provided together by so-called 'blossom showers' at the end of a dry season (Rees, 1964; Browning, 1975b; Magalhaes and Angelocci, 1976; Reddy, 1979). Irrigation, mists, immersion in water, immersion of cut ends of detached branches in water, spraying the buds with water, or a sudden drop in temperature in a dry glasshouse, can all provide the stimulus for regrowth (Mes, 1957; Went, 1957; Alvim, 1958; Mathew and Chokkanna, 1961; Rees, 1964). Exogenously applied GA can replace the environmental stimulus, but applied auxins have little effect, and applied ABA and inhibitors of GA biosynthesis inhibit regrowth (Alvim, 1958; Veen, 1968; Browning, 1975b). During the 3-4 days after the stimulus has been received, meiosis occurs, and there is an increase of almost two orders of magnitude in levels of endogenous, active GA in the buds, possibly released from the bound form (Browning, 1973a). This increase in GA levels occurs before the buds increase abruptly in fresh weight, and is

Physiology of the Coffee Crop 125

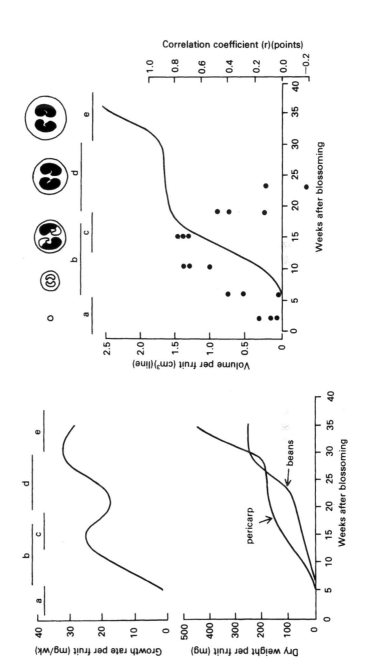

Figure 5.9: Growth of Arabica Coffee Fruits. Left: Mean fruit dry weight increase. Right: Mean fruit volume, and correlation coefficients (points) between the number of rainy days at different stages of fruit growth and the proportion of large beans at harvest (i.e. proportion of beans retained on a 6.75mm sieve). a = pinhead stage, b = rapid expansion and pericarp growth, c = bean (endosperm) formation, d = bean dry matter accumulation, e = fruit ripening. Data taken from Wormer (1964), Cannell (1971e) and Cannell (1974), supported by Leon and Fournier (1962) and Ramaiah and Vasudeva (1969).

thought to overcome the inhibitory effects of ABA, which remains in the buds at the same levels as before (Browning, 1975b).

Regrowth to Anthesis

During the next 6-12 days the flower buds grow very rapidly in fresh and dry weight, their cytokinin levels increase (Browning, 1973b) their metabolic rates increase (Janardhan, Raju and Gopal, 1977), they increase in length 3-4 fold, and develop to blossoming and anthesis. The rate, and hence the duration, of this regrowth is temperature dependent (Mes, 1957; Barros *et al.*, 1978) and seems to occur more readily if leaves are present at the flowering nodes (Raju *et al.*, 1975; Magalhaes and Angelocci 1976).

The possible adaptive advantages of this remarkable mechanism are discussed on page 128.

Fruit Growth and Effects of Fruiting on Vegetative Growth

Stages of Fruit Growth

For the first 6-8 weeks after fertilisation the ovaries of coffee undergo cell division, but the fruits grow very little in weight or volume, remaining as so-called 'pinheads'. This is apparently a true dormancy, which, like flower bud dormancy, is associated with high levels of endogenous ABA and low levels of active GA (Opile, 1979).

From about 6 to 16 weeks after blossoming the fruits grow rapidly in volume and dry weight, mostly owing to pericarp growth (Figure 5.9). During this period there is rapid cell expansion, and the fruits attain a large water content (80-85 per cent). Most importantly, the two fruit locules, which will contain the beans, swell to full size, and the endocarps, which line the locules, lignify, so that during this swelling stage the maximum volume of the beans is determined. The size to which the locules swell depends greatly on the water status of the trees. Fruits that expand during wet weather become larger, with larger locules, which are subsequently filled with larger beans, than fruits which expand during hot, dry weather. In Kenya, the size of the beans is overwhelmingly determined by the amount of rainfall during this fruit expansion stage (Figure 5.9; Cannell, 1974). Rainfall at other times has little effect on bean size, and bean size is not much affected by yield level, partial defoliation, or cultural treatments other than irrigation (Abruna, Silva and Vicente-Chandler, 1966; Cannell, 1974). This is in marked contrast to fleshy fruits, whose size is greatly dependent on yield level.

Between about 12 and 18 weeks after blossoming the beans are formed; these are the seed endosperms. There is, at this stage, a marked increase in endogenous GA levels (Opile, 1979), but a decrease in the rate of fruit dry weight increase (Figure 5.9). Once the beans fill the locules they begin to act as priority sinks for assimilates and minerals. The beans increase rapidly in dry weight, with little increase in fruit size. About 30-35 weeks after blossoming the fruits ripen, losing chlorophyll, producing ethylene, and turning red. During ripening the pericarps increase greatly in dry weight and volume (Figure 5.9).

The total respiration rate per fruit follows approximately the same course as the rate of fruit dry weight increase, which is bimodal (Figure 5.9), whereas the respiration rate per unit of fruit dry weight decreases throughout the period of fruit development (Cannell, 1974). The developing beans themselves have very low respiration rates, so that dry matter accumulated in them imposes a low respiratory burden on the trees, which is perhaps an additional reason why fruiting trees have high rates of net carbon fixation.

'Peaberries' result when one of the two ovules aborts, due invariably to poor pollination (see page 00) and a variety of bean malformations can occur as a result of incomplete endosperm development or filling (e.g. Wormer, 1966; Gopal and Venkataraman, 1974).

Effects of Fruiting on Vegetative Growth

The growing fruits can draw assimilates from all but the terminal leaves on the same branch, and also from leaves on lateral branches (Cannell and Huxley, 1969; Cannell, 1970; Cannell, 1971a, b, f). When there is a heavy crop, or when Ea values are low, fruiting greatly reduces the growth of the vegetative parts. The first effect is a depletion of stored carbohydrate reserves in the wood (Wormer and Ebagole, 1965; Janardhan, Gopal and Ramaiah, 1971), so that the dry weight increase of structural parts of fruiting trees (i.e. the branches, stems and thick roots) represents a smaller proportion of the total increment of vegetative parts than it does on nonfruiting trees (Cannell, 1972b). Secondly, the fruits restrict the supply of assimilates to the roots (Cannell, 1971b), and, as mentioned, inhibit the outgrowth of lateral branch buds (Clowes and Wilson, 1977). Thirdly, the shoots begin to 'die-back' at the tips. Bearing in mind that the young expanding leaves at the shoot apices must be favoured sinks for their own assimilates, it seems likely that the shoots die mostly because of a shortage of mineral nutrients and root-produced factors. In fact, ring-barked, heavily bearing branches do not die-back unless the rest of the tree is also heavily bearing (Clowes, 1977b; Clowes and Wilson, 1977). Lastly, if there is less than about $10\,cm^2$ of leaf per fruit, assimilates are diverted even from the fruit pericarps, so that the fruits, but not the beans, are reduced in size (Cannell, 1974).

Heavy bearing is a typical, not an unusual, phenomenon on unshaded coffee trees grown at conventional spacings. In Kenya, trees bearing a typical on-year yield of about 2,500 kg/ha of sun-dried beans produced about 1 kg dry weight of fruits each, which represented 36 per cent of the total tree dry weight including roots. During their later stages of growth the fruits took 72 per cent of the current dry matter increment, and over 95 per cent of the current total uptake of potassium, phosphorus and nitrogen (Cannell, 1971a, b; Cannell and Kimeu, 1971).

Conclusions: Adaptive Characteristics

Many of the physiological characteristics of arabica coffee described can be most readily understood by recalling that the species evolved in Ethiopia, in cool, shady habitats with one annual dry season, and it has recently been brought into open, sunny environments, including regions with two annual dry seasons.

The photosynthetic characteristics of coffee are clearly those of a shade-adapted species. Shade-grown leaves are photosynthetically more efficient than sun-grown ones, and in full tropical sunlight even sun-grown leaves (i) receive irradiances 3-5 times above saturation, (ii) can be 5-20°C above their optimum temperature, and (iii) can suffer physical damage. To maximise dry matter production per hectare, coffee needs to be grown in hedges or at close-spacings so that radiation and heat loads are spread over a large leaf area. In fact, coffee allocates a large proportion of its dry matter to leaves (like other shade plants) and is well able rapidly to produce deep, well-structured canopies. Alternatively, it may be possible to select cultivars that are more sun tolerant (Nunes, 1976); indeed the poor performance of many recent Ethiopian acquisitions in Kenya suggests that selection in this direction has occurred over the last 100 years.

But floral initiation is also light dependent. In its native habitat, in deep shade, coffee produces very few flowers, whereas in full tropical daylight it can produce so many that the trees suffer 'overbearing die-back' and bear irregularly. Teleologically, it could be argued that, because coffee produces so few flowers in its native habitat, it therefore failed to evolve satisfactory mechanisms to keep its fruiting load in balance with carbohydrate and mineral resources. Certainly, coffee's inability to shed 'excess' fruits is unusual among woody perennials. The trees become committed to filling all the beans that are formed after the fruit expansion stage. It is left to the grower to regulate the leaf/fruit ratio, either physically, by pruning or deblossoming, or by managing the trees so that mutual shading suppresses floral initiation. Note that, according to this argument, the causes of overbearing have to do with profuse floral initiation in the absence of shade, little compensatory fruit shedding, and the sink 'strength' of the seed endosperms, not with low photosynthetic rates.

Coffee beans are, in fact, relatively heavy seeds, which ripen within sweet, red fruits that are dispersed naturally by birds and animals. In shady, colonised habitats within tropical forests it is likely to be important to produce large seeds, with a large reserve for seedling establishment. The large sink 'strength' of coffee seed endosperms is one mechanism that helps to ensure that the trees do not 'waste' energy by producing light seeds. But many features of coffee phenology serve this same purpose. Bearing in mind that bean size is determined during the period of fruit expansion, it may be argued that in Ethiopia, flower and fruit development are phased to maximise the likelihood that the fruits will expand during the rains and after a 'flush' of new leaves. Thus, floral initiation occurs during the cool, dry 'winter' period, the flower buds remain dormant during the dry season, they blossom after the first showers, which invariably precede the main rains, and the 'pinhead' fruits then remain dormant for 6-8 weeks before expanding, by which time the rains will have begun, and a new 'flush' of leaves, triggered by the same blossom showers, will have expanded. This phenological cycle is followed very well in most non-Equatorial regions, but in Kenya and Colombia fruit expansion can often coincide with a drought.

The remarkable synchronous blossoming mechanism may confer other adaptive advantages (Browning, 1975b). First, flower bud dormancy may protect the sexual structures from water stress during the dry period, in the same way that winter dormancy protects plant tissues from frost damage. Secondly, synchronous flowering may increase the chances of outcrossing, which would otherwise be very rare in self-compatible species like arabica coffee. Thirdly, the need of a rehydration/temperature drop stimulus for blossoming increases the likelihood that anthesis will occur in dry sunny weather because 'blossom showers' invariably occur before the main rains; this will increase the chances of successful pollination and fertilisation.

Acknowledgements

I am grateful to Dr G. Browning, who suggested some alternative explanations of observed phenomena, and to Dr M.H. Unsworth for editorial comments.

References

Abruna, F., Silva, S. and Vicente-Chandler, J. (1966) 'Effects of yields, shade and varieties on size of coffee beans', *Journal of Agriculture of the University of Puerto Rico*, 50, 226-30

Altmann, P.L. and Dittmer, D.S. (1968) *Biology Data Book*. Federation of American Societies of Experimental Biology, Washington DC., USA, pp. 213-14

Alvim, P. de T. (1958) 'Estimulo de la floración y fructifición del cafeto por aspersiones con ácido giberélico', *Turrialba*, 8, 64-72

Alvim, P. de T. (1960) 'Moisture stress as a requirement for flowering of coffee'. *Science*, 132, 354

Alvim, P. de T. (1973) 'Factors affecting flowering of coffee', *Journal of Plantation Crops*, 1, 37-43 (Also in *Indian Coffee* (1977), 41, 218-24)

Avery, D.J. (1970) 'Effects of fruiting on the growth of apple trees of four rootstock varieties', *New Phytologist*, 69, 19-30

Barros, R.S. and Maestri, M. (1974) 'Influencia dos factores climaticos sobre a periodicidade de crecimento vegetativo do café (*Coffea arabica* L.), *Revista Ceres*, 21, 268-79

Barros, R.S., Maestri, M. and Coons, M.P. (1978) 'The physiology of flowering in coffee: a review, *Journal of Coffee Research*, 8, 29-73

Barua, D.N. (1956) 'Light intensity, assimilation rate and yield of young tea plants, *Annual Report of Tocklai Experimental Station Tea Association, 1956*, 34-7

Beaumont, J.H. and Fukunaga, E.T. (1958) 'Factors affecting the growth and yield of coffee in Kona, Hawaii', *Hawaii Experimental Station Bulletin*, 13

Bierhuizen, J.F., Nunes, M.A. and Ploegman, C. (1969) 'Studies on productivity of coffee. II. Effect of soil moisture on photosynthesis and transpiration of *Coffea arabica*', *Acta Botanica Neerlandica*, 18, 367-74

Bjorkman, O. (1968) 'Further studies on differentiation of photosynthetic properties in sun and shade ecotypes of *Solidago virgaurea*', *Physiologia Plantarum*, 21, 84-99

Blackman, G.E. and Black, J.N. (1959) 'Physiological and ecological studies on the analysis of plant environment XII. The role of the light factor in limiting growth, *Annals of Botany*, 23, 131-45

Borchert, R. (1973) 'Simulation of rhythmic tree growth under constant conditions', *Physiologia Plantarum*, 29, 173-80

Boyer, J. (1969) 'Etude experimentale des effets du régime d'humidité du sol sur la croissance vegetative, la floraison et la fructification des cafeiers Robusta', *Café, Cacao, Thé*, 13, 187-200

Browning, G. (1973a) 'Flower bud dormancy in *Coffea arabica* L. I. Studies of gibberellin in flower buds and xylem sap and abscisic acid in flower buds in relation to dormancy release, *Journal of Horticultural Science*, 48, 29-41

Browning, G. (1973b) 'Flower bud dormancy in *Coffea arabica* L. II. Relation of cytokinins in xylem sap and flower buds to dormancy release, *Journal of Horticultural Science*, 48, 297-310

Browning, G. (1975a) 'Shoot growth in *Coffea arabica* L. I. Response to rainfall when the soil moisture status and gibberellin supply are not limiting, *Journal of Horticultural Science*, 50, 1-11

Browning, G. (1975b) 'Environmental control of flower bud dormancy in *Coffea arabica* L.' in J.J. Landsberg and C.V. Cutting (eds.) *Environmental Effects on Crop Physiology*, Academic Press, London, pp. 321-32

Browning, G. and Cannell, M.G.R. (1970) 'Use of 2-chloroethane phosphonic acid to promote abscission and ripening of fruit of *Coffea arabica* L.' *Journal of Horticultural Science*, 45, 223-32

Browning, G. and Fisher, N.M. (1975) 'Shoot growth in *Coffea arabica* L. II. Growth flushing stimulated by irrigation', *Journal of Horticultural Science*, 50, 207-18

Browning, G. and Fisher, N.M. (1976) 'High density coffee: yield results for the first cycle from systematic plant spacing designs', *Kenya Coffee*, 41, 209-17

Butler, D.R. (1977) 'Coffee leaf temperatures in a tropical environment, *Acta Botanica Neerlandica*, 26, 129-40

Cannell, M.G.R. (1970) 'The contribution of carbohydrates from vegetative laterals to the growth of fruits on the bearing branches of *Coffea arabica* L.', *Turrialba*, 20, 15-19

Cannell, M.G.R. (1971a) *Effects of Season and Fruiting on Accumulation and Distribution of Dry Matter in Coffee Trees in Kenya East of the Rift*, Unpublished PhD thesis. University of Reading. 176 pp

Cannell, M.G.R. (1971b) 'Production and distribution of dry matter in trees of *Coffea arabica* L. in Kenya as affected by seasonal climatic differences and the presence of fruits, *Annals of Applied Biology*, 67, 99-120

Cannell, M.G.R. (1971c) 'Seasonal patterns of growth and development of Arabica coffee in Kenya. Part III. Changes in the photosynthetic capacity of the trees', *Kenya Coffee*, 36, 68-74

Cannell, M.G.R. (1971d) 'Use of gibberellic acid to change the seasonal fruiting pattern of Arabica coffee in Kenya, *Journal of Horticultural Science*, 46, 289-98

Cannell, M.G.R. (1971e) 'Changes in the respiration and growth rates of developing fruits of

Coffea arabica L.', *Journal of Horticultural Science, 46,* 263-72

Cannell, M.G.R. (1971f) 'Effects of fruiting, defoliation and ring-barking on the accumulation and distribution of dry matter in branches of *Coffea arabica* L. in Kenya', *Experimental Agriculture, 7,* 63-74

Cannell, M.G.R. (1972a) 'Photoperiodic response of mature trees of Arabica coffee', *Turrialba, 22,* 198-206

Cannell, M.G.R. (1972b) 'Primary production, fruit production and assimilate partition in Arabica coffee: a review' in *Coffee Research Foundation, Kenya, Annual Report for 1971/72,* pp. 6-24

Cannell, M.G.R. (1973) 'Effects of irrigation, mulch and N-fertilizers, on yield components of Arabica coffee in Kenya', *Experimental Agriculture, 9,* 225-32

Cannell, M.G.R. (1974) 'Factors affecting Arabica coffee bean size in Kenya', *Journal of Horticultural Science, 49,* 65-76

Cannell, M.G.R. (1975) 'Crop physiological aspects of coffee bean yield: a review', *Journal of Coffee Research, 5,* 7-20

Cannell, M.G.R. and Kimeu, B.S. (1971) 'Uptake and distribution of macro-nutrients in trees of *Coffea arabica* L. in Kenya as affected by seasonal climatic differences and the presence of fruits', *Annals of Applied Biology, 68,* 213-30

Cannell, M.G.R. and Huxley, P.A. (1969) 'Seasonal differences in the pattern of assimilate movement in branches of *Coffea arabica* L.', *Annals of Applied Biology, 64,* 345-57

Capot, J. (1964) 'La pollinisation artificielle des cafeiers allogames et son role dans leurs amelioration', *Café, Cacao, Thé, 8,* 75-88

Carvalho, A., Ferwerda, F.P., Frahm-Leliveld, J.A, Medina, D.M., Mendes, A.J.T. and Monaco, L.C. (1969) 'Coffee' in F.P. Ferwerda and F. Wit (eds.) *Outlines of Perennial Crop Breeding in the Tropics,* Misc. Papers 4. Landbouwhogeschool, Wageningen, The Netherlands. pp. 189-241

Castano, F.J.O. and Robledo, A.J. (1978) 'Efecto del deficit de humedad en el suelo sobre la temperatura del suelo y de lan hojas en plantas de *Coffea canephora* y *C. arabica*', *Cenicafé, 29,* 121-34

Castillo, J.Z. and Lopez, A.R. (1966) 'Nota sobre el efecto de la intensidad de la luz en la floracion del cafeto', *Cenicafé, 17,* 51-60

Clowes, M. St J. (1977a) 'The effects of ethrel on ripening *Coffea arabica* L. fruits at different stages of maturity', *Rhodesian Journal of Agricultural Research, 15,* 79-82

Clowes, M. St J (1977b) 'A study of the growth of *Coffea arabica* L. fruits in Rhodesia', *Rhodesian Journal of Agricultural Research, 15,* 89-93

Clowes, M. St J. (1980) *Growth and Development of Coffee,* Unpublished PhD thesis. University of Zimbabwe. 259 pp.

Clowes, M. St J. and Wilson, J.H.H. (1977) 'The growth and development of lateral branches of *Coffea arabica* L. in Rhodesia', *Rhodesian Journal of Agricultural Research, 15,* 171-85

Decker, J.P. and Tio, M.A. (1959) 'Photosynthetic surges in coffee seedlings', *Journal of Agriculture of the University of Puerto Rico, 63,* 50-5

Dublin, P. (1960) 'Biologie florale du *Coffea dewevrei* de Wild race *excelsa* A. Chevalier. I. Floraison. II. Pollinisation. III Fecondation', *Agronomie Tropicale, 15,* 189-212

Fisher, N.M. and Browning, G. (1979) 'Some effects of irrigation and plant density on the water relations of high density coffee (*Coffea arabica* L.) in Kenya. *Journal of Horticultural Science, 54,* 13-22

Franco, C.M. (1940) 'Fotoperiodismo em cafeeiro (*C. arabica* L.). Instituto de Café do Estado de Sao Paulo, Brasil', *Revista, 27,* 1586-92

Gebre-Egziabher, T.B. (1978) 'Some vegetative parameters of coffee, *Coffea arabica* L., proportional to yield, *Ethiopian Journal of Science, 1,* 51-7

Gomez, L.G. (1977) 'Influencia de los factores climaticos sobre la periodicidad de crecimiento del cafeto', *Cenicafé, 28,* 3-17

Gomez, L.G. and Robledo, A.J. (1974) 'Temperaturas en arboles de cafe al sol', *Cenicafé, 25,* 61-2

Gopal, N.H. (1971) 'Preliminary studies on the control of fruit drop in Arabica coffee', *Indian coffee, 25,* 1-7

Gopal, N.H. (1976) 'Hastening of fruit ripening in Robusta coffee', *Indian Coffee, 40,* 23-4

Gopal, N.H. and Vasudeva, N. (1973) 'Physiological studies on flowering in Arabica coffee under South Indian conditions. I. Growth of flower buds and flowering', *Turrialba, 23,* 146-53

Gopal, N.H. and Venkataraman, D. (1974) 'Studies on black bean disorder in coffee', *Indian Coffee, 38,* 259-67

Heath, O.V.S. and Orchard, B. (1957) 'Temperature effects on the minimum intercellular space carbon dioxide concentration', *Nature, 180,* 180-1

Hollies, M.A. (1967) 'Effects of shade on the structure and chlorophyll content of Arabica coffee leaves', *Experimental Agriculture, 3,* 183-90

Huerta, H.S. and Alvim, P. de T. (1962) 'Indice de area foliar y su influencia en la capacidad fotosinthetica del cafeto', *Cenicafé, 13,* 75-84

Huxley, P.A. (1967) 'The effects of artificial shading on some growth characteristics of Arabica and Robusta coffee seedlings. I. The effects of shading on dry weight, leaf area and derived growth data', *Journal of Applied Ecology, 4,* 291-308

Huxley, P.A. (1970) 'Some aspects of the physiology of Arabica coffee — the central problems and the need for a synthesis', in L.C. Luckwill and C.V. Cutting (eds.), *Physiology of Tree Crops,* Academic Press, London, New York and San Francisco, pp. 253-68

Huxley, P.A. and Cannell, M.G.R. (1969) 'Some crop physiological factors to be considered in intensification', in P.A. Huxley (ed.) *Proceedings of a Seminar on Intensification of Coffee Growing in Kenya,* Coffee Research Foundation, Kenya. Mimeo. pp. 45-50

Huxley, P.A. and Ismail, S.A.H. (1969) 'Floral atrophy and fruit set in Arabica coffee in Kenya', *Turrialba, 19,* 345-54

Huxley, P.A. and Turk, A. (1975) 'Preliminary investigations with Arabica coffee in root observation laboratory in Kenya', *East African Agriculture and Forestry Journal, 40,* 300-12

Jackson, J.E. and Palmer, J.W. (1972) 'Interception of light by model hedgerow orchards in relation to latitude, time of year and hedgerow configuration and orientation. *Journal of Applied Ecology, 9,* 341-58

Jahnke, L.S. and Lawrence, D.B. (1965) 'Influence of photosynthetic crown structure on potential productivity of vegetation based primarily on mathematical models', *Ecology, 46,* 319-26

Janardhan, R.V., Gopal, N.H. and Ramaiah, P.K. (1971) 'Carbohydrate reserves in relation to vegetative growth, flower and fruit formation, and crop levels in Arabica coffee', *Indian Coffee, 35,* 145-8

Janardhan, K.V., Raju, K.I. and Gopal, N.H. (1977) 'Physiological studies on flowering in coffee under South Indian conditions. II. Changes in growth rate, indoleacetic acid and carbohydrate metabolism during flower bud development and anthesis', *Turrialba, 27,* 29-36

Jarvis, P.G. and Jarvis, M.S. (1964) 'The growth of woody plants', *Physiologia Plantarum, 17,* 654-66

Jones, M.B. and Mansfield, T.A. (1970) 'A circadian rhythm in the level of the carbon dioxide compensation point in *Bryophyllum* and *Coffea, Journal of Experimental Botany, 21,* 159-63

Kuguru, F.M., Fisher, N.M., Browning, G. and Mitchell, H.W. (1977) *The Effect of Tree Density on Yield and some Yield Components of Arabica Coffee in Kenya.* Mimeographed report. Coffee Research Station, Ruiru, Kenya

Kumar, D. (1978) 'Investigations into some physiological aspects of high density planting of coffee', *Kenya Coffee, 48,* 263-72

Kumar, D. (1979) 'Morphactin changes orientation of branches of *Coffea arabica* L. and in this way facilitates vegetative propagation', *Naturwissenschaften, 66,* 113-14

Kumar, D. (1982) 'Preliminary investigations into some flowering abnormalities of coffee in Kenya', *Kenya Coffee, 47,* 16-24

Kumar, D. and Tieszen, L.L. (1980a) 'Photosynthesis in *Coffea arabica* I. Effects of light and temperature', *Experimental Agriculture, 16,* 13-19

Kumar, D. and Tieszen, L.L. (1980b) 'Photosynthesis in *Coffea arabica.* II. Effects of water stress', *Experimental Agriculture, 16,* 21-7

Leon, J. and Fournier, L. (1962) 'Crecimiento y desarrollo del fruto de *Coffea arabica* L.',

Turrialba, 12, 65-74
Maestri, M. and Barros, R.S. (1977) 'Coffee' in F. de Alvin and T.T. Kozlowski (eds.) *Ecophysiology of Tropical Crops,* Academic Press, London, New York and San Francisco. pp. 249-78
Magalhaes, A.C. and Angelocci, L.R. (1976) 'Sudden alterations in water balance associated with flower bud opening in coffee plants', *Journal of Horticultural Science, 51,* 419-23
Magambo, M.J.S. and Cannell, M.G.R. (1981) 'Dry matter production and partition in relation to yield of tea', *Experimental Agriculture, 17,* 33-8
Maggs, D.H. (1961) 'Changes in the amount of distribution of increment induced by contrasting watering, nitrogen and environmental regimes', *Annals of Botany, 25,* 353-61
Mathew, P.K. and Chokkanna, N.G. (1961) 'Studies on the intake of water and nutrients during development of flower buds to blossoms in coffee', *Indian Coffee, 25,* 264-72
Mes, M.G. (1957) *Estudos Sobre o Florescimento de Coffea arabica L.* (Translated by C.M. Franco; *Studies on Flowering of Coffea arabica L.*). IBEC Research Institute, Bulletin 14, New York
Mitchell, H.W. (1976) 'Research on close spacing systems for intensive coffee production in Kenya' in *Coffee Research Foundation, Kenya, Annual Report for 1974-75* pp. 13-58
Monaco, L.C., Filho, H.P.M., Sondahl, M.R. and Alves de Lima, M.M. (1978) 'Efecto de dias longos no crescimento e florescimento de cultivares de café', *Bragantia, 37,* 25-31
Montoya, C.A., Sylvain, P.G. and Umana, R. (1961) 'Effect of light intensity and nitrogen fertilization upon growth differentiation balance in *Coffea arabica* L.', *Coffee, 3,* 97-115
Mulinge, S.K. and Griffiths, E. (1974) 'Effects of fungicides on leaf rust, berry disease, foliation and yield of coffee', *Transactions of the British Mycological Society, 62,* 495-507
Nogueira, P.N., Carvalho, A. and Antunes, H.F. (1959) 'Efeito da exclusao dos insetos polinazadores na producao do cafe bourbon', *Bragantia, 18,* 441-68
Nunes, M.A. (1976) 'Water relations of coffee. Significance of plant water deficits to growth and yield: a review', *Journal of Coffee Research, 6,* 4-21
Nunes, M.A., Bierhuizen, J.F. and Ploegman, C. (1968) 'Studies on productivity of coffee. I. Effect of light, temperature and Co_2 concentration on photosynthesis of *Coffea arabica* L.', *Acta Botanica Neerlandica, 17,* 93-102
Nunes, M.A., Bierhuizen, J.F. and Ploegman, C. (1969) 'Studies on productivity of coffee. III. Differences in photosynthesis between four varieties of coffee', *Acta Botanica Neerlandica, 18,* 420-4
Nunes, M.A., Brumby, D. and Davies, D.D. (1973) 'Estudo comparativo do metabolismo fotossintético em folhas de cafeeiro, beterraba e cane-de-acúcar', *Garcia de Orta, Sér Estudos Agronómicas (Lisboa), 1,* 1-14
Nutman, J.F. (1937) 'Studies on the physiology of *Coffea arabica* L. I. Photosynthesis of coffee leaves under natural conditions', *Annals of Botany, 1,* 353-62
Opile, W.R. (1979) 'Hormonal relations in fruit growth and development of *Coffea arabica* L.', *Kenya Coffee, 44,* 13-21
Opile, W.R. and Browning, G. (1975). 'Regulated ripening of *Coffea arabica* L. in Kenya: studies on the use of 2-chloroethyl-phosphonic acid', *Acta Horticulturae, 49,* 125-36
Ostendorf, F.W. (1962) 'The coffee shade problem: review article', *Tropical Abstracts, 17,* 577-81
Oyebade, T. (1976) 'Influence of pre-harvest sprays of ethrel on ripening and abscission of coffee berries', *Turrialba, 26,* 86-9
Piringer, A.A. and Borthwick, H.A. (1955) 'Photoperiodic response of coffee', *Turrialba, 5,* 72-7
Porteres, R. (1946) 'Action de l'eau, apres une periode seche, sur le declenchement de la floraison chez *Coffea arabica* L.', *Agronomie Tropicale, 1,* 148-58
Raju, K.I. Venkataramanan, D. and Janardhan, K.V. (1975) 'Physiological studies on flowering in coffee under South Indian conditions. II Flowering in relation to foliage and wood starch', *Turrialba, 25,* 235-42
Ramaiah, P.K. and Vasudeva, N. (1969) 'Observations on the growth of coffee berries in South India', *Turrialba, 19,* 455-64
Rao, G.S., Venkataramanan, D. and Rao, K.N. (1977) 'Changes in fruit growth and pectic content in *Coffea canephora* (Robusta) in relation to the exogenous application of ethylene', *Zeitschrift fur Pflanzenphysiolie, 83,* 459-61

Raw, A. and Free, J.B. (1977) 'The pollination of coffee (*Coffea arabica*) by honeybees', *Tropical Agriculture*, 54, 365-70

Reddy, A.G.S. (1979) 'Quiescence of coffee flower buds and observations on the influence of temperature and humidity on its release', *Journal of Coffee Research*, 9, 1-13

Reddy, G.S.T. and Srinivasan, C.S. (1979) 'Variability for flower production, fruit set and fruit drop in some varieties of *Coffea arabica* L.', *Journal of Coffee Research*, 9, 27-34

Rees, A.R. (1964) 'Some observations on the flowering behaviour of *Coffea rupestris* in southern Nigeria', *Journal of Ecology*, 52, 1-7

Reffye, Ph. de. (1981) 'Modèle mathématique aléotoire et simulation de la croissance et de l'architecture du cafeier Robusta', *Café, Cacao, Thé*, 25, 83-104

Reffye, Ph. de. and Snoeck, J. (1976) 'Modèle mathématique de base 1 étude et la simulation de la croissance et de l'architecture du *Coffea robusta*', *Café, Cacao, Thé*, 20, 11-32

Robledo, A.J. and Aristizabel, G.V. (1980) 'Los elementos climaticos y el desarrollo de *Coffea arabica* L. en Chinchina, Colombia', *Cenicafé*, 31, 127-43

Robledo, A.J. and Santos dos, J.M. (1980) 'Balance de radiacion solar en *Coffea arabica* L. variedades Catui y Borbon Amarillo', *Cenicafé*, 31, 86-104

Snoeck, J. (1977) 'Essais de groupement de la récolte des fruits du caféier Canephora à l'aide de éthéphon', *Café, Cacao, Thé*, 21, 163-78

Sondahl, M.R., Crocomo, D.J. and Sodek, L. (1976) 'Measurements of ^{14}C incorporation by illuminated intact leaves of coffee plants from gas mixtures containing $^{14}CO_2$', *Journal of Experimental Botany*, 27, 1187-95

Stemmer, W.P.C., Adrichen, van J.C.J. and Roorda, F.A. (1982) 'Inducing orthotropic shoots in coffee with the morphactin chloroflurenolmethylester'. *Experimental Agriculture*, 18, 29-35

Suryakantha, R.K., Srinivasan, S. and Vishveshwara, S. (1974) 'Vegetative-floral balance in coffee. II. Effect of crop thinning on the growth of axillary vegetative shoots, extension of nodes, defoliation and fruit drop', *Journal of Coffee Research*, 4, 21-4

Taschdijan, E. (1932) 'Beobachtung über Variabilitat, Dominanz und Vizinismus bei *Coffea arabica*', *Zeitschrift fur Zuchtung*, 17, 341-54

Tio, M.A. (1962) 'Effect of light intensity on the rate of apparent photosynthesis in coffee leaves', *Journal of Agriculture of the University of Puerto Rico*, 46, 159-66

Trojer, H. (1968) 'The phenological equator for coffee planting in Colombia'. *UNESCO Natural Resources Research*, 7, 107-13 UNESCO, Paris

Valencia, A.G. (1973) 'Relacion entre el indice de area foliar y la productividad del cafeto', *Cenicafé*, 24, 79-89

Vasudeva, N. and Ramaiah, P.K. (1979) 'The growth and development of Arabica coffee under South Indian conditions', *Journal of Coffee Research*, 9, 35-45

Vasudeva, N. and Ratageri, M.C. (1981) 'Studies on leaf to crop ratio in two commercial species of coffee grown in India', *Journal of Coffee Research*, 11, 129-36

Veen, van der R. (1968) 'Plant hormones and flowering in coffee', *Acta Botanica Nederlandica*, 17, 373-6

Vossen, van der, H.A.M. and Browning, G. (1978) 'Prospects of selecting genotypes of *Coffea arabica* L. which do not require tonic sprays of fungicide for increased leaf retention and yield', *Journal of Horticultural Science*, 53, 225-33

Went, F.W. (1957) *The Experimental Control of Plant Growth*, Chronica Botanica Co., Waltham, MA USA

Willey, R.W. (1975) 'The use of shade in coffee, cocoa and tea', *Horticultural Abstracts*, 45, 791-8

Wormer, T.M. (1964) 'The growth of the coffee berry', *Annals of Botany*, 27, 47-55

Wormer, T.M. (1966) 'Shape of bean in *Coffea arabica* L. in Kenya', *Turrialba*, 16, 221-36

Wormer, T.M. and Ebagole, H.E. (1965) 'Visual scoring of starch in *Coffea arabica* L. II. Starch in bearing and non-bearing branches', *Experimental Agriculture*, 1, 41-53

Wormer, T.M. and Gituanja, J. (1970) 'Floral initiation and flowering in *Coffea arabica* L. in Kenya', *Experimental Agriculture*, 6, 157-70

Yamaguchi, T. and Friend, D.J.C. (1979) 'Effect of leaf age and irradiance on photosynthesis of *Coffea arabica*', *Photosynthetica*, 13, 271-8

6 MINERAL NUTRITION AND FERTILISER NEEDS

C. Willson

Nutrient Requirements

The coffee crop (green beans) includes mineral nutrients which are therefore removed from the plantation system. Compared with some crops (for example, sugar cane) the quantities are not large. Catani and de Moraes (1958) estimated that the major nutrients removed in 1 tonne of arabica green beans amounted to 34.0 kg N, 5.2 kg P_2O_5 and 47.8 kg K_2O. However the crop is harvested as cherry which includes pulp and parchment in addition to the beans. In many cases these are not returned to the field so that the nutrients therein are lost to the system. Using data published by Ripperton, Goto and Pahau (1935: 55) the nutrients removed in the bean, pulp and parchment equivalent to 1 tonne of arabica green beans are: in bean, 45.5 kg N, 7.67 kg P_2O_5, and 37.9 kg K_2O; in parchment, 2.27 kg N, 0.3 kg P_2O_5 and 1.87 kg K_2O; in pulp, 15.33 kg N, 3.67 kg P_2O_5 and 27.4 kg K_2O. Ripperton et al. (1935: 47) showed that the concentration of nitrogen, phosphorus and potassium in the constituents of the cherry varied according to the soil and fertiliser applications. Roelofsen and Coolhaas (1940) reported that the total losses of nutrients from the plantation equivalent to 1 tonne of robusta green bean were: 35 kg N, 6 kg P_2O_5, 50 kg K_2O, 4 kg CaO, 4 kg MgO, 0.3 kg Fe_2O_3, 0.02 kg Mn_3O_4. Malavolta, Graner, Sarruge and Gomes (1963) reported the concentrations of macro- and micro-nutrients in pulp and beans of arabica coffee. Samples were taken from three coffee soils in Brazil. They found a significant variation in the concentration of nitrogen, potassium, iron and molybdenum between the soils; variations in other nutrients were not significant. No significant variations were found between the three varieties sampled. The figures quoted for nutrient losses can only, therefore, be an approximate guide to nutrient requirements.

Nutrients are also incorporated into vegetative growth, above and below ground. The major part of the root system does not decay when the tree is healthy so the nutrients therein are lost permanently. If leaf-fall and prunings remain in the plantation they decompose, releasing mineral nutrients back to the soil. This recycling of nutrients is important in minimising the requirements of fertiliser. The humus formed is valuable in maintaining soil structure and helps to keep nutrients in an available form. Trees increase in size with age until a complete canopy is formed and will continue to increase unless a very strict pruning system is in operation. Nutrients will therefore be permanently removed from the system in stems and leaves.

Removal of material from the field, particularly pruned stems which are often used as firewood, can cause heavy loss of nutrients. The author weighed and analysed large stems which were removed from a plantation for fuel after pruning. All primaries, leaves, etc. were cut off and remained on the ground. The average weight of the stems was 2 kg. A sample was analysed and found to contain 0.6 per cent N, 0.05 per cent P and 0.4 per cent K in dry matter. The dry matter content was 91.4 per cent. With one stem only removed from each of 1350 trees per hectare the gross weight of timber removed is 2700 kg per hectare, which would carry with it 14.8 kg N, 2.8 kg P_2O_5 and 11.9 kg K_2O.

De Castro and Rodriguez (1955) reported that losses of nitrogen by erosion from unprotected areas exceeded the amount removed by the coffee crop. Leaching of soluble nutrients causes further losses. Catani and de Moraes (1958) reported that an arabica tree removed from the soil during its fifth year 118 g N, 16 g P_2O_5, 120 g K_2O, 76.5 g CaO and 23.4 g MgO. Where trees are planted at a density of 1350 per hectare the total loss of nutrients amounts to 277 kg N, 37.6 kg P_2O_5, 282 kg K_2O, 180 kg CaO and 55 kg MgO.

For robusta, Forestier (1969) reported that, without returning pulp and parchment to the plantation, one tonne of green bean removes 30 kg N, 3.75 kg P_2O_5 and 36.5 kg K_2O from the plantation.

The figures above show that nitrogen and potash are the major nutrient requirements.

The root system of coffee explores the soil extensively to a depth of at least 3 metres in suitable soil (Nutman, 1933). De Castro (1960) reported that the proportion of roots in all sizes below 30 cm depth in the soil increased with the age of the tree. Of the roots of seven-year-old trees 61 per cent were at depths below 30 cm. Cuenca, Aranguren and Herrera (1983) investigated the root system of arabica coffee var. *mundo novo* planted 25 years earlier under shade and managed without fertilisers. Of the fine roots which are mainly responsible for absorption of nutrients 33 per cent were found in the top 10 cm of soil, which consisted largely of litter. Of the fine roots 73 per cent were in the upper 30 cm of soil. Root production was measured at 660 g m^{-2} yr^{-1} in the upper 7.5 cm of soil which is largely litter and 10g m^{-2} yr^{-1} in the surface layer of fresh litter. The major purpose of the deeper roots is to absorb water; most absorption of nutrients is carried out by the fine roots in the top 30 cm of the soil profile. It is therefore important that the upper soil layers retain a good structure and a high organic content.

In addition to the nutrients required to maintain the trees in good health, yield is stimulated by the application of nitrogen. Vegetative growth is also increased. Higher yield and vegetative growth require correspondingly higher amounts of all other nutrients.

Nutrients are absorbed from the soil. A fertile soil with high levels of available nutrients maintains a higher growth rate and crop yields than a poor soil, unless fertiliser is applied. In the absence of fertiliser applications the nutrient reserves in the soil can be used up. Coffee is a long-lived crop and a significant reduction in the amounts of nutrients available in the soil occurs over tens of years. Table 6.1 exemplifies this.

Arabica coffee is prone to overbearing. The physiological reasons for this are discussed by Cannell in Chapter 5. This overbearing leads to exhaustion of nutrients within the tree which restricts vegetative growth, thereby reducing the number of buds available to flower the following year. Vegetative shoots often die back when the nutrients in the leaves and stems have been transferred to the fruit. Production therefore becomes biennial. Die-back of stems is accompanied by die-back of roots, which restricts the ability of the trees to recover. There may be a permanent reduction of the cropping potential of the tree due to loss of feeder-bearers (Nutman, 1933). Improvement of the nutrient supply by application of fertiliser can reduce the degree of overbearing and ensuing die-back, but the physical capacity of the roots to absorb nutrients often prevents the uptake of sufficient nutrients to prevent overbearing, which must then be controlled by other means. The effect of fertilisers on the incidence of die-back has been studied in several countries. Pereira and Jones (1954) reported a significant reduction in die-back following application of nitrogen fertiliser. Montoya and Umana (1961) found that the incidence of die-back decreased with the application of nitrogen. Malavolta, Gomes and Coury (1958) found that die-back was inversely correlated with the levels of nitrogen and potassium in the leaves.

The degree of shading interacts with nutrition. Cannell discusses the physiological effects of shade in Chapter 5. Apart from the physiological effect of different light intensities on the coffee trees, shade trees supply nutrients in leaf-fall; leguminous trees will make a larger contribution by fixation of atmospheric nitrogen. Shading affects soil conditions; organic matter decomposes more slowly at lower temperatures. The net effect of shade, in general, is that unfertilised shaded trees yield at a higher level than unfertilised unshaded trees. However, unshaded trees have a higher rate of response to fertiliser so that the highest yields are given by heavily fertilised unshaded coffee. Triana (1957) demonstrated the effect of shade on yields and fertiliser responses, and Montoya and Umana (1961) showed that die-back increased with light intensity whilst applications of nitrogen reduced the disorder; the effect of nitrogen increasing as the light intensity increased.

Roskoski (1980) found that shade tree leaf litter contributed over 80 kg N per hectare per year; the total of = returned to the soil was greater than the loss of nutrients in the crop. Leguminous trees (*Inga jinicuil*) fixed about 40 kg N ha^{-1} yr^{-1} (Roskoski, 1982).

Nutrients are recycled in a tree-crop ecosystem. Willson (1978) pointed

Table 6.1: Comparison of Soils Under Different Ages of Coffee

Age of coffee (years)	pH	P (μg/ml)	K (meq %)	Ca (meq %)	Mg (meq %)	Na (meq %)	CEC (meq %)	K (% CEC)	Ca (% CEC)	Mg (% CEC)	Na (% CEC)
1	5.9	12	0.67	6.5	2.16		24	2.8	27	9.1	
11	5.0	16	0.92	7.8	3.1	0.08	20	4.5	38	15	0.4
30	4.0	21	0.52	0.5	0.2	0.11	16	3.3	3	1.3	0.7
30 (limed)*	4.6	16	0.40	1.8	1.3	0.10	15	2.6	11	8.5	0.7

*Three tonnes dolomitic limestone applied during two years prior to sampling.
Source: Author's records for Western Highlands, Papua New Guinea (unpublished).
CEC: cation exchange capacity

out that large quantities of nutrients are involved therefore losses from this system must be minimised; also considerable ecological advantages follow from proper management. The cycling of nitrogen in coffee plantations has been discussed by Bornemisza (1982) and Aranguren, Escalante and Herrera (1982).

For every fertiliser nutrient reports can be found of responses varying from a reduction in yield through no effect to an increase in yield. Fertiliser applications to any area must be related to local soil conditions. The results must be monitored continuously using leaf and soil analysis in addition to yield and climatic data, and fertiliser applications varied accordingly.

Individual Nutrients

Nitrogen

Nitrogen is essential for the development of stems, leaves and fruit. Under heavy shade where growth rate and crop yield are low, the low nitrogen requirement will be supplied from the soil, as shown by Bornemisza (1982). When the yield is increased by reduction or removal of shade, pruning and other cultural measures nitrogen will be in short supply.

Applications of nitrogen to coffee which is not deficient in nitrogen will usually increase vegetative growth and yield, provided deficiencies of other nutrients do not restrict production. Montoya, Sylvain and Umana (1961) found that, in arabica coffee, var. *Bourbon*, nitrogen increased the number of nodes per branch without affecting the internodal distance. Snoeck and de Reffye (1980) showed that, in robusta coffee, the increased yield was a consequence of an increase in the number of fruiting nodes which was significantly related to the application level of nitrogen. Snoeck (1981) reported that nitrogen also increased the number of flowers per node and the sucessful setting rate.

Quantities of nitrogen fertiliser applied to coffee vary from 50 kg N per hectare annually, upwards to very high levels (400 kg N per hectare). Low levels may be applied in a single application, the higher levels are almost invariably split into a number of applications.

Investigations in Kenya (Oruko, 1977) showed that greater responses were obtained when nitrogen was applied at periods when demand was high and supply low. There was therefore an optimum time of application during the two periods (long and short rains) when the growth rate was at its maximum and supply low due to leaching. Further benefit was obtained by splitting the application in each rain period but the months of application had a critical effect on crop yields.

Various straight nitrogen fertilisers are used on coffee. Sulphate of ammonia is the most common but has the disadvantage of acidifying the soil with consequent reduction in availability of base nutrients. Less acidifying materials, such as calcium ammonium nitrate (CAN), are used. The response may vary depending on the form of nitrogen fertiliser used. Continuous use of CAN can apply sufficient calcium in time to affect adversely the balance of bases. Urea is commonly used, but must be applied under moist conditions to minimise loss of ammonia. The biuret concentration in urea must be low; Vaidyanathan, Cokkana and Narayanan (1956) described leaf symptoms of biuret toxicity, but reported no adverse effect on the growth rate of affected seedlings.

Where other nutrients are applied also, nitrogen is often applied as a compound fertiliser of which there are many of varying complexity.

Nitrogenous fertilisers can be applied as a foliar spray. Urea is commonly applied in this manner, other totally water-soluble materials can be used. Responses follow quickly after spraying. Malavolta et al. (1959) reported that 95 per cent of urea applied in a foliar spray was absorbed within 9 hours of application.

Phosphate

The amount of phosphate removed in the crop is small. In the more acid soils phosphate fixation may prevent the trees from absorbing sufficient phosphate. Direct responses to experimental applications of phosphate to mature coffee are rare. However phosphate is often applied to mature coffee in small quantities.

The benefits of phosphate are more clearly seen in nurseries and on young coffee plants. Admixture of phosphate fertiliser with nursery soil and with planting-hole soil before field planting are common practices. In many nurseries soluble phosphate fertiliser is sprayed on to young plants at frequent intervals. M'Itungo and Van der Vossen (1981) reported an investigation into the nutrient requirements of polybag nurseries. Farmyard manure and NPK foliar feeds of varying composition all improved the growth rate of seedlings. By mixing superphosphate and farmyard manure with the potting soil good seedlings could be obtained without foliar feeding; the major effect on the seedlings can therefore be ascribed to the phosphate.

A wide variety of phosphate fertilisers are applied to coffee and this nutrient is usually included in complex fertilisers where these are used. The proportions are often varied in accordance with the age of the coffee; for example the N: P_2O_5: K_2O ratio would be 1: 2: 1 on young coffee but 2: 1: 2 on older trees. Even where heavy applications of other nutrients, particularly nitrogen and potash are given, applications of phosphate do not usually exceed 100 kg P_2O_5 per hectare per annum. A significant response was obtained in Kenya to one application at a very high level (268 kg P_2O_5 per hectare) (Chawdhry, 1974). However there was no response to this level of application in subsequent years (Kabaara, 1976).

Foliar application of phosphate fertiliser is much more effective than ground application. Malavolta *et al.* (1959) reported the uptake of phosphorus from superphosphate applied in various ways, as follows:

1. Top-dressed   10.2 per cent
2. Applied in a circular furrow   2.4 per cent
3. Applied in a semi-circular furrow   1.7 per cent
4. Sprayed directly on to the leaves   38.0 per cent

Haag and Malavolta (1960) reported leaf symptoms caused by excess phosphorus.

Potassium

In terms of losses in the crop this element is slightly more important than nitrogen. It holds a key position in the nutrition of the plant and the development of the fruit. Along with some other nutrients, but to a greater extent than any other, it is moved from leaves and other vegetative parts to the fruits as they develop. If the crop needs more potassium than is available from the soil, this element is transferred from the leaves. The level in the leaves may be reduced to a deficiency; then vegetative growth will stop and the leaf deficiency will result in leaf-fall and ultimately in die-back.

On deep, fertile soils the coffee can extract sufficient potassium from the soil. Regular mulching maintains potassium availability at a high level. Under such conditions no significant experimental response to potash fertiliser is obtained; for example, in Kenya and Tanzania (Pereira and Jones, 1954; Robinson, 1964: 73; Kabaara, 1976). However, continuous production over many years will ultimately exhaust soil reserves and potassium deficiency is usually the first consequence of this process. On less fertile or denuded soils significant responses have been obtained to potash fertilisers; for example, in Brazil (Malavolta *et al.*, 1958) and in Hawaii (Ripperton *et al.*, 1935: 42-46).

Potassium and magnesium are antagonistic. High levels of potassium in the soil or excessive application of potash fertilisers can cause magnesium deficiency; potassium deficiency can arise from high levels of magnesium in the soil or excessive magnesium applications. High calcium levels in the soil will also restrict potassium uptake.

Experiments often show significant positive interactions between nitrogen and potassium (for example, Malavolta *et al.*, 1958). This interaction occurred in Kenya only in the presence of phosphate (Kabaara, 1976). This indicates the need to keep the nutrients in balance. The ratio of potash to nitrogen (K_2O/N) in the total annual fertiliser application is a convenient measure of the potash fertiliser required at a particular location. From the many publications advising on fertiliser applications or reporting on fertiliser requirements, this ratio is seen to vary from zero to two. The application rate of potash therefore varies from nil to over 500 kg K_2O per hectare per annum.

Young coffee which is not producing crop requires less potassium in relation to nitrogen but a deficiency must be avoided or this will retard the development of trees with high crop potential.

Muriate of potash is commonly used where a straight potash fertiliser is applied. The effects of the chloride in this material have not been studied extensively; this is discussed in greater detail later. Sulphate of potash is used in some locations; this fertiliser provides sulphur which has to be applied to some soils. Compound fertilisers containing nitrogen, phosphate and potash are convenient where all three nutrients are required. Many compounds, of which there is a wide range of composition, are in common use.

Haag and Malavolta (1960) reported toxicity symptoms from excessive potassium applications.

Calcium

Little calcium is removed in the crop, most soils can provide adequate amounts. Calcium deficiency symptoms have been described, but they are rarely seen. Many fertilisers which are used primarily as a source of other nutrients (for example, calcium ammonium nitrate, superphosphate) include calcium. Normal applications of such materials will usually provide sufficient calcium to prevent a deficiency.

This element has a greater importance in relation to soil acidity; the application of lime to counter soil acidity is recommended in many countries. Experiments have shown that lime is effective in reducing the acidity of the soil; improvement in coffee yields has not been reported (Valencia-Aristizabal and Bravo-Grijalba, 1981). The quantities are usually large (tonnes per hectare) but the interval between applications often lengthy (up to five years).

Coffee planted on soils derived from pure limestone or other highly calciferous soils will absorb excessive quantities of calcium. This does not produce a toxicity as such but restricts the uptake of potassium and/or magnesium, depending on the amounts of these nutrients available. Heavy applications of potash or magnesium fertilisers are then necessary.

Magnesium

The characteristic deficiency symptoms of magnesium are often seen. Because magnesium is translocated freely within the tree, the symptoms affect older leaves initially. The degree to which the deficiency affects yield depends on the stage in the annual cycle at which the deficiency occurs. If the symptoms appear at a late stage in the development of the fruit, it is probable that the developing harvest has adequate magnesium and will not be reduced. The earlier the symptoms, or deficiency as indicated by leaf analysis, arise, the more likely is it that crop yield will be reduced. If plenty of new foliage has grown during the current year, the loss of some old leaves from this deficiency will not significantly affect the crop for the following year.

Magnesium deficiency may arise from high applications of potash fertilisers or other materials high in potassium. Robinson and Chenery (1958) found that repeated mulching with napier grass (*Pennisetum purpureum*), which has a high potassium content, induced magnesium deficiency.

Many authors have described experiments which cured the deficiency symptoms without increasing yield; for example, Anon. (1961/63: 40-46). Others have reported significant yield increases (e.g. Hernandez, 1962).

A variety of chemicals have been applied in order to relieve this deficiency. Magnesium sulphate (epsom salt or kieserite) is commonly used for ground application. Magnesium oxide (calcined magnesium carbonate) is also used. Where lime is applied also, a dolomitic limestone is often used. Recovery after soil applications is often slow; Muller (1959) reported six to twelve months, Fiester (1961) three to five years. It is common practice when applying a compound NPK fertiliser to use one also containing magnesium.

Foliar applications produce quicker responses. Magnesium sulphate is used most frequently, in solution of strength 0.5 to 0.75 per cent, with addition of 0.25 to 0.35 per cent calcium carbonate (Anon., 1957). Concentrations up to 1.2 per cent have been used; combined with copper spray concentrations up to 12 per cent are effective, as are magnesium sulphate/urea solutions with equal amounts of each chemical at concentrations up to 6 per cent. Five applications during the period of cherry development at a rate of 375 to 1,150 litres per hectare are recommended in Kenya (Anon., 1961/63). Magnesium nitrate and magnesium chloride have also been recommended. Magnesium nitrate was the most effective, absorption of 90 per cent of the material was claimed (Fiester, 1961) and deficiency symptoms disappeared after two applications within three months (Chanchay, 1964). However this chemical is phytotoxic; therefore the concentration should not exceed 0.75 per cent; leaf scorch has been reported from a concentration of 1.2 per cent. Magnesium sulphate was reported as the least effective foliar treatment (Chanchay, 1964).

Toxicity symptoms from excess magnesium have not been reported. Haag and Malavolta (1960) attempted to induce toxicity without success. However, the interaction between potassium and magnesium may result in potassium deficiency after excessive application of magnesium fertiliser. Mehlich (1967) reported yield reductions following application of magnesium oxide, more frequent in the absence of mulch.

Sodium

Concentrations of sodium in leaves up to 0.065 per cent have been reported following application of sodium nitrate (Robinson, 1961). Symptoms closely resembling potassium deficiency appeared in leaves containing

2.75 per cent potassium and 0.033 per cent sodium, also on leaves with higher sodium levels (Espinosa, 1960). No work has been reported on saline soils.

Sulphur

Sulphur deficiency occurs occasionally. Growth and fruit-set are sometimes reduced (Cibes and Samuels, 1955). In other cases growth is not affected (Franco and Mendes, 1949; Loue, 1957). It is common practice to ensure that the fertiliser programme includes some sulphur (as sulphates or in compound fertilisers containing sulphur) to prevent the occurrence of this deficiency.

Accorsi and Haag (1959) reported symptoms of sulphur toxicity. Haag and Malavolta (1960) reported a reduction in leaf dry weight and leaf sulphur content of 0.25 per cent following application of excess sulphur in a pot experiment with arabica coffee. However, application of large quantities of sulphates in the field does not lead to excessive absorption of sulphur by the trees. This is in contrast to chlorine, as was shown by Gonzalez, Lopez, Carvajal and Briceno (1977). They reported sulphur contents of arabica leaves between 0.1 and 0.26 per cent. Forestier and Beley (1966) reported leaf contents between 0.1 and 0.4 per cent for robusta; leaf chlorosis was present at sulphur levels of 0.12 per cent and lower.

Zinc

Zinc deficiency is common and the visual symptoms are easily recognised. Foliar spraying of a small quantity (up to 5 kg per hectare) of a zinc compound (commonly either sulphate or oxide) to trees showing the typical 'miniaturisation' symptoms are quickly followed by growth of normal sized internodes and leaves (Muller, 1959; Fiester, 1961). Fruits grow to normal size also, with corresponding benefit in yield (Perez, 1958). At other locations foliar sprays of zinc have not been followed by yield increases (Uribe-Henao and Salazar-Arias, 1981). Reduction of yield has also been reported (Anon. 1969; Fenner and Adamson, 1982). Such inconsistencies arise probably either by interaction between minor nutrients, by application of large quantities which produce temporary excesses, or by application at unduly lengthy intervals. The author found that a foliar spray of 5 kg zinc sulphate per hectare raised a low leaf content to an excessively high level. The leaf content fell to a deficient level within three months. An annual or bi-annual spray may therefore not improve yield as zinc deficiency is present for several months of the year. A better effect is obtained if the same total quantity of zinc compound per year is split over more frequent applications, ideally every two months. No toxicity symptoms have been reported from high leaf zinc contents following foliar application of excessive quantities (Ananth, Iyengar and Chokkana, 1965).

Ground applications of zinc fertilisers are not effective unless much larger quantities (up to ten times the foliar application levels) are made. Malavolta *et al* (1959) demonstrated the very small proportion of soil-applied zinc which was absorbed by the plants.

Malavolta *et al.* (1959) also showed that excessive amounts of manganese, copper and molybdenum restricted the uptake of zinc. Iron did not have any inhibiting effect in the experiment reported. Forestier and Beley (1966) reported a zinc-manganese interaction in robusta.

Copper

Copper deficiency has only been described from pot experiments (Malavolta, Haag and Johnson, 1961). Application of copper fungicides is necessary in many areas; this automatically provides sufficient copper by leaf absorption to avoid a deficiency. Where fungicides are not necessary a foliar spray of a copper compound provides sufficient copper to prevent a deficiency.

The 'tonic' effect of copper fungicides which increases yield is well authenticated. (See Chapter 9, page 222 and Figure 3.5.)

The continuous use of fungicidal copper sprays could possibly lead to a nutritional problem due to excess copper. Aduayi (1977) investigated the effects of varying application levels on copper in soils of a range of pH values. Low concentrations of copper improved growth, at the higher concentrations toxicity symptoms appeared and growth was reduced. The effects were more severe at the lower pH values. The effects of copper in relation to phosphorus were also investigated (Aduayi, 1978). Growth, as measured by number of leaves and flower-buds, increased as very low levels increased to optimum values, above which growth decreased and toxicity symptoms, or deficiency symptoms of calcium and magnesium due to imbalance, appeared. Vasudeva and Ratageri (1980) reported that copper sulphate was deposited in leaves following heavy applications of this material.

Iron

Iron deficiency often occurs on more alkaline soils or soils with high organic content (Robinson, 1959) and it is often associated with high phosphorus levels or caused by high manganese levels (Medcalf and Lott, 1956). Apart from correction of soil conditions application of an iron chelate will correct the deficiency (Robinson, 1959). Iron deficiency in liberica coffee in Malaysia is cured by an application of 20 g ferrous sulphate per tree (Willson, 1984). Foliar applications are not normally absorbed (Muller, 1959), but solid injection is effective (Robinson, 1959).

Manganese

Manganese deficiency is rare, but sometimes occurs on alkaline soils. Manganese toxicity occurs on very acid soils and can be cured by application of lime, or other means to reduce soil acidity. Both deficiency and toxicity of manganese are known in Costa Rica, where the toxicity effects are known as 'Cafe macho' (Perez, 1957). Loue (1960) reported deficiencies in robusta coffee in the Ivory Coast.

Aluminium

This element is often highly available in acid soils and is absorbed in excessive quantities by some plants under these conditions. Catani *et al.* (1967) reported 346 ppm of aluminium in leaves of coffee growing on a latosol (pH not recorded).

Aluminium is believed not to be essential for the coffee plant. Pavan, Bingham and Pratt (1982) reported toxicity effects on arabica coffee and established critical levels for aluminium in soil and leaves, above which growth was reduced (100 and 62 ppm Al in leaf, for effects on shoot and root growth respectively).

Boron

Boron deficiency is common and excessive amounts of boron are toxic. Leaves irreversibly deformed by deficiency can absorb more boron and ultimately show toxicity symptoms also. Boron deficiency is often associated with high calcium in the soil but low boron restricts the uptake of calcium. Because of the relationship between potassium and calcium nutrition, application of potash fertilisers may affect the boron status of the plants (Malavolta *et al.*, 1962). Perez, Chaverri and Bornemisza (1956) found that the calcium/boron ratio was important.

Ground application of 30-60 g of borax to each tree per annum will correct this deficiency. A foliar spray of a 0.4 per cent solution of borax is equally effective.

Molybdenum

Catani *et al.* (1967) reported molybdenum concentrations in arabica coffee leaves, trunk, branches and fruits. Malavolta *et al.* (1961) reported molybdenum deficiency in arabica coffee grown in nutrient solution, and suggested a minimum critical level of 0.9 ppm. Deficiency conditions may arise on very acid soils.

Forestier and Beley (1966) reported molybdenum levels in robusta leaves and proposed a minimum critical level of 0.5 ppm.

Chlorine

Catani *et al.* (1967) noted very high concentrations of chlorine in leaves (6,815 ppm) and in fruits (3,380 ppm). No information was given as to fertiliser applications which might have been the source of the chlorine.

Arana (1967) reported that toxicity symptoms appeared when the chlorine content of leaf reached 3,400 ppm, becoming more severe as the level of chlorine increased. He suggested a tentative tolerance limit for chlorine of 2,000 ppm. Furlani, Catani, de Moraes and Franco (1976) reported toxicity symptoms when leaf content exceeded 7,600 ppm in arabica coffee *var. Catuai.*

Gonzalez *et al.* (1977) showed that the chlorine concentration in the leaves rose rapidly after the application of potassium chloride to the tree to over 2,000 ppm, remaining above that value for three months before declining to the original level.

Morillo, Nieto, Linares and Roman (1978) investigated the effect of potassium chloride on arabica coffee plants in the nursery, using repeated applications at a low concentration. The highest foliar application level produced a significant reduction in growth; soil applications had no effect. The chlorine concentration in leaves was 201 ppm two months after treatment, 467 ppm six months after treatment. No deficiency symptoms were reported.

The possibility that application of muriate of potash to coffee may be harmful under certain conditions has not been considered by the industry. Very few data are available for the chlorine content of leaves of producing coffee. The results above were obtained using arabica coffee, there are no publications concerning robusta. Problems which might be expected to arise on saline soils have not been reported. It is unlikely that arabica would be planted on such soils. However robusta, which will grow at sea-level, may be planted close to the sea. The author has seen robusta coffee planted close to the sea and which was suffering severely from the combined effects of saline soil and salt spray. The leaves contained 0.06 per cent sodium and 0.80 per cent chlorine.

Organic Manures and Mulch

Organic manures are used on coffee where they are available. Fernie (in Robinson, 1964: 32) recommended the use of compost or cattle manure in Tanganyika. Acland (1971: 67) repeated this recommendation, for East Africa, Anon. (1975: 13-ii-2) recommended composting coffee pulp. The use of bio-gas digester sludge was reported by Hutchinson (1968). Coffee pulp is sometimes returned to the field; excess water is drained off and the moist pulp carted to the field. Alternatively the water/pulp mixture leaving the pulper is pumped into the field.

As mulch material decays, the mineral nutrients it contains are released to the soil in an available form. Robinson and Hosegood (1963) reported the benefits of mulch on a latosol. Organic manures and mulch can provide significant quantities of nutrients and organic matter which benefit the soil. Manures containing unbalanced amounts of nutrients can, if used in sufficient quantity, create problems of imbalance in the coffee. Mehlich (1965) reported the large increase in potassium in the soil as a result of mulching with elephant grass (*Pennisetum purpureum*), and following application of cattle manure.

The great benefit of organic material in stimulating root growth was shown by Cuenca *et al* (1983) who reported that the production rate of roots was $660 \text{ g m}^{-2} \text{ yr}^{-1}$ in the top 7.5 cm of soil which consisted largely of decomposed litter compared with $10 \text{ g m}^{-2} \text{ yr}^{-1}$ in fresh litter.

Nutrition and Crop Quality

The effects of nutrients on the quality of the crop have been studied in several places.

Mehlich (1967) reviewed investigations in Kenya. The criterion for quality in this case was the proportion of grade 'A' coffee in the product. The coffee is sorted into grades by screening (see also Chapter 3 p. 57 and 10; p. 243). Gordian (1963) and Northmore (1965) found that the higher quality coffee beans had a higher mean weight; weight was therefore used as a criterion in several experiments. It was found that nitrogen reduced quality by this criterion although yield was increased. The reduction was greater in the absence of other nutrients or mulch, or on low fertility soil. Irrigation improved quality; the reduction caused by nitrogen was less severe on irrigated coffee.

If the balance of other nutrients was moved away from the optimum, quality tended to decline. For example, application of magnesium to unmulched coffee reduced quality, whereas application of magnesium to mulched coffee (the mulch would provide potassium) improved quality. The application of copper and 'tonic' sprays had no effect on quality. Northmore (1965) had earlier reported that unduly high levels of potassium or calcium in the bean reduced quality. Iron deficiency on high pH soil leads to the production of 'amber beans' (Robinson, 1960). Clarke and Walker (1974) reported the potassium content of green beans from a range of sources and investigated the relationship between these data and extraction yield during soluble coffee manufacture. (See also Chapter 13; p. 307.)

Oyejola (1975) investigated the effect of NPK fertiliser on the wet and dry weights of robusta beans, pulp and parchment, and leaves. The moisture content was not changed significantly, but there were significant increases in weights following fertilisation.

Quijano-Rico and Spettel (1975) reported the concentrations of 26 elements in eight species or cultivars of coffee. The elements assayed included a number not reported elsewhere. Significant differences of some elements between species were noted. No attempt was made to relate these differences to quality of the coffee product.

The effect of some nutrients on the amount of caffeine and chlorogenic acids in robusta coffee was investigated by Chassevent, Gerwig and Bouharmont (1973). Various levels of nitrogen, phosphate, potassium, calcium and magnesium caused no significant variations, but see also Chapter 13; p. 334.

Malavolta *et al.* (1959) reported differences in the concentrations of amino acids, carbohydrates and carboxylic acids between normal and nitrogen-deficient leaves. They investigated the changes in concentration following foliar application of urea to deficient leaves; the urea was labelled with radioactive carbon so that the mechanism of production of these compounds could be studied.

Georgier and Vento (1975) investigated the effects of NPK fertilisation and shade on the carbohydrate and organic acid content of the leaves of arabica var. *caturra*. The production of sugars and organic acids was enhanced in sunlight whereas starch production was favoured in shade. Nitrogen and phosphorus both increased the production of starch and hemicellulose.

Foliar Analysis

Foliar analysis is being used increasingly for assessment of the nutritional status of coffee and as a guide to fertiliser programmes. Muller (1966) reviewed the nutrition of the coffee plant and discussed the use of foliar analysis to diagnose nutrient abnormalities. The third or fourth leaf pair is normally sampled and the results are sufficiently reproducible for production diagnosis and interpretation of some research results. Nutrient composition can vary to some extent throughout the year and be affected by variation of climatic and shade conditions. For the most accurate research, age of leaf, time of year and environmental conditions need to be specified in detail.

Numerous authors have reported leaf nutrient levels for various locations and suggested critical levels. For example Cooil (1954), Fahmy (1977) and Valencia-Aristizabal and Arcila-Pulgarin (1977). Robinson (1969) discussed the detail of a nutrition advisory service based on foliar analysis and compared published critical values for unshaded arabica coffee. Bould, Aduayi and Kimeu (1971) described the leaf analysis service for coffee growers established by the Coffee Research Foundation, Kenya. For robusta coffee, data have been presented by Loue (1960), Forestier and Beley (1966) and Verliere (1973a, b). Colonna (1964) reported an investigation into critical nutrient levels in excelsa coffee. Critical data for arabica and robusta which have been used successfully by the author are given in Table 6.2

Table 6.2: Critical Levels of Nutrients in Leaves

Arabica	N %	P %	K %	Ca %	Mg %	S %	Cl %	Fe ppm	Mn ppm	Zn ppm	Cu ppm	B ppm	Al ppm	Mo ppm
Deficient	2.00	0.10	1.50	0.40	0.10	0.10		40	25	10	3	25		0.5
Subnormal	2.60	0.15	2.10	0.75	0.25	0.15		70	50	15	7	40		0.8
Normal	3.50	0.20	2.60	1.50	0.40	0.25	0.2*	200	100	30	20	90	60	
Excess														

Robusta	N %	P %	K %	Ca %	Mg %	S %	Cl %	Fe ppm	Mn ppm	Zn ppm	Cu ppm	B ppm	Al ppm	Mo ppm
Deficient	1.80	0.10	1.20	0.40	0.20	0.12		40	20	10	13	20		0.3
Subnormal	2.70	0.13	1.80	0.80	0.30	0.18		70	35	15	20	35		0.5
Normal	3.30	0.15	2.20	1.50	0.36	0.26		200	70	30	40	90		
Excess														

Source: Author's values, based on reconciliation of published data.
*Tentative value

Deficiency and Toxicity Symptoms

Most nutrient deficiencies and toxicities produce a characteristic disorder of the leaves and stems. These are listed in Table 6.3.

Coloured photographs of leaves showing deficiency or toxicity symptoms have appeared in a number of publications. For example: Anon. (1961/63), (1975), Cibes and Samuels (1955), van Dierendonck (1959), Franco and Mendes (1954), Loue (1957), Malavolta et al. (1962), Robinson and Chenery (1958), Robinson (1964) (black and white only), Southern and Hart (1969) and Southern and Dick (1969).

Table 6.3: Visual Symptoms of Nutrient deficiency and Toxicity in Coffee. Key for the identification of mineral deficiencies in coffee. This is essentially a translation of the description given by Müller (1959) with additional information from various sources.

	I. Leaves without chlorosis
Boron deficiency	Leaves are smaller, narrow, twisted, with irregular edges and rough surface. Death of the terminal bud causes branching in a fan-like pattern. The young deficient leaves keep the symptoms even after getting old.
	II. Leaves with chlorosis
	A. More or less uniform chlorosis over the leaf blade
Manganese deficiency	1. The younger leaves, usually till the fourth pair, pale green with numerous small, yellowish points. The older leaves, when exposed to the sun, show a lemon-yellow colour.
Nitrogen deficiency	2. The older leaves first lose their green colour, becoming sometimes yellowish, almost white. The shaded portions of the leaves are greener. The leaves fall and the branches start to wither from tip to bottom (die-back).
Sulphur deficiency	3. The younger leaves show a yellowish-green colour. The chlorosis starts as a band which comprises the main vein and extends towards the middle of the blade. The lower surface of the leaf is much lighter in colour than the upper one. Some mottling is apparent near the margins.
Copper toxicity	4. Slight chlorosis leading to defoliation.
	B. Non uniform chlorosis, with different patterns
	(i) Leaves are smaller in size
Zinc deficiency	(a) The veins form a green network against the pale green or yellowish tissues. In acute cases the leaf is almost whitish-yellow, shaped as a spear-head, easy to break. Shortening of the internodes. Rosette, die back of the branches is frequent.
Low temperature	(b) Severely affected leaves are reduced in size. Irregular white or yellow chlorosis, usually starting at leaf margins. Mottling may develop. Leaves may scorch and fall, excessive branching may occur and shoot tips blacken, distort and shrivel.
	(ii) Leaves are normal in size
	(a) Green veins on a chlorotic background mottling
Iron deficiency	1. Young leaves pale green, yellowish or almost white. The veins, including the very fine ones keep their green colour, except in very acute cases (whitish leaves).

Table 6.3 continued

Manganese deficiency	2. Main and secondary veins, including a narrow (1-3mm) band on both sides of them are green, producing a coarse mottled appearance. The background is pale green or yellow. When the deficiency is severe, the number of green veins is small. The symptoms appear in young leaves.
Boron toxicity	3. General chlorosis with scattered necrotic spots.
	(b) Chlorosis begins at the tip or margins of the leaf
Nitrogen deficiency	1. Young leaves show yellow band, 2-4mm wide (or more) around the edges. The rest of the leaf blade is brown or has a brown tint. Leaves show a tendency to curl along the principal vein taking a 'V' shape. In the unshaded leaves the chlorosis is persistent.
Calcium deficiency	2. In the young leaves the chlorosis appears on the margins, in an irregular shape, leaving area with saw-like edges in both sides of the main veins; the leaf blade shows a convex shape. In severe cases, the chlorosis begins at the tip of the older leaves. Frequently the formation of cork in the larger veins, especially on the lower surface of the leaves, is observed, Sometimes death of the bud occurs (die-back).
Aluminium toxicity	3. Marginal necrosis of younger leaves leading to leaf curling because of faster growth of leaf centre.
	(c) Chlorosis between the main lateral veins
Magnesium deficiency	1. The yellowing begins near the mid-rib and proceeds inside the area between the main lateral veins. A 3-5mm wide band remains green on both sides of the mid-rib (which is yellow) and of the secondary veins. The chlorotic areas are orange-yellow or brown, and in advanced cases they cover nearly all of the leaf blade. The affected areas lose their shiny appearance. The older leaves, when deficient, fall off easily. The symptoms are easier to find in leaves near the berries.
Biuret toxicity	2. Chlorosis of marginal interveinal regions and outward curling of leaves ('cupping'). (d) Chlorosis in the form of irregular spots.
Phosphorus deficiency	1. Leaves are mottled, with yellow spots showing a reddish or dark red tint. In case of serious deficiency the entire leaf is chlorotic. The older leaves, which first show the deficiency, soon fall off.
Aluminium toxicity	2. Marginal chlorosis and small necrotic spots on older leaves, starting at leaf apex and progressing to the centre.
	III. Leaves with necrotic areas
	A. Tip and margins
Potassium deficiency or sodium toxicity	The older leaves appear scorched at their tips and margins. The affected leaves fall off so that only a few remain attached to the branches. Die-back of the branches is the last stage.
	B. Margins
Copper deficiency	The younger leaves are distorted, having an S-shape for lack of growth of the veins. They lose their green colour. Necrotic patches, rather large, appear on the margins.
	C. Near the margins
Molybdenum deficiency	First, yellow spots develop near the margins. They become necrotic, in the centre, first. Downward curling of the leaf blade occurs from the mid-rib, so that the opposite edges touch each other underneath. Subterminal leaves are affected first.

References

Accorsi, W.R. and Haag, H.P. (1959) 'Alteracoes morfologicas e citologicas do cafeeiro (*Coffea arabica* L., var. *Bourbon (B. Rodr.) Choussy*) cultivado em solucao nutritiva, decorrentes des deficiencias e excessos dos macronutrientes', *Revista Cafe Portuguesas*, 6, 5-19

Acland, J.D. (1971) *East African Crops*, Longmans/FAO, Rome

Aduayi, E.A. (1977) 'Relationship between varying levels of copper and soil pH on the growth and mineral composition of arabica coffee plants', *Turrialba*, 27, 7-16

Aduayi, E.A. (1978) 'Role of phosphorus and copper in growth and nutrient composition of coffee grown in sand culture', *Turrialba*, 28, 105-11

Ananth, B.R., Iyengar, B.R.V. and Chokkana, N.G. (1965) Widespread zinc deficiency in coffee in India', *Turrialba*, 15, 81-7

Anon. (1957) 'Deficiencias visuales en elementos menores', *El Cafe de Nicaragua*, 12, 22-6

Anon. (1961/63) *An Atlas of Coffee Pests and Diseases; Illustrations of the Common Insect Pests, Diseases and Deficiency Syndromes of Coffea arabica in Kenya*, Coffee Board of Kenya, Nairobi

Anon. (1969) 'Caution in use of zinc foliar sprays and herbicides', Information Circular No. 4. Department of Agriculture, Stock and Fisheries, Papua New Guinea

Anon. (1975) *Coffee Handbook*, Rhodesia Coffee Growers Association, Salisbury, Southern Rhodesia

Arana, M.L. (1967) 'Fertilizacion con cloruro y sulfato de potassio en plantaciones de cafe. Verificacion de la absortion de iones K, Cl y S por medro de analisis foliares', *Cenicafe*, 18, 47-54

Aranguren, J., Escalante, G. and Herrera, R. (1982) 'Nitrogen cycle of tropical perennial crops under shade trees. 1. Coffee', *Plant and Soil*, 67, 247-58

Bornemisza, E. (1982) 'Nitrogen cycling in coffee plantations', *Plant and Soil*, 67, 241-6

Bould, C., Aduayi, E.A. and Kimeu, B.S. (1971) 'A leaf analysis service for coffee-growers', *Kenya Coffee*, 36, 37-9

de Camargo, T. (1937) 'Physiologia vegetal influencia da relacao potassio-azoto sobre a desenvolvimiento do cafeeiro durante o primeiro periodo de vegetacao', *Boletin Technico Instituto Agronomia Campinas*, 5

de Castro, F.S. (1960) 'Distribucion de las raices del cafeto (*Coffea arabica* L.) en un suelo de El Salvador', *El Cafe de El Salvador*, 30, 421-49

de Castro, F.S. and Rodriguez, A. (1955) 'Nutrient losses by erosion as affected by different plant covers and soil conservation practices', *National Coffee Growers Association of Colombia, Technical Bulletin*, 2, 1-24

Catani, R.A. and de Moraes, F.R.P. (1958) 'A composico quimica do cafeeiro. Quantidade e distribucao de N, P_2O_5, CaO e MgO em Cafeeiro de la 5 anos de idade', *Revista de Agricultura, Brazil* 33, 45-52

Catani, R.A., Pellegrino, D., Bittencourt, V.C., Jacintho, A.O. and Graner, C.A.F. (1967) 'A concentracao e a quandidade de micronutrientes e de aluminio no caffeeiro, *Coffee arabica* L. variedade *Mundo Novo* (B. Rodr.) Choussy, aoz des anos de idade', *Anals Escola Superior de Agricultura Luiz de Queiroz*, XXIV, 97-106

Chanchay, C.A.G. (1964) *Efecto de Tres Fuentes de Magnesio Aplicadas al Suelo y a las Hojas de Cafetos Sobre la Concentracion Foliar de este Elemento*. Unpublished Ph.D. thesis, Inter-American Institute of Agricultural Sciences, Turrialba, Costa Rica. Quoted by Muller (1966)

Chassevent, F., Gerwig, S. and Bouharmont, M. (1973) Influence eventuelle de diverses fumures sur les teneurs en acides chlorogeniques et en caffeine de grains de cafeirs cultives, in *6th international Colloquium on the Chemistry of Coffee*, ASIC, Paris, 57-60

Chawdhry, M.A. (1974) in *Annual Report, Coffee Research Foundation, Kenya, 1973/74*, 34-5

Cibes, H. and Samuels, G. (1955) Mineral Deficiency Symptoms Displayed by Coffee Trees Grown under Controlled Conditions, *Agriculture Experiment Station Technical Paper No. 14*, University of Puerto Rico

Clarke, R.J. and Walker, L.J. (1974) 'Potassium and other mineral contents of green, roasted and instant coffees', *Journal of the Science of Food and Agriculture*, 25, 1389-404

Colonna, J.P. (1964) 'Contribution a l'etude pratique du diagnostic foliaire du cafeier Excelsa' *Café, Cacao, Thé, VIII*, 264-74

Cooil, B.J. (1954) *Leaf Composition in Relation to Growth and Yield of Coffee in Kona*, Hawaii Coffee Information Exchange

Cuenca, G., Aranguren, J. and Herrera, R. (1983) 'Root growth and litter decomposition in a coffee plantation under shade trees', *Plant and Soil, 71*, 477-86

van Dierendonck, F.J.E. (1959) *The Manuring of Coffee, Cocoa, Tea and Tobacco*, Centre d'Etude de L'azote, Geneva

Espinosa, F.M. (1960) 'El analisis foliar en el diagnostico del estado nutricional del cafeto', *Instituto Salvadoreno de Investigaciones del Cafe. Boletin Information Suplemento No. 2* Santa Tecla, El Salvador

Fahmy, F.N. (1977) 'Soil and leaf analyses in relation to the nutrition of tree crops in Papua New Guinea' in *Proceedings of International Soil Conference. Classification and Management of Tropical Soils*, Kuala Lumpur, Malaysia, pp. 309-18

Fenner, R.J. and Adamson, E. (1982) 'The response of coffee to the foliar application of zinc oxide', *Zimbabwe Agricultural Journal, 79*, 131-2

Fiester, D.R. (1961) 'Nutricion del cafeto', *Revista Cafetelera (Guatemala), Oct. Nov. Dec.*, 9-15

Forestier, J. (1969) *Culture du Cafeier Robusta en Afrique Centrale*, Institut Francais du Cafe, du Cacao et autres Plantes Stimulantes, Paris

Forestier, J. and Beley, J. (1966) 'Teneurs en soufre et en oligo-elements des feuilles du cafeier robusta en Lobaye (Republique Centrafricaine)', *Café, Cacao, Thé, X*, 17-27

Franco, C.M. and Mendes, H.C. (1949) 'Sintomas de deficiencias minerais no cafeeiro', *Bragantia, 9*, 165-73

Franco, C.M. and Mendes, H.C. (1954) 'Deficiencia de zinco em cafeeiro', *Boletin da Superintendencia dos Servicos do Cafe, 29*, 34-9

Furlani, A.M., Catani, R.A., de Moraes, F.R.P. and Franco, C.M. (1976) 'Efectos de aplicacao de cloreto e de sulfato de potassio na nutricao de cafeeiro', *Bragantia, 35*, 349-62

Georgier, G.H. and Vento, H. (1975) 'Effect of light intensity and mineral nutrition on carbohydrate and organic acid content in leaves of young coffee plants', *Fiziologiya na Rasteniyata, 1*, 41-51

Gonzalez, M.A., Lopez, C.A., Carvajal, J.F. and Briceno, J.A. (1977) 'Efecto de la fuente de potassio en el acumulamiento de cloruros y sulfatos en el cafeto', *Agronomia Costa Rica, 1*, 31-7

Gordian, M.R. (1963) *Green and Roasted Coffee Tests*, Gordian Publishing House, Hamburg

Haag, H.P. and Malavolta, E. (1960) 'Estudos sobre a alimentacao do cafeeiro IV. Efeito dos excessos de macronutrientes mo crescimento e na composicao quimica do cafeeiro (*Coffea arabica* L. var. Bourbon) cultivado em solucao nutritiva', *Revista Cafe Portugueses, 7*, 5-12

Hernandez, E. (1962) 'El magnesio aumenta los rendimientos del cafe' *Agricultura al Dia (Puerto Rico), 36-7*

Hutchinson, T.H. (1968) 'The methane system of coffee culture' in P.A. Huxley (ed.), *Proceedings of a Seminar on Intensification of Coffee Growing in Kenya*, Coffee Research Foundation, Nairobi, 85-90

Kabaara, A.M. (1976) in *Annual Report, Coffee Research Foundation, Kenya, 1975/76*, pp. 59-60

Loue, A. (1957) *La nutrition minerale du Cafeier en Cote d'Ivoire*, Gouvernment General de l'A.O.F. Centre de Recherches Agronomiques, Bingerville, Cote d'Ivoire

Loue, A. (1960) 'Nouvelles observations sur les oligoelementes dans la nutrition du cafeier', *Café Cacao Thé, IV*, 133-49

Malavolta, E., Graner, E.A., Sarruge, J.R. and Gomes, L. (1963) 'Estudos sobre a alimentacao mineral do cafeeiro. XI. Extracao de macro e micro nutrientes, na colheita, pelas variedades '*Bourbon Amarelo*', '*Caturra Amarelo*' e '*Mundo Novo*'. *Turrialba, 13*, 188-9

Malavolta, E., Haag, H.P. and Johnson, C.M. (1961) 'Estudos sobre a alimentacao mineral do cafeeiro. VI. Efeitos das deficiencias de micronutrientes em *Coffea arabica* L. var. *Mundo Novo* cultivado em solucao nutritiva'. *Anais Escola Superior de Agricultura Luis de Queiroz, 18*, 147-67

Malavolta, E., Haag, H.P., Mollo, F.A.F. and Brasil Sobro, M.O.C. (1962) '3. Coffee' in *On the Mineral Nutrition of some Tropical Crops*, International Potash Institute, Berne, pp. 43-67, 150-1

Malavolta, E., Menard, L.N., Arzolla, J.D.P., Crocomo, O.J., Haag, P.J. and Lott, W.L. (1959) 'Tracer studies in the coffee plant (*Coffea arabica* L.)', *Anais. Escola Superior de Agricultura Luis de Queiroz*, 16, 65-78

Malavolta, E., Pimentel Gomes, F. and Coury, T. (1958) 'Estudos sobre a alimentacao mineral do cafeeiro (*Coffea arabica* L. variedade *Bourbon Vermellio*) 1. Resultados preliminares' *Boletin Superior Servizio Cafe, San Paulo*, 33, 10-24

Medcalf, J.C. and Lott, W.L. (1956) 'Metal chelates in coffee', *Research Institute Bulletin No. 11*. International Basic Economic Co-operation, New York

Mehlich, A. (1965) 'Soil analyses', *Annual Report, Coffee Research Foundation, Kenya 1964/65*, 46-52

Mehlich, A. (1967) 'Mineral nutrition in relation to yield and quality of Kenya coffee. 1. Effect of nitrogen fertilisers, mulch and other materials on yield and grade 'A' quality of coffee'. *Kenya Coffee*, 32, 399-407

M'Itungo, A.M. and van der Vossen, H.A.M. (1981) 'Nutrient requirements of coffee seedlings in polybag nurseries: the effect of foliar feeds in relation to type of potting mixture', *Kenya Coffee*, 46, 181-7

Montoya, L.A., Sylvain, P.G. and Umana, R. (1961) 'Effect of light intensity and nitrogen fertilisation upon growth differentiation balance in *Coffea arabica* L.' *Coffee (Turrialba)*, 3, 97-104

Montoya, L.A. and Umana, R. (1961) 'Effect of three light intensities and three levels of nitrogen (urea) on incidence of dieback' *Coffee (Turrialba)*, 3, 1-3

Morillo, R.A., Nieto, E.P., Linares, J.A. and Roman, G.N.C. (1978) 'Efecto de la fertilizacion foliar y al suelo con cloruro de potassio a plantas de cafe en vivero' *Agronomia Tropical*, XXIX, 319-25

Müller, L.E. (1959) *Algunas deficiencias minerales communes en el cafeto (Coffea arabica L.) Boletin Tecnico No. 4*, Instituto Interamericano de Ciencias Agricolas, Turrialba, Costa Rica

Müeller, L.E. (1966) 'Coffee nutrition' in *Temperate to Tropical Fruit Nutrition*, Rutgers State University, pp. 685-774

Northmore, J.M. (1965) 'Some factors affecting the quality of Kenya coffee', *Turrialba*, 15, 184-93

Nutman, F.J. (1933) 'The root system of *Coffea arabica*', *Empire Journal of Experimental Agriculture*, 1, 271-95

Oruko, B.A. (1977) 'Yield responses of arabica coffee to fertilisers in Kenya', *Kenya Coffee*, 42, 227-39

Oyejola, B.O. (1975) 'The effect of NPK fertiliser on the quality of robusta coffee (*Coffea canephora*) berries during their development stages', *Turrialba*, 25, 67-71

Pavan, M.A., Bingham, F.T. and Pratt, P.F. (1982) 'Toxicity of aluminium to coffee in ultisols and oxisols amended with $CaCO_3$, $MgCO_3$ and $CaSO_4 2H_2O$', *Soil Science Society of America*, 46, 1201-6

Pereira, H.C. and Jones, P.A. (1954) 'Field responses by Kenya coffee to fertilisers, manures and mulches', *Empire Journal of Experimental Agriculture*, 22, 23-36

Perez, S.V.M. (1957) *Algunas deficiencias minerales del cafeto en Costa Rica. Informacion Tecnico. No. 2*. Ministerio de Agricultura e Industrias — STICA. San Jose, Costa Rica

Perez, S.V.M. (1958) *Labores del Projecto No. 23 en el mejoramiento del cultivo del cafe. Informacion Tecnico No. 4*. Ministerio de Agricultura e Industrias — STICA. San Jose, Costa Rica

Perez, S.V.M., Chaverri, R.G. and Bornemisza, S.E. (1956) *Algunas aspectos del aboniamento del cafeto con boro y calcio en las condiciones de la Meseta Central de Costa Rica. Information Tecnico No. 1* Ministerio de Agricultura e Industrias STICA. San Jose, Costa Rica

Quijano-Rico, M. and Spettel, B. (1975) 'Determinacion del contenido en varios elementos en muestras de cafes de diferentes variedades', in *7th International Colloquium on the Chemistry of Coffee* ASIC, Paris, pp. 165-74

Ripperton, J.E., Goto, Y.B. and Pahau, R.K. (1935) 'Coffee cultural practices in the Kona district of Hawaii', *Bulletin of Hawaii Agriculture Experiment Station. No. 75*

Robinson, J.B.D. (1959) 'Iron deficiency control. The control of chronic iron deficiency in coffee (*Coffea arabica* L.) growing on localised native hut or boma sites', *Kenya Coffee*, 24, 16-19

Robinson, J.B.D. (1960) 'Amber beans' *Kenya Coffee*, 25, 91-5

Robinson, J.B.D. (1961) 'Mineral nutrition of coffee. Preliminary results with the leaf analysis technique' *East African Agriculture and Forestry Journal*, 27, 1-9

Robinson, J.B.D. (1964) *A Handbook on Arabica Coffee in Tanganyika*, Tanganyika Coffee Board, Moshi, Tanganyika

Robinson, J.B.D. (1969) 'Defining and improving the mineral nutrition of bearing arabica coffee' in Huxley, P.A. (ed.) *Proceedings of a Seminar on the Intensification of Coffee Growing in Kenya*, Coffee Research Foundation, Nairobi, 75-84

Robinson, J.B.D. and Chenery, E.M. (1958) 'Magnesium deficiency in coffee with special reference to mulching', *Empire Journal of Experimental Agriculture*, 26, 259-73

Robinson, J.B.D. and Hosegood, P.A. (1963) 'Effects of organic mulch on fertility of a latosolic coffee soil in Kenya.' *Experimental Agriculture*, 1, 67-80

Roelofsen, P.A. and Coolhaas, C. (1940) 'Waarnemingen over de periodiciteit in de chemische samenstelling van de takken van de produceerenden koffieboom en over de samenstelling van den produceerden oogst', *Archief der Koffiecultur in de Nederlands-Indies*, 14, 133-58

Roskoski, J. (1980) 'The importance of fixation in the nitrogen economy of coffee plantations' in *1st Symposium on Ecological Studies in the Coffee Agrosystem*, National Institute of Biological Resources, Xalapa, Mexico

Roskoski, J. (1982) 'Nitrogen fixation in a Mexican coffee plantation', *Plant and Soil*, 67, 283-91

Snoeck, J. (1981) 'Facteurs du rendement influences par les apports d'azote chez le cafeier robusta en Cote d'Ivoire', *Café, Cacao, Thé*, XXV, 173-80

Snoeck, J. and de Reffye, Ph. (1980) 'Influence des engrais sur l'architecture et la croissance du cafeier robusta', *Café, Cacao, Thé*, XXIV, 259-66

Southern, P.J. and Dick, K. (1969) 'Trace element deficiencies in tropical tree crops in Papua and New Guinea', *Research Bulletin No. 7*. Department of Agriculture, Stock and Fisheries, Port Moresby, Papua New Guinea

Southern, P.J. and Hart, G. (1969) 'Nutritional studies of coffee in the Territory of Papua and New Guinea', *Research Bulletin No. 1, Crop Production Series*. Department of Agriculture, Stock and Fisheries, Port Moresby, Papua New Guinea

Triana, J.V. (1957) 'Informe preliminar sobre un estudio de "Modalides del cultivo del cafeto"', *Cenicafe*, 8, 156-68

Uribe-Henao, A. and Salazar-Arias, N. (1981) 'Efecto de los elementos menores en la produccion de cafe', *Cenicafe*, 32, 122-42

Vaidyanathan, L.V., Cokkana, N.G. and Narayanan, B.T. (1956) 'Leaf symptoms of biuret toxicity in coffee', *Indian Journal of Horticulture*, XIII, 163-4

Valencia-Aristizabal, G. and Arcila-Pulgarin, J. (1977) 'Efecto de la fertilizacion con N, P, K a tres niveles en la composicion mineral de las hojas del cafeto', *Cenicafe*, 28, 119-38

Valencia-Aristizabal, G. and Bravo-Grijalba, E. de J. (1981) 'Influencia del encalamento en la produccion de cafetales establecidos' *Cenicafe*, 32, 3-14

Vasudeva, N. and Ratageri, M.C. (1980) 'Upward translocation and deposition of copper sulphate in arabica coffee seedlings', *Journal of Coffee Research*, 10, 36-8

Verliere, G. (1973a) 'La nutrition minerale et la fertilisation du cafeier sur sol schisteux en Cote d'Ivoire. I. Etude de la nutrition minerale' *Café, Cacao, Thé*, XVII, 97-124

Verliere, G. (1973b) 'La nutrition minerale et la fertilisation du cafeier sur sol schisteux en Cote d'Ivoire. II. Influence de la fertilisation minerale sur les rendements. III. Relation entre les rendements et la composition minerale des fruits', *Café, Cacao, Thé*, XVII, 211-22

Willson, K.C. (1978) 'Nutrient recycling in tree crops' in *Plant Nutrition, 1978, 8th International Colloquium on Plant Analysis and Fertiliser Problems, Auckland, New Zealand*. New Zealand Department of Scientific and Industrial Research. Information Series no. 134, Wellington

Willson, K.C. (1984) Verbal communication from staff of the Malaysian Agricultural Research and Development Institute (MARDI), Kuala Lumpur

7 CULTURAL METHODS

K.C. Willson

Choice of Cultivar

The location of a new coffee development will determine which of the two most important species, arabica and robusta, will be planted. There will be a few sites in marginal areas where both could be planted, perhaps with arabica at the higher elevations and robusta lower down. In areas suitable for robusta, liberica and excelsa are possible alternatives but the market for these is limited. Van der Vossen discusses in Chapter 3 the populations of coffee.

Within arabica there are many cultivars. Some of these are universally known. Others are local selections. Two cultivars which have been recognised for many years are bourbon and typica. These were listed by Chenney (1925) as distinct varieties: *Coffea arabica* var. *bourbon* (B. Rodr.) Choussy and *Coffea arabica* var. *arabica* (syn. var. *typica* Cramer). All other cultivars in widespread use are believed to have derived from these. They can be distinguished visually: bourbon primaries grow initially upwards at 45° to the stem and young leaves are green; typica primaries are initially almost horizontal and the young leaves have a bronze colouration. In identical conditions one of these varieties may yield more crop than the other. Triana (1957) showed that bourbon yielded more than typica and its yield in unshaded conditions was proportionally higher than under shade compared to bourbon, but typica gave a greater percentage yield response to applied fertiliser. Such differences may not apply in other locations.

Some cultivars are known by different names in different countries. For example, typica is known as nyasa in Malawi and Uganda. Other cultivars are named from their place of origin and are usually either a hybrid of bourbon and typica or a mutant from an established cultivar. Some were discovered by chance, for example kent (from Mr Kent's estate in India) and maragogipe (from the town of Maragogipe in Brazil). Many have been found during planned selection of good trees and are often identified only by numbers.

In addition to the cultivars of arabica which grow to a 'normal' size there is at least one, maragogipe, which has a larger habit of growth and produces very large beans. Such reports as there are of its yield indicate that several normal cultivars are more productive.

There are a number of dwarf cultivars, some of which have been planted widely. Caturra is perhaps the best known; others include san ramon, mokka and san bernardino. Caturra has been included in several experimental investigations. Spacing experiments are discussed on page 167. Kiara (1981) reported that caturra was very prone to biennial bearing under Kenya conditions. The quality of the coffee produced, as measured by the size of the beans, was lower than that from a normal-sized cultivar.

Within robusta there are two main types which have been classified as botanical varieties. *Coffea canephora* var. *nganda* has a spreading habit. It is indigenous to Uganda, and grown extensively there but rarely grown elsewhere. The similar kouilou (or quillou) type has spread more widely, including Brazil. *Coffea canephora* var. *canephora* has a more upright habit. Robusta is very variable because it is self-sterile. Many species names have been proposed for groups of trees growing in various locations but it is most probable that the minor differences represent varying degrees of crossing between the two varieties and local selection. When looking for robusta for propagation as seedlings the best that can be achieved is to obtain seed from a population known to produce good progeny.

Liberica and excelsa are also self-sterile, so numerous forms are known. Breeding programmes are increasing the already large number of hybrids available. Van der Vossen in Chapter 3 discusses hybridisation. Hybrids must be propagated vegetatively.

The choice of planting material for a particular location may be limited by availability. When the choice is made account must be taken of the method of propagation which is necessary. Other cultural variables, particularly spacing, will depend on the cultivar. Snoeck (1963) showed that some clones of robusta branched more freely than others; this will have a bearing on the method employed to establish multiple-stem trees.

Propagation

A variety of propagation methods can be used satisfactorily on coffee. The choice of method in a particular situation will depend on the species, cultivar, material available and the purpose for which the coffee is being planted.

Coffea arabica is predominantly self-pollinated. The progeny from seed is therefore very uniform and a good variety can be multiplied by seed and remain true. Careful selection of the best and most uniform material in the nursery should reject most plants which are not true to type. Propagation as seedlings is therefore adequate for most plantings for crop production.

Coffea canephora, Coffea excelsa and *Coffea liberica* are largely self-sterile. Seed is therefore cross-pollinated and the progeny very variable. Vegetative propagation is therefore of much greater interest to growers of these species. Hybrids must be propagated vegetatively.

Advice and instruction on the preparation and management of nurseries has been published in most coffee-growing countries. The advice is normally related to local conditions and may therefore require modification before being applied elsewhere. Useful examples of such advice are in Forestier (1969: 34-37) (for robusta) and Robinson (1964: 22-30) (for arabica).

Location of Nurseries

The chosen site needs to be convenient for the area to be planted, to minimise transport costs. It should be reasonably level but with no risk of waterlogging in the heaviest rain. The soil should be chemically suitable with a good structure if it is to be dug to provide the beds. A deep topsoil ensures a good depth of fertile soil in the beds. If plants are to be raised in containers a source of soil suitable for the type of propagating material needs to be conveniently placed.

A supply of water is necessary. Choosing a site under large trees which provide a light shade avoids the need to provide shade, but young plants which have not been hardened off by exposure to full sunlight for the last few months in the nursery suffer shock on field planting and the success rate is likely to be lower than with plants which have been exposed.

Production of Seedlings

Seed. Seed loses its viability quickly unless stored under special conditions; this is discussed by Van der Vossen in Chapter 3. Seed should be collected from selected trees, which are known to give a high yield of large seeds giving good quality coffee. Van der Vossen discusses the stability of the inheritance of various characteristics, and the selection of good material.

Cherries from the selected trees should be pulped, fermented and washed, but not dried. Van der Vossen (1979) showed that the germinative energy was better preserved in seed with a moisture content of 41 per cent. However, dry seed can be preserved at a temperature of 15°C. Plantations raising seedlings from their own seed are best advised to use wet seeds immediately after washing. The seeds should be carefully inspected after washing and all small, misshapen and damaged seeds rejected.

Germination. The planter has the choice of germinating the seed in a bed or other situation reserved for this operation, or planting seed directly into the bed or container in which the plants will grow until planted out.

A germinating bed is prepared by digging the soil and producing a fine tilth at the surface. Seeds are planted closely, 2 cm by 2 cm square, 1 to 1.5 cm deep. It does not matter whether the seeds have the flat or convex surface upwards, as the hypocotyl lifts the cotyledons out of the soil. The bed must be kept moist but not waterlogged. In dry periods a grass mulch helps to retain moisture. During long cloudless periods a light shade is beneficial but is unnecessary if not harmful in overcast weather. The best time to lift the seedlings is when the two cotyledon leaves have grown to their full extent.

Seeds can be germinated away from soil; for example by spreading them on sacking or a black plastic sheet and covering with sacking. The sacking must not be allowed to dry out but seeds must not stand in water. Seeds can be mixed with damp vermiculite (3 parts seed to 1 part vermiculite) and placed in a sealed plastic bag, which is stored at room temperature in the shade. Seeds should be inspected regularly from 10 to 14 days later, germination can be expected at around three weeks from mixing (Robinson, 1964: 23). Seeds should be planted in their final nursery location as soon as the hypocotyl has emerged. This requires them to be inspected at least once each day, germinated seeds being removed immediately.

Separate germination enables inferior material to be discarded at an early stage, and seeds which do not germinate do not occupy valuable space in the main nursery. Weed control problems are reduced slightly, but a greater labour input is required.

Seedling Nurseries. Seedlings can be raised in beds, or in separate containers. Beds should be well-dug to a depth of at least one metre. It is preferable for the beds to be raised above the original soil level to avoid waterlogging. For convenience, beds should not be more than one metre wide, with a pathway at the original soil level between each bed. It is often beneficial to incorporate some compost, farmyard or other organic manure, or phosphate fertiliser, into the bed, which should preferably be made of a good topsoil. Mestre-mestre (1977) reported the benefit of mixing coffee pulp into a potting mixture. Other fertilisers may be incorporated if analysis shows a specific need. If cut-worm is a problem treat the beds with an appropriate insecticide before planting. Weed control is assisted if a crop of weeds is allowed to grow on the beds, and then removed, before planting the coffee seed.

Ungerminated seeds, or those which have been pre-germinated, are placed in the beds, or in individual containers. Spacing in the beds depends on the species grown and the size at which seedlings are normally transplanted to the field: for arabica, 12.5 cm × 15 cm rectangular (Robinson, 1964: 22), 20-25 cm triangular (Wellman, 1961: 158), 23 cm × 15 cm (Haarer, 1962: 116); for robusta, 23 cm × 23 cm (Haarer, 1962: 116) 15 cm × 15 cm (Forestier, 1969: 37); for liberica and excelsa, 30 cm × 30 cm (Haarer, 1962: 116).

Containers can be woven from a variety of fibrous materials but polythene tubes are now most commonly used, 9 cm diameter by 35 cm deep, or material close to these measurements, is a suitable size. Containers should be filled with soil of the same quality as would be used in a bed.

Containers should be held upright in 'beds' by wires attached to short posts which go around the bed.

Adjustable shade is necessary over each bed. This can be provided at a low level, not more than one metre above bed soil level, by woven panels, loose poles or roller-blinds made from stems about 2 cm in diameter. These rest on a frame each side of the bed and are easily removed when required. Shade can be constructed at a higher level, about 2 metres above ground level, over the whole nursery area. Alteration of the level of shade is not so easy with this method.

Vegetative Propagation

Vegetative propagation was first used in Java from 1887 (Cramer, 1934). This work was initiated as a means to reduce the crop losses from nematodes, primarily *Tylenchus coffeae*, which attacked the roots of arabica. At a later stage the aim was to improve the overall yield by reducing the variability between individual trees. Trees of known high yield were propagated. Grafting was the method used almost invariably. Experiments included trials of rootstocks, selected trees of several species and some natural hybrids.

The experimental work in Java was reported in some detail by Ferwerda (1934). It was found that the polarity of coffee was unchanged by grafting; grafts from plagiotropic stems produced low spreading trees. Such trees would occasionally produce an orthotropic stem. Grafts had to be made from orthotropic stems to produce upright trees. Also, monoclonal plantings gave a low yield due to partial self-sterility; mixed clonal plantings were more productive. The problems which are to be expected with this method of propagation, such as the varying degree of success from one clone to another and occasional incompatibility of scion and rootstock, were investigated.

Shield budding was used successfully in the Philippines (Wester, 1917). Hardwood cuttings were tried in the Ivory Coast but the success rate was low (Porteres, 1935). A high success rate was achieved in Tanganyika with softwood cuttings (Fernie, 1940) using an effective propagator devised in Trinidad.

Cuttings. These are preferably taken from young orthotropic stems. Forestier (1969: 37) recommends the fourth to sixth leaf-pairs as most suitable, they should be taken when the stem is growing actively and the stem is still green at the cuttings. Fernie (1940) recommended retention of the younger leaves and tips. Multi-node and single-node cuttings have been used successfully. A high proportion of successful rooting can be obtained using single-node cuttings, which are cut longitudinally to give single-leaf cuttings. Leaves are cut to reduce their area. Rooting is improved by using a hormone (Feilden and Garner, 1940; Forestier, 1969: 37).

Cuttings should be planted in a suitable medium, for example decomposed sawdust (Forestier, 1969: 37), equal parts sand and peat (Fernie, 1940). After rooting, which will take up to three months, they are transferred to polythene bags containing a good forest soil (Forestier 1969: 37) or a rich compost with white sand (Fernie, 1940).

Cuttings must be kept cool and moist during the rooting phase. Shade is essential, with frequent watering or a misting installation. Sealing the bed under a polythene sheet maintains a humid atmosphere without frequent watering. Techniques of rooting cuttings have been discussed and described by Fernie (1940) for arabica, Capot (1966) for robusta and Dublin (1964) for excelsa.

Budding and Grafting. Experimental work has been carried out on both methods of uniting coffee plants. In Java (Cramer, 1934) and Tanganyika (Gilbert, 1936) it was found that grafting was more successful. Budding was successful in the Philippines (Webster, 1917). Gilbert (1936) reported that 50-90 per cent of buddings were successful in Tanganyika irrespective of the time of year. In Costa Rica 95 per cent were successful in the dry season; a much lower proportion succeeded in the wet season (Ulate, 1946).

Marshall (1936) reported work on grafting and budding in Tanganyika. A success rate of 92 per cent was achieved from bark and side grafting. Budding was equally successful but the union was much weaker so that wind would break the tree back to the original union. Callusing was slow.

Buds should be taken from young wood, and inserted using the T or inverted T method. After two months it is usually possible to tell if the bud has taken, and shoot growth will start one or two months later. Wester (1917) and Feilden and Garner (1940) discussed the practical technique of budding.

Coffee has been grafted successfully by cleft, saddle, side, notch and rind methods. Grafting techniques have been discussed by Wester (1922) Ferwerda (1934) and Feilden and Garner (1940). Grafting onto seedlings has been described by Reyne (1966) at the two-leaf stage and by Velasco and Rodriguez (1974) at the cotyledon stage.

Problems arise from incompatibility of rootstock and scion. The relative suitability of rootstock and scion must be established before a large-scale programme of propagation is instituted.

Maintenance of Nurseries

The moisture status of the beds must be maintained. Shading reduces moisture loss, but regular watering is essential in dry weather. The water requirement will increase as plants grow more leaves and the shade is reduced.

Shade should be reduced in stages as the trees grow; the plants should have been in full sunlight for several weeks before removal for field planting.

Mulch can be re-applied if necessary, between the rows, before the plants become too large (two or three pairs of true leaves).

Application of fertilisers will increase the rate of growth. When the soil of the bed is fertile, nitrogen alone will have a significant effect. Phosphate is also effective in most situations. Other nutrients will only be useful where a specific need has been established by analysis and experiment. Foliar application of fertilisers is more convenient and effective. Small quantities can be evenly applied at frequent intervals and the risk of damaging young plants by excessive quantities of fertilisers is removed. M'Itungo and Van der Vossen (1981) showed that fertiliser application gave no benefit when the potting mixture contained farmyard manure and superphosphate. However, foliar fertilisation improved the growth of seedlings in less ideal soil, or without farmyard manure.

Weeds can be a problem when the plants are small; as the plants grow they shade the soil which discourages most weeds. Application of a pre-emergent herbicide, for example simazine will usually check weed growth until the coffee plants shade the soil completely.

Applications of insecticides and fungicides will depend on the local pests and diseases which create problems. It is common practice to spray a fungicide regularly, every two weeks is a common frequency. Two fungicides are often used alternately. If one is a copper preparation, the 'tonic' effect is beneficial by increasing leaf retention.

Root pruning in beds was advised in Tanganyika (Robinson, 1964: 24). It stimulates production of a dense clump of roots around the stem, the plants are easier to lift and root damage is minimised. Plants in containers are automatically restrained from spreading laterally but the tap root will anchor the plant and container if allowed to grow unchecked. Moving each container a short distance at appropriate intervals breaks the tap root whilst it is weak and prevents anchorage.

Rotation of Nurseries

It is inadvisable to use a nursery for more than two consecutive years, three at the extreme limit, due to exhaustion of the soil and build-up of pests and diseases.

Nurseries can be rehabilitated by planting a green manure, preferably a legume. Robinson (1964: 29) suggests pigeon pea (*Cajanus cajan*) or *Glycine javanica*, but any easily available legume which grows well in the local conditions will do. *Cajanus cajan* is host to some of the nematodes which attack coffee, and is therefore not suitable where there is a nematode problem. Two years of green manure, which is then ploughed in, should be sufficient to return soil to a condition suitable for a nursery.

Land Clearance and Preparation

The method of clearing land and the amount of work involved depend on the initial vegetation. When planting in an area of forest, useful timber trees are usually selectively removed as a first step. Remaining large trees are poisoned if time is available for their slow death and root disease is present which would affect the coffee. With or without poisoning the trees must be knocked down; it may be necessary to remove some or all of the stumps. If the coffee is to be planted under shade and some of the forest trees are suitable they can be left. The vegetation and unusable timber should next be burnt. This operation is most effective during the dry season. An overall burn, leaving felled trees where they lie, is suitable for robusta. Deuss (1969) showed that robusta yields in the first few years were considerably higher when the vegetation was burnt than when the land was cleared mechanically. For arabica, which is more sensitive to high soil pH values, the timber should be stacked on the lines of the coffee inter-rows before burning.

The land should next be ploughed and disced to destroy the roots of small plants, particularly the perennial grasses. The operations may need repeating several times to effect complete clearance. The whole area can be worked if all the timber has been removed; if not, operations are restricted to the coffee rows, leaving the timber in the inter-rows.

In savannah, small trees and bushes should first be knocked over. If conditions are suitable, and robusta is to be planted, the vegetation can then be burnt. Otherwise the soil can be ploughed and disced so that the vegetation is broken up and worked into the soil.

On sloping land, terraces and drains need to be made for erosion control; this operation can be combined with soil preparation. On flat land, drains may be necessary to speed the flow of excess water and prevent waterlogging.

As soon as possible after land preparation and terracing, and before the rains, a cover crop should be sown. This is preferably a legume which is locally available and is known to grow well. This should reduce erosion during wet weather and smother such weeds as may germinate. The legume must be inoculated.

Roadways should have been lined and smoothed as soon as possible after the start of the operation because they will be required for access by machinery. The final operation, after sowing the covercrop, is to line the plantation and mark the site for each tree with a stake.

Shade

Because the natural habitat of most coffee species is in the understorey of a tall forest, it was natural that early plantings were made under shade. Residual forest trees, which were often made use of, were largely replaced by especially planted trees. These were generally more suitable, could be planted in a regular pattern and controlled by selective pruning to give an even shade of the required density. Cannell has discussed the physiological effects of shade in Chapter 5.

Sturdy (1935) showed that arabica coffee under artificial shade suffered less severely from die-back than unshaded coffee. Maidment (1948) found that robusta coffee under *Ficus* shade outyielded unshaded coffee and coffee under *Cordia* or *Croton* shade. Boyer (1968) studied the effects of artificial shade on robusta coffee. Leaf area was increased under shade but the thickness of the tissues was reduced. Flowering was improved on some clones where abnormalities occurred in direct sunlight, and fruiting was improved.

It was found that coffee could grow well without shade, under some, if not all, conditions. Many investigations into the effect of shade on coffee have been carried out. As a broad generalisation, unfertilised shaded coffee yields at a higher level than unfertilised unshaded coffee, but responses to fertiliser are greater on unshaded coffee so that the highest yields are given by fertilised unshaded coffee. The higher yields from unshaded coffee were clearly demonstrated by Triana (1957).

In addition to the physiological effects of shade on coffee, shade trees modify the plantation ecosystem in various ways. They compete with the coffee for moisture. In areas with low or marginal rainfall the water used by shade trees reduces the amount available for the coffee which then suffers (Franco, 1952). In such conditions coffee should be grown without shade trees, which are the only economic way to provide shade. The damaging effect of high insolation on coffee leaves must be accepted. However such effects are limited to the outermost leaves and shading within the canopy will ensure that lower leaves are not affected. Provided that the canopy includes sufficient leaves the cropping potential of the tree will not be reduced.

Shade modifies the micro-climate around the coffee. This can be beneficial where high insolation rates can be reduced. Wind speed and temperature are lower around the coffee, reducing water loss; such a climatic modification can be important in areas subject to hot, dry winds.

Shade trees may extract nutrients from the soil at levels not explored by coffee roots and leguminous trees fix nitrogen. The additional nutrients reach the soil surface as leaf fall where they become available to the coffee; the leaf fall adds to the mulch.

Shade trees are sometimes foci for root diseases and lightning. The timber from shade trees is often an important source of fuel or constructional material, particularly for smallholders, and justifies a reduction in coffee yield.

Several species of tree have been widely used for shade. Many are legumes; *Acacia* spp., *Albizzia* spp., *Cassia* spp., *Erythrina* spp., *Gliricidia* spp., *Inga* spp. and *Leucaena* spp. are most common. Among non-legumes *Grevillea robusta* is widespread; *Casuarina* spp., which also fix nitrogen, are used in some places although they are not ideal as a shade tree.

A good shade tree needs to compete at a minimum level with the coffee for nutrients and water. The growth habit should be such that it can easily be controlled to give an even canopy at a reasonable height (about four metres). Leaves should not be large; small and feathery leaves give a dappled shade.

Other economic plants are used as shade for coffee. Bananas are common, but they compete strongly for water and nutrients. Robusta is grown under coconuts.

Windbreaks

Where a plantation site is regularly affected by strong winds, it is desirable to protect the coffee by planting windbreaks across the line of the wind. An effective windbreak should not be too dense; the aim is to reduce air speed as it filters through the canopy. With a very dense canopy the airstream flows over the top and produces excessive and damaging turbulence on the lee side.

The effect of a windbreak persists for a downstream distance of approximately ten times the height. On large plantations, therefore, windbreaks are needed at intervals related to the height.

Many species of tree have been planted as windbreaks. Those used for shade will also form windbreaks if planted close together. Larger trees, such as *Eucalyptus* spp. and *Pinus* spp., can also be used. Most windbreaks adversely affect coffee planted too close, mainly by competing for water and nutrients; some excessive shading may occur also. A space should therefore be left between a windbreak and the coffee.

Temporary Shade, Nurse Crops and Intercropping

Temporary shade or a nurse crop is often planted to shelter young coffee for a period, up to about two years, from planting in the field. This is often done even when the mature plantation will be unshaded.

Short-lived legumes are often planted as temporary shade, in lines midway between the coffee lines. *Crotalaria* spp., *Gliricidia* spp., *Sesbania* spp. and *Mimosa invisa* var. *inermis* have all been used at various sites. Trees which are also used for permanent shade can be planted closely along inter-row lines and later thinned out completely or to leave a normal shade-tree pattern. If planting and removal are correctly timed there will be no significant competition with the coffee, whilst there will be benefits from shelter, leaf-fall and nitrogen fixed by legumes.

Spreading cover-crops can be used as nursecrops. Those with an erect habit provide some shelter, which may be sufficient in locations where taller species give too much shade. Such spreading crops are more effective in controlling weeds and reducing soil erosion, but need regular attention to prevent them smothering the coffee. Legumes such as *Desmodium* spp., *Flemingia* spp., *Calopogonium* spp., *Pueraria* spp. and *Stylosanthes* spp. have all been used.

An economic crop can be planted as temporary shade or nurse-crop. Bananas are often used but compete severely with the coffee if not removed before the banana and coffee roots meet. Tall annual crops such as maize are good nurse-crops. Less erect crops have been grown, for example ground-nuts, soya bean, sweet potato and various vegetables. The benefits of the nurse-crop together with an early economic return are achieved. No competition will occur provided that spreading crops such as sweet potato are controlled.

Grasses can be used as nurse-crops. Guatemala grass, *Tripsacum laxum*, has been used effectively in arabica coffee but spreads if not ruthlessly controlled, as do other grasses.

Oladokun (1980) reported an investigation of the effect of various cover crops, mulch and banana shade on soil moisture and pH level, and plant height and nutrient contents. This was carried out using the quillou form of robusta coffee.

Some species die out when the coffee shades the ground well. Others will not but can be removed at the start of weeding.

Spacing of Coffee

Theoretical aspects of plant population and crop yield have been discussed by Holliday (1968).

A complete canopy is desirable with any tree crop growing under ideal conditions so that maximum use is made of sunlight. Therefore, a triangular arrangement of trees with the distance between each tree equal to the diameter of the canopy of the mature tree, is as close to ideal as can be achieved. Such an arrangement also allows for maximum root development. However, in coffee, a number of practical considerations constrain the arrangement so that the ideal is unusual, although it is common in other crops, for example oil palm. The ideal distance between trees depends on the species and cultivar.

The pruning which is essential for coffee is also a means of controlling the size and density of the canopy. By pruning, therefore, the trees can be shaped to suit a wide range of planting arrangements. It is often necessary to allow for mechanical operations (cultivation, spraying etc.) along inter-rows. Access for harvesting and other operations is important even when all operations are carried out manually. In areas of low rainfall, water demand is limited by planting with wide inter-rows.

Close spacings yield a higher crop in the early harvest years. This can be important where a large early return is required on the investment in planting. Such plantings may be too dense at a later stage and will require heavy pruning or thinning by removal of a proportion of the trees. Thinning does not necessarily ensure high yields in later years (see below).

The increased yield in early years from higher-density planting was shown clearly by Basagoita (1981). These results showed a significant correlation between total yield over the first seven cropping years and number of trees per hectare. The length of the primaries and the diameter of the stem decreased as the distance between trees was reduced. This experiment was carried out with arabica, variety bourbon, at densities from 2994 trees per hectare to 7,128 per hectare. Uribe-Henao and Salazar-Arias (1981) experimented with the dwarf variety caturra at densities from 2,500 to 6,410 trees per hectare at several locations. At each site the total crop over the first four, or five, years' harvest increased in line with the number of trees. Awatrami (1982) planted arabica selection 795 at six densities, from 2,058 to 12,345 per hectare. The denser treatments were thinned in stages extending over nine years by removal of trees starting after the third harvest. Over the first three years' harvests the crop increased with the density. Over the next three years the crop from the denser plots fell, that from 3,086 per hectare remained constant and the yield from 2,058 per hectare increased to outyield all others. In the following years the lowest density continued to give the highest yield. Yield from the plots which had been thinned from original high densities continued to fall. The highest yield over 14 years was given by coffee at 3,086 trees per hectare. Uribe-Hénao and Mestre-Mestre (1980) found the reverse.

Bouharmont (1981) reported two experiments with arabica. In the first using the normal-size variety known as java in West Africa he obtained a significant correlation between increasing yield and increasing density over the range 1333 to 3333 trees per hectare. With the dwarf variety caturra, over the range 1,250 to 10,000 trees per hectare, yield over the first two harvests reached its maximum at 7500 trees per hectare. In this case yields were smaller at the highest densities.

Cestac and Snoeck (1982) found that the highest gross yield from robusta over 16 years was given by a density of 2,194 trees per hectare. Two experiments reported included densities ranging from 952 to 2,666 trees per hectare; yields at the highest densities were lower than at the optimum spacing. In all spacings the distance between the lines was three metres. The reduction of the inter-tree space within the line from 2 to 1.25

metres reduces tree efficiency and thus total yield despite a greater density of stems.

Hatert (1958) studied the root systems of robusta coffee grown at a number of densities. He found that the system was consistently circular and did not exceed 3.2 m in diameter. Roots of adjoining trees touched but rarely crossed. The roots of closely planted trees covered circular areas which expanded until they touched the roots of the closest neighbour. The root systems thereafter remained static and circular, not expanding to explore soil beyond the circle. This would explain disappointing long-term yields from closely planted trees and the inability of such trees to increase yield after thinning. Deuss and Borget (1964) found that double-density planting of robusta, thinned after three years' production, produced no more crop, cumulated over five years, than the trees planted at the lower density and not thinned. Trees planted initially at very close spacing in a line gave a lower cumulative yield than trees planted initially in lines evenly spaced at half the normal distance. The latter conclusion would follow from Hatert's observations.

Browning and Fisher (1976) investigated yields of arabica from fan plantings which compared a number of plant densities between 2200 and 75,000 per hectare. The highest yields on each of several experimental sites were obtained from densities between 3600 and 8500 plants per hectare with a mean at 5600 for the normal-size variety SL28. The dwarf variety caturra gave a lower optimum density. Yields tended to fall as the coffee aged.

A high density of individual stems can be achieved at an early stage in plantation development by planting more than one tree at each site. This is common practice in Brazil, where the group at each site is known as a 'Cova'. Four trees per cova is the most common arrangement, but numbers from two to six can be found. Trials of this arrangement have been carried out elsewhere. Kiara (1981) reported work with arabica in Kenya. Using SL28, a variety of normal size, there was no significant difference between yields from one, two and three plants per site; the yield from four per site was significantly lower. The dwarf variety caturra showed a tendency to increase yield from one to four trees per site, but this was not significant. Root development was found to be restricted to a degree which increased as the number of trees per site increased from two to four. Cestac and Snoeck (1982) tried three robusta trees per site. There was no advantage over one tree per site at 1111 trees per hectare and a small advantage at 1333 trees per hectare but the yield from this arrangement was much lower than that from single trees at 1666 per hectare.

The effects of spacing upon overall yield can be modified by pruning. It is sometimes argued that the key factor is the number of stems per hectare rather than the number of trees per hectare. This might be valid if yield per stem were constant, but it has been found in most experiments that yield per tree falls as numbers of trees increase, although total yield increases. The nunber of stems per tree is constant throughout most experiments.

A wide range of planting distances and arrangement is found in plantations. Traditionally, arabica coffee is planted at a density of the order of 1350 trees per hectare. Typical spacings are 2.7 m × 2.7 m square or triangular; 3.0 m × 2.4 m rectangular. Hedge planting, with trees much closer, in rows which are wider apart, is common, particularly when large machines are used in the inter-rows for cultivation, mulching, spraying or other cultural operations. A higher density will be used when a dwarf cultivar is used, for example 2.0 m × 2.0 m, giving 2,000 plants per hectare when planted square, using cultivars such as caturra or san ramon.

Robusta is a larger tree and lower densities are usual, 1,000 to 1,350 trees per hectare. For example, Forestier (1969: 38) recommends 2.0 to 2.25 m within lines which should be 3.7 to 4.0 m apart (1,110 to 1,372 trees per hectare). In some areas growth can be so vigorous that the density has to be reduced to just over 900 trees per hectare (3.3 m × 3.3 m gives 918 trees per hectare).

Shade, spacing, climatic conditions, pruning and nutrition interact in a complex manner. The optimum spacing for a particular set of conditions has to be established by trial.

During the last 25 years there has been much interest in increasing the coffee crop by raising the density of planting. Huxley and Cannell (1968) discussed crop physiological factors which should be considered in intensification of planting. Mitchell (1975) reviewed work in Kenya on intensification of arabica coffee planting.

Field Planting

This operation is preferably carried out early in the rains, although planting at other times has been successful where conditions have been modified locally, for example by temporary shade, or where irrigation can be carried out. Trees planted earlier in the rains often establish more certainly and show a significant advantage in growth in the first year, compared with those planted later.

In most coffee-growing areas trees are raised in nurseries. However, planting seed directly into the field (seed-at-stake) is practised in some areas. In particular, mulch planting in Brazil is done on the cova system (see p. 00). Planting a single seed per site gives no opportunity for selection of the better plants and a failure to germinate is inevitably a missing tree. The cova system partly overcomes these difficulties. Maintenance of the plantation after seed-at-stake planting is difficult, particularly in very wet areas, and there is a high risk of damage to small seedlings.

In East Africa it is recommended that the holes be dug several months before planting to allow the soil to weather. The holes are refilled two to three weeks before planting with topsoil mixed with cattle manure, compost and/or phosphate fertiliser. This must be firmed down to minimise settlement after planting. The hole is then partly re-opened shortly before planting (Acland, 1971: 67; Robinson, 1964: 32). In areas with slow-draining soil and high rainfall it is not possible to dig holes until immediately before planting, as they will fill with water overnight. Large holes (60-100 cm diameter and depth) are recommended in East Africa and must be beneficial if refilled with a mixture containing manure. Smaller holes are often used. If holes are not dug in advance, ploughing along the planting line breaks the soil up in advance; this method is improved by ploughing lines in two directions across the field with the lines crossing at the plant sites.

The age at which plants are removed from a nursery varies, but is usually between 6 and 18 months. The exact age is usually determined by the availability of seed and the wet season for planting. Sometimes plants are allowed to remain in the nursery for another year. They should then be large enough to be stumped; the stumps are used to infill where plants have died. They have sufficient reserves to compete with the existing plants.

Plants raised in a nursery bed may be planted either with a ball of soil enclosing the roots, or with bare roots. The former is preferable, but is not possible unless the soil is suitable and nursery technique develops a tight mass of roots (see page 163). When planting bare roots care should be taken to spread the roots evenly around the stem; ensuing root development will make the most efficient use of the soil around the tree.

Plants in containers retain the soil around the roots. The containers must be removed unless they are made of material which decomposes quickly and does not obstruct root development.

Although planting with the stem vertical is usual, planting at an angle, about 30° to the horizontal, is practised. This enables a multiple-stem tree to be formed more quickly and with less labour input than vertical planting. The root system is, however, inevitably shallow and largely to one side of the tree. The anchorage is relatively poor and large trees have been known to fall over by rotation about the axis of the original stem. The tap root may not develop properly; if it does it is bent which is likely to restrict the flow of sap.

Trees should be planted so that the soil level on the stem is the same as it was in the nursery. Settlement of the soil in the planting-hole is inevitable so the tree after planting should be at the top of a small mound.

Unless there is almost continuous cloud cover during the wet season it is desirable to shade the young trees after planting. If temporary shade is to be used (see page 166) it must be planted in advance of the coffee if it is to be effective. The same applies to a nurse crop (see page 167) which can provide shade from the side in addition to sheltering the plants from wind. In the absence of temporary shade or a nurse-crop, individual plants can be shaded by appropriate vegetation, for example banana leaves, palm leaves or grass on sticks.

If cutworm or other insect is known to attack a significant proportion of the young plants, a suitable insecticide can be sprayed around the base of each tree.

Where application of mulch is standard practice, this should be spread over the planted area as soon as possible. A space of about 10 cm radius should be left uncovered around each plant.

Pruning

In many coffee-growing areas, pruning is a major part of coffee husbandry. Most of the coffee crop is grown on two-year-old plagiotropic stems (primaries*) and one important objective of pruning is to maximise the amount of such wood. Others are:

(a) to shape the tree to make the best use of the space between trees, whilst allowing the necessary access;
(b) to remove dead, diseased and over-age wood. Arabica coffee does not shed over-age laterals naturally, but robusta does;
(c) to provide an environment within the tree which is conducive to maximum crop production but minimises the spread of diseases and pests;
(d) to have the crop formed where it can be harvested easily and cheaply;
(e) to enable sprays of fungicides and insecticides to reach all of the tree;
(f) to minimise biennial bearing and consequent risk of die-back.

All pruning systems can be grouped as either single-stem or multiple-stem systems. Trees on either system require regular attention to meet the objectives listed above.

Variety, environmental conditions and other factors affect the measures necessary to establish and maintain particular systems. The angle and degree of branching varies with the variety or clone, as shown by Snoeck (1963). Low temperatures restrict branching, as shown by Abrego (1962) for arabica and by Snoeck (1963) for robusta.

Many variations of pruning are seen in practice. Bouharmont (1968) described a range of formation and pruning systems which were in use in the east of the Cameroons (Figure 7.1). These are on robusta, but the systems shown apply equally to arabica. The diversity in handling arabica in various countries is shown by Mitchell (1974) who reported as follows:

---

*In this chapter primary refers to a plagiotropic stem which originates on an orthotropic stem. A secondary branch from a primary.

Coffee is allowed to grow freely in Brazil so that old coffee with four to six trees per cova will carry up to 25 stems; younger coffee, more closely planted with two or three trees per cova, has six to twelve stems. Capped single-stem is the normal system in Colombia. In Costa Rica the usual system was 'Candelabra' formation (p. 177) followed by free growth; 'Agobio' to form three stems followed by free growth was used as an alternative. The current recommendation is to plant three seeds in a pot, and allow a single uncapped stem on each tree in the cova. Older trees in El Salvador and Guatemala were formed by an extended 'Agobio' or 'Parras' system in which the suckers were also bent over. Modern plantings use a closer spacing and allow a succession of stems to grow without bending ('Multiples Verticales'). In Jamaica seedlings are planted at an angle of 45° and three suckers selected. These are capped once to produce six stems. In Malawi the older coffee is multiple-stem with newer plantings single-stem.

Robinson (1964) describes only multiple-stem pruning for Tanganyika. Anon. (1975) for Zimbabwe states that both single- and multiple-stem have been proven commercially; both systems are described. Ripperton, Goto and Pahau (1935) describe one single-stem and five multiple-stem methods used in Hawaii.

For robusta Bouharmont (1968) describes single- and multiple-stem systems in use in East Cameroons. There is no advice to use one system in preference to the other. Ways to make the best use of either system are described. Robusta in Madagascar does not produce additional stems freely due to low temperatures, nevertheless multiple-stem pruning is recommended (Snoeck, 1963). Forestier (1969: 63-82) describes only multiple-stem systems for the Central African Republic. Cestac and Snoeck (1982) showed that the multiple-stem method recommended for the Ivory Coast outyielded by a substantial margin single-stem and three alternative multiple-stem systems, when applied to trees at a density of 1333 per hectare. This advantage persisted up to a density of 2500 per hectare; at 3333 per hectare single-stem gave the highest yields. In Malaysia, liberica is grown as a single stem and capped at 2 m. No further pruning is necessary during the 30 year life of the tree.

Single-stem Systems

Formation. Few trees from a nursery grow more than one orthotropic stem, although they may produce plagiotropic primaries if kept for sufficiently long in the nursery. The only variations in formation practices relate to shortening the stem (capping) and treatment of primaries.

Fernie, in Robinson (1964: 37), recommends capping at three levels: 53 cm, 114 cm and 168 cm (see Figure 7.1 no. 4). Figure 7.2 shows the steps in tree formation. The same applies for Zimbabwe (Anon., 1975). In

Hawaii single-stem arabica is allowed to grow unchecked, often until two crops have been harvested, before capping at 2 to 2.3 metres (Ripperton *et al.*, 1935: 20, 29). Triana (1957) in Colombia found no significant difference in crop yield between capped and uncapped single-stem coffee. Figure 7.3 (from Fernie, in Robinson, 1964) shows how primaries should be kept free of secondaries close to the stem; this process must start during the formation period.

Bouharmont (1968) recommends for the Cameroons the formation of single-stem robusta by capping at three levels: 0.7 m, 1.2 m and 1.7 m. In Indonesia, single-stem trees are capped at 1.8-2.0 metres (Poetiray, 1981).

Maintenance Pruning. In arabica it is possible to maintain a canopy of reasonable density by stimulating new secondaries and removal of old ones. The original primary will remain and should ideally carry not more than two secondaries from any one node. Harder pruning will be required at the top of the tree to prevent concentration of the primaries there and consequent formation of an umbrella. Figure 7.4 illustrates the secondaries to be removed in maintenance pruning.

Ultimately, progressive loss of lower primaries will restrict the depth of the productive part of the trees to a short section at the top of the tree, to form an umbrella. At this stage the tree must be renewed from the base. The trees can be stumped and one new stem selected from the regrowth. It is also possible to change to a multiple-stem system by selecting more than one new stem. Stumping is followed by a period of two years during which little or no crop is produced. The loss of crop can be reduced if a new stem can be well advanced prior to stumping. It may be necessary to raise the skirt of the canopy, or remove the central part, to stimulate formation of new stems (see Figure 7.1 nos. 5 and 6).

Multiple-stem Systems

Formation. There are four main ways of producing multiple-stem trees. These are cutting, bending, angle-planting and multiple planting.

Cutting the stem of a coffee tree will, in most conditions, be followed by the growth of new stems from below the cut. Commonly, two stems will arise from the axils of the pair of leaves immediately below. Others may grow from a lower level, depending on variety, local conditions and the tree configuration, particularly whether there is foliage below the cut which shades the stem. The stems required are selected from the new growth; others are removed.

The earliest cut can be made in the nursery. Trees then go to the field with the beginning of a multiple-stem formation. This operation will lengthen the period in the nursery; the trees will be bulkier when removed which may create transport problems.

Otherwise, the first cut is made in the field. The stages recommended by

Figure 7.1: Pruning Systems in Use in East Cameroons. 1. Single-stem tree which has formed an umbrella. 2. and 3. Cylindrical pruning of single-stem tree. 4. Formation of single-stem tree by sequential cropping. 5. and 6. Development of a new single-stem following partial pruning. 7. Development of a new single-stem following stumping. 8. Formation of a multiple-stem by sequential capping ('Candelabra'). 9. Formation of multiple-stem tree by a single capping. 10. Formation of a multiple-stem tree by angle planting. 11. Formation of a multiple-stem tree by bending a single stem ('Agobio'). 12. Annual pruning of one stem of a multiple-stem tree. 13. Pruning of a multiple-stem tree by stumping after several years. 14. and 15. Pruning of a multiple-stem tree by stumping, leaving one stem as a 'lung'. (See also Figure 7.12.)

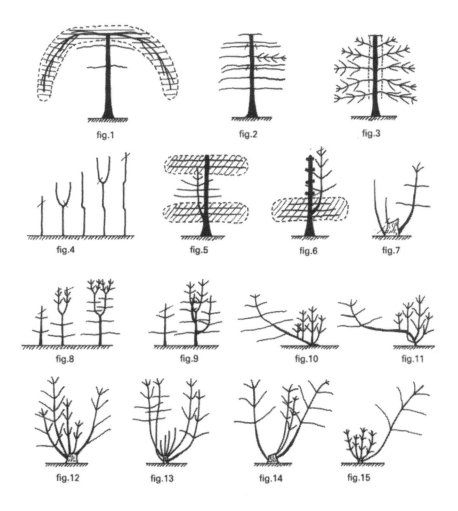

Source: Bouharmont (1968) Café, Cacao, Thé, *XII*, 16. By courtesy of IRCC, Paris

176 *Cultural Methods*

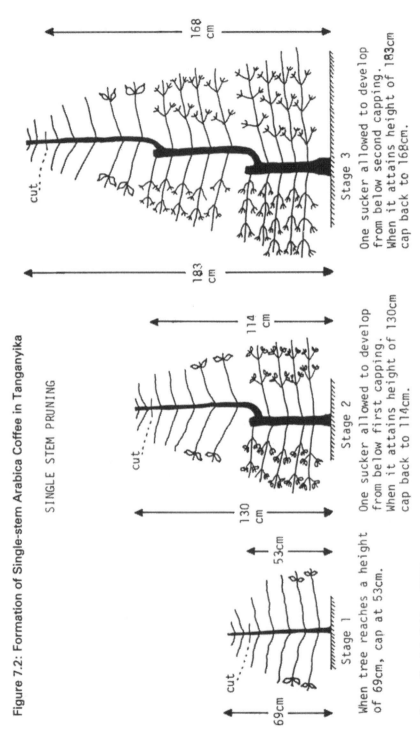

Figure 7.2: Formation of Single-stem Arabica Coffee in Tanganyika

SINGLE STEM PRUNING

Stage 1
When tree reaches a height of 69cm, cap at 53cm.

Stage 2
One sucker allowed to develop from below first capping. When it attains height of 130cm cap back to 114cm.

Stage 3
One sucker allowed to develop from below second capping. When it attains height of 183cm cap back to 168cm.

Source: Fernie in Robinson (1964).

Figure 7.3: Single-stem Pruning: Ideal Formation of Primaries and Secondaries. Diagram looking down on top of a tree showing how all primary branches are kept free of growth within 6in of the main stem.

Source: Fernie, in Robinson (1964).

Fernie (in Robinson, 1964) are illustrated in Figures 7.5, 7.6 and 7.7. Two stems only are recommended for the first cycle.

Bouharmont (1968) describes how a single cut at a height of 40 cm can produce a multiple-stem tree with more than two stems by the growth of stems below the first node under a cut at 40 cm (Figure 7.1, no. 9). In cases where no suckers arise below the first node both new stems are cut at an unstated height, to give another two stems from each. This produces a 'candelabra' tree (Figure 7.1, no. 8).

Forestier (1969: 63) states that robusta tends to produce many stems without any stimulation by cutting or other means. This effect is dependent on light; more stems are produced in unshaded coffee. Low temperatures inhibit suckering, as in Madagascar. Varietal differences arise, some clones in particular will not sucker easily. Snoeck (1963) reports that in Madagascar where temperatures fall to a monthly average of 20-22°C for four

178 *Cultural Methods*

Figure 7.4: Single Stem Pruning: Maintenance.

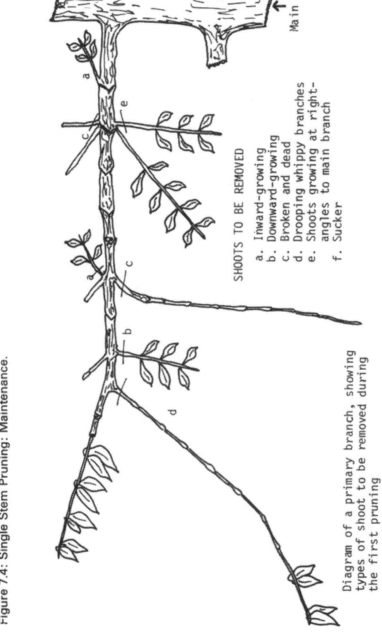

SHOOTS TO BE REMOVED

a. Inward-growing
b. Downward-growing
c. Broken and dead
d. Drooping whippy branches
e. Shoots growing at right-angles to main branch
f. Sucker

Diagram of a primary branch, showing types of shoot to be removed during the first pruning

Source: Fernie, in Robinson (1964).

months of the year and new plantings are of clonal cuttings, 96 per cent of the trees produce no suckers. Forestier (1969: 64) states that, even where suckers are relatively numerous, there is a wide variation between trees, which makes establishment of uniform trees difficult. The example quoted is 32 per cent of uncut trees formed three or more suckers, 20 per cent two suckers, 21 per cent one sucker, 27 per cent no sucker.

Bending of coffee plants to stimulate production of main stems was originally developed in Central America; hence such trees are known as 'agobio' and the method as 'agobiada'. When a stem is bent over, the light stimulates the production of new stems along the upper surface. The required number of stems in the most favourable situations at the lower end of the original stem are selected, others are removed (see Figure 7.1, no. 11). The original stem needs to be held down firmly until growth and the weight of new stems prevents the lower end from returning to a vertical position. The unwanted section of the original stem can then be cut off. Snoeck (1963) detailed the technique: Figure 7.8 illustrates this description.

Planting at an angle (Figure 7.1, no. 10) achieves the same effect as bending, but avoids the work of bending and risk of straightening if the fixing is disturbed. Figure 7.9 illustrates the method. Using this method, root depth is initially restricted and to one side only of the trees. It is not uncommon for trees to fall over. Snoeck (1963) compared bending with angle-planting at 30° and found bending significantly better; 73 per cent of bent trees produced four or more stems compared with only 61 per cent of angle-planted trees.

Single-stem trees can be converted into multiple-stem by stumping. This operation may give the tree an excessive shock which would kill it. It is advantageous to remove the primaries from one part of the canopy (for example, Figure 7.1 nos. 5 and 6) which will stimulate the growth of at least one new stem. This will act as a lung and minimise the shock to the tree. More stems will grow after the main stem is removed. To form a multiple-stem tree a number of these will be selected, they will not be cut off as shown in Figure 7.1, no. 7. Robinson (1964: 46) suggests removing all primaries from one side of the main stem. He suggests leaving one primary as a lung about 1 metre above soil level; all other primaries will be removed and the main stem can be cut back to just above the lung. The lower section of the main stem can be cut back to just above the point of divergence of the new main stems when the latter are large enough to maintain the tree.

During the formation process, primaries must be selected so that the desired shape of tree is achieved. To this end, surplus primaries should be removed at as early a stage as possible. Particularly, primaries which are too close to the ground, those which pass too close to another stem, those which cross the centre of the tree and those which grow at a bad angle, should be removed. The number of primaries needs to be restricted to

180 *Cultural Methods*

Figure 7.5: Formation of Multiple-stem Arabica Coffee in Tanganyika (Part 1)

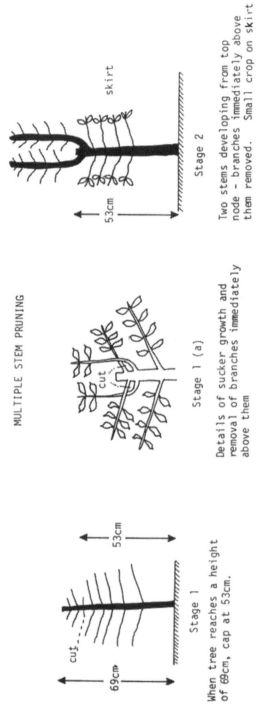

Source: Fernie, in Robinson (1964).

Cultural Methods 181

Figure 7.6: Formation of Multiple-stem Arabica Coffee in Tanganyika (Part 2)

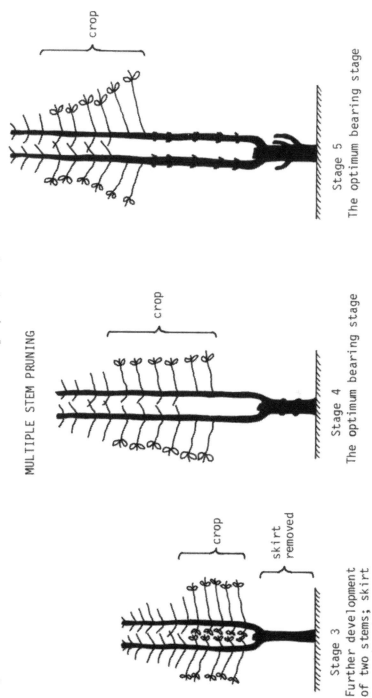

Source: Fernie, in Robinson (1964).

182 *Cultural Methods*

Figure 7.7: Formation of Multiple-stem Arabica Coffee in Tanganyika (Part 3)

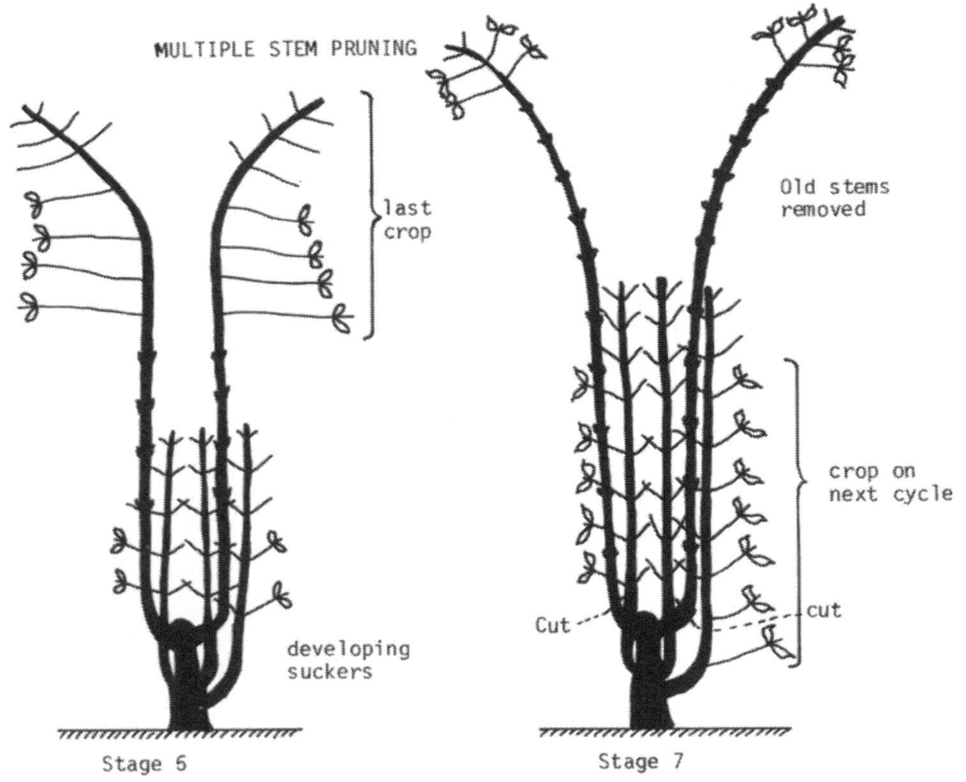

Source: Fernie, in Robinson (1964).

Cultural Methods 183

Figure 7.8: Formation of Multiple-stem Tree by Bending

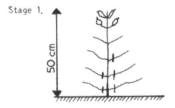

Stage 1.

Removal of lower primaries before bending, when tree is 50 cm tall.

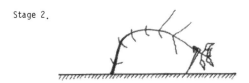

Stage 2.

Tree bent over and held down by a strong peg.

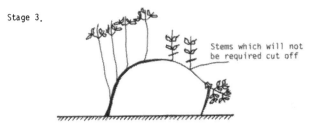

Stage 3.

New stems growing from upper surface of original stem.

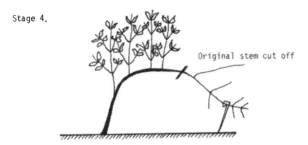

Stage 4.

Original stem cut to leave 4 new stems close to base of tree. Original stem has now set into curved position.

Source: Expanded from Snoeck (1963), by courtesy of IRCC, Paris.

184 *Cultural Methods*

Figure 7.9: Formation of Multiple-stem Trees by Angle Planting

Stage 1.

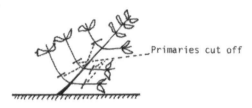

Planting at an angle of about 30° to horizontal. Preferably, lower primaries should be cut off, unless there is a long section of lower stem without primaries.

Stage 2.

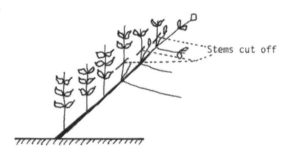

Removal of new stems too far from base of tree.

Stage 3.

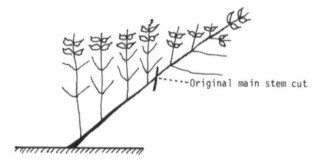

Original main stem cut off leaving four main stems.

minimise overbearing. Fernie (in Robinson, 1964: 42) recommends the use of a measuring stick; this is used to measure downwards from the top of each stem. All primaries below the lower end of the stick are removed. The length of the stick is suggested as 1.2 to 1.4 m normally; reduced to 1 m after an overcropping season. In very good conditions the stick length could be increased to 1.5 m in second and subsequent cycles. Inward-growing primaries should be removed over a further 30 to 50 cm above the base of the head, so that stems will bend outwards to help close the canopy and facilitate harvesting.

Maintenance Pruning. Main stems have to be removed at intervals, as with single-stem pruning. There are two approaches to the renewal of stems: a cyclic system whereby one or more stems is removed (Figure 7.1, no. 12) and replaced each year, or stumping at longer intervals (Figure 7.1. nos. 13, 14 and 15).

In order to minimise the loss of crop arising from the removal of a stem, it is desirable to have the replacement growing before removal of the old stem. New stems may arise without any special measures being necessary; the necessary light intensity will arise, particularly in unshaded coffee, when the bearing head of primaries on the old stem is well above the ground and the skirt of lower primaries has been removed. Where suckering does not occur freely, it will be necessary to increase the light intensity at the base of the tree. To do this, the depth of the head can be reduced by removing primaries to a greater height above the ground. This action may not be sufficient, in which case all primaries should be removed from one side of the stem or the tree.

Fernie (in Robinson, 1964: 42) recommends removing all primaries lower than about 1 m below the top of each stem. Allow all suckers to grow to about 30 cm long before selecting the required number in the best positions and removing all others. New suckers should arise between 15 and 45-60 cm above ground level. Figure 7.10 shows a tree on a four-stem system; the oldest stem will be removed after harvest. The new stem will flower the following year for the first time. Figure 7.11 shows a tree with five stems where, in alternate years, two stems and one stem are replaced. This figure shows the tree before removal of lower primaries to stimulate growth of suckers. The number of stems removed in any one year is not necessarily restricted to one. Where the spacing justifies more than four stems per tree it may well be necessary to remove more than one stem on occasions to ensure that the crop is accessible.

When pruning by annual removal of one or more stems it is necessary to remove dead and diseased wood, unnecessary suckers and badly placed primaries. The lengths of the heads can also be adjusted to the required measurement.

Stumping may involve removing all stems (Figure 7.1, no. 13); alter-

186 *Cultural Methods*

natively one stem may be left as a lung (Figure 7.1 nos. 14 and 15). The latter minimises the shock to the trees, which may kill a proportion of the planting. Where a lung is left, this must be a vigorous, healthy stem capable of producing a good crop for one further year. This stem should be as far to one side of the tree as is possible so that there is minimum shading of the centre of the tree. New stems will arise from the central stump, there will usually be more than the number required so some thinning will be necessary. Stems retained should be evenly spaced around the centre of the tree. The lung is removed after harvest the following year; a small crop is

Figure 7.10: Multiple-stem Pruning: Four-stem Tree with One Stem Replaced Each Year

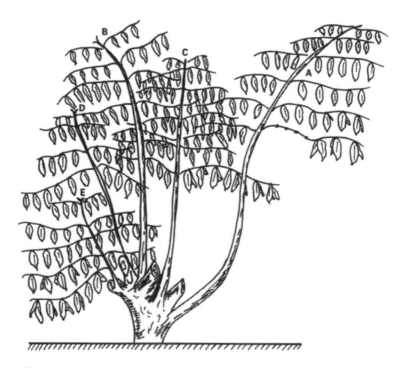

A. Oldest stem; to be removed after current harvest
B. Four-year-old stem
C. Three-year-old stem
D. Two-year-old stem on which first harvest is maturing
E. Replacement stem, less than one year old, which will flower the following year for the first time.
All other suckers have been removed.

Source: Forestier (1969), by courtesy of IRCC, Paris.

thereby obtained whereas no crop would be given if all stems had been removed. Figures 7.12 and 7.15 illustrate stumping leaving a lung. Some faults which can occur are shown in Figures 7.13, 7.14, 7.16, 7.17 and 7.18. Stumps should be cut about 30 cm above ground level. A cut which is too high at 50 cm is shown in Figure 7.16. When old stems are removed, a clean cut which is sloping, not horizontal, should be made. Figure 7.12 shows the ideal. Figure 7.17 shows an incorrect horizontal cut and 7.18 ragged cuts made by a knife or axe. Horizontal or ragged cuts facilitate the entry of rainwater and disease: this applies equally to all variations of pruning. Rough saw cuts should preferably be smoothed with a sharp knife.

Stumping can be carried out either by lines or by blocks. Sometimes when one line is stumped adjacent lines create too much shade and growth of new suckers is delayed or may not occur at all. This effect can be minimised by stumping lines in order, for example, 1, 4, 2, 5, 3 on a five-year cycle. This approach may not suffice; in that case, stumping by blocks is essential.

Figure 7.11: Multiple-stem Pruning: Five-stem Tree with Stems Replaced Alternately Two and One. New sucker not yet growing

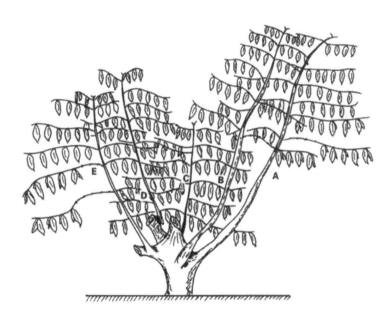

A, B. Four-year-old stems
C.   Two-year-old stem
D, E. Three-year-old stems.

Source: Forestier (1969), by courtesy of IRCC, Paris.

188 *Cultural Methods*

Figure 7.12: Multiple-stem Pruning: Stumping Leaving a Lung: Ideal

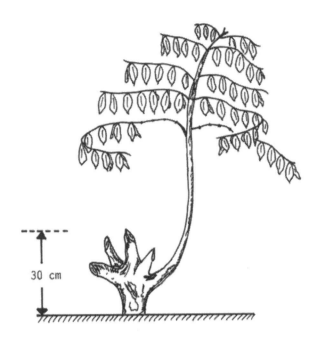

Source: Forestier (1969), by courtesy of IRCC, Paris.

Choice of Pruning System

The pruning system chosen for any particular plantation must be related to the cultivar planted, the location, spacing, the need to obtain an early return on capital invested and the quality and quantity of labour and supervision.

In mature coffee, there needs to be an even but not overdense canopy. The number of stems per hectare and thence per tree can have an effect on this, but it can be modified by the degree of effort put into control of primaries.

Early crop is largely dictated by spacing. However, if trees are allowed to grow in the nursery without any check by pruning or bending there is likely to be more fruiting nodes when the first harvest is formed. This would be expected to produce a larger crop. Such an approach would lead to uncontrolled single stem trees at close spacing; multiple-stem trees would arise if suckers arose spontaneously which occurs, particularly with robusta, under certain conditions. Lack of control may overtax young trees leading to die-back at an early age, low second-year crop, biennial bearing and possibly permanent damage to cropping potential. Uncontrolled

Figure 7.13: Multiple-stem Pruning: Lung in Incorrect Central Position which Shades Centre

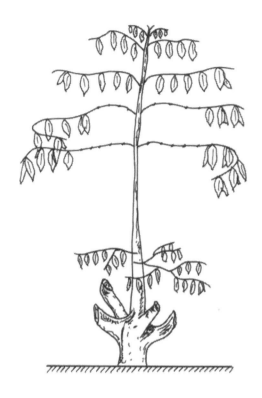

Source: Forestier (1969), by courtesy of IRCC, Paris.

growth may also lead to excessively dense foliage in the centre of the tree, which will restrict crop in the second year.

Pruning must therefore be carefully controlled from the start and an appropriate system maintained.

Single-stem pruning requires more skill to control the primaries and secondaries and delay the development of umbrella trees. Formation by capping requires a labour input but may delay the development of an umbrella.

Multiple-stem formation requires labour for capping or bending. Angle-planting reduces the labour requirement but an inferior root system is formed. Unless some skill is exercised in selecting stems from bent or angle-planted trees, the stems will arise from the same (upper) side of the original stem. They will therefore have no tendency to produce a circular, spreading tree, and there may be a restriction on sap flow due to the closeness of the stems.

190 *Cultural Methods*

Figure 7.14: Multiple-stem Pruning: Stumping Leaving a Lung: Incorrect Examples

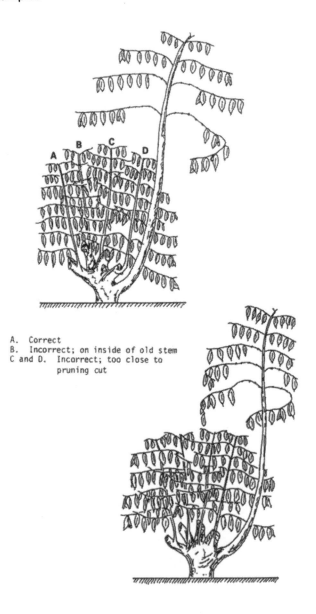

A. Correct
B. Incorrect; on inside of old stem
C and D. Incorrect; too close to pruning cut

Arrangement of stems incorrect: too close to one another and all on one side of the tree.

Source: Forestier (1969), by courtesy of IRCC, Paris.

Figure 7.15: Multiple-stem Pruning: Tree after Removal of Lung: Five New Stems Evenly Arranged

Source: Forestier (1969), by courtesy of IRCC, Paris.

Multiple-stem maintenance pruning requires less skill, stumping probably less than annual removal of one or more stems, because there is no need to remove unwanted primaries. However, with all the systems it is necessary to inspect at intervals to remove surplus suckers and other unwanted growth. With single-stem pruning there are inevitably years with little or no crop; stumping multiple-stem coffee has a similar effect, which is minimised by leaving a lung. However, by stumping a part of the plantation each year total production is smoothed. In spite of the large drop in production every fourth or fifth year this system may give the highest yields, as reported by Cestac and Snoeck (1982).

Consistent application of an appropriate system is probably more important than the details of the method. Sudden changes in pruning system, also spacing which interacts with pruning, may have much more severe effects on crop yields than regular stumping of part of the plantation each year.

192 *Cultural Methods*

Figure 7.16: Multiple-stem Pruning: Stumping at Too Great a Height Above Ground Level

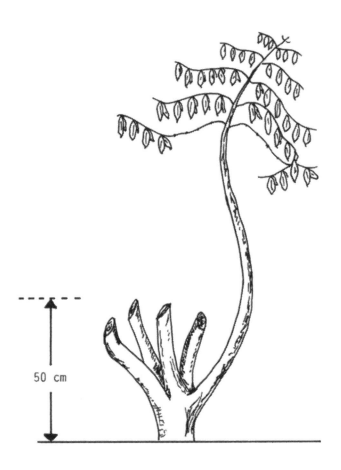

Source: Forestier (1969), by courtesy of IRCC, Paris.

Rehabilitation of Neglected Plantations

The part which pruning can play in rejuvenating coffee which has been neglected has been studied. Capot (1966) recommended stumping robusta leaving one stem as a lung, after which four stems only are permitted to remain on the tree. At the same time the ground should be hoed around each tree and weeds removed. Shade should be thinned and excess shade trees removed. A leguminous cover crop should be planted. Bouharmont (1977a, for arabica and 1977b, for robusta) investigated the effects of a

Figure 7.17: Multiple-stem Pruning: Incorrect Horizontal Stumping Cut

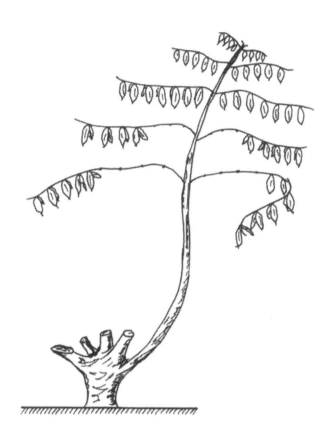

Source: Forestier (1969), by courtesy of IRCC, Paris.

variety of pruning methods, both the initial cutting-back and subsequent treatment of new stems. The main conclusion from this work was that there was a substantial loss of production from all the heavy pruning systems. This underlines the importance of establishing an appropriate system and ensuring that it is adhered to. Okelana (1982) studied the effect of stumping height and the number of stems permitted to remain on robusta coffee. Total production in the first 20 months after stumping was highest when there were four stems per tree, but the quality of the coffee produced, as measured by size of beans, was lowest from the four stem trees.

Figure 7.18: Multiple-stem Pruning: Incorrect Ragged Cut

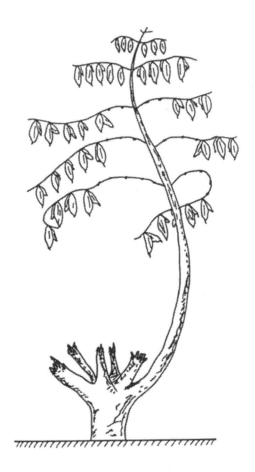

Source: Forestier (1969), by courtesy of IRCC, Paris.

Irrigation

Augmentation of the amount of water available to coffee is valuable at times of water shortage, to prevent damage from lack of moisture. Boyer (1969) found that growth of two robusta cultivars would continue without interruption throughout the year if soil water reserves did not fall below one-quarter of the maximum storage. Changes in humidity and insolation are sufficient to create the internal water deficit necessary to induce flowering.

The physiological effects of changes in moisture regime in coffee have been discussed by Cannell in Chapter 5. The dormancy of flower buds is broken by a reduction of a water deficit. Boyer (1969) reported that floral initiation in robusta occurred provided that 1 to 1.5 mm of growth was achieved daily on average during the preceding vegetative growth period in the wet season. Therefore, when planning irrigation for coffee it is important to relate frequency and amount of water applications to the growth cycle of the crop. Irrigation can also be used to regulate crop distribution.

The water requirement to prevent wilting has been found to vary with the cultivar. Lemee and Boyer (1960) reported that a kouilou cultivar had a greater resistance to lack of moisture than a robusta cultivar. They showed that this resistance arose from several factors. The kouilou could tolerate a higher internal deficit, had a lower transcuticular rate of moisture movement and had a larger root system in relation to the size of the tree. These factors enabled the stomata of kouilou to remain open at a higher moisture deficit thus permitting growth to continue to a later stage than with the robusta during a dry period.

Müller (1975) showed that irrigation or arabica early in the dry season when insolation was at a high level stimulated intense vegetative growth. This ensured that there were plenty of nodes on young stems which would support a heavy crop the following year. Flowering was also induced and the early stages of fruit development occurred when temperatures and insolation were high. When this irrigation was carried out every year it ensured a high level of crop which was consistent from year to year. An appropriately high level of fertiliser application was essential. This routine minimised the problems caused by coffee berry disease (*Colletotrichum coffeanum*) but attacks by leaf rusts (*Hemileia vastatrix* and *Hemileia coffeicola*) were stimulated. Treatment with fungicides was therefore essential.

The amount of water required to restore the soil to field capacity can be calculated from the water lost by evapotranspiration (E_t). This can be determined by simple calculation from a measured water loss from an open evaporation pan (E_o) and a record of rainfall. See Chapter 4 (page 97) for a fuller discussion. The evaporation pan is a simple open-topped box, two metres square and about 60 cm deep, sunk in the ground with its rim 7.5 cm above ground level. The tank can be made of 30 mm sheet steel and needs strengthening by internal cross struts at a depth of 30 cm. It should be painted black inside. A scale must be fixed to or painted on the inside so that the height of the water can be measured. The tank, and the raingauge, should be sited close to the centre of a flat area at least 30 metres square which must be covered with either clean gravel or grass which is kept cut short and fenced to keep animals out. Adjust the water level in the pan to a fixed level each day, note the level before refilling or removing surplus water; the difference in levels is the amount by which E_o has exceeded rainfall, or vice versa. E_o can be calculated from several meteorological measurements (see Chapter 4 page 97). The pan is cheaper than the necessary instruments and avoids a complex calculation.

Evapotranspiration from plants is lower than open-pan evaporation. For a complete canopy of foliage assume that $E_t/E_o = 0.09$. Where there are wide inter-rows, not shaded by coffee, within which there is bare soil or dry mulch, assume that $E_t/E_0 = 0.7$. Measurements should be entered in a table (Table 7.1). Use a separate table for each field. At the end of each week the net gain or loss of moisture from the soil can be calculated as shown in Table 7.1.

The water content of the soil must never be allowed to fall below the wilting-point; irrigation should be done well before this level is reached. Measurement of field capacity and wilting-point requires special equipment in a laboratory so these data are unlikely to be available. A local research station may be able to advise on particular soils; otherwise as a first approximation assume that water equivalent to 60 mm of rain is available in the soil. Irrigation will therefore be necessary before the moisture deficit reaches 60 mm. If these data are available for a number of years it is possible to calculate the requirement of irrigation water. The requirement of water must be sufficient for the driest year on record; the quantity should then be doubled to allow for losses and more extreme conditions.

Fisher and Browning (1978) investigated the water requirements of high density coffee. Fears had been expressed that the trend towards higher planting densities might increase the water requirement to a level where

Table 7.1: Data and Calculation Sheet for Irrigation Control

Week commencing .... Field ....

Day	Change in pan level		Evaporation $-R + D$ or F^1 (mm)	Rainfall (mm) R
	Increase (mm) (I)	Decrease (mm) (D)		
Monday				
Tuesday				
Wednesday				
Thursday				
Friday				
Saturday				
Sunday				
Weekly total	.... (I)	.... (D)	.... (E_0)	.... (R)
Moisture deficit at end of previous week			.... mm (P)	
Irrigation applied			.... mm (W)	
Moisture loss by evapotranspiration^2 = B × E_o =			.... mm (E_t)	
Moisture deficit at end of week = P + E_t − (R + W)^{3,4} =			.... mm (F)	

Notes:
1. When pan level increases, rainfall has exceeded evaporation. In this case use for evaporation loss an average value for rainless days during an earlier dry period.
2. For a full canopy of coffee use B = 0.9. With wide inter-rows, not shaded by coffee or shade trees, which are bare ground or dry mulch, use B = 0.7.
3. If R + W greater than P + E_t deficit F = 0. Surplus water is lost by drainage.
4. Irrigation is due when F is greater than 60mm, or a value determined locally.

problems would arise in low-rainfall areas. They found no evidence that water demand increased at higher densities.

Akunda and Kumar (1982) showed that the colour change of a cobalt chloride disc could be used to indicate the internal water status of coffee leaves. They determined that growth stopped when the internal water potential fell to −20 bar. They described a simple test based on the foregoing which would enable growers to determine when irrigation was required.

Where irrigation is necessary a reliable supply of water is essential. Unless there is a river nearby which maintains a good flow throughout the dry season some storage is essential. Distribution can be by any convenient method, using gravity where possible, otherwise pumping is necessary. Application of water to trees can be by any method which is convenient, but distributors which emit water with sufficient force to damage flowers or leaves must not be used. Sprinklers of the smaller sizes are probably the most common method of distribution, but trickle or flood irrigation is equally acceptable.

Mulch

In Kenya, mulching is a very important part of the normal husbandry of coffee. Elsewhere the practice is less common. Mulch reduces the moisture loss from the soil surface in dry conditions; this is of particular importance in the drier coffee areas east of the Rift Valley in Kenya.

The organic matter content of the soil benefits from the breakdown of mulching materials. This provides nutrients in available form and stimulates the growth of roots (see page 148). Unfortunately most mulching materials do not include nutrients in the proportions which coffee requires; continuous mulching with the same material produces an imbalance of nutrients which must be corrected by application of an appropriate fertiliser. The nutritional aspects of mulching are discussed in Chapter 6, p. 147.

By shading the ground, mulch minimises weed growth. Loss of soil by erosion in heavy rainstorms is greatly reduced by a covering of mulch.

The results of investigations into the effect of mulch on crop yields over 10 years on arabica coffee in Kenya and over 20 years on robusta coffee in Uganda, were summarised by Haarer (1962: 221). Substantial yield increases were achieved. Bouharmont (1979) reported a significant yield increase from arabica coffee in the Cameroons. Deuss (1967) reported experimental results with robusta in the Ivory Coast. Here also mulch gave significant increases in crop yields.

Any foliage material can be used as a mulch. Those commonly used include: natural savannah grass, elephant or napier grass (*Pennisetum pur-*

pureum), guatemala grass (*Tripsacum laxum*), maize stalks and leaves (stover) (*Zea mays*), banana trash (*Musa* spp.), sisal waste (*Agave sisalana*) and other *Agave* spp.) and a wide variety of legumes.

It is preferable for mulch material to be obtained without the effort of growing it on the plantation. Waste from sisal factories, maize stalks or banana trash from farms or plantations merely require transportation, savannah grass requires cutting and transportation. Wallis (1960) discussed the properties and effects of various grasses used for mulch. Where such material is not available a mulch material must be grown on the plantation. Napier grass is commonly grown for this purpose. A substantial area of land is needed; up to an area equal to the area planted with coffee. If land unsuitable for coffee is available this can be used satisfactorily for this purpose. It has an adverse effect on plantation economics if the land could otherwise support coffee.

A major risk with mulch is that of fire, particularly as it is usually easier and more convenient to apply mulch in the dry season. Often alternate inter-rows are mulched each year, this minimises the risk of a fire spreading rapidly over a large area. Wet mulch materials, such as banana trash, are a lower fire risk.

Bouharmont (1979) and Deuss (1967) calculated the profitability of mulching and showed that it was not profitable in the areas which each was discussing. In some other areas the benefits of mulch justify the necessarily large handling costs.

Cover Crops and Intercropping

There can be advantages from cover crops in perennials. Some are the same as those from mulch: suppression of weeds and protection of the soil. Unless the cover crop dies back in the dry season it will not restrict moisture loss from the soil at that time. A cover crop which maintains a large area of green leaf throughout the dry season will compete strongly for moisture.

The mineral nutrition of the cover crop may compete with the coffee, but if the foliage remains in the field when it either dies naturally or is cut, the nutrients therein are recycled. The continuing nutrient demand will be small after the cyclic system has taken up the required nutrients. There may be a benefit in this process in that the nutrients are converted into an organic form in which they are easily available. The litter stimulates root development (see page 148). A legume cover crop, which has been properly inoculated, will provide nitrogen by fixation of atmospheric nitrogen. Burke (1975) evaluated a series of leguminous cover crops.

Under a dense canopy of coffee, or coffee under shade trees, there may be insufficient light to maintain a cover crop. Such conditions do have some effect in restricting weed growth and soil movement in heavy rain. In young coffee, or widely spaced mature coffee, a cover crop will grow satisfactorily and may have a beneficial effect on the coffee crop.

Planting a cover crop requires a labour input and some maintenance work is necessary. Most of the suitable legumes will spread as vines and climb into the coffee so clearing along the tree lines is essential. It is common practice to cut cover crops at the beginning of the dry season; the cut foliage is left as mulch and may be moved onto the tree lines. There can be a fire hazard as this material dries.

Intercropping is attractive, particularly to smallholders, because it creates a second income. Intercropping when a plantation is young can be beneficial, effectively providing a nurse crop (see page 166). In older coffee there may be more competition for nutrients because the product from the intercrop is removed from the field. This may remove the major part of the intercrop, as, for example, with maize. Such losses can be reduced if the effort is made to return the stalks to the field. Some smallholders will regard some loss of coffee crop as a small price to pay for additional food crops.

Investigation of the effects of cover crops and intercropping on coffee have been made by a number of workers. With arabica in the Cameroons, Bouharmont (1979) reported that *Flemingia congesta* was used successfully among unshaded coffee in two research stations.

Experimental comparison of four legumes showed that *Mimosa* gave a substantial increase in yield, *Pueraria* and *Flemingia* a modest increase. *Pueraria* and *Mimosa* were cut and the vegetation spread as mulch along the interlines at the beginning of each dry season. If the foliage was not cut there was no beneficial effect on yield. *Mimosa* dies back naturally in the dry season which intensifies the fire hazard. *Stylosanthes* had no effect on yield. *Flemingia* did not increase yield to as high a level as application of nitrogen fertiliser up to 167 kg N per hectare to clean soil. *Flemingia* reduced the responses to nitrogen; nitrogen application at rates above 167 kg N per hectare produced only small yield increases. Bouharmont concludes that, in contrast to robusta, there is little to be gained from cover crops in arabica in the Cameroons.

Balasubramanian (1970) reported trials with *Mimosa invisa* in India. Savings in weeding costs and a control of soil erosion on sloping land in a high rainfall area are reported.

Bouharmont (1978) reported investigations in robusta coffee in the Cameroons. Comparisons were made between *Pueraria, Stylosanthes, Mimosa* and *Tithonia.* *Flemingia* and *Mimosa* stimulated substantial improvements in yield from unfertilised coffee; *Pueraria* and *Stylosanthes* increased coffee yield, to a smaller extent and on fewer experiments. Nitrogen fertiliser was applied to one experiment involving *Pueraria*; there was a small increase in yield with *Pueraria* and a low nitrogen application and a depression of yield at a higher nitrogen level. *Tithonia* is not a legume and yields were not significantly different from control (natural vegetation). Deuss (1967) compared mulch, *Leucaena glauca* (planted as a cover crop), sweet potato, clean weeding and natural vegetation. Mulch gave the highest coffee yield. *Leucaena glauca* promoted yields a little above those from clean weeding. Yields were lower with sweet potato and natural vegetation than with clean weeding.

Intercropping is not of great interest in mature coffee although it can be useful in the first few years after planting. Little experimental investigation has been carried out. Bouharmont (1978) reported that sweet potato reduced yields of robusta coffee.

Chemical Modification of Fruit Abscission and Ripening

Overbearing could be minimised if a proportion of fruit could be removed at an early stage when a very heavy crop is set. Cannell (page 128) has pointed out that coffee, unlike many fruit trees, has not developed a regulating mechanism which discards fruit beyond the capacity of the tree to ripen. Hand thinning of fruit is practised in some plantations, when labour is available. Cannell (pages 124-7) has mentioned the effects of gibberellic acid and 2-chloroethane phosphonic acid (Ethephon) on floral initiation and ripening. The potential value of controlling ripening is such that further discussion of some experimental results is justified. Browning and Cannell (1970) investigated the use of Ethephon for this purpose. They found that a solution of Ethephon, concentration 1,400 ppm, removed 60-65 per cent of expanding berries, four to six days after application. Some leaves abscissed but no other effect was seen.

Ethephon has also been applied in attempts to accelerate the ripening process. This could shorten the harvest period and ensure that a higher proportion of ripe cherries were present at the time of harvest. The latter could be valuable if mechanical harvesting was introduced. Monroe and Wang (1968) observed that, 'despite efforts to obtain satisfactory selectivity, the amount of immature cherries in mechanically harvested fruit is still from 15 to 40 per cent. Therefore until some method of controlling ripening is developed we must separate immature fruit from ripe in order to have a marketable product.'

In arabica coffee Rodriguez and Jordan-Malero (1970) reported that a solution containing 4,000 ppm Ethephon ripened 92 per cent of the cherries in shaded coffee; in unshaded coffee 500 and 1,000 ppm were effective to a similar degree. A high proportion of leaves on unshaded trees abscissed when Ethephon concentrations exceeded 1,500 ppm; leaf abscission was much less severe on unshaded coffee. Oyebade (1971) reported significant differences in the proportion of ripe cherries 11 and 17 days after treatment with 25, 50 and 100 ppm concentrations of Ethephon; the differences diminished to insignificance by day 25. Ethephon had no significant effect on the size of the berries. It has since been reported that under some conditions Ethephon ripens the pulp but not the bean. Immature beans were found after processing and in very small cherries the bean was liquid although the flesh was red (private communication).

Snoeck (1977) reported on a comprehensive investigation into the effect of Ethephon on polyclonal plantings of robusta at two stations in the Ivory Coast. Solutions of strength 480 and 720 ppm active ingredient were sprayed at a rate of 500 ml per tree at the time when most fruit were close to maturity. The flesh was reddened but the beans were not brought any closer to maturity. The degree of maturity varied from clone to clone. Labour could be saved by harvesting on fewer occasions but at the expense of smaller bean size due to immaturity and corresponding reduction in yield. There were no other effects on the trees. An excessive proportion of immature beans (more than 30 per cent) caused an appreciable loss of cup quality. The variability between clones made timing of an application difficult.

Alvim (1958) reported uniform ripening following application of gibberellic acid in concentration from 10 to 100 ppm. Adenikinju and Poole (1967) reported that gibberellic acid, 2, 4, 5-T, and two proprietary products (Alar and Jiffy Grow) produced significant increases in the number of ripe cherries on arabica coffee four to six weeks after treatment.

Adenikinju (1977) investigated the effects of gibberellic acid and 2, 4, 5-T on robusta coffee cherries. Low concentrations (10 and 50 ppm) of 2, 4, 5-T produced a significant increase of ripe cherries 14 days after treatment. Higher concentrations produced fewer ripe cherries. Gibberellic acid delayed the ripening process, and there were no significant differences after 51 days between the treated and control plots.

Weed Control

The adverse effect of weeds in perennial crops is well known. As an example, Bouharmont (1979) showed that adventitious vegetation, mainly *Graminae*, which was cut down at intervals, reduced crop yield by about 20 per cent.

Some cultural operations can assist in minimising weed growth. Some nurse crops will discourage or smother weeds; this effect may last until shade from a full canopy discourages both nurse crop and weeds. Shading of the ground discourages many weeds. A complete cover of the ground by an even canopy therefore helps in weed control. Some cover crops continue the effect of the nurse crop in discouraging and smothering weeds.

Mulch discourages weeds by shading the soil surface. The mulch may be material carried in, or may form from fallen leaves and other litter from the coffee and from shade trees where they are planted. The mulch builds up and should be almost free of weed seeds. This is another reason why litter, including woody stems which break down, should be, as far as possible, left undisturbed on the ground.

Gaps in the canopy where trees have died lose the litter layer quickly and weeds grow to give a dense cover if not removed. Such gaps should be kept clean, and replacement trees planted as soon as is possible.

The best cultural programme cannot completely prohibit the growth of weeds. Weeds can be removed mechanically or by use of herbicides. Mechanical removal can be by machine or by manual labour. Where interrows are wide enough for machinery to pass, cultivation (disc, harrow, rake or other suitable implement) by a self-propelled or tractor-drawn implement is quick and easy. Where machinery is not available, or cannot move between closely spaced trees, some form of manual cultivation must be carried out. This is usually a hoeing operation, but forking, and cutting very close to the ground using some form of knife are also practised.

Unless the operation is restricted to cutting close to but always above ground level, the litter layer will be disturbed. It is impossible to shave soil so accurately that no disturbance results, either manually or mechanically. The coffee roots active in nutrient absorption proliferate in the litter. These inevitably are damaged and the capacity of the trees to absorb nutrients is reduced. Many weeds are not damaged sufficiently to be killed, they are in effect merely transplanted with some check in growth. Weed seeds are not affected and will germinate after a weeding operation. The production of seeds can be minimised by cutting before flowering. Plants are weakened by frequent regular removal of foliage, but many weeds withstand such treatment over a long period. Deep-rooted, rhizomatous and stoloniferous weeds are particularly difficult to remove; digging them out causes major damage to coffee roots.

The use of herbicides is essential in order to master any weed problem. Cutting, or cultivation, carried out only once or very infrequently, can assist herbicides in controlling a difficult situation. As herbicides are used the weed flora will change; species resistant to the herbicides in use will become predominant. The chemicals used will need to be changed as soon as there are signs of such a development. As with all control methods, success is easier to achieve when infestations are low. If herbicides can be used sufficiently frequently in the early stages to prevent the weeds flowering there will be a reduction in the population of viable seeds, provided that the soil is not disturbed; seeds too deep to germinate will not be brought to the surface. The degree of infestation will therefore diminish and the cost of control will fall. It is almost inevitable that when a programme of herbicidal weed control is introduced, costs are initially unpleasantly high. However it is vital to treat weedy areas initially at a high frequency to prevent regeneration by suppression of seeding. Soil disturbance must be avoided although it may be necessary to pull out or dig up the occasional deep-rooted plant. If the programme is carried through conscientiously costs will fall as weed regrowth becomes progressively lighter.

There is a large range of herbicides available. Many have severe effects on coffee and must never be used. Those that are in general use act in various ways. Some destroy most green foliage which they contact, a few weed species are fairly resistant. These chemicals generally lose their activity on the ground and have no persistence. Paraquat has this character and is still widely used after over 20 years. Other chemicals are absorbed by the weeds and translocated; such materials are essential to remove rhizomatous materials such as 'couch' grass: the name is applied to different species in various locations. These materials do not necessarily persist in the soil. Dalapon has been widely used for this purpose. There are many herbicides which persist for long periods and will thus prevent regrowth. Diuron is widely used in coffee. Some have no effect on existing growth but only control seedlings. Simazine is often used to control weed seedlings in nurseries; it does not affect coffee seedlings. The use of glyphosate has spread in recent years; it has a strong effect on *Graminae* and has some advantages over paraquat and dalapon. The above chemicals are quoted as examples in common use; the many materials available must be evaluated for any situation by reference to research investigations and careful study of the manufacturers' literature.

The effects of chemicals can often be enhanced by applying more than one at the same time. The spread of resistant species can be checked, and a broader spectrum of weeds controlled, by such mixing. The suppliers must be consulted before any such mixing as some chemicals react with one another.

All herbicides must be treated with respect, although their toxicity to humans and coffee trees varies. Application methods must avoid the chemical reaching coffee leaves, and the labour carrying out the work must be properly instructed and protected as may be necessary. Machinery must be maintained in good condition so as to apply only the required quantity in the right place. Leaks can give rise to toxic levels or reach coffee leaves, either of which event could reduce coffee yield. Manual or tractor application can be used as is convenient. Manufacturer's instructions must be followed.

Harvesting

Coffee is harvested either as ripe cherry or as cherry which has ripened and then dried out. Ripe cherry must be harvested from the tree. Dried cherry can be harvested from the tree or from the ground when it has dried and abscissed.

Harvesting of ripe cherry is the practice in most coffee-growing countries. This is usually followed by wet processing, but in Brazil the cherry is sun-dried (see Chapter 10). In Uganda, some robusta coffee is allowed to dry on the trees and is harvested either from the tree or the ground below. The product is known as 'mbuni'.

Harvesting is usually a manual process, employing seasonal labour. Wages are usually paid according to weight harvested. Control is needed to ensure that the amount of unripe cherry harvested is kept to a minimum and that damage to coffee trees is limited.

Mechanical harvesting has not hitherto been of much interest. However Van der Vossen (Chapter 3) refers to breeding trees suited to mechanical harvesting and interest will grow as profit margins tighten and labour costs rise. This process has been under way in the tea crop for some years. Little work has been published concerning harvesting machines. However, Wang (1965), Wang and Shellenberger (1967) and Monroe and Wang (1968) carried out a lengthy study and operated experimental machines in Hawaii. These machines operated by shaking the trees to dislodge the ripe cherry, which was collected by a suitable device close to the ground.

Watson (1980) described two coffee harvesters which were developed in Brazil. These machines employ vibrating fingers mounted on rotating vertical cylinders. It was possible selectively to remove only ripe cherry by adjustment of the vibration amplitude and frequency. The machine was not fully developed at the time of this report but was already in commercial use.

References

Abrego, L. (1962) 'Cambie la epoca de podar sus cafetales en fineas de altura', *Boletin informativo, Instituto Salvadoreno de Investigaciones del cafe, 33*, 6-7

Acland, J.D. (1971) *East African Crops*, Longmans/FAO, Rome

Adenikinju, S.A. (1977) 'Attempts at inducing uniform ripening in robusta coffee (*Coffea canephora* Pierre ex-Froehner) with 2, 4, 5-T and GA. *Acta Horticulturae, 53*, 297-303

Adenikinju, S.A. and Poole, R.T. (1967) *An attempt to induce uniform ripening and to delay abscission of ripe berries in Coffea arabica L. with growth regulators.* Unpublished, Department of Horticulture, University of Hawaii. Quoted by Adenikinju (1977)

Akunda, E.M.W. and Kumar, D. (1982) 'Using internal plant water status as a criterion for scheduling irrigation in coffee east of the Rift Valley in Kenya', *Kenya Coffee, 47*, 281-4

Alvin, P. de T. (p958) 'Use of growth regulators to induce fruit set in coffee (*Coffea arabica* L.)', *Turrialba, 8*, 64-72

Anon. (1975) *Coffee Handbook*, Rhodesia Coffee Growers Assocation, Zimbabwe

Awatrami, N.A. (1982) 'Spacing for arabica coffee', *Indian Coffee, 46*, 120-6

Balasubramanian, M.A. (1970) '*Mimosa invisa* Linn: a possible cover crop for coffee', *Indian Coffee, 34*, 47-8

Basagoita, C.R. (1981) 'Efecto de cuatro distanciamentos de siembra en el desarollo y produccion del cafeto', *Boletin Tecnico, Instituto Salvadoreno de Investigaciones del cafe, 7*, 11-22

Bouharmont, P. (1968) 'La taille du cafeier robusta dans la zone cafeicole de l'est du Cameroun' *Café, Cacao, Thé, XII*, 13-27

Bouharmont, P. (1977a) 'Experimentation sur le renouvellement de l'appareil vegetatif du cafeier par recepage des anciennes tiges. Premiere partie: le cafeier arabica', *Café, Cacao, Thé, XXI*, 91-8

Bouharmont, P. (1977b) 'Experimentation sur le renouvellement de l'appareil vegetatif du cafeier par recepage des anciennes tiges. Deuxieme partie: le cafeier robusta', *Café, Cacao, Thé, XXI*, 99-110

Bouharmont, P. (1978) 'L'utilisation des plantes de couverture dans la culture du cafeier robusta au Cameroun', *Café, Cacao, Thé, XXII*, 113-36

Bourharmont, P. (1979) 'L'utilisation des plantes de couverture et du paillage dans la culture du cafeier arabica au Cameroun', *Café, Cacao, Thé, XXIII*, 75-102

Bouharmont, P. (1981) 'Experimentation sur les dispositifs et les densites, de plantation du cafeier arabica au Cameroun', *Café, Cacao, Thé, XXV*, 243-62

Boyer, J. (1968) 'Influence de l'ombrage artificiel sur la croissance vegetative, la floraison et la fructification des cafeiers Robusta', *Café, Cacao, Thé, XII*, 302-20

Boyer, J. (1969) 'Etude experimental des effets du regime, d'humidite du sol sur la croissance vegetative, la floraison et la fructification des cafeiers robusta', *Café, Cacao, Thé, XIII*, 187-200

Browning, G. (1975) 'Perspectives in the hormone physiology of tropical crops responses to water stress', *Acta Horticulturae, 49*, 113-23

Browning, G. and Cannell, M.G.R. (1970) 'Use of 2-chloroethane phosphonic acid to promote the abscission and ripening of fruit of *Coffea arabica* L.', *Horticultural Science, 45*, 223-32

Browning, G. and Fisher, N.M. (1976) 'High density coffee: yield results for the first cycle from systematic plant spacing designs'. *Kenya Coffee, 41*, 209-17

Burke, R.M. (1975) 'Evaluation of leguminous cover crops at Keravat, New Britain', *Papua New Guinea Agricultural Journal, 25*, 1-9

Capot, J. (1966) 'Regeneration et tailles des cafeiers en Cote d'Ivoire'. *Café, Cacao, Thé, X*, 3-16

Capot, J. (1966) 'La production de boutures de clones selectionnees de cafeiers Canephora' *Café, Cacao, Thé, X*, 219-27

Cestac, Y. and Snoeck, J. (1982) 'Les essais de densites, de dispositifs de plantation et de taille de cafeiers robusta en Cote d'Ivoire', *Café, Cacao, Thé, XXVI*, 183-98

Chenney, R.H. (1925) *Coffee; a Monograph of the Economic Species of the Genus Coffea L.* New York University Press, New York

Cramer, P.J.S. (1934) 'Early experiments on grafting coffee in Java', *Empire Journal of Experimental Agriculture, 2*, 200-4

Deuss, J. (1967) 'Protection de la fertilite du sol et modes de couverture utilises en culture cafeiere en Republique Centrafricaine', *Café, Cacao, Thé, XI*, 312-20

Deuss, J. (1969) 'Influence du mode d'ouverture de plantation, avec ou sans brulis, sur le fertilite du sol et la productivite des cafeiers robusta en zone forestiere centrafricaine', *Café, Cacao, Thé, XIII*, 283-9

Deuss, J. and Borget, M. (1964) 'Resultats d'un essai de plantation du *Coffea robusta* a densite double effectue au Centre de Recherches Agronomiques de Boukoko', *Café, Cacao, Thé VIII*, 17-21

Dublin, P. (1964) 'Le bouturage du cafeier Excelsa. Progres realises au Centre de Recherches Agronomiques de Boukoko', *Café, Cacao, Thé, VIII*, 3-16

Feilden, G., St. Clair and Garner, R.J. (1940) 'Vegetative Propagation of Tropical and Sub-tropical Crops', *Technical Communication Bureau of Horticulture, East Malling, no. 13* Maidstone, Kent, UK

Fernie, L.M. (1940) 'The rooting of softwood cuttings of Coffea arabica', *The East African Agricultural Journal, 5*, 323-9

Ferwerda, F.P. (1934) 'The vegetative propagation of coffee', *Empire Journal of Experimental Agriculture, 2*, 189-99

Fisher, N.M. and Browning, G. (1978) 'The water requirements of high density coffee: 1. Responses to irrigation and plant water stress measurements', *Kenya Coffee, 43*, 43-6

Forestier, A. (1969) *Culture du Cafeier Robusta en Afrique Centrale*, Institut Francais du Café et du Cacao. Paris

Franco, C.M. (1952) 'A agua do solo e sombreamento dos cafezais em Sao Paulo', *Boletin Superintendencia Servicio Cafe, Sao Paulo*, 27, 10-19

Gilbert, (1936) *Report of Coffee Research Station, Moshi*

Haarer, A.E. (1962) *Modern Coffee Production*, Leonard Hill, London

Hatert, J. (1958) 'Premieres observations sur le systeme radiculaire du cafeier robusta'. *Bulletin agricole du Congo-belge*, 49, 461-82

Holliday, R. (1968) 'Theoretical aspects of plant population and crop yield', *Proceedings of a seminar on Intensification of Coffee in Kenya, Nairobi, 1968*, 19-26

Huxley, P.A. and Cannell, M.G.R. (1968) 'Some crop physiological factors to be considered in intensification'. *Proceedings of a Seminar on Intensification of Coffee in Kenya, Nairobi, 1968*, 45-54

Kiara, J.M. (1981) 'Response of two coffee cultivars to spatial arrangements at high density planting', *Kenya Coffee*, 46, 277-81

Lemee, G. and Boyer, J. (1960) 'Influence de l'humidite du sol sur l'economie d'eau et la croissance des cafeiers du groupe *Canephora* cultives en Cote d'Ivoire', *Café, Cacao, Thé*, IV, 55-63

Maidment, W.T.O. (1948) 'Report on minor experimental work. Mulch and shade demonstration plots of Uganda coffee at Bulindi', *Annual Report of Department of Agriculture, Uganda*, April 1947-March 1948, Part II Experimental, pp. 90-1

Marshall, T.H. (1936) 'Coffee grafting and budding', *The Planters' Chronicle*, 31, 568-70

Mestre-Mestre, A. (1977) 'Evaluacion de la pulpa do cafe como abono para almacigos', *Cenicafe*, 28, 18-26

Mitchell, H.W. (1974) 'A study tour of coffee producing countries', *Kenya Coffee*, 39, 24-32, 108-15, 140-7, 164-9, 194-201, 233-7, 283-91

Mitchell, H.W. (1975) 'Research on close spacing systems for intensive coffee production in Kenya', *Coffee Research Foundation, Kenya, Annual Report, 1974-75*, 13-58

M'Itungo, A.M. and van der Vossen, H.A.M. (1981) 'Nutrient requirements of coffee seedlings in polybag nurseries: the effect of foliar feeds in relation to type of potting mixture', *Kenya Coffee*, 46, 181-7

Monroe, G.E. and Wang, J.K. (1968) 'Systems for mechanically harvesting coffee', *Transactions of the American Society of Agricultural Engineers*, 11, 270-2, 278

Muller, R.A. (1975) 'L'irrigation precoce, assurance pour une production reguliere de haut niveau du cafeier arabica', *Café, Cacao, Thé*, XIX, 95-122

Okelana, M.A.O. (1982) 'Rehabilitation of robusta coffee, *Coffea canephora* Pierre. Influence of stumping height and multiple shooting on flushing and cherry production', *Café, Cacao, Thé*, XXXI, 273-8

Oladokun, M.A.O. (1980) 'Legume crops mulch and associated conditions, and plant nutrient content for establishing quillou coffee', *Horticultural Science*, 15, 305-6

Oyebade, I.T. (1971) 'Effect of preharvest spray of Ethrel (2-chloroethane phosphonic acid) on robusta coffee berries (*Coffea canephora* Pierre)', *Turrialba*, 442-4

Poetiray, P. (1981) *Improvements in the Production of Robusta Coffee. A Review of Past and Current Experience*. FAO/DANIDA National Seminar on the improvement of small scale cash crop farming in the Philippines, Davao City, August 1981. FAO, Rome

Porteres, R. (1935) 'Multiplication vegetative des Cafeiers en Cote d'Ivoire', *Revue Botanique Appliquée, et d'Agriculture Tropicale*, 15, 682-94

Reyne, E.H. (1966) 'Un nuevo metodo de injertacin en Cafe', *Boletin Tecnico Dericcion General de Investigacion y Control Agropecuario*, Ministerio de Agricultura, Guatemala 21

Ripperton, J.C., Goto, Y.B. and Pahau, R.K. (1935) *Coffee Cultural Practices in the Kona District of Hawaii. Bulletin N. 75*, Hawaii Agricultural Experiment Station, Honolulu, Hawaii

Robinson, J.B.D. (1964) *A Handbook on Arabica Coffee in Tanganyika*, Tanganyika Coffee Board, Moshi, Tanganyika

Rodriguez, S.J. and Jordan-Malero, J. (1970) *ETHREL: a Potential Coffee Ripener*. Unpublished, University of Puerto Rico, Agricultural Experiment Station, Rio Pedras, Puerto Rico. Quoted by Amchem Products Inc., Ambler, Pennsylvania, Information sheet no. 55

Snoeck, J. (1963) 'La taille du cafeier robusta a Madagascar', *Café, Cacao, Thé*, VII, 421-32

Snoeck, J. (1977) 'Essais de groupement de la recolte des fruits du cafeier *Canephora* a l'aide de l'Ethephon', *Café, Cacao, Thé, XXI.* 163-78
Sturdy, D. (1935) 'Observations on coffee under artificial shade at Selian Coffee Estate, Arusha, 1931-35', *East African Agricultural Journal, I*, 135-9
Triana, J.V. (1957) 'Informe preliminar sobre un estudio de "modalides del cultivo del cafeto"', *Cenicafe, 6*, 156-68
Ulate, R. (1946) 'El injerto en el cafeto'. *Revista de Agricultura,* 569-77
Uribe-Henao, A. and Mestre-Mestre, A. (1980) 'Efecto de la densidade de poblacion y su sistema de manegio sobre la produccion de café *Cenicafe, 31*, 29-31
Uribe-Henao, A. and Salazar-Arias, N. (1981) 'Distancias de siembra y dosis de fertilizante en la produccion de cafe', *Cenicafe, 32*, 88-105
Van der Vossen, H.A.M. (1979) 'Methods of preserving the viability of coffee seed in storage', *Kenya Coffee, 45*, 31-5
Velasco, J.R. and Rodriguez, M.C. (1974) 'Grafting of Coffee', *The Philippines Agriculturalist, 58*, 156-65
Wallis, J.A.N. (1960) 'Note on grasses for mulching coffee', *Kenya Coffee, 25*, 366-7
Wang, J.K. (1965) 'Mechanical coffee harvesting, (Part A)', *Transactions of the American Society of Agricultural Engineers, 8*, 400-5
Wang, J.K. and Shellenberger, F.A. (1967) 'Effect of cumulative damage due to stress cycles on selective harvesting of coffee', *Transactions of the American Society of Agricultural Engineers, 10*, 252-5
Watson, A.G. (1980) 'The mechanization of coffee production', in *9th International Colloquium on the Chemistry of Coffee*, ASIC, Paris, pp. 681-6
Wester, P.J. (1917) 'Shield budding of coffee', *Tropical Agriculturalist, 49*, 179
Wester, P.J. (1922) 'Graftage of coffee', *Tropical Agriculturalist, 58*, 62-3
Wellman, F.L. (1961) *Coffee: Botany, Cultivation and Utilization*, World Crops Books; Leonard Hill, London and Interscience Publishers, New York

8 PEST CONTROL

R. Bardner

Introduction

The most serious pests of coffee are insects, and over 900 species are known to infest the crop. Other invertebrates attacking coffee include nematodes, mites and molluscs, whilst amongst the vertebrates, birds and mammals are occasionally troublesome. The literature on all these pests is extensive and in a chapter of this length it is impossible to give a detailed account of even the most important species. Instead, the principles and problems of pest management in coffee are surveyed, with appropriate examples and references. The discussion is mostly concerned with field problems, for fortunately the pests of stored coffee are few and much less important.

A very comprehensive textbook dealing with all aspects of coffee pests is that of Le Pelley (1968) and this lists over a thousand references to other publications. Since then several hundred more references have been abstracted by the Commonwealth Agricultural Bureaux, many being published in the *Review of Applied Entomology*, Series A. There have been relatively few important developments since Le Pelley's book was published, though some cosmopolitan pests have continued to extend their range and, as mentioned later, some new pesticides have come into general use.

Information on world-wide losses in major crops caused by pests, diseases and weeds was collated and summarised by Kramer (1967). His estimates for losses in coffee attributable to insects were: Africa, 15 per cent; Asia, 10 per cent; America, 12 per cent and Oceania, 10 per cent.

These are gross estimates and conceal much variation between individual holdings. Arabica varieties are generally more susceptible to pests than are the robustas, but climatic factors and management practices all affect pest incidence. Fortunately there are a number of factors making for stability within a coffee plantation and with proper management arthropod pests can usually be kept within bounds without routine 'blanket' spraying of the crop. Nevertheless there are some examples of catastrophic losses such as those that were caused by the pseudococcid *Planococcus kenyae* (Le Pelley) in Kenya in the 1920s and 1930s (Le Pelley, 1943).

Of the insect pests, roughly 400 are Ethiopian, 250 are Oriental and 200 Neotropical (Le Pelley, 1973); these zoogeographical regions are roughly comparable to Kramer's divisions of Africa, Asia plus Oceania and America. There are also a small number of species which are now pan-tropical; these are mostly those that can infest stored coffee thus facilitating accidental transport around the world. Coffee originated in Africa so it is to be expected that this continent also has the greatest range of pests, but the plant family to which coffee belongs (Rubiaceae) is distributed throughout the tropics and many insects feeding on these plants have little difficulty in attacking introduced coffee.

Of the 900 or so species of insect pests, 34 per cent are Coleoptera, 28 per cent Hemiptera, 21 per cent Lepidoptera, 4 per cent Hymenoptera, 6 per cent Orthoptera, 3 per cent Diptera, 3 per cent Thysanoptera and 1 per cent Isoptera. Other invertebrates include nine species of Acarina, two myriapods, several snails and slugs and about 20 nematodes, these latter including both endo- and exoparasites. Birds sometimes eat ripe cherries, and coffee trees are occasionally undermined by burrowing mammals, grazed by herbivores or used as rubbing posts by large mammals, but such damage is infrequent. Some rodents will eat coffee beans.

Coffee as a Habitat for Pests

Coffee provides three main growth stages for colonisation by pests. These are the young seedlings from germination to establishment in the plantation, the mature bush and ripe cherries, including the coffee bean.

Seedlings are normally grown under shade for about 18 months before transplanting. In these nurseries the young plants are crowded together and if grown in seed beds there is ample opportunity for attack by soil-inhabiting pests such as nematodes, cutworms and slugs. This danger is lessened by the modern practice of growing each seedling in its individual plastic container, and can be decreased still further by soil sterilisation or soil pesticides. The young tender shoots and leaves attract a wide range of sucking and leaf-eating pests. When newly planted out the seedlings are subject to similar pests and as they grow larger they may be attacked by shoot-miners and boring beetles.

Mature coffee trees can continue to bear satisfactory crops for many decades, and as the tree is always in leaf the habitat is very stable, although there are annual or bi-annual cycles of growth, leaf production, flowering and fruiting in response to rainy seasons. Methods of growing coffee differ greatly, not only between regions, but often between neighbouring holdings. Pruning systems, presence or absence of shade trees, mulch and irrigation are all reflected in the range of pests found in coffee and their relative abundance, quite apart from the obvious effects resulting from the choice of pesticides and the frequency of their application. It is important to remember that the wide variety of arthropod pests to be found in small numbers in most coffee plantations have numerous parasites and predators. In a well-managed plantation there is a rough balance between pests and their enemies without the need for routine 'blanket' spraying of the trees. Such spraying can often exacerbate pest problems, as explained later.

Some Important Pests

Each part of a coffee plant has its own spectrum of pests. Roots are attacked by nematodes. Among the most serious endoparasitic species are the root knot nematodes such as *Meloidogyne africana* Whitehead, *M. exigua* Goeldi, *M. coffeicola* Lordello and Zamith, *M. decalineata* Whitehead and other *Meloidogyne* spp. Root lesion nematodes of the genus *Pratylenchus* are also troublesome, especially *Pratylenchus coffeae* (Zimmerman), as are the burrowing nematode *Radolphus similis* (Cobb) and the reniform nematodes *Rotylenchus* spp. As with insects, arabica seems to be generally more susceptible than robusta. Many ectoparasitic nematodes have been recorded, but they are usually less troublesome. Some nematodes are widely distributed in more than one continent, e.g. *Pratylenchus coffeae*, others such as *Meloidogyne africana* are more local. For general accounts of nematodes attacking coffee see Whitehead (1968, 1969) Flores and Yepez (1969) and Chapter 9.

Larvae of several beetles also attack roots. Those which are borers are more frequently found in stems and branches, but sometimes reach the larger roots after an initial entry in the lower stem, often known as the collar. Larvae of melolonthid beetles are free-living in the soil and sever smaller roots. Amongst the Hemiptera, coccids (e.g. *Planococcus citri* (Risso)) are particularly troublesome, being associated with a webbing fungus on the roots. They are usually protected by ants. Cicada nymphs also pierce and suck root tissues and are a long-standing problem in Brazil (Le Pelley, 1968: 306-7).

Stems and woody branches are often attacked by larvae of boring beetles, and such pests as *Anthores leuconotus* Pascoe and *Dirphya nigricornis* (Olivier) caused the abandonment of attempts to grow coffee in some parts of Africa (Crowe, 1963; Tapley, 1966), and the cerambycid *Xylotrechus quadripes* Chevrolat (the shot-hole borer) is notorious in parts of the Orient. Borers can cause two types of damage. Some girdle the stem before entering it, destroying the phloem, whilst boring within the xylem weakens the tree mechanically, resulting in stem or branch breakage. Some ants excavate nests in branches and others make nests of leaves, and the bites and stings of a few species are often a nuisance to labourers picking or pruning, but the main problem is that many species protect aphids, scales and pseudococcids feeding on foliage and young stems. Young stems and the unlignified tips of branches are attacked by many of the pests which also feed on foliage or the immature berries. They are particularly attractive to sucking pests such as the scales and pseudococcids. Some Heteroptera can cause multiple branching, an example being the well-known Antestia bug (*Antestiopsis* spp.) of Central and East Africa. Similar damage can be caused by chewing insects feeding on the shoot-tip; this sometimes happens with attacks by larvae of the giant looper, *Ascotis selenaria reciprocaria* Walker. There are also larvae which bore into the tips of young shoots, such as the moth *Eucosoma nereidopa* Meyrick.

The foliage of coffee trees supports the greatest number of pests. As well as scales and pseudococcids (mealy bugs) there are many lepidopterous larvae which eat all the leaf lamina, leaving little except cuticle (skeletonisers) or mine it. Amongst the Hemiptera the aphid *Toxoptera auranti* (Boyer) is most often found on young foliage but rarely in large numbers. Tingids (lace bugs) are more frequent especially in dry conditions which also encourage thrips infestation. Such conditions also favour mites, but usually these are only a minor problem.

Lepidoptera are undoubtedly the most serious pests of foliage, over one hundred species having been recorded. Among the best known are the lyonetiid leaf miners of the genus *Leucoptera*. Several species are among the commonest pests of arabica coffee in East Africa; robusta coffees are less heavily attacked. The new world has its own species, *Leucoptera coffeella* Guerin-Menevill. A characteristic of attacked leaves is that although only a small part of the lamina is destroyed the whole leaf is likely to be shed prematurely. Most of the smaller Lepidoptera eat only the leaf lamina, but as was mentioned in the previous section, larger species can also feed on young shoots. The family Limacodidae have larvae with urticating hairs, and these can be a considerable nuisance to pickers and others working in coffee.

Flower buds can also be damaged by caterpillars more often found on the foliage. *Antestiopsis* will feed on flower buds, preventing any further growth and mirids will attack developing anthers. Young berries can also be attacked by some of these pests and are also colonised by scales and mealybugs. Large green berries are attacked by a variety of boring caterpillars and by sucking insects like *Antestiopsis* — the attacks of the latter can introduce the pathogen *Nematospora* and often result in the production of 'Zebra beans'. As the berries mature and the beans harden they are colonised by a different fauna. The flesh of the berry is utilised by the larvae of several trypetid flies, among them being the Mediterranean fruitfly, *Ceratitis capitata* (Wiedemann). These are now thought to cause little damage when the ripe cherry is infested but infestation while the berry is still green can cause premature berry drop. Many of the pests attacking beans are akin to those found in stored products. Prominent among them is the berry borer *Hypothenemus hampeii* (Ferrari) sometimes known as *Stephanoderes coffeae* Hagedorn. This beetle originated in Africa, but is now common in America and the East Indies and is still extending its range. In Guatemala, for example, it was first noticed in 1971 (Hernandez

Paz and Penagos Dardon, 1974). Although harvested beans are often infested by larvae, the young larvae can only bore into the softer beans found in large green berries so the pest does not usually breed in stored coffee unless it is exceptionally moist. A pest more adapted to storage is the anthribid beetle *Araecerus fasciculatus* (Degeer). In the field it apparently oviposits in ripe berries, but it will breed freely in stored coffee if damp enough and it is now widely distributed. Other storage pests are of very minor importance, with the possible exception of the lamiid beetle, *Sophronica ventralis* Aurivillius.

Pest Assessment

Earlier, world estimates for losses in coffee were quoted, but it is difficult to find data based on sound experimental evidence for even the commonest pests. Losses are obvious if trees are killed, for they are not easy to re-establish and they are not usually close enough together for neighbouring trees to compensate for the loss, but in most areas it is uncommon for established trees to be killed by pests. More often a stem may be killed — typically by boring larvae of beetles, causing complete loss of yield for that stem until it is replaced by a new shoot from the stump — perhaps for 18 months to 2 years. Leaf-eating insects can be important, especially those whose attacks cause leaves to be shed, for premature shedding can result in yield loss. Recent results from Kenya suggest that controlling moderate attacks of leaf miner has increased yields by 6-8 per cent (Waikwa and Dooso, 1980). It should be noted that the effects of foliage insects may take more than one season to become apparent. The fruits are extremely strong sinks for photosynthesis and their growth is often unaffected by defoliation during their development (Cannell, 1972 and Chapter 5, p. 127), because they will draw on other reserves of the plant, even though this may decrease yields the following season.

Flowers and young berries are normally produced in excess, so there are possibilities for compensation if there is any loss at this stage. Indeed, some loss may prevent plants overbearing. Larger berries are not replaced, so attacks by *Antestiopsis* spp. or berry-boring larvae can be very serious — up to 80 per cent of the crop in parts of Brazil has been lost because of infestation by *Hypothenenus hampeii* (Bondar, 1929) and in Uganda, an average population of two *Antestiopsis* per tree will decrease yields by half (McNutt, 1979). This is one of the few coffee pests for which the thresholds for treatment are actually founded on adequate data relating population to subsequent losses, though thresholds based on intuition and experience are used for several other pests, as described below.

Harvested coffee beans marked by earlier attacks of such pests as *Antestiopsis* or *Hypothenemus* may require much labour to remove the damaged beans, and the sale price can decrease through loss of quality. Storage losses are generally more of quality than of the actual weight of the beans.

Principles for the Control of Arthropod Pests in Coffee

The stable environment of a coffee plantation with its diverse and resident population of pests and their parasites and predators makes it particularly favourable for the practice of integrated control techniques, nowadays often called integrated pest management. This seeks to utilise and cordinate appropriate methods of chemical, cultural and biological control to keep pest populations below levels which decrease yields. Frequent foliar sprays of the broad-spectrum insecticides to large blocks of coffee or entire plantations are unwise but there is a temptation to simplify management by mixing such materials with the fungicide sprays that may have to be applied as a routine throughout much of the year. This is often a disastrous policy, because often the pests are less susceptible to insecticides than are their enemies, and instead of attacks diminishing they may increase. Some examples are the use of DDT on dieldrin increasing outbreaks of *Leucoptera* spp. in Tanzania (Tapley, 1961), the appearance of *Ascotis selenoria reciprocaria* in Kenya when routine parathion sprays were used (Wheatley, 1963) and the increase of *Leucoptera* spp. in South America when persistent insecticides were used against *Hypothenumus hampeii*.

Integrated control techniques for coffee have usually been developed on an *ad hoc* basis, utilising keen observation but without any detailed analysis of the population dynamics involved (but for an exception see Bigger, 1973). Those used in East Africa on an arabica coffee are a good example, having been in continuous development ever since the crop came into widespread cultivation in the first decades of the century. The necessity for any treatment (whether cultural or chemical) is determined by crop inspection, and in estates it is usual to employ specially-trained full-time personnel for this purpose. Considerable efforts have also been made to educate small holders so that they also can recognise the major pests and estimate their incidence. As has been mentioned, thresholds for insecticide treatments are used for a number of pests. With *Antestiopsis* this is 1-2 per tree, depending on the locality. Numbers are estimated by test-spraying a few trees with pyrethrum, sheets being placed beneath the trees to catch the falling insects (Wheatley, 1962). A simple sequential sampling system has been devised to increase accuracy and speed (Rennison, 1962). A similar knock-down technique is used to estimate mirids. *Leucoptera* species have about eight well-marked cycles a year brought about by interaction with their parasites. Here the technique is to shake the trees when moths are most abundant. If more than 35 moths fly out insecticides should be considered, the optimum time for foliage sprays being one week later when the population consists mainly of newly-hatched larvae or eggs about to hatch. There are other thresholds for mirids, *Ascotis*, leaf skeletonisers and thrips (Bardner, 1975).

Manipulation of the crop can also decrease pest incidence. *Antestiopsis* prefers dense tangled bushes, so pruning to open out the centre of the bush decreases populations. *Hypothenumus hampeii* prefers shaded coffee, and removal of shade trees is beneficial. Unfortunately the possibilities for this sort of manipulation are often limited because other considerations are more important. Copper sprays are used to control leaf rust. *Hemileia vastatrix* B. et Br. and to increase leaf retention (see also Chapter 9, p. 225), and grass mulches between the rows of coffee are used to decrease evaporation from the soil and competition from weeds; both techniques have favoured *Leucoptera*, the former because there are more older leaves which the pest prefers, and the latter because the mulch is an excellent pupation site.

Two other methods of cultural control are also of interest because they favour natural enemies at the expense of the pest. Both white waxy scale, *Gascardia brevicauda* Hall and brown scale, *Saissetia coffeae* Walker can be discouraged by pruning out infested branches and leaving them under the bushes. Parasites emerging from twigs on the ground will reinforce the attack on pests still on the foliage and berries. The star scale, *Asterocalanium coffeae* Newstead infests bushes near dusty roads, because the scale resists dust more than its parasites. Preventing the dust by treating the road with sump oil or planting a hedge as a dust barrier usually results in the gradual disappearence of the infestations.

Ant-attended scales and mealy bugs can be controlled by a combined chemical and cultural technique. A band of persistent insecticide such as dieldrin is painted round the trunk and below the lowest branches to prevent ants climbing the tree. At the same time judicious pruning will prevent any of the branches forming a bridge for ants between the plant and the ground or weeds. The main parasite of *Planococcus kenyae* is *Anagyrus kivuensis* Compere, and this is one of the classic examples of biological control, for it was not until the parasite was introduced into Kenya from Uganda, the original home of the 'Kenyan Mealybug' that attacks were brought within bounds (Le Pelley, 1943). Attempts have also been made to control *Hypothenenus hampeii* in South America by parasites introduced from Africa but these seem to have had little effect on populations of the pest.

An alternative method of biological control is the development of coffee varieties resistant to pest attack. There is evidence that cultivars differ in resistance to *Leucoptera* spp. (Van der Graaff, 1981) but as yet coffee breeders have considered it more important to select for resistance to disease.

Pesticides and their Application

Previous sections have emphasised that the use of pesticides in coffee should be selective, having the maximum effect on the pest and the minimum effect on beneficial organisms, but the methods used must minimise risks to the operator and the local population, and also be compatible with available application equipment. (See also Chapter 9, p. 222). Most of the latter is designed primarily for the application of fungicides, as these need to be used much more frequently. Large tractor-mounted or tractor-trailed machines are utilised by estates and these have the advantage of speed, essential for pests that have synchronous generations, especially *Leucoptera* spp. Aerial spraying is sometimes done, but the risk to local inhabitants makes the technique inadvisable for the more toxic insecticides, and in any case it may be impracticable in hilly country or if there are shade trees in the coffee. The small farmer usually has to rely on knapsack sprayers or bucket pumps. Even these may be too expensive for individual growers, and often applications are made by spray gangs making the rounds of all the holdings in the district. This facilitates the supply and handling of pesticides and spray coverage is likely to be more uniform, but correct timing of applications is difficult.

Another problem for the small farmer is the safe storage and handling of toxic chemicals. (See also Chapter 9, p. 223.) Lock-up storage, supplies of protective clothing and good handling and hygiene practices are more likely to be found on large farms where there are adequate capital resources and educated managements, and it is desirable to avoid recommending and supplying the more toxic pesticides to the small farmer.

Pesticide applications are of two types — those used as overall treatments, and those applied selectively to smaller areas, but it is important to ensure that even the overall treatments should be made as selective as possible. Hence the broad spectrum organochlorine insecticides have fallen into general disuse as foliage sprays and a range of organophosphorus or carbamate sprays are commonly used, those with the lowest mammalian toxicity being most suited to the small farmer. It is important to test new materials and ensure that they have no greater toxicity or residual activity against common beneficial insects than those already in use, and quite simple tests may be adequate for this. Dosage rates can sometimes be manipulated to decrease the risk to beneficial insects — *Antestiopsis* for example is killed by lower concentrations of insecticides than many other pests. The influence of timing has already been mentioned in connection with *Leucoptera* spp. Real selectivity against some Lepidopterous pests is possible with the anti-feedant fentin hydroxide (Bardner and Mathenge, 1974). Earlier, lead arsenate was found to control many leaf-eating caterpillars with minimal effects on beneficial insects; though today this material might be thought undesirable on the grounds of possible residues. *Bacillus thuringensis* has also been shown to be selective against *Ascotis* larvae (Waikwa and Mathenge, 1977).

Quite a number of pests can be controlled by the selective applications of persistent or moderately persistent insecticides. Dieldrin-banding of trees is a common technique to prevent soil-nesting ants climbing trees and protecting aphids, scales and mealybugs. An injection of the same material into the galleries of wood-boring beetles is also an effective method of control. Adults of the tip-boring caterpillar *Eucosoma nereidopa* rest on the trunks of shade trees about 1.5-2 m above the ground and they can be killed by spraying the trunks of these trees from ground level to 3 m with a persistent organochlorine insecticide.

In recent years a new method of application has been developed against foliage pests, especially *Leucoptera* species. Granules of persistent systemic carbamates or organophosphates are scattered round the roots of the tree, preferably near the start of a rainy season. Subsequent absorption and translocation render leaves toxic to insects feeding on them without jeopardising the beneficial fauna (Wanjala 1976, Almeida, Arruda and Pereira, 1977). Some of these compounds also have nematocidal properties and have been used in soil treatments for nematode control, especially in nurseries (Baeza-Aragon, 1977; Curi, Silveira and Elias, 1975; Echavez Badel and Ayala, 1980; Moraes, 1977; Waikwa, Waikenya and Gichure, 1978; See also Chapter 9: p. 228). Formerly it was impossible to control nematodes in plantations without the use of soil sterilants and fumigants which could only be applied to the soil before the coffee was planted. Soil insects are somewhat easier to control as there is wide range of suitable soil insecticides, both of the older chlorinated hydrocarbon type, and the newer carbamates and organophosphates, but treatment is easier in seed beds than in established plantations, where it is usually necessary to scrape the soil away from the upper roots before treatment.

The control of insect infestations in stored coffee requires close attention to the conditions of storage. A low moisture content is necessary to discourage pests, but care must also be taken to ensure that possibility of cross-contamination between different lots of coffee is minimal. Fumigation is often required, particularly in moist, hot climates, but the tablet method of generating phosphine has made such treatments much easier than the older method using methyl bromide. Synthetic pyrethroids have been used to protect harvested coffee against *Araecerus* (Bitran, Campos and Oliveira, 1977; 1980).

Conclusions and Future Development

Effective principles of integrated pest management were developed in coffee long before such an approach became fashionable among most agricultural entomologists. These techniques could be refined still further if more was known about pest intensity/crop loss relationships and pest population dynamics, but such work cannot be done without adequate resources of scientific manpower and technical facilities. Both are in very short supply in many coffee-producing countries. Changes which are likely are the gradual replacement of persistent organochlorine insecticides with semi-persistent carbamates and organophosphates, and materials with a high mammalian toxicity may gradually fall into disuse. Importers of green coffee are becoming aware of the possibility of pesticide residues, slight though these are, and it may become increasingly important to avoid the use of some pesticides on ripening coffee.

So far there has been little sign of the development of insecticide resistance, but this remains a possibility where insecticide use is frequent. Another risk is further movement of insect pests between continents, especially as many developing countries are having difficulty in establishing effective plant quarantine and plant health inspection schemes.

References

Almeida, P.R. de, Arruda, H.U. and Pereira, L. (1977) 'Efficiency of some granulated systemic insecticides for the control of the coffee leaf miner *Perileucoptera coffealla*', *Biologico, 43*, 29-31, *Horticultural Abstracts, 48*, 8658

Baeza Aragon, C.A. (1977) 'Evaluation of nematicides for the control of *Meloidogyne exigua* in seedlings of the coffee cultivar Caturra.' *Cenicafe, 28*, 108-16, *Horticultural Abstracts, 49* 7141

Bardner, R. (1975) 'Integrated control of coffee pests'. *Kenya Coffee, 40*, 62-6

Bardner, R. and Mathenge, W.M. (1974) 'Organo-tin compounds against caterpillars on coffee leaves', *Kenya Coffee, 39*, 257-9

Bigger, M. (1973) 'An investigation by Fourier analysis into the interaction between coffee leaf-miners and their larval parasites', *Journal of Animal Ecology, 42*, 417-34

Bitran, E.A., Campos, T.B. and Oliveira, D.A. (1977) 'Residual action of insecticides for the control of the coffee beetle *Araecerus fasciculatus*', *Biologico, 43*, 111-17, *Review of Applied Entomology (A), 66*, 5113

Bitran, E.A., Campos, T.B. and Oliveira, D.A. (1980) 'Evaluation of insecticides for the protection of maize and coffee during storage II — pyrethroids', *Biologico, 46*, 45-57, *Review of Applied Entomology (A), 69*, 4978

Bondar, G. (1929) 'Relatorios das viagens', *Boletim do Laboratorio de Pathologia vegetal do estado da Bahia, 6*, 1-64, *Review of Applied Entomology, A* 17, 462

Cannell, M.G.R. (1972) 'Primary production, fruit production and assimilate partition in arabica coffee: a review,' Report for 1971/72. *Coffee Research Foundation, Ruiru, Kenya*

Crowe, T.J. (1963) 'The biology and control of *Dirphya nigricornis* (Olivier), a pest of coffee in Kenya', *Journal of the Entomological Society of South Africa, 25*, 304-12

Curi, S.M., Silveira, S.G.P. de and Elias, E.G. (1975) 'Preliminary results with the chemical control of the coffee nematode *(Meloidogyne incognita)* in the field.' *Biologico, 41*, 67-72, *Horticultural Abstracts, 46*, 3953

Echavez Badel, R. and Ayala, A. (1980) 'Chemical control of coffee nematodes in seed beds in Puerto Rico'. *Journal of Agriculture of the University of Puerto Rico, 64*, 474-81, *Horticultural Abstracts, 51*, 4105

Flores, J.M. and T. Yepez, G. (1969) '*Meloidogyne* in coffee in Venezuela' in J.E. Peachy (ed.) *Nematodes in Tropical Crops*. Technical Communication No. 40, Commonwealth Bureau of Nematology, St Albans, UK pp 251-56

Graaff, N.A. Van der. (1981) *'Selection of Arabica Coffee Types Resistant to Coffee Berry Disease in Ethiopia'*, Mededlingen Landbouwhoogeschool, Wageningen

Hernandez Paz, M. and Penagos Dardon, H. (1974) 'Evaluation of the system of low-volume spraying for the control of the coffee berry borer *Hypothenemus hampeii*', *Revista Cafetalera* No. 134, 15-21, *Review of Applied Entomology, (A) 64*, 4987

Kramer, H.H. (1967) (trans J.H. Edwards) 'Plant protection and world crop production', *Pflanzenschutz-Nachrichten "Bayer", 20*

Le Pelley, R.H. (1943) 'The biological control of a mealybug on coffee and other crops in Kenya'. *Empire Journal of Experimental Agriculture 11*, 78-88

Le Pelley, R.H. (1968) *Pests of Coffee*, Longmans, Green & Co, London

Le Pelley, R.H. (1973) 'Coffee Insects', *Annual Review of Entomology, 18*, 121-42

McNutt, D.N. (1979) 'Control of *Antestiopsis* spp. on coffee in Uganda', *Pesticides Abstracts & New S Service* (PANS), *25*, 5-15

Moraes, M.V. de (1977) 'Systemics for nematode control in coffee nurseries,' *Trabalhos apresentados a II reunion de nematologia, Piricaba, Brazil, Horticultural Abstracts, 48* 10145

Rennison, B.D. (1962) 'A method of sampling *Antestiopsis* in Arabica coffee in chemical control schemes', *East Agrican Agricultural & Forestry Journal, 27*, 197-200

Tapley, R.G. (1961) 'Coffee leaf miner epidemics in relation to the use of persistent insecticides', *Research Report for 1960, Coffee Research Station, Lyamangu, Tanganika*, 43-55

Tapley, R.G. (1966) 'The White Coffee Borer, *Anthores leuconotus* Pasc. and its control', *Bulletin of Entomological Research, 51*, 279-301

Waikwa, J.W. and Mathenge, W.M. (1977) 'Field studies on the effects of *Bacillus thuringiensis* (Berliner) on the larvae of the giant coffee looper, *Ascotis selenaria reciprocaria* and its side effects on the larval parasites of the leaf miner *(Leucoptera sp)*', *Kenya Coffee, 42*, 95-101, *Horticultural Abstracts, 48*, 2881

Waikwa, S.W., Waihenya, W.W. and Gichure, E. (1978) 'Control of root-knot nematode, *Meloidogyne africana*, in a coffee nursery using systemic nematicides.' *Kenya Coffee*, August, 233-37

Waikwa, J.W. and Dooso, B.S. (1980) 'Assessment of crop loss due to leaf miner (*Leucoptera* spp.) damage,' *Report for 1978-79. Coffee Research Foundation, Kenya*, pp. 28-30

Wanjala, F.M.E. (1976) 'Disyston (disulfoton) granules against leaf miner, *Leucoptera meyricki* Ghesq applied at different periods in a rainy season'. *Kenya Coffee, 41*, 277-80

Wheatley, P.E. (1962) 'Antestia testing'. *Kenya Coffee, 27*, 405

Wheatley, P.E. (1963) 'The giant coffee looper, *Ascotis selenaria reciprocaria*', *East African Forestry and Agriculture Journal, 29*, 143-6

Whitehead, A.G. (1968) 'Nematoda' in Le Pelley, R.H. (ed.) *Pests of Coffee*, Longmans, Green & Company, London, pp. 407-22

Whitehead, A.G. (1969) 'Nematodes attacking coffee, tea and cocoa and their control'. in J.E. Peachy (ed.) *Nematodes in Tropical Crops*, Technical Communication No. 40, Commonwealth Bureau of Nematology, St Albans, UK, pp. 238-50

9 CONTROL OF COFFEE DISEASES

J.M. Waller

Coffee diseases are caused by pathogenic microfungi and occasionally by bacteria and some viruses; they affect different plant organs resulting in debility, deformity and sometimes the death of the whole plant. Appropriate measures are often necessary to prevent diseases developing to a level that would reduce the productivity or quality of the crop. The need to undertake disease control depends upon the effects of particular diseases and for this purpose they can conveniently be grouped according to the part of the coffee plant they attack.

The Effects of Diseases on Coffee Production

Berry Diseases

The coffee berry is the harvestable portion of the plant so that diseases affecting berries can cause direct loss of yield even though they may not reduce the vegetative vigour of the plant or its subsequent productive potential. Several fungal pathogens can attack coffee berries; coffee berry disease (CBD) caused by *Colletotrichum coffeanum* is a particularly destructive disease which infects growing berries causing them to rot or to be shed from the plant before the beans are formed inside. CBD only occurs in Africa, but a less virulent form of the same fungus occurs worldwide. This only attacks ripening berries causing 'brown blight' and the mature beans inside are not destroyed. Because this disease causes the pulp to stick to the bean, it makes wet processing more difficult and can reduce the quality of the coffee produced. Similar problems can be caused by other fungi which attack ripe berries. The normal resistance of healthy green berries is reduced when the tree is under physiological stress and these fungi can then infect immature fruit resulting in light or empty beans, thus reducing both the quality and quantity of the harvested crop. Some yeast-like fungi (*Nematospora* spp.) infect berries attacked by *Antestiopsis* spp. (Hemipteran bugs) and can induce an internal rot of the bean. Both lights and *Antestiopsis* damaged beans can produce off-flavours when coffee is wet processed, due to continuing microbiological decay inside the bean. (See Chapter 8.)

Dieback Diseases

Several pathogens infect immature coffee branches causing a 'dying back' of the young stems that will bear the following seasons crop. This symptom may also be caused by root and trunk disease and by foliage diseases which reduce the photosynthetic capacity of the tree. Dieback diseases may not be sufficiently severe to have much effect on the crop already on the tree (which is carried on growth produced in the previous season), but they reduce the cropping potential of the tree so that the following seasons' crop can be drastically reduced. *Corticium* spp., causing pink disease and web blight, can infect bearing branches and destroy the crop that they carry.

Root and Trunk Diseases

These cause general debilitation of coffee trees by reducing their capacity to absorb water and minerals and by disrupting the translocation of substances between roots and shoots. Initially, these diseases cause leaf wilting, shedding or chlorosis. On trees showing these early symptoms the crop may fail to mature properly producing many light or empty beans. Cessation of growth and shoot dieback then occurs and diseased trees, or large parts of them, are killed. These diseases usually occur sporadically on individual trees or groups of trees and do not cause rapidly spreading epidemics. Some are greatly influenced by soil and climatic conditions, only causing severe damage if trees are under climatic stress. Others may be restricted to younger roots; these are not fatal but restrict tree vigour and can be widespread.

Foliage Diseases

Leaf diseases reduce the photosynthetic capacity of the plant and eventually its productive potential, but only those which damage large areas of leaf tissue or cause leaf shedding, such as South American leaf spot and coffee rust, have major effects on the plant. Photosynthesis has to provide carbohydrates for both the developing berries and vegetative growth. Developing berries provide the strongest physiological sink for carbohydrates, so that any reduction in photosynthesis will result in slower vegetative growth. Because the following seasons' crop is carried on the current season's new growth, the main effect of foliage diseases is to reduce the next season's crop. Where major leaf diseases continue unchecked over a number of seasons, progressive decline in vigour and yield occurs. Severe leaf disease on trees carrying a heavy crop may result in photosynthesis being unable to meet the demands of the growing crop; carbohydrates are then withdrawn from remaining leaves and young vegetative tissue resulting in complete leaf loss, overbearing stress and dieback of young shoots and roots. Often, a large proportion of the crop on such trees fails to mature properly; the berries appear dull rather than glossy and are particularly prone to berry diseases. Yellow ripening is another characteristic symptom and there is a large proportion of empty or light beans and accompanying loss of quality.

Strategies for Controlling Coffee Diseases

Wherever possible, it is best to avoid disease problems by preventing the initial introduction of pathogens to new areas through appropriate plant quarantine regulations. It is the infectious nature of diseases which makes them so damaging to plant populations, this is due to the ability of pathogenic organisms to reproduce on their hosts and to spread to infect other individuals. The strategies used to control diseases are related to the different stages of this biological cycle. Cultural and some chemical methods act on the pathogen before it infects the crop, either by reducing or eliminating the source from which it spreads, or by making conditions for spread and infection particularly unfavourable. Most chemical methods act by protecting susceptible parts or stages of the plant from infection. Resistant coffee cultivars and some systemic fungicides act by stopping or reducing the growth of the pathogen after it has infected the crop. Recent advances in the epidemiology of crop diseases have helped to improve the efficiency of many control measures. Further information on the principles and methods used to control diseases of tropical crops can be found in Hill and Waller (1982).

Cultural and Agronomic Practices

General plantation hygiene is one of the most important aspects of controlling root and trunk diseases. These diseases tend to spread fairly slowly and locally; eliminating the source of the pathogen, usually a diseased tree or branch, as soon as possible will prevent the pathogen spreading to neighbouring trees. With root pathogens such as *Armillaria*, excavation of the main roots, from which fungal rhizomorphs can spread to infect neighbouring trees is important, but with *Ceratocystis* or *Corticium*, cutting out the diseased parts of the crown is usually adequate.

Plants suffering from environmental stress are often predisposed to infection by certain types of pathogen, e.g. drought stress and infection by *Fusarium* spp. Adequate pruning and mulching at the start of the dry season helps to prevent *Fusarium* infections of roots and trunks in Africa. The use of shelter belts and light shade in higher altitude plantations prevents coffee being affected by 'hot and cold' disease, a physiological effect which causes a chlorotic deformation of young leaves. These can become infected with *Phoma* spp. resulting in a dieback of young growing shoots. Conversely some diseases, especially those caused by pathogens which require water for dispersal and infection can be favoured by dense shaded canopies. Adequate spacing and pruning to allow air circulation in coffee canopies assists more rapid drying of plant surfaces so that infection is restricted; spray penetration is also facilitated, thus improving the efficiency of chemical control.

Generally, optimum cultural conditions to ensure good growth of the tree help to offset the ill effects of disease by making the tree physiologically more tolerant. Sometimes it may be necessary to remove some of the developing crop if this is very heavy and the trees begin to show signs of overbearing stress.

Chemical Control Methods

Fungicides are widely used on coffee particularly to control foliage and berry diseases. Chemical control methods are well suited to perennial cash crops because their value allows the costs to be economic and other methods of control which rely on changing crop varieties, crop rotations or cultivation practices cannot be used to any great extent. Even so, economic constraints have stimulated the search for increasingly efficient chemical usage and higher yields to offset costs. There are three major aspects of chemical control; which substance and how much to use, how to apply it to the tree, and when should it be used. The first depends largely on which pathogen is being controlled but there are basically two types of fungicide.

Protectants act by killing spores or other pathogen propagules on the plant surface before they can infect; they have to be applied by methods which achieve good coverage of the target (susceptible part of the plant) and fairly frequently to ensure protection of new growth and to replace that removed by rain. Copper-based fungicides, captafol and chlorothalonil are used widely on coffee as they are fairly tenacious, but are redistributed throughout the tree canopy as rainwater percolates through it. Redistribution of fungicides in rainfall is a major factor assisting the control of diseases such as coffee rust and berry disease which require rainwater for infection.

Systemic fungicides enter the plant tissues and kill the pathogen during or after infection. They are also redistributed in the plant to some extent so that maximum spray coverage is not so essential, but they will not move from leaves to roots or leaves to berries. Generally, systemic fungicides are more expensive than protectants, but the lower doses used can offset this. A major snag with most systemic fungicides is that many pathogens have developed resistance to them.

A wide variety of spraying machines have been used on coffee, and the type used depends very largely on the terrain and suitability of the plantation for mechanisation. On fairly level estates with well-spaced coffee, tractor-drawn sprayers are suitable. Hydraulic sprayers using high volumes of spray have now largely been replaced with medium or low volume mistblowers which carry spray droplets into the tree canopy in a flow of air; this disturbs the canopy and gives good spray coverage of most plant surfaces. Ultra-low volume machines producing even-sized spray droplets from a spinning disc have also given good control of coffee rust in Brazil, but the lower dosages applied by these machines makes them more suited to the application of systemic fungicides. Because good control of coffee rust and berry disease can be achieved by redistribution of protectant fungicides in rainwater, aerial spraying from fixed-wing aircraft, which apply chemicals mostly to the top and edges of the canopy, has been used successfully in East Africa and Brazil.

The large areas of smallholder coffee in many tropical areas and plantations on steep land, e.g. in Colombia and Central America, are only suitable for spraying by hand. Motorised knapsack mistblowers give good coverage of most plant surfaces and use fairly low volumes, but they are heavy to use in hilly terrain and require investment and maintenance which is often beyond the capacity of smallholders in isolated areas. Manually operated knapsack sprayers of the pressure retaining type or with a hydraulic hand pump are the most widely used type of sprayer. With all hand-directed equipment, even coverage of the canopy is important particularly at the top of the tree which can carry a reservoir of fungicide for subsequent redistribution through the canopy. Hand directed ultra-low volume, spinning disc sprayers, are now being tested for use on coffee. These have the advantage that they are much lighter to use, requiring minimal volumes of spray liquid, and apply smaller doses of chemical more efficiently with a consequent saving on costs.

The most common problem with spraying machines is blocked or worn nozzles; using clean water in which to mix chemicals, and cleaning the filters built into the sprayer after use will help maintain them in good working order. None of the fungicides currently used on coffee have caused problems with processing, residues in beans or tainting of processed coffee and although they present no special hazards to the operator, the appropriate precautions should always be taken. These are:

(1) Avoid splashes and spillages when mixing chemicals and filling machines;
(2) Do not eat or smoke while handling chemicals and avoid contamination of skin or clothing;
(3) Use the correct dose and application method as recommended by the manufacturer;
(4) Dispose of old containers, excess spray liquid, etc. in a safe place — usually a pit from which substances cannot seep to contaminate waterways;
(5) Clean equipment after use and wash off any spray material from hands and other areas.

The timing of fungicide applications depends on the epidemiology of the pathogen and the persistence of the chemical being used. Weather conditions, particularly rainfall, are important in determining when pathogens spread and infect crops so that in general, protection of the crop is only needed during the rainy season. Furthermore, only some stages of the plant may be susceptible to infection so that fungicide application can be limited to those times when these susceptible stages occur. Fungicide application may need to be more frequent during periods of rapid growth or when heavy rainfall results in rapid erosion of fungicide deposits. Protectant fungicides act by preventing infection so that timely applications are essential; systemic fungicides can eradicate existing infections so that timing is not so critical (see also Chapter 8: pp. 215-6).

Disease Resistance in Coffee

The use of resistant varieties is an ideal method of controlling crop diseases as it does not require the recurrent effort and costs needed by chemical and cultural methods. Resistance is a genetically determined characteristic which enables the plants natural defence mechanisms to 'recognise' pathogens and react appropriately. Resistance based on single genes having major effects (vertical resistance) is easily manipulated by plant breeders, but is often only temporarily effective as it can be countered by matching pathogen virulence genes which can become predominant in the pathogen population through selection. However, a more durable form of resistance (horizontal resistance) is usually expressed by the effect of many interacting genes but is much more difficult to manipulate by recognised breeding techniques. In perennial crops such as coffee, which require a major initial investment at planting and which must continue to give satisfactory yields over many years, disease resistance which may only be temporary is a disadvantage. This has prevented genetic resistance from being used against coffee rust as major resistance genes in *Coffea arabica* can be overcome by virulence genes in the pathogen. However, there are other genes in coffee which can confer a more durable resistance to rust. Some of these occur in other *Coffea* spp. and research on coffee breeding and interspecific hybrids has now produced hybrid populations which possess partial but apparently durable resistance to rust (Eskes, 1983).

Resistance to other diseases such as CBD appears to be durable. Several selections of *Coffea arabica* are known to possess resistance to CBD, and have been successfully used to combat the disease in Rwanda, for example. When CBD reached Ethiopia, the centre of diversity of *Coffea arabica* FAO established a programme to select resistant varieties from the diverse semi-wild population which existed there. Despite the availability of reliable resistance to CBD, the disease is still a problem in many parts of Africa, requiring expensive spray programmes to control it. This is because there are many other characters which a variety must possess before it can be considered acceptable to growers and consumers. Of particular importance are quality, yield and other desirable agronomic traits. The incorporation of resistance to several diseases together with these other characteristics can only be produced by long-term breeding programmes requiring considerable investment by governments or commodity boards,

Coffee Diseases

Individual diseases are recognised by the symptoms they produce and by the identification of the causal organism based on its biological characteristics. There is no single text which adequately describes all diseases but the Kenya Coffee Board (1979) illustrates those which occur in East Africa.

Coffee Rust

This is caused by the fungus *Hemileia vastatrix* and is recognised by the orange powdery spots on the undersides of coffee leaves. Diseased leaves are shed prematurely reducing the amount of vegetative growth and consequently the following seasons yield. Heavy defoliation may cause overbearing stress and dieback which reduces the yield and quality of the current seasons crop.

Coffee rust is a classic among plant diseases, being one of the first to be studied scientifically when it destroyed the Ceylon coffee industry between 1860 and 1880. The disease spread across Asia to the Pacific Islands and across Africa. It reached the New World as late as 1970 and now affects most Latin American coffee-producing countries. Spores of the fungus, produced from the leaf lesions, are dispersed by wind, rain and passive carriers such as man and machinery. Infection requires rainwater so that disease spread and development is limited to the rainy season, but because of the relatively long incubation phase (2-6 weeks depending on temperature) maximum incidence occurs early in the dry season. Cool conditions in coffee at high altitudes ($>1,700$ m) prevent severe epidemics.

Control is based on the application of fungicidal sprays during the rainy season. Copper-based protectant fungicides applied at 4-6 week intervals using 3-5 kg/ha are the standard treatment. Lower doses are effective when low volume, controlled droplet sprayers are used. Systemic fungicides such as pyracarbolid, oxycarboxin and triadimefon have also been used, but seem to be less reliable than copper. Sources of resistance in *C. arabica* have been overcome by virulent races of the pathogen, but hybrids between *C. arabica* and *C. canephora* show resistance to all known pathogen races. Backcrosses using these hybrids, e.g. 'Catimor', are now under trial, and have reached commercial status in Colombia where the multiline 'Colombia' is being released. Waller (1982) provides the most recent review of this disease.

Grey rust, caused by *Hemileia coffeicola*, occurs only in West Africa where it can be serious on some cultivars of *C. canephora* grown in very wet areas. It is favoured by hot humid conditions which are unsuitable for arabica coffee.

Other Foliage Pathogens

Brown eye spot, caused by the fungus *Cercospora coffeicola*, is the most common and widespread leaf disease but rarely causes detectable yield loss so that specific control measures are not justified. The circular brown spots with a greyish centre sometimes have a yellow halo but seldom grow to more than 5 mm diameter. The disease flourishes in warm humid conditions, and where coffee is overcrowded or under nutrient stress; it can reduce the vigour of very young plants so that chemical control with protectant fungicides may be justified in coffee nurseries (Siddiqui, 1970).

South American leaf spot, caused by the fungus *Mycena citricola*, only occurs in Latin America where it can seriously damage sheltered coffee in cool, wet areas. The disease produces buff to white circular lesions up to 15 mm diameter which may occur on stems and berries as well as leaves. Heavy infection, which is often very localised, causes defoliation with subsequent yield loss. Phytosanitary measures such as the removal of badly diseased coffee, shade or hedge plants which can act as alternative hosts is effective but control by fungicides is necessary in some mountainous areas. Lead arsenate has been widely used but copper-based compounds and tridemorph are also effective.

Irregular dark necrotic lesions can be produced by *Phoma* spp. and *Colletotrichum* spp. on old or damaged leaves. A virus disease producing chlorotic rings and spots on leaves occurs sporadically in Latin America. The bacterium *Pseudomonas syringae* can also cause leaf lesions and blighting; this is described under 'Dieback diseases'.

Coffee Berry Disease (CBD)

This disease, caused by a virulent form of the fungus *Colletotrichum coffeanum*, is limited to east, west and northern parts of central Africa. It attacks young expanding berries (5-20 weeks old) producing black sunken anthracnose lesions which cause fruit rotting and shedding. Ripe berries are also susceptible and the overlapping of successive crops can increase disease severity. The disease is favoured by wet conditions as rainfall is required for spore production, dispersal and infection.

Durable resistance to CBD exists and is used in Rwanda, Ethiopia and some other countries, but the SL varieties in Kenya are very susceptible and fungicides are necessary for control. These are applied during the rainy seasons when berries, at susceptible stages, are on the trees. Copper-based fungicides have mostly been replaced by captafol or chlorothalonil applied at about 3 kg/ha on a monthly schedule. Benomyl and carbendazim are widely used, but the pathogen developed resistance to these fungicides after a few years intensive use. *C. coffeanum* can also infect leaves, flowers and maturing stems where it exists as a major component of the bark microflora. (See Firman and Waller (1977) for further information.)

Other fungi can attack berries when they ripen and earlier when they are under stress. *Botrytis cinerea* can cause 'warty berry' in very wet conditions, *Cercospora coffeicola* causes berry blotch, and *Fusarium stilboides* can rot the pulp of mature berries. These pathogens may cause substantial damage on overbearing trees.

Root Diseases

The fungus *Armillaria mellea* is a widespread root pathogen of many woody plants and occurs frequently on coffee, particularly on recently cleared land or where shade trees have been removed. Moribund stumps or roots provide a food source from which the fungus can spread to infect coffee. Symptoms usually appear as a rapid debilitation, wilting and death of affected trees. Creamy white mycelial strands can be seen beneath the bark and clusters of the characteristic pale brown mushroom-like sporophores occur on the base of recently killed trees. Control measures are restricted to disease prevention by adequate land clearance, ring barking of shade trees to starve and kill the roots before they are cut down, and the prompt removal and destruction of diseased coffee trees and their roots.

Rosellinia spp., causing black root rot of many tropical crops, can attack coffee in Latin America especially on recently cleared land. Symptoms are broadly similar to those caused by *Armillaria*, except that the fruiting bodies are minute spherical structures. Control measures are also similar. Another lethal root disease, caused by *Fusarium solani*, occurs sporadically in Africa, especially in dry areas.

Several nematode species attack the roots of coffee causing debilitation of the tree when severe, *Meloidogyne exigua* causes knots or galls which deform the roots and *Pratylenchus coffeae* causes lesions on the fine feeder roots; both have caused damage to plantations in parts of Latin America.

Wilt and Dieback Diseases

Fusarium bark disease, caused by *F. stilboides*, has restricted coffee growing in Malawi and is important in other central African countries. The fungus is widely distributed but only damages susceptible varieties grown under warm, dry conditions. The characteristic symptom is a scaling of the bark leading to stem cankers and a progressive dying back of the whole tree; green vegetative shoots are also killed. Symptoms are most pronounced at the collar region and can be alleviated by the application of fungicidal paints and optimum cultural conditions (Baker, 1970). Recent selections of geisha and agarro types of *C. arabica* show good resistance. Another related species, *Fusarium xylarioides*, affects the vascular system of the coffee tree causing a wilt and dieback and is of sporadic occurrence in Africa.

Coffee wilt in Latin America is caused by the fungus *Ceratocystis fimbriata* which usually gains entry to the plants through wounds at the base of the trunk. The pathogen causes dark sunken necrotic cankers and can kill trees when the stem is girdled. *C. canephora* and *C. liberica* are more resistant. Apart from the early removal of diseased trees, no control measures are practised.

Bacterial blight or dieback of arabica coffee is caused by the bacterium *Pseudomonas syringae* which infects young tissues during cool, wet conditions to produce a dark necrotic leaf blight and dieback of vegetative shoots. A recent upsurge of the disease in Kenya was due to the replacement of copper fungicides, which are bactericidal, with organic fungicides which have no bactericidal effect. The disease is most severe at high altitudes. *Phoma* spp. can cause a similar, but less severe dieback of young shoots in high altitude areas.

Corticium spp. causing pink disease and web blight or koleroga can infect mature branches of coffee in wetter areas of South and Central America. The mycelium spreads over the surface of stems, leaves and fruits causing the death of individual branches. Removal and destruction of diseased branches is usually sufficient to control the disease, but copper-based fungicides are also effective.

Nursery Diseases

Damping off of young coffee seedlings can be a problem in the coffee nursery. This is usually caused by the fungi *Rhizoctonia solani*, *Fusarium solani* and *F. stilboides* and can be controlled by applying thiram-based fungicides to seed or soil. Nematodes can also cause problems in the nursery; these live on the roots and can cause debilitation of mature coffee trees in the field, especially when spread from the nursery on young plants. Sterilising soil with heat or fumigant chemicals such as methyl bromide or methyl isothiocyanate (applied as dazomet granules which release the gas when mixed with soil) will kill nematodes, but soil must be left for a week or two to allow dispersion of the fumigant before seeds are planted. Granular systemic nematicides can also be applied to soil after planting; they are very toxic and should be handled with care using gloves. (See also Chapter 8; p. 216).

References

Baker, C.J. (1970) 'Coffee bark diseases in Kenya', *Kenya Coffee*, 35, 226-8
Coffee Research Foundation (1979) *Atlas of Coffee Pests and Diseases*, Coffee Board of Kenya.

Eskes, A.B. (1983) 'Incomplete Resistance to Coffee leaf rust' in F. Lamberti, J.M. Waller and N.A. Van der Graaff (eds.), *Durable Resistance in Crops*, Plenum, New York

Firman, I.D. and Waller, J.M. (1977) 'Coffee berry diseases and other *Colletotrichum* diseases of coffee', *Phytopathological Paper* No. 20, Commonwealth Mycological Institute

Hill, D.S. and Waller, J.M. (1982) 'Pests and diseases of tropical crops,' Vol I *Principles and Methods of Control*, Longman, ITAS

Siddiqui, M.A. (1970) 'Incidence, development and symptoms of *Cercospora* diseases of coffee in Malawi' *Transactions of the British Mycological Society, 54*, 415-21

Waller J.M. (1982) 'Coffee rust — epidemiology and control' *Crop Protection*, 1, 385-404

10 GREEN COFFEE PROCESSING
R.J. Clarke

Introduction

Following harvesting, green coffee is prepared from the cherries (or berries) of the coffee tree by a relatively complex series of process steps carried out entirely within the producing countries. Technically, the basic problem to be overcome is to take out the coffee seeds from within the cherry, by removing the various covering layers (see Figure 10.1) in the most efficacious manner; and to provide the coffee as far as possible as marketable green coffee beans, with a moisture content of less than 12-13 per cent. Two main processes have been used for many years, if not centuries, traditionally distinguished as the dry process and the wet process, which differ fundamentally in their approach. Apart from differentiation by species and geographical origin, green coffee prepared by these processes is known as either dry process ('natural') or wet process ('washed') coffee. In addition, certain cleaning and polishing processes may be additionally adopted, together with various grading and sorting techniques to provide the finished coffees by specification for export.

Numerous accounts of these processes have of course appeared in the lay and technical literature, with especially detailed and authoritative texts by Haarer (1962) (with particular emphasis on African practices), Foote Sivetz (1963) and Sivetz and Desrosier (1979) (with greater reference to Central and South America), and earlier by Wilbaux (1952) with particular reference to French-speaking regions. Much additional information

Figure 10.1: A. Cross Section of Coffee Fruit; B. Longitudinal Section of the Coffee Fruit

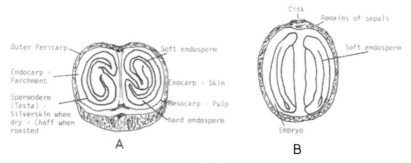

is to be found in the proceedings of the Association Scientifique Internationale du Café, Paris (ASIC), whilst an excellent review by Kulaba (1979) on green coffee processing in Kenya should be consulted. It is of particular interest that much of the engineering equipment involved has been supplied the world over by old-established British specialist engineering companies, such as Bentell's, Gordon's, Mackinnon's and others, and has continued in reliable usage over a large number of years. There are of course, also US companies, such as Squiers and Marcus and Mason, and French companies supplying reliable equipment. Green coffee processing is in fact, another field of application for chemical engineering (Clarke, 1976). The various steps in these processes have been investigated technically and scientifically by many workers as will be seen from published papers and reports from the many coffee research stations throughout the world. A listing of such institutes would be too lengthy for this review; but we should certainly mention the Institut Français du Café et cacao (IFCC) in Paris, with its offshoots in the Ivory Coast, Cameroons and Togo; the research station at Ruiru, Kenya (and KIRDI, formerly EAIRO, the East African Industrial Research Organisation); the Lyamungu station in Tanzania; those in South America, Turrialba in Costa Rica, Campinas in Brazil and Chinchina in Colombia, and those in Asia, and India. It would be difficult to provide a simultaneous and up-to-date global view of all the precise green coffee processes in use. Nevertheless this chapter will provide an overall outline, together with modifications of steps that have been studied, retained or dropped.

Some distinction should be made between native-grown and plantation-grown coffee (i.e. from large estates), in respect of processing. In both cases, however, the coffee is often sent to central coffee 'curing works' which complete the late stages of green coffee processing and prepare the coffee for export.

Although the composition of the green coffee beans is fully described in Chapter 13, a knowledge of the composition of the various layers in Figure 10.1 is desirable, though their removal is primarily mechanical in nature. The mucilage layer is, however, deliberately subjected to biochemical changes for its removal, (see pages 235-40). The fleshy pulp represents about 29 per cent of the weight of the coffee cherry on a dry basis and when fresh, contains about 77 per cent water. The composition of pulp and parchment has been examined by a number of workers as reported by Braham and Bresani (1979).

Harvesting

Agricultural practices in the cultivation of coffee have been described in Chapters 6 and 7, when harvesting is the final stage. Harvesting is an important element in the overall green coffee processing, in the selection of coffee cherries that are made.

Traditionally sequential hand-picking is employed as for example in Colombia, where selection of properly ripe cherries (i.e. those with red skins) can be made, though as it is dependent upon the weather conditions, it may be difficult to organise. Coffee trees in different locations will differ in respect of the range of maturity of its crop on one tree, ranging from immature yellow or green cherries to soft overripe and even hard dried; some cherries may contain 'black beans'. A proportion of the crop may also be found on the ground, or on matting designed to collect any falls, accidental or deliberate. At this stage also, some beans will have insect damage consisting of cavities made by the coffee berry borer (see also Chapter 8). In Brazil, where in the states of Saõ Paulo and Parana a long dry period usually coincides with the harvest season, strip picking is generally practised with the coffee tending to be overripe but variable. In Brazil also, considerable progress has been made in the use of mechanical harvesters (Watson, 1980) and in appropriate pruning and spacing of the trees (see also Chapter 7).

The Dry Processing Method

This process is the oldest and simplest, consisting of three basic steps. The harvested cherries may be subjected to some form of classification, but then are dried in their entirety, most usually in the sun. The dried cherry coffee ('café en coque, Fr.; cafe em coca, Port.) is then subjected to a milling operation (or 'hulling', or rather 'dehusking') to separate out the green beans. The initial drying of the cherries also means that the final beans are sufficiently dry for subsequent storage and bagging for export.

This process method in particular, is adopted for nearly all (about 95 per cent) Brazilian arabica coffee; and is also a feature of preparation of robusta coffee in most parts of the world where it is produced. Whilst the process has a somewhat lower quality connotation than the wet process (see pages 234-43), this is not necessarily so, and indeed the prized Mocha-type coffees are produced in this way.

Classification

Ideally, classification, that is the removal of unripe and damaged cherries, and therefore underlying inferior seeds, should take place primarily as a result of the harvesting. As already mentioned, stripping is general in Brazil, but some classification on the way to the next step of drying is and can be practised by flotation and separation with flowing water during conveyance to the drying areas. This step also serves to clean the coffee of dirt and other material. Sieving can be used to remove dirt and soil, whilst stones, twigs and leaves can be picked up from the top side of screens.

Drying

Coffee cherries are laid out in the sun to dry, though machine drying is also practised. In Africa, on small plots, cherries are often laid out on matting supported by cane sticks or staging. Solid concrete (or other hard material) surfaces on the ground are more general, especially in the larger patio installations in Brazil, fed by water fluming methods. Though the cherries will be heaped on to the surfaces, attention must be paid to the thickness of the drying layers adopted, and furthermore, the cherries need to be turned during their drying period of some eight to ten days in good weather. The recommended level of loading is about 20 kg/m^2, though in practice higher levels will be found. Coffee cherry is hygroscopic, so that it is possible for moisture eliminated during the day to be re-absorbed at night. The final moisture content required is 13 per cent w/w, and preferably rather less. As with all coffee moisture values, the value is somewhat arbitrary, depending upon the particular oven drying method used for determination (typically 105°C for 16-24 hours). Moisture meters, calibrated by an oven method, giving rapid readings are increasingly popular; locally, tactile methods will be used to indicate adequate drying (see also Chapter 13).

The machine drying of coffee cherry is particularly used in Brazil, but also in larger installations in Africa for total or finish drying. Foote and Sivetz (1963) and Sivetz and Desrosier (1979) give a good account of the type of drying machines in use, together with an examination of the various process parameters of temperature and flow rates. Drying by this method only takes some three days from the original moisture content of 65-70 per cent.

Clearly sun-drying is subject to the vagaries of atmospheric conditions, together with the possibilities of growth of both desirable and undesirable micro-organisms generating substances from the drying pulp, affecting subsequent flavour of the coffee brew made from the coffee after roasting. Related to this type of drying, is the occurence of so-called 'hard' coffee, or Rio type in Brazil, characterised by a 'medicinal' type of flavour, especially notable in certain areas. Some work has been carried out on this subject (Haarer, 1962; Amorim, 1975), but such flavours are not inherent to the process. Indeed, Brazilian coffee is classified in the degrees of 'softness'; whereas the 'hard' coffee is separately sold.

A comparison of machine-dried with sun-dried Brazilian coffee was reported by Teixeira (1982), who showed the importance of keeping the air drying temperature down to 30°C to minimise the incidence of defective beans, including 'blacks'.

Hulling (Dehusking)

The finally dried coffee cherries must next have their dried husks removed to release the coffee beans. Traditional African native practices would involve the use of large wooden mortars with pole-like pestles; though now a large number of powered machines available in different sizes are, of course, in use. The general tendency is, however, to have native grown-and-dried, and also plantation coffee, sent to a central 'coffee curing' works, of which a good modern example is that in the Ivory Coast at Toumbokro, described by Richard (1977). Deliveries of dried robusta cherries are made and paid for on a weight basis, but with discounts made against high moisture contents, as determined on a sample taken from the delivery. The Control Laboratory also assesses the defective bean count, especially in respect of black beans, after hulling and electronically sorting a sample, when again there is a scale of discounting the basic rate of payment.

The machines for dehusking dried coffee cherries are usually somewhat similar to those for milling off dry parchment in the wet process (see page 242), but they need to be able to dispose of the larger quantity of non-coffee bean debris. It should be noted that the layer closest to the beans themselves, i.e. now-called the silverskin ('testa' in the wet cherry), is difficult to remove from coffee prepared by the dry process, so that considerable silverskin may be found adhering.

The average colour of beans prepared by the process ranges from 'greenish' to 'brownish'.

The Wet Processing Method

This process is more sophisticated than the dry process, and by general consent leads to better quality coffee and commands a higher price. The process carried out correctly when applied to arabica coffee provides so-called parchment coffee (café en parche, Fr.; pergamino, Sp.) and finally hulling (decorticage, Fr.) provides the dry green beans.

This method is generally used for arabica coffee, except in Brazil; but is also occasionally used for robusta coffee, and expected for general use with arabusta coffee. It requires the use of substantial quantities of water. The method is expertly used in Kenya and Colombia in particular, but also numerous other countries, in Central American countries, India, Tanzania, Indonesia and so on.

The product should flow in the most convenient and economic way from a high to low ground level; the movement through the pulpery section to the wet parchment stage is assisted by the gravity flow of water. The location of the water supply is, therefore, important, and it will be noted that a hillside is a convenient means of providing this flow. Perhaps the key operation (and difference from the dry process) is the pulping, i.e. in which the soft pulpy part of the cherry together with the skin is 'torn off' as soon as possible. However, the fermentation and washing process to remove residual pulp and mucilage before drying is particularly important.

Classification

Special care is usually taken in the provision of ripe fruit only, which may be sorted out by hand at picking or immediately after picking. Payments to pickers can be arranged to maximize the provision of ripe cherries. In the subsequent operation of pulping, the pulpers will not handle satisfactorily hard, partially dried cherries, nor are unripe and overripe cherries desirable in the process. Separation by flotation in water may also be carried out in two or more stages, first to remove stones and dirt, and second to separate cherries by means of large tanks filled with flowing water. The heavy cherry sinks and may be syphoned off, whilst the lighter cherry (essentially mainly hard partially dried coffee) floats away through an escape to a draining table. The separation is not clear cut between truly ripe and unripe cherries, and between small and large cherries, though screens and additional stages help. A typical classification flow diagram is given by Sivetz and Desrosier (1979).

Pulping

There should be as little delay as possible in the next stage of pulping, since the juicy fruit may start to ferment. It is recommended that this stage should be completed within 12-24 hours after picking, otherwise the cherries should be kept under water for another day. Taints can arise from rotting pulp or unclean water. Different qualities of cherry for whatever reason should be processed separately.

The cherry coffee arrives at the hoppers of the pulpers with water, where its flow is regulated. The pulping machine needs to be adjusted, for the average size and ripeness of the coffee cherry being presented. The waste pulp should be thrown out at the rear of the machine, and conveyed well away; but some may inevitably go forward with the beans. The pulped beans therefore need to go next to some form of vibrating screens when the larger pieces of pulp and imperfectly pulped cherries are sent to another smaller pulper (or so-called 'repasser'). Descriptions of these machines are given by Sivetz and Desrosier (1979) and Haarer (1962), including such machines as the Aquapulper, which also appear able to divest the cherries of the mucilaginous material, otherwise removed by the next step of fermentation.

The separated pulped beans should then pass in short water-washing channels (usually made of concrete) where a further flotation separation can be arranged by means of sluice gates.

Fermentation

This stage is necessary to remove any residual adhering pulp, and importantly to remove a very adhesive mucilaginous layer (some 0.5-2 mm thick, according to species and maturity), leaving only the parchment covering, enclosing the green beans, or seeds (as they still are, strictly speaking, called). The composition of the mucilaginous layer has been investigated by a number of workers, as reported by Braham and Bresani (1979). By percentage weight, it is typically 84.2 per cent water, 8.9 per cent protein, 4.1 per cent sugar (about 60 per cent of which are reducing sugars), 0.91 per cent pectic acid (substances) and 0.7 per cent ash, but its composition over a range of different types of cherry is not totally clarified. Mucilage contains various pectin-degrading and hydrolysing enzymes, not totally defined which are important to the fermentation that is required to loosen this gel-like material. Wilbaux (1956) as reported by Braham and Bresani (1979) believes that some of these enzymes are not inherent to mucilage itself, but have migrated from the pulp or even the endosperm by osmosis and diffusion.

Fermentation in large scale-practice is a batch process taking place in rectangularly sectioned concrete tanks of which a number will be needed in a typical pulpery, for both capacity reasons, and the need to process separately and simultaneously any different qualities of coffee in process at that stage. As pulping proceeds, so the water outlet from a tank is closed until the tank is filled with its complement of pulped coffee. Fermentation takes place whilst the coffee lies in a sticky mess (say one metre thick) with the water drained away — so-called 'dry fermentation'. The time of fermentation allowed varies according to the location, with temperature being an important factor, from about six hours to as long as 80 hours, though a time of 24 hours may be regarded as typical in a pulpery at say 900 metres altitude. The finish of fermentation is empirically assessed by feel, and when adjudged complete, the coffee should be immediately washed with water. Washing may be conducted in further concrete channels (say 100 feet long with suitable slope), and the water drained off. Washing machines are also available for this operation, with a washing time of some ten minutes.

During fermentation, variations in pH have been found to have an important influence on enzyme activity, and to indicate the presence of various generated acids. An initial typical pH of 6.8-6.7 drops to 4.2-4.5 when the end-point is reached. It is also clear that pulped coffee is a good substrate for other enzymes present in air-borne moulds, yeasts and bacteria.

Fermentation Studies

This process has been especially examined by a number of workers over the years. Wootton of the East African Industrial Research Organization, Nairobi, published a series of papers and reports (1965, 1967, 1971) on the subject of coffee quality as affected by the fermentation step, following earlier studies also in Kenya. This work conflicted with the conventional views that the care demanded in this process is necessary merely to avoid deterioration of the raw bean appearance and the development of taints; though this fermentation is not of the kind associated with cocoa to develop specific flavour. Wootton found that various aliphatic acids are produced during fermentation; acetic and lactic acids being dominant, with butyric and especially propionic acids developing in the later stages of the fermentation. As the time of removal of mucilage becomes longer, so the rate of acid production is greater, but lower pH values are inhibitory to pectic enzyme activity in respect of its main object of mucilage loosening. The incidence of 'onion flavour' a known flavour defect, as suggested by Wootton, is connected closely with the presence of propionic acid in intimate contact (and therefore potential absorption) with the fermented beans immediately before washing. The use of 'under-water' fermentation as opposed to 'dry' accentuates the formation of acids.

Wootton also pointed out that the fermentation process itself contributed to an up-grading of the final appearance of the green beans (obtained after subsequent hulling and drying). When the mucilage was removed either carefully by means of a scalpel knife or by rapid treatment with 0.1 M sodium hydroxide (followed by quick washing), he found that the resultant air-dried coffee had a very poor appearance (e.g. yellowish green with a brown centre cut and associated silverskin), whereas with conventional fermentation and washing, this was not so. Kenyan coffees are sold by auction, and in addition are graded by expert liquorers who report in quality standards, numbered from 1-10 (low). Raw bean appearance, roast bean appearance and brewed flavour quality are noted, and all contribute to the final assessment. In his preliminary experimental work, the flavour quality when compared, was essentially unchanged, but the liquorers would on average advance the grading standard by one or two units because of the better appearance of the conventional product. Wootton went on to suggest that it would be advantageous to promote separately the two desired effects of fermentation (i.e. removal of mucilage/ enhancement of appearance).

There are a number of large-scale methods which can be used to accelerate mucilage degradation, of which one is the use of pectic enzyme preparations, e.g. pectinases as described by Ehlers (1980), and which function almost as well in the 'dry' state as in under-water fermentation. Though their usage cost is stated to be low (typically 0.5 per cent of coffee value), Wootton commented that in 1963, usage tended to be restricted to emergency situations, whilst Sivetz and Desrosier (1979) commented that although they had been in commercial use in Brazil and Central America for two decades, their use is declining in favour of attrition methods, which are available. The use of sodium hydroxide however presents various problems in practice. After intermediate washing to remove the fermentation degradation products, Wootton then proposed a second stage of soaking under water for a period of 24 hours, in which the desirable eventual raw bean appearance would be further enhanced. Preliminary laboratory work indicated that this was indeed so.

Wootton (1967a) reported further on the subject, following a collaborative study in different wet processing methods arranged on behalf of the Association Scientifique Internationale du Café. In this report he was only able to comment that the quality of coffee was indeed very dependent upon the overall process, and that his early observations were largely confirmed.

This subject was taken up again by Chassevent, Vincent, Hahn, Pougneaud and Wilbaux, of the IFCC, and reported in 1969. Trials in the Cameroons on the same type robusta coffee, carefully picked, with replicates at each of two plantations, were conducted. Comparison was made between coffees prepared by:

(1) a standard dry process;
(2) a wet process, with either:
   (a) pulping, fermentation (24 hours), washing/draining and air drying — conventional;
   (b) pulping, mucilage removal by sodium hydroxide, washing, draining and air-drying;
   (c) pulping, mucilage removal by sodium hydroxide, soaking in water for 24 hours, draining and air drying — as per Wootton.

No comments in this work were made on differences of bean appearance, whilst the total numbers of defective/undesirable beans on average for each treatment showed no statistically determined differences; though actual numbers were relatively high for the 300g samples.

In assessing flavour quality of brews of the roasted coffee prepared in the same way, it was evident that there were, however, marked differences, when the data was subjected to statistical tests. In overall numerical subjective quality grading, the dry process robusta gave an inferior cup compared to the standard wet process (trials 2a) sample, whilst samples 2c, with water soaking, showed a marked improvement. This improvement appeared to be associated with the absence of particular notes of 'poivre' (peppery), 'puant' (stinking) and fermenté (fermented), though all samples had a woody ('ligneux') note, and none had the good acid taste characteristic of arabica coffee. The different processing treatments had not affected astringence and bitterness characteristics. This paper also demonstrated negligible differences in caffeine, trigonelline and total chlorogenic acid (Lehman method) contents, which suggests that the parchment layer minimises diffusion of these components out of the beans even during water soaking. The numerical figure for chlorogenic acid is, however, lower for the dry process coffee (see also Chapter 13).

Wootton (1971) reported experimental work in determining weight losses during dry fermentation (for 48 hours) and found these to be only about 1 per cent dry weight (based on the pulped coffee), about half of which was total sugar, the greater part of this being the reducing sugars. The loss during a 24 hour water soak is similar, when the expected loss for a two-stage process with fast removal of mucilage is 1.0-1.5 per cent dry basis. The loss he believed to be due to diffusion rather than respiration. Certain other workers have reported losses of 3 per cent or more. Vincent *et al.* (1977) followed up the subject of losses by soaking green beans of various origins both with and without silverskin.

They found that the loss of components such as caffeine and chlorogenic acid was quite rapid, certainly up to 48 hours. This paper also then reported on a comparative study of quality and other factors of green coffees, prepared by the wet process but otherwise differentiated by 'dry' fermentation (for 12, 24 and 39 hours). Comparison was also made with green coffee prepared by the dry process. This work again suggested that the dry process gave an inferior product on a flavour quality scale (1-10), whilst within the wet process, it was marginally preferred to have underwater fermentation (optimally at 24 hours) for robusta coffee, which appeared to be associated with a reduction of bitterness. Kulaba (1979) reported that the soaking process was now general in Kenya.

Drying

Drying the wet parchment coffee from which mucilage has been removed, is accomplished by either sun drying or by hot air drying in a suitable mechanical dryer, or indeed by a combination of the two.

Sun drying is carried out in a number of different ways; for example (Haarer, 1962) the parchment coffee may be placed on permanent three-foot-wide staging of wire mesh covered with loose strips of strong hessian, which may be grasped at the corners and carried under cover as necessary, or have loose rainproof covers that can be placed over. Protection is necessary from rain showers or any heavy dews at night. No matter how the coffee is laid, i.e. on matting, trays or terraces, it should be turned or stirred frequently during the day time, to enable even and quick drying. Unlike cherry coffee, wet parchment coffee is more uniform in moisture content and initially has a lower moisture content of around 57 per cent. Haarer (1962) suggests that the coffee be laid initially about one inch thick, though the thickness can be increased later; whilst Sivetz and Desrosier (1979) gives a figure of 2-4 inches on terraces. Sun drying should take from eight to ten days, though may be shorter depending upon ambient temperature and humidity; up flow of air through the coffee serves to shorten the time compared with that on a terrace. Latent heat of evaporation is, however, supplied essentially by radiation. Absorption and desorption of moisture during drying is undesirable, and may result in 'spotty' beans, hence the protection needed, already mentioned. The parchments should not be broken or cracked as this will lead to dried beans which are white. The finish of drying can be empirically assessed by hand tests, in the field, since the beans shrink within their parchment shells and the silverskin loosens. The final moisture content to be achieved is again a figure of less than 13 per cent as determined by an oven drying method or calibrated moisture meter (Wootton, 1967b). The colour of the beans inside will be a bluish-green for arabica coffee when drying has been properly conducted.

Machine dryers are particularly necessary on large estates or even smaller ones, to accommodate the peaks in a harvest or problems during spells of rainy weather. Dryers are generally of the rotating drum type (e.g. the Guardiola dryer), when again Sivetz and Desrosier (1979) should be consulted for further details of the machines that are available, and the effect of operating parameters. There will inevitably be controversy on the relative merits of sun-drying, and machine drying in respect of quality factors. Machine drying needs of course to be carefully controlled, whilst the drying times will still be of the order of days. Initial sun-drying to about 44 per cent moisture content or lower, is often practised, followed by machine drying. Fluidised bed drying though not commercial apparently enables shorter drying times, and has been the subject of a number of papers discussed in the next section.

Drying Studies

It is generally recognised that the conditions of drying, especially in machine dryers, are important to the quality of the final coffee, in respect of flavour and appearance. Like the drying of foodstuffs generally, the effect of temperature on these quality factors will be related to moisture content at any time in the drying, and to the time of drying. Parchment coffee is somewhat different, in that the outer shell is subsequently removed. It has to be remembered that it is the temperature of the coffee as it is being dried, rather than the air temperature itself, that is important. So long as the surface of the coffee is wet (the so-called constant rate period of drying), the temperature will only rise to the wet-bulb temperature of the air, but as soon as there is a falling rate, the coffee will reach temperatures approaching that of the outlet drying air. The rate of drying generally is dependent upon both the relative humidity and the temperature of the drying air. High air flows will help to keep the relative humidity nearer to its entering humidity, though velocity of itself will not assist the rate of drying in the falling rate period. Drying has a complex interaction of variables, not least in its effect on any product or quality changes. These matters will have been considered in less available reports in that coffee (even at less than 20 per cent moisture) will tolerate a temperature of 40°C for a day or two,

50°C for a few hours, and 60°C for less than one hour without damage. Rolz, Menchu and Arimany (1969) state that coffee will be completely damaged by using air at above 80°C for more than five minutes. It seems that the general consensus is that coffee (like other foodstuffs) is even less resistant, to damage by heat, the higher the moisture content. Many workers in Africa and elsewhere, quoted by Rolz, recommend that the air temperature in the initial stage should not exceed 30°C. From an engineering and economic point of view, it is desirable to use high air temperature in the initial stage, though so long as constant rate drying is occurring, the coffee will be at the wet-bulb temperature, which could be below 30°C, a figure assessable from a pyschrometric chart. Examination of data from moisture content of parchment coffee against time however suggests that the constant rate period of drying (calculated as kg water/hour) is likely to be short, perhaps only in the first few hours if at all. Kulaba has obtained water sorption isotherms for parchment coffee, and finds that a number of different drying stages can be identified. It is of interest to note that Haarer (1962) reports that it is usually considered that a starting air temperature of 80°-85°C is quite high enough and that this should be lowered to 65°C after six hours and remain so to the end, though he states correctly that actual coffee temperatures are not given and are important. Further evidence on the effect of temperature comes from Rolz's studies in Guatemala on the fluidised bed drying of parchment coffee, relating temperature of operation to cup quality. Using a range of inlet air temperatures from 100°C to 50°C in short contact times (from a few minutes to 3 hours), even just drying to moisture contents of 30-40 per cent (followed by finish sun-drying to 10 per cent) gave poor quality coffees, at best barely acceptable. He concluded that the drying had to be carried out in two stages, first using drying air at ambient temperatures; but at around 43 per cent moisture in the coffee, air temperature could be increased to 60°C and drying continued to 10 per cent moisture, when a drying time of 8-9 hours can be expected. An especially favourable coffee was obtained by initial sun drying to 15.5 per cent followed by fluidised bed drying at 60°C for 50 minutes.

A desirable feature of wet process coffee is the occurrence of a bluish-green appearance in the final beans; and of course, the general absence of discoloured beans, such as yellowish, brownish and reddish. Apart from the particular methods of grading Kenya coffee already described, there is general agreement that overall bean colour is correlatable with final beverage quality. Studies have been conducted as to the nature and development of this colour in processing.

Gibson (1971) and other workers of EAIRO have indicated the importance of sunlight irradiation on coffee quality, at least in respect of raw bean appearance, so that coffee dried in the absence of sunlight will not achieve the same quality. Sivetz and Desrosier (1979) however, without particular reference to this work, dismisses the need for sun drying for flavour quality. Gibson provided a number of ground rules for drying to good quality; (1) rapid drying of the surface parchment; (2) slow shade drying for the next 48 hours, to allow slow moisture loss without breaking the parchment (so called white stage coffee); (3) sun-drying of the resultant parchment coffee down to 10-11 per cent moisture, allowing some 40 hours of strong sunlight irradiation, within the range of moisture content of 30% to 23 per cent (so-called 'soft black stage'). It is apparent therefore that a finish machine drying from about 20 per cent moisture content would be equally acceptable, and also in the earlier stage. Gibson rationalised these observations by claiming that the cafestol and kahweol esters were only synthesised in the coffee oil, from an unknown precursor, as the white stage has been approaching 30 per cent moisture content. Ultra-violet radiation destroys these esters, hence the need for shade. It is not clear however whether this also is meant to be true of the free cafestol and kahweol. In the soft black stage however, we have the development of blue, blue-green, green and the less-desirable yellow pigments (though also generatable in fermentation) in the surface of the bean. The green and blue pigments are produced by some mild catalysis of these esters by bean acids present, though the blue component was unidentified. Silverskin was shown to contain chlorophyll, which has a light-filtering effect, to give only 600-650 nm wavelengths, so that if a considerable amount is present, any bleaching effect of greens and yellow is slow. In the absence of these silverskin chlorophyll pigments, the light will consist of 350-650 nm wavelengths which tend to cause rapid bleaching of all the bean pigments. A correlation was found between the chlorophyll content of silverskin integument and the final raw bean appearance. Whilst the need for shade at one stage of the drying is understandable on this basis, the rationale behind the positive effect of irradiation during the 'soft, black stage' is less clear, especially since McCloy (as cited by Gibson, 1971) had stated that it was light between 480 and 400 nm which gave the desirable results. In particular respect of the blue component, other explanations have been advanced, notably that of Northmore (1967), who related it to a reaction product with chlorogenic acid. Gibson states that a very blue component is very rare, and is also related to the genetic make-up of the particular coffee (see also Chapter 13).

Hulling

Hulling, with synonymous usage of the terms 'milling', 'peeling' or 'shelling', is the operation in which the dried parchment coffee is divested of its parchment shell to release the green coffee beans. Hulling is the first in a sequence of operations known in East Africa as 'curing' and including subsequent grading and sorting, which are often conducted in large premises or 'curing works', to which dried parchment coffee is delivered in loads from different areas. As in Kenya, the loads may be kept in their separate identity through the curing, and subsequent auctioning. Smaller estates and individual plots may have their own small hulling equipment.

There are a number of different machines available for hulling, such as those based on the original Smout peeler. A typical machine uses a long rotating screw with helical pitch increasing towards the discharge and the parchment coffee is crushed by friction in its movement along the screw, whilst an adjustable weight-loaded discharge gate regulates the back pressure on the beans as they issue. Broken parchment is removed either by suction applied to the underside of the trough which is perforated, and/or by a counter-flow of air as the beans are discharged. Though cast iron machines may be used, it is considered that the working parts should be made of phosphor-bronze, as this enchances the blue-green appearance of the beans.

Sivetz and Desrosier (1979) gives a figure for the average hulling weight loss for arabica coffee, in East Africa and South America at around 20 per cent.The coffee can get undesirably hot after several hours working especially if damp, and excessive back pressure is employed. As well as the parchment, the silverskin is also generally removed by the abrasive actions involved, though a further machine for polishing may be used. Such machines can have a capacity of some 2-3 tons per hour of parchment coffee, requiring about 50 horse power. Machines based on impact milling, such as the Jackson huller are also available, requiring less horse power.

Grading and Sorting

Size Grading

Most green coffee, whether prepared by the wet or dry process, is marketed according to bean size specifications, which can be assessed by a conventional laboratory test screen analysis. Perforated plate screens are used, with apertures of specified dimensions, round holes for normal so-called 'flat beans', and lesser used slotted holes for peaberries. A set of screens for 'flat beans' will consist of individual screens, ranging from a so-called No. 20 screen down to No. 10 in whole unit steps (also sometimes half units). These screens are almost universally used, believed to have been devised by Jabez Burns in the USA for use in exports from Central and South American countries, whilst Gordon's of England have been and are a major supplier. These particular numbered sizes in fact, represent 1/64 to 10/64 (with the half-sizes at 1/128 inch) intervals; these holes need to be accurately drilled, otherwise screening result differences will occur. Gordon's in fact drill to the nearest metric equivalent (in mm) to several places of decimals. In countries, with entirely metric systems of measurement, numbering is sometimes in mm diameter size, at these same intervals but with occasional small differences; but the general tendency is to retain the basic

244 *Green Coffee Processing*

English numbering system. An International Standard (ISO, 1980) was published, harmonising existing practices and setting out nominal aperture diameters (in mm) and allowable tolerances for each numbered screen (except half-sizes). This standard also sets out other characteristics of standard screens (e.g. spacing and pitch, overall size and surround depth), in line with specifications for perforated plate screens in other ISO standards for such screens used generally in the food and other industries. In addition, the Standard details a procedure for using the screens, which is usually manual (i.e. shaking side to side) for some five minutes, and a format for reporting results. Not more than four screens are used at a time, and are selected from the top, middle or bottom of the available range, according to the average bean size of the sample. It will be noted that the results of a test screen analysis of commercial sample will fall as a very good straight line (at least centrally) on a plot of aperture size (in mm) on a linear scale, against percentage cumulative amount by weight, held at each aperture size on a probability scale (see Figure 10.2). The value of the aperture size at which there is a 50 per cent cumulative amount, is an indication of the average size of the coffee beans, whilst the slope of the line is an indication of the size distribution. It is often particularly also required to know the largest screen size at which all the beans pass through, and also the smallest size at which all beans are retained (or say only 1 per cent passes through).

Different producing countries have different size grade numbering, lettering or word descriptive systems, and quantitative specifications (including allowable tolerances in terms of screen analyses). Fuller details may be found by consulting Clarke (1972), Jobin (1982), or occasional articles in the journal 'World Coffee and Tea', or best, direct from the marketing organisations of the different countries, or the New York Coffee Exchange. Brazilian coffee is marketed in a large number of different size grades, by word description, which say at 'Medium' represents a green coffee, specified as a minimum of 20 per cent on a No. 15 screen, 50 per cent on a No. 14 and a maximum of 20 per cent on a No. 13 screen, with a tolerance of 10 per cent below minimum screen for each bean size. These figures are not however expressed cumulatively.

A particular bean shape characteristic is the incidence or otherwise of mechanically broken beans. In particular, the broken beans of less than half the original size (brisures, Fr.), which though they may be assessable in amount by test screen analyses, their quantitative level is determined and specified by limit after hand picking from a separate standard 300 g laboratory sample.

Separate grades of 'peaberries' (caracoli Fr.) which are sometimes marketed separately, are rounded beans occasionally found singly within a parchment shell. Robusta beans often tend however to be rounded, as opposed to fully 'flat beans' (i.e. one flat surface with an open cleft or centre-cut in the middle). True peaberries are size graded in test screen

Green Coffee Processing 245

Figure 10.2: Typical Screen Analysis for Ivory Coast Robusta Coffee Beans (Grade 2)

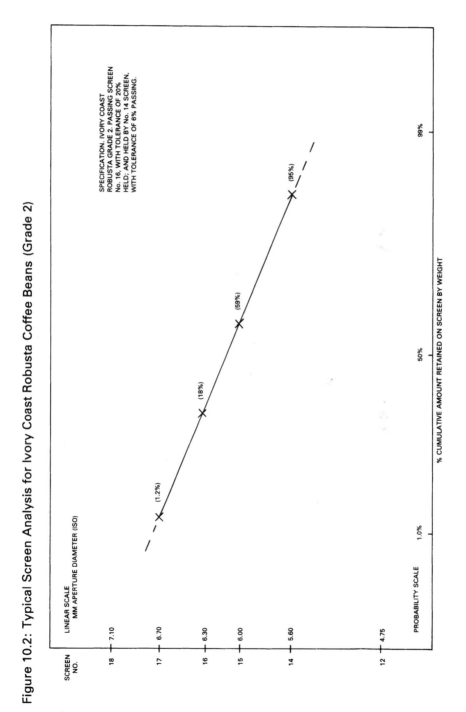

analyses, using screens with slotted holes, numbered No. 12 to No. 6, and these are also described in the International Standard already mentioned. Round holes essentially determine thickness of flat beans, whilst slotted holes determine width/length. There are also extra large size beans such as Elephants and Maragogipes in Brazil and some green coffees are sold unscreened.

To meet these marketing requirements, the hulled green coffee is put through rotating drums fitted with detachable screens, with bars or perforated holes of different dimensions in detachable sections, to permit the recognised sizes of beans to fall through as they pass along the drum or drums. These drums are often referred to as reel graders, but other size grading machines may also be used.

Sorting — Airlifting

Green coffee, especially from the dry process, and coffees which have been 'strip-picked' by whatever process will inevitably contain, even after size-grading, numbers of green beans which are defective in a number of different ways. Such coffee will also contain both foreign matter and matter extraneous to the beans themselves, such as whole dried cherries, pieces of parchment and husk, even pieces of stalk ('twig') and stones. Most of the bean defectives, usually characterised by non-standard colour, have already been discussed under the different stages of the green coffee processes.

Making use of the air lifting or blasting principle, some of these impurities (whether bean or extraneous) which have markedly lower density may be removed, though some sound beans may generally be removed as well. This operation is known as triaging (triage Fr.); the residues from such operations are sometimes exported, as triage or air blast grades. A particular type of machine working on this principle is the 'Catador' in Brazil.

Apart from the size grades already described, green coffee is also marketed in different grades or types, determined by the maximum allowable number of defects allowed in each type. For fuller details, reference should be made to the citations under size-grading already given. It should however be emphasised that wet process arabica coffee, especially from Kenya and Colombia, contains very few imperfections (e.g. Colombia by the New York Coffee Exchange not more than 13 defects by counting, including complex systems of equivalancy, in a 1 lb sample, of some 2,000 beans). The Federal Drug Administration in the US proscribes import above 10 per cent (by direct count) of insect-damaged beans. Dry robusta coffees may carry percentages up to this amount, according to source. There is no mechanical means of their separation; apart from hole damage to cherries whilst on the coffee tree, subsequent damage by other insects may take place during storage/transportation (see also Chapter 8).

Sorting — Manual

The removal of defective beans by hand picking of green coffee laid on travelling belts has been a traditional practice for wet process arabica coffee. Clearly considerable labour is involved, but justified by the upgrading of quality achieved when concentrated upon particular defectives, i.e. discoloured beans especially those such as sour beans which affect the flavour of the brew even when present in a small number, though these are often difficult to recognise. Black beans, likely to be very few in number in wet process arabica coffee, are of course, fairly easily recognised. Hand-picking has been developed to a fine art in Colombia, and often two hand-picking sequences are carried out on request, as in the so-called European preparation of Colombian (MAMS) excelso coffee.

Sorting — Electronic

Since about 1960, the use of so-called electronic sorters in producing countries has become quite usual. In principle, the coffee beans pass at high speed along chutes, and are assessed by reflected light beams, which can actuate a mechanism whereby those beans of an undesired colour are ejected separately. Machines on this principle, manufactured by Gunson's Sortex of the UK and by Icore in the USA and by others have now reached considerable sophistication. The earlier machines, based upon a monochromatic system were first used to separate out actual black beans from coffee, particularly for upgrading dry process robusta. Batteries of these machines, which are necessary for capacity reasons, can be seen in the Ivory Coast for robusta coffee, but also for arabica, in Kenya, Colombia, Tanzania, Brazil and elsewhere. Later machines using a bichromatic principle with pattern recognition can according to pre-inserted computer programs be used to separate a range of discoloured beans. The use of this kind of system is well described by Illy, Brumen, Mastro-pasqua and Maughan (1982), and by Maughan, Milo and Roarzi (1980) and Milo (1980), for use in an importing country, resulting from joint investigations by Illycaffè in Italy and Gunson's.

Sorting — UV Excitation

It has already been mentioned that sour beans and 'stinker' beans (fève puante, Fr.) are very undesirable beans, but sometimes difficult to detect by eye, though often possible by laboratory dissection of the individual beans. Gibson and Butty (1975) examined the fluorescence of beans under ultraviolet light excitation, and set out the different and complex types of fluorescence they had encountered, especially from defective beans of the 'stinker' type found in wet process arabica. Whilst the correlation is not always clear cut, the age of bean playing some role in degree of fluorescence, machines based on this principle have been manufactured by Gunson's Sortex and are used commercially, particularly in large batteries, in a curing mill in Nairobi, Kenya. The use of UV excitation after electronic sorting along the lines already described has eliminated hand-picking.

Carter (1980) has also described experimental machines based on UV excitation, but did not find any particular upgrading as a result, though this may well be due to their use in an importing country on necessarily older coffee. This subject is also discussed by Kulaba (1979).

Polishing

The removal of silverskin, except that retained in the centre-cut of beans which is only really subsequently released as chaff on roasting and grinding is desirable to give the green beans an attractive appearance. However, wet process arabica coffee is usually free from this material, as a result of combined hulling/polishing. The silverskin in dry process coffee, particularly robusta coffee is very tenacious, but may be removed by a polishing operation in which some water is also used. Such robusta coffee is referred to as 'washed and cleaned', though the word 'washed' has no connection with the 'washing' of the wet process.

Storage and Transportation

Green coffee is sold in sacks or bags (usually of jute), generally at 60 kg net weight, and may need to be stored either in warehouses of the producing country or at its overseas destination, and is necessarily stored in ship transit. Coffee may be held in silos for short periods in the producing country, and again in the premises of roasters in the consuming country. Green coffee is susceptible to changes affecting appearance and flavour, but especially at elevated temperatures (say above 25°C) and in conditions of high relative humidity where the moisture content can rise above 12-13 per cent. Under the latter conditions, mould growing on beans is a particular hazard, creating a new defect of 'mouldy' beans; additional black beans may arise. Insect damage, resulting in holes in the beans may also occur during unsanitary storage. This particular subject is large, and the object of considerable investigations by particular workers at different times. Wet process arabica coffee kept in parchment is less susceptible to change under unfavourable storage conditions (Stirling, 1980). At one time, shipments of parchment coffee were made, with hulling taking place in the host country despite the additional 15 per cent shipping weight. The most recent development is that of containerised shipment, which however has its own separate technical problems, e.g. the release of moisture within the containers, which has been described by Jouanjan (1980) and in a series of articles in Coffee International by various authors (1983).

References

Amorim, H.V. (1975) 'Relationship between some organic compounds of Brazilian green coffee with the quality of the beverage' in *6th International Colloquium on the Chemistry of Coffee*, ASIC. Paris, pp. 113-23

Braham, J.E. and Bresani, R. (1979) *Coffee Pulp*, International Development Research Centre, Ottawa, Canada

British Standards Institution (1980-83) Green coffee — Methods of Test. Parts 1-5, B S 5752

Carter, H.W. (1980) 'Coffee sorting with UV excitation' in *9th International Colloquium on the Chemistry of Coffee*, ASIC. Paris, pp. 219-26

Chassevent, F., Vincent, J.C., Hahn, D., Pougneaud, S. and Wilbaux, R. (1969) 'Study of the relationships, gustatory and chemical as a function of the preparation of Robusta coffee at the first stage' (in French) in *4th International Colloquium on the Chemistry of Coffee*, ASIC. Paris, pp. 179-85

Clarke, R.J. (1972) 'Grading Green Coffee', *Process Biochemistry*, October, pp. 18-20

Clarke, R.J. (1976) 'Food engineering and coffee', *Chemistry and Industry*, pp. 362-5

Ehlers, G.M. (1980) 'Possible applications of enzymes in coffee processing', in *9th International Colloquium on the Chemistry of Coffee*, ASIC. Paris, pp. 267-72

Foote, H.E. and Sivetz, M. (1963) *Coffee Technology*, Volume 1, AVI, Westport, Connecticut, USA

Gibson, A. (1971) 'Photochemical aspects of the drying of East African coffees', in *5th International Colloquium on the Chemistry of Coffee*, ASIC. Paris, pp.246-58

Gibson, A. and Butty, M. (1975) 'Overfermented coffee beans', in *7th International Colloquium on the Chemistry of Coffee*, ASIC. Paris, pp. 141-52

Haarer, A.E. (1962) *Modern Coffee Production*, 2nd ed, Leonard Hill, London

Illy, E., Brumen, G., Mastro-pasqua, L. and Maughan, W. (1982) 'Study on the characteristics and industrial sorting of defective beans', in *10th International Colloquium on the Chemistry of Coffee*, ASIC. Paris, pp. 219-26

International Standard No. 4650 (1980) Green coffee. Methods of Test — Manual Sieving. (BS 5752 Part 5, 1981)

Jobin, P. (1982) *The Coffees Produced Throughout the World*, (4th edn.), P. Jobin, Le Havre

Jouanjan, F. (1980) 'Maritime transport of coffee in containers' (in French) in, *9th International Colloquium on the Chemistry of Coffee*, ASIC. Paris, pp. 177-88

Kulaba, G.W. (1979) 'Coffee processing research; a review', *Kenya Coffee*, 44, 23-34

Maughan, W.S., Milo, S. and Roarzish (1980) 'Instrumentation system for the analysis of coffee beans', in *9th International Colloquium on the Chemistry of Coffee*, ASIC. Paris, pp. 201-10

Milo, S. (1980) 'Pattern recognition in sorting green coffee', in *9th International Colloquium on the Chemistry of Coffee*, ASIC. Paris, pp. 211-18

Northmore, J.M. (1967) 'Raw bean colours and the quality of Kenya Arabica coffee', in *3rd International Colloquium on the Chemistry of Coffee*, ASIC. Paris, pp. 405-14

Richard, M. (1977) 'Progress in the treatment and storage of green coffee' (in French) in *8th International Colloquium on the Chemistry of Coffee*, ASIC. Paris, pp. 187-96

Rolz, C., Menchu, J.F. and Arimany, E. (1969) 'The fluidized bed drying of coffee' in *4th International Colloquium on the Chemistry of Coffee*, ASIC, Paris, pp. 166-73

Sivetz, M. and Desrosier, N.W. (1979). *Coffee Technology*, AVI, Westport, Connecticut, USA

Stirling, M. (1980) 'Storage research on Kenya Arabica coffee', in *9th International Colloquium on the Chemistry of Coffee*, ASIC. Paris, pp. 189-200.

Teixeira, A.A. (1982) 'Effect of temperature in drying in relation to defects in coffee' (in Portuguese) in *10th International Colloquium on the Chemistry of Coffee*, ASIC. Paris, pp. 73-80

Various (1983) 'Containerisation of green coffee', *Coffee International*, 5, 13

Vincent, J.C.,Guenot, M.C., Perriott, J.J., Hahn, H. and Guele, D. (1977) 'The influence of different technical treatments on the chemical and organoleptic characteristics of Robusta and Arabusta coffee' (in French) in *8th International Colloquium on the Chemistry of Coffee*, ASIC. Paris, pp. 271-84

Watson, A.G. (1980) 'The mechanisation of coffee production', in *9th International*

Colloquium on the Chemistry of Coffee, ASIC. Paris, pp. 681-6

Wilbaux, R. (1952) *'The Treatment of Coffee'* (in French), FAO, Rome

Wootton, A.E. (1965) 'The importance of field processing to the quality of East African coffee', in *2nd International Colloquium on the Chemistry of Coffee*, ASIC. Paris, pp. 247-58

Wootton, A.E. (1967a) 'A report by the working group on the wet processing of coffee', in *3rd International colloquium on the Chemistry of Coffee*, ASIC. Paris, pp. 398-404

Wootton, A.E. (1967b) 'A comparison on the methods of measurement of moisture content in parchment and green coffee', in *3rd International Colloquium on the Chemistry of Coffee*, ASIC. Paris, pp. 92-100

Wootton, A.E. (1971) 'The dry matter loss from parchment and green coffee', in *5th International Colloquium on the Chemistry of Coffee*, ASIC. Paris, pp. 316-24

11 WORLD COFFEE TRADE

C.F. Marshall

The world trade in coffee is an important contributor to the income of some fifty or more producing countries on widely differing scales from Brazil with over 30 million bags annually to an equally famous producer, Jamaica, with a few thousand bags only. The trade mostly talks and thinks in terms of 'bags' so far as quantity is concerned, although occasionally 'tonnes' are traded. Bags vary in weight in fact, with the majority at 60 kg each. All statistics and agreement provisions are calculated on the basis of a 60 kg bag, but in most of Latin-America 69 kg bags are the rule, and Brazil uses 60 kg and Colombia 70 kg. Traders prefer natural fibre bags (sisal, jute, kenaf, etc.) but there are some bags in use made from man-made filament. These are not popular as they are considered prone to slip when piled. Prior agreement from a potential buyer must be sought before such bags are tendered in commerce.

The first assumption in all coffee trading is that the contents of each bag is uniform throughout as to quality and grade and that each bag identically marked is also identical. This uniformity is vital to commerce and especially to the eventual user. The principle is taken for granted throughout; so much so that in fact it is not even mentioned in the various contract forms.

World international trade is over 60 million bags each year valued at about 12,000 million dollars at 1981 prices. Some 19 million bags are consumed by producers themselves. To a few countries coffee is the biggest earner; Burundi and Rwanda for example depend on coffee for 80 per cent of foreign exchange income.

Prices fluctuate a great deal and often rapidly. There are also wide quality differentials which, themselves, fluctuate according to supply and demand. Change in every aspect is continuous to an extent even greater than in most other commodities.

International Coffee Agreement

It was the economic importance of coffee that brought importing countries into cooperation with producers in the first International Agreement of 1962. Advancing prices in the 1950s had brought a big response in production, especially in Africa, and it was by then clear that demand would be overwhelmed and cause the utter collapse of prices. This Agreement was amended in 1968, 1976 and 1983. Current members are listed in Table 11.1.

Table 11.1: Members of the International Coffee Organisation in 1983

Exporting Members (47)
Angola
Benin [a]
Bolivia
Brazil
Burundi
Cameroon [a]
Central African Republic [a]
Colombia
Congo [a]
Costa Rica
Dominican Republic
Ecuador
El Salvador
Ethiopia
Gabon [a]
Ghana
Guatemala
Guinea
Haiti
Honduras
India
Indonesia
Ivory Coast [a]
Jamaica
Kenya
Liberia
Madagascar [a]
Malawi
Mexico
Nicaragua
Nigeria
Panama
Papua New Guinea
Paraguay
Peru
Philippines
Rwanda
Sierra Leone
Sri Lanka
Tanzania
Thailand
Togo [a]
Trinidad and Tobago
Uganda
Venezuela
Zaire
Zimbabwe

Importing Members (25)
Australia
Austria[b]
Belgium/Luxembourg
Canada
Cyprus
Denmark
Fiji
Finland
France
Germany, Federal Republic of
Greece
Ireland
Italy
Japan
Netherlands
New Zealand
Norway
Portugal
Singapore
Spain
Sweden
Switzerland
United Kingdom
United States of America
Yugoslavia

European Economic Community

Note: a. OAMCAF members
b. Provisional, pending deposit of official ratification by legislation in Austria
Source: International Coffee Organisation, London.

The Agreement applies to all forms of coffee including roasted coffee, and 1 kg soluble powder is equated to 2.6 kg of green coffee, although earlier Agreements used a ratio of 1 to 3. The Agreement also regulates movements of all forms of coffee between importing members, so that quota or non-quota status of individual lots is maintained.

'Members' of the Agreement are governments, both producers and consumers. Individual exporters, traders and importers carry out their functions in the controls system to regulate the quota in collaboration with their governments. In the consumer states, the trade has influence on practical details and is consulted by government, but there appears to be far less trade participation in most of the producing countries. Major decisions of Agreement members are made by the Council consisting of delegations of all members and usually sitting twice a year with the most important meeting in September, but an Executive Board of eight countries on each side is elected to make minor decisions at some seven or eight meetings through the year. Voting power is regulated by the volume of interest of each country in the trade; but in practice most decisions are agreed and voting is comparatively infrequent. The big interests are Brazil and Colombia on the producer side, with the USA and the European Community on the other, though it should be noted that at present Community members vote separately, and figure separately in the Board.

A new 1983 Agreement text has been negotiated to enter into force on 1 October 1983, assuming that by then a sufficient number of powers has ratified or legalised participation. It will remain in force until 30 September 1989. The Agreement is negotiated and monitored by the International Coffee Organisation. The expenses of the Organisation, with its HQ in London, are subscribed by all members, importers and exporters, in proportion to their trade interest. The machinery of the Organisation and its working are very complex and would require a whole book to themselves for a full explanation. Here we can only outline its function in a rudimentary fashion.

The ICO, as the Organisation is known all over the world, has no knowledge of, or control over traded prices. The object is to control the flow of coffee to the market, and to monitor the market-sharing implied by the quota. Within the limits so imposed prices must find their own levels according to the normal commercial factors of quality, reliability and supply against competing growths. The contribution of importers is that of policing the imports and, of course, paying prices higher than might be the result of unrestricted competition. To be fair, the consuming governments recognise that ultimately this is to the advantage of buyers as much as sellers.

An overall quota or export entitlement (see Table 11.2) for the coffee year (October to September) is set each September by the Council based on the calculated 'disappearance' figure supplied by the Secretariat of the

Figure 11.1: Indicator Prices 1976. Monthly averages since 1973 in current and constant terms[a]

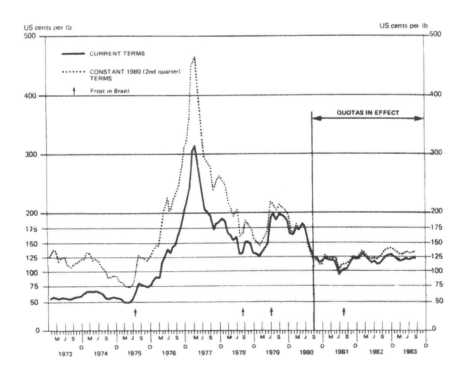

Note: a. Prices in current terms deflated by the UN index of unit values of exports of manufactured goods from developed market economies.

Source: International Coffee Association, London.

International Coffee Organisation, and with a nominal price 'bracket' in mind. In other words, the aim is to keep the 'price' within certain limits for the ensuing year. To do this, provision is made for the quota to be reduced if prices fall, or increased if they rise beyond the limits. Recently, the price 'bracket' has been 120 to 140 US cents per pound (see Figure 11.1).

The price referred to is not the price of any particular coffee but is a nominal 'indicator' price for the purpose of measuring the rise and fall of the innumerable commercial prices in the market. To do this, certain typical descriptions have been selected and daily quotations are supplied from the markets to the ICO in London, which calculates from them indicators for certain main groups:

1. Colombian Milds (comprising Colombia, Kenya and Tanzania).
2. Other Milds (most other producers of washed arabica).
3. Brazil and Other Arabica (unwashed arabica).
4. Robusta.

The countries are classified according to their main product — many produce a small proportion of some other type of green coffee. Of the four group indicators, two — the Other Milds and Robusta — are used to calculate a daily composite average and a running average for 15-day periods. It is this 15-day running average that is in fact used to 'trigger' quota adjustments and to judge the performance of the market. Note that the Brazil and Colombian group indicators are at present excluded from inclusion in the running average — for reasons explained later (see page 278).

Under the 1962 and 1968 Agreements when there was a considerable accrued surplus there was also an element known as 'selectivity', which permitted a limited possibility for consumers to influence the distribution of quota between the four groups. This is why there are four groups, each with its own indicator and in those days each group could gain, or lose, by quota adjustments independently of the other groups. At the same time, there were also the overall adjustments as at present. Consumers, by increasing their demand for, say, robusta and forcing prices higher could bring the robusta group limited increases in quota. This is what happened, but unfortunately, the Other Milds tended to suffer.

Ultimately, the quota system was suspended and coffee became much scarcer, until in 1980 it became necessary to re-institute the quota. Even then stocks were too small to make selectivity really practicable. Now, many members are calling for selectivity to be resumed but there are practical difficulties and much opposition from producers. So, at present, all adjustments and quota decisions are concerned with the overall quota regulated by the composite 15-day running average.

The overall figure has to be allocated to individual countries; their market share has to be decided, in fact. Therein lies the supreme difficulty, and probably no single country feels it is justly dealt with. They would almost all like a bigger share. There are elaborate rules included in the Agreement and, in fact, these rules to govern the larger producers have at this moment still to be decided, though a basis for 1983/84 only, has been agreed. The present rules do provide for a certain limited flexibility in that 30 per cent of the quota for the larger producers depends on the stocks, as verified by the ICO or its agents, at the previous 30 September or 31 March according to the crop times. The quota is divided into four quarterly portions which, except for the very smallest producers, are intended to be equal; though there is a limited facility for the Board to vary the quarters. This is to ensure an even flow of coffee to the market over the whole year.

Control and regulation of the system is achieved by the fact that con-

256 World Coffee Trade

Table 11.2: Initial Annual and Quarterly Quotas (60kg bags) for all Exporting Members — Coffee Year 1983/84

Exporting member	Initial annual quota (1)	Quarterly quotas[a]			
		Oct-Dec 1983 (2)	Jan-March 1984 (3)	April-June 1984 (4)	July-Sept 1984 (5)
TOTAL	56,200,000	14,049,996	14,050,007	14,049,996	14,050,001
A. Sub-total: Members entitled to a basic quota	53,633,086	13,408,270	13,408,276	13,408,270	13,408,270
Colombian milds	10,466,344	2,616,586	2,616,587	2,616,586	2,616,585
Colombia	8,468,791	2,117,198	2,117,198	2,117,198	2,117,197
Kenya	1,290,086	322,521	322,522	322,521	322,522
Tanzania	707,467	176,867	176,867	176,867	176,866
Other Milds	12,151,777	3,037,944	3,037,946	3,037,944	3,037,943
Costa Rica	1,123,623	280,906	280,906	280,906	280,905
Dominican Republic	494,186	123,546	123,547	123,546	123,547
Ecuador	1,128,825	282,206	282,206	282,206	282,207
El Salvador	2,330,478	582,619	582,620	582,619	582,620
Guatemala	1,805,080	451,270	451,270	451,270	451,270
Honduras	775,092	193,773	193,773	193,773	193,773
India	645,043	161,261	161,261	161,261	161,260
Mexico	1,898,715	474,679	474,679	474,679	474,678
Nicaragua	665,851	166,463	166,463	166,463	166,462
Papua New Guinea	603,427	150,857	150,857	150,857	150,856
Peru	681,457	170,364	170,364	170,364	170,365
Brazilian and other arabicas	17,400,557	4,350,139	4,350,140	4,350,139	4,350,139
Brazil	16,037,643	4,009,411	4,009,411	4,009,411	4,009,410
Ethiopia	1,362,914	340,728	340,729	340,728	340,729
Robustas	13,614,408	3,403,601	3,403,603	3,403,601	3,403,603
Angola	450,000	112,500	112,500	112,500	112,500
Indonesia	2,366,892	591,723	591,723	591,723	591,723
OAMCAF	6,915,030	1,728,757	1,728,758	1,728,757	1,728,758
Philippines	470,000	117,500	117,500	117,500	117,500
Uganda	2,309,670	577,417	577,418	577,417	577,418
Zaire	1,102,816	275,704	275,704	275,704	275,704
B. Sub-total: members exempt from basic quotas (other than OAMCAF)	2,566,914	641,726	641,731	641,726	641,731
Sub-total: members exporting 100,000 bags or *less* (without OAMCAF)	477,272	119,317	119,319	119,317	119,319
Ghana	50,512	12,628	12,628	12,628	12,628
Jamaica	17,467	4,366	4,367	4,367	4,366
Malawi	23,368	5,842	5,842	5,842	5,842
Nigeria	73,408	18,352	18,352	18,352	18,352
Panama	65,855	16,464	16,464	16,464	16,463
Sri Lanka	54,053	13,513	13,513	13,513	13,514
Trinidad and Tobago	34,226	8,556	8,557	8,556	8,557
Venezuela	80,2554	20,063	20,064	20,063	20,064
Zimbabwe	78,129	19,532	19,532	19,532	19,533

Sub-total: members exporting more than 100,000 bags (without OAMCAF)	2,089,642	522,409	522,412	522,409	522,412
Bolivia	109,759	27,440	27,440	27,440	27,439
Burundi	450,000	112,500	112,500	112,500	112,500
Guinea	100,317	25,079	25,079	25,079	25,080
Haiti	401,032	100,258	100,258	100,258	100,258
Liberia	130,294	32,573	32,574	32,573	32,574
Paraguay	108,814	27,203	27,204	27,203	27,204
Rwanda	450,000	112,500	112,500	112,500	112,500
Sierra Leone	234,624	58,656	58,656	58,656	58,656
Thailand	104,802	26,200	26,201	26,200	26,201

Note: a. The quota shown for each quarter represents 25% of the corresponding annual quota of each Member shown in column (1).
Members exporting 100,000 bags or less, including Members of OAMCAF, are not subject to quarterly quotas. Quarterly quotas in this table in respect of those Members are shown for illustrative purposes.

Source: International Coffee Organisation London.

suming countries will not admit coffee unless it is accompanied by a quota certificate. Coffee not accompanied by the correct certificate is not admitted by the Customs into a member country. The Customs collect the certificates and forward them to the ICO in London where the quota statistics are maintained. In order to ensure a country does not over-issue certificates, they have to be validated by special adhesive stamps issued from London strictly in accordance with the volume of quota allocated to each country. Thus, when the stamps are used up the quota is exhausted. In the first three quarters of the year under-shipments may be carried forward and used later, but any unused stamps/quota at 30 September are not carried forward, but lost.

There are arrangements to deal with sales between importing members, changed destinations and for other contingencies. In addition, consumer countries have agreed to limit their imports from producing countries not members of the Agreement to figures based on historical data. The total dwindles as producers join the Agreement and is now only about 500,000 bags.

In some cases, producers know they will be unable, usually because of some climatic irregularity, to make use of their quota and the rules provide that the 'shortfall' may be re-allocated to other products. Brazil and Angola have, in recent years, made very substantial shortfall declarations — in 1980 the total was 7 million bags.

The exports to members of the Agreement are covered by the quota, but exports to countries outside the Agreement are unregulated and completely free of quota. In theory, they must not be sold at prices lower than for quota coffee but in practice this cannot be controlled. As a result, sales at much lower prices have been made. There are two results. First, there is a temptation for economically hard-pressed consumer states to fall out as Israel and Hungary have done, and secondly, it affords great scope for fraud or forgery to smuggle cheap coffee into the quota area. Undoubtedly this happens, though the scale is very much a controversial issue. One substantial forgery of quota stamps has been detected before too much damage was done and some other cases of illegal entry by one method or another have been discovered. The really successful case is that which is not discovered so, in the nature of the thing, we cannot know just how much leaks from the cheap non-quota area into the relatively expensive quota area. But, certainly, it is at present much less than it was in the sixties.

Efforts are being made to monitor non-quota shipments to ensure they do indeed reach the declared destination but the problem is that non-member states have no obligation to cooperate. For that matter, producers can ship coffee secretly without declaring it as such — just as they did in the sixties — so that the ICO will not even know it has happened.

Imports to the European Community

The Lome Convention, regulating the relations between the European Community and so-called ACP (African, Caribbean and Pacific) countries, concerns many commodities of which coffee is only one. A Common External Tariff of 5 per cent ad val. is imposed but ACP origin coffee has duty-free entry when accompanied by the privilege certificate (EUR. 1). Coffee is also one of the articles included in the STABEX Scheme which sets out to cushion the impact of price fluctuations for primary producers. The downward fluctuations have, however, proved to be large enough in recent years to cause some serious problems but it is a matter between the Commission and the ACP states and has no relevance to normal trade relations.

Coffee Exporting

Historical statistics on exports and imports are presented in Tables 11.3 to 11.9 inclusive.

Quota Exports to Member Countries

The two biggest producers in the world are Brazil and Colombia, together usually responsible for between 40 and 50 per cent of world supply. Brazil is notorious for fluctuating crops due to adverse weather conditions which occur more frequently, or so it appears, than in the past. Cold weather, even frost is dramatic and well known but drought can also be serious.

Through the sixties there were a number of frost incidents (see Figure 11.1) culminating in the worst of all in 1975. These gradually caused the erosion of the stock of over 50 million bags which existed in the early sixties; a good deal must have also been lost or spoiled through poor storage. As a result we had the great rise in prices through 1976 to about April 1977. No quota was in operation at this time and, in fact, no Agreement existed in theory as the 1976 Agreement was then being negotiated or in course of ratification.

Other nations were therefore able to sell as much coffee as they wished and some took the opportunity to allow production to grow even though they had been warned that Brazil intended fully to resume her 'rightful' place in the market. Colombia in particular increased production enormously but has since been forced to restrain herself. For the most part, the increases in production were achieved through higher yields assisted by high prices rather than expansion of planting. At the present time the world surplus is growing year by year, though not quite as dramatically as some would have us think. There is always, too, the risk that a dramatic frost or drought in Brazil could reverse the trend for a year or two.

Most of the coffee in the world is grown by smallholders, though there are big plantations, too, in some areas. It reaches the international market through local merchants, cooperatives of grower/processors or government agencies. These organisations eventually pass on the coffee to exporters, who may be government agencies or monopolies, as in Uganda for example, or in other cases private enterprise organisations. A monopoly corporation can deal with a crop which is highly standardised but is ill-suited to deal with a crop where there is a large range of grades and qualities such as Brazil or Kenya. Private enterprise has the advantage that it can adapt to the detailed needs of users in consuming areas and supply even small quantities efficiently in a way a monopoly organisation handling a whole crop seems unable to manage. So we find Brazil combining private enterprise with a very far reaching and thorough control over the whole business. The exporters are responsible for selecting qualities and making up bulks to suit the requirements of their clients but in the realm of price, foreign exchange and administration the exporters are regimented by the Instituto Brasileira do Cafe (IBC), a government institution. In Kenya, all the production is sold through auction by the semi-official Coffee Board to private exporters who have the task of making up the shipments to the grade and quality standards required by buyers. Other countries, particularly in Latin-America work similarly but in the African robusta countries there is a greater tendency to use government organisations throughout.

Although an arabica producer, Colombian quality is more uniform than in Brazil and though there is a dual system of private exporters and the Government Federacion Nacional de Cafeteros (FNC), it is the latter which controls the system and prices. It also sells and exports directly to

Table 11.3: Exports (60kg bags × 1000) by Exporting Members to All Members October-September 1977/78 to 1982/83

Exporting Member	October-September					
	1977/78	1978/79	1979/80	1980/81	1981/82	1982/83[a]
Total	45,892	57,250	54,268	51,506	54,432	55,216
Sub-total: members entitled to a basic quota	44,181	55,398	52,289	49,540	52,359	52,604
Colombian Milds	9,343	12,637	12,608	10,168	10,065	10,462
Colombia	7,141	10,706	10,678	8,329	8,044	8,462
Kenya	1,300	1,195	1,317	1,117	1,301	1,291
Tanzania	902	736	613	722	719	709
Other Milds	13,527	16,233	13,802	12,458	12,327	11,990
Costa Rica	1,164	1,480	1,221	1,326	1,220	1,126
Dominican Republic	703	364	585	403	575	488
Ecuador	1,399	1,163	813	1,033	1,021	1,106
El Salvador	2,427	3,379	2,913	1,983	2,163	2,328
Guatemala	2,181	2,528	2,010	1,800	1,779	1,769
Honduras	938	1,051	978	955	905	759
India	546	645	529	967	905	615
Mexico	1,701	2,839	2,253	1,884	1,808	1,861
Nicaragua	893	927	809	696	655	652
Papua New Guinea	760	808	835	635	566	603
Peru	814	1,046	856	776	730	682
Brazilian & Other Arabicas	8,988	12,289	13,374	14,568	16,449	17,056
Brazil	8,216	11,237	12,383	13,375	15,230	15,715
Ethiopia	782	1,052	991	1,192	1,219	1,341
Robustas	12,314	14,240	12,505	12,345	13,518	13,097
Angola	725	366	289	241	498	350
Indonesia	3,141	3,836	3,570	3,058	2,080	2,363
OAMCAF	(5,575)	(6,588)	(5,482)	(5,821)	(6,714)	(6,624)
Benin	2	2	1	35	36	53
Cameroon	1,335	1,554	1,582	1,490	1,481	1,397
Central African Republic	152	102	167	209	253	266
Congo	48	105	46	42	38	40
Gabon	3	5	10	14	14	40
Ivory Coast	3,129	4,002	2,587	3,023	4,008	3,787
Madagascar	813	694	940	799	629	749
Togo	94	124	149	209	256	293
Philippines	202	244	246	291	399	462
Uganda	1,488	2,157	2,057	1,935	2,708	2,194
Zaire	1,182	1,048	862	1,000	1,119	1,103
Sub-total: members exempt from basic quotas	1,711	1,852	1,978	1,966	2,073	2,612
Arabicas	1,309	1,295	1,478	1,510	1,525	1,820
Bolivia	82	69	60	94	106	114
Burundi	331	415	333	400	400	446
Haiti	316	237	411	247	350	408
Jamaica	14	12	17	13	19	23
Malawi	3	3	2	12	16	23

Panama	41	37	51	66	66	76
Paraguay	0	0	5	97	103	140
Rwanda	313	341	557	480	376	446
Venezuela	209	180	41	26	18	47
Zimbabwe	0	0	0	75	71	96
Robustas	401	558	501	456	548	792
Ghana	26	16	15	31	20	30
Guinea	0	4	46	18	16	70
Liberia	167	141	168	126	125	150
Nigeria	15	53	12	3	51	73
Sierra Leone	117	218	170	155	153	234
Sri Lanka	32	40	22	9	58	72
Thailand	23	55	53	90	95	144
Trinidad and Tobago	22	31	15	25	31	19
Brazil	8,216	11,237	12,383	13,375	15,230	15,715
Colombia	7,141	10,706	10,678	8,329	8,044	8,462
All other arabicas	17,820	20,510	18,201	17,000	17,092	17,149
Robustas	12,715	14,798	13,006	12,801	14,066	13,889

Note: a. Due to rounding the totals may not always reflect the sum of the relevant components
Final annual quotas
Source: International Coffee Organisation, London

foreign markets, particularly in Europe as opposed to the USA, where private exporters are more important.

In Brazil, Kenya and certain other arabica exporters the matter of beverage flavour is extremely important and it is in the exporting houses where the organoleptic expertise is found. It is their task to discover and agree with potential buyers exactly what standards are required as to quality and grade. They then set up 'types' or samples to illustrate the desired features and will offer their clients suitable quantities 'equal to type'. To work successfully in this way, ideally there should be some community of understanding with the buyers and this is, once again, usually easier for private enterprise than it is for governments. Shipments against types are rarely absolutely identical as time goes by, and the existence of a confidential relationship between buyers and sellers is important. With a natural product, vulnerable in so many ways during processing and transit, mutual understanding is necessary.

In fact, though, probably most business is not transacted by exporters directly with users. In many markets, dealers in green coffee are more important and play a significant part. It is often thought that the dealer function is unncessary and that he just takes a profit for nothing. If this were the case, he would surely have disappeared long ago even if he had come into existence at all.

Dealer Trading

International trade is complicated enough for this bridge between exporter and importer to have its uses. A financial function is obvious enough;

exporters require prompt payment and users often pay only later when the goods reach them (or even later still). Users often wish to buy only on 'landed terms' in their own country, whereas exporters prefer to sell Free on Board vessel at their own port. Sellers wish to buy when prices look high, users when they seem low. Users may want to secure a guaranteed supply well in advance; shippers dare not sell until they know the supply of coffee is sure. All these 'gaps' can be covered by dealers using the facilities afforded by terminal markets, finance houses and foreign exchange markets in the USA or in Europe.

The dealer, too, can often perform a useful publicity function. Where a country has only a modest production it cannot afford to take up direct relations with too many possible buyers. It would not be economic to do so. But a dealer can, for example, buy a couple of thousand bags of Burundi coffee and go on including them in his daily offers to clients together with other coffees, so that they are brought to the notice of a much wider range of buyers than would otherwise be possible. Burundi ships around 400,000 bags per year, often less, and it pays her to work through a limited number of dealers knowing quite well they will in turn circulate their offers much further afield.

The dealer has also another task — that of keeping his contacts informed. Because dealers are usually in important trade centres like New York, London, Hamburg or elsewhere they have access to a range of world wide information. Buyers and sellers can rely on them to give a picture of the market and all its features which is not possible for a seller operating in a place like Bujumbura (to return to our Burundi example), or for a buyer, in say, Reykjavik.

In short, dealers are a very definite necessity and though they have become fewer over recent years, and there have been many, many adaptations, disappearances of some names and appearance of new ones, they seem likely to continue.

Although business is usually based on type samples, often tailor-made for specific buyers, there are other methods. First, there is the simple stocklot where a sample is drawn from a particular identifiable parcel of coffee and the buyer buys that lot having approved the sample submitted. Here he is not buying equal to the sample; he is buying that parcel and no other. The system is known as 'as per sample' as that is usually the phrase adopted in the paper contract, and the sample is customarily identified by a number or name. Sometimes, by the time the buyer gives his approval the particular lot has been sold and in such cases it is not unusual, where confidence exists, to buy equal to the stocklot; in other words, not that specific group of bags but some closely matching it.

Buying 'as per sample' is for the buyer who demands precise knowledge of what he is to receive — when commencing relations with a new origin, perhaps — but at the opposite end of the 'precision scale' is the 'fair aver-

World Coffee Trade 263

Table 11 4: Net Imports (60kg bags × 1000) by Importing Members from all Sources July-June 1977/78 to 1982/83

Importing Member	July-June					
	1977/78	1978/79	1979/80	1980/81	1981/82	1982/83
Total	40,377	54,392	55,331	53,387	53,192	55,489[a]
USA (Subtotal)	13,933	19,289	19,463	16,896	15,957	17,761
EEC (Subtotal)	17,508	22,190	22,526	23,471	24,047	23,882
Belgium/Luxembourg	956	1,294	1,309	1,372	1,341	1,288[b]
Denmark	760	941	935	989	976	941
France	4,442	5,281	5,188	5,344	5,250	5,399
Germany, FR of	5,645	6,567	6,742	7,414	7,379	7,457
Greece	250	339	408	401	439	473[b]
Ireland	32	66	59	67	60	60
Italy	2,928	3,508	3,751	3,737	4,080	3,971
Netherlands	1,327	1,870	1,861	2,007	2,187	2,057
United Kingdom	1,168	2,324	2,273	2,140	2,335	2,236
Other Members (Subtotal)	8,936	12,913	13,342	13,020	13,188	13,846
Australia	368	478	593	567	635	627
Austria	478	684	828	838	959	1,039
Canada	1,411	1,715	1,798	1,916	1,843	1,763
Cyprus	18	28	29	30	28	32
Fiji	3	3	3	3	3[b]	3[b]
Finland	724	1,049	1,147	862	1,179	990
Japan	1,670	2,994	3,241	3,171	3,481	3,720
New Zealand	63	91	108	99	104	119
Norway	527	669	691	683	751	706
Portugal	148	229	163	198	249	261
Singapore	46	1	−38	−23	−379	−188
Spain	1,306	1,711	1,736	1,377	1,550	1,983
Sweden	1,113	1,684	1,635	1,699	1,715	1,591
Switzerland	487	626	581	710	584	631
Yugoslavia	574	951	827	844	486[b]	569[b]

A negative sign indicates net exports
Note: a. Preliminary
b. Estimated
Source: International Coffee Organisation, London.

age quality' concept used where there are relatively large volumes of a roughly standardised grade and quality as in the large robusta producing areas and in many places in Latin-America. This means that the buyer expects to receive the average standard of the crop but implies that if the crop is poor, his delivery will be poor but not significantly worse than the average from the area in question. This is a fairly common practice in commodities and in coffee one finds two sub-divisions. The first is 'fair average quality of the season' where the standard is the average of the whole crop up to the moment of shipment. But it can be 'fair average quality at time and place of shipment', which means precisely what it says and takes no account of deliveries some months or weeks earlier, for example. In case of

Table 11 5: Exports (60kg bags Green Bean Equivalents) of Soluble Coffee Powder by Exporting Member Countries

Origin	1973	1974	1975	1976	1977	1978	1979	1980	1981	1982
Grand Total	2,217,361	2,085,725	1,847,229	2,483,688	1,964,209	2,624,379	3,248,526	2,526,783	2,781,424	2,955,340
Colombian Milds (subtotal)	15,202	45,965	46,922	105,261	174,959	86,154	138,424	102,171	126,323	104,863
Colombia	15,192	45,965	46,922	105,261	174,959	86,154	138,424	102,171	126,323	104,863
Tanzania	10	0	0	0	0	0	0	0	0	0
Other Milds (subtotal)	133,791	89,916	144,247	84,180	87,974	141,913	238,402	311,076	351,624	392,283
Ecuador	8,319	14,730	16,740	19,243	18,338	31,371	65,231	82,567	185,600	246,632
El Salvador	50,482	43,531	60,690	17,102	26,400	44,268	52,378	45,810	37,592	25,252
Guatemala	30,726	17,591	29,541	11,630	518	5,459	6,191	8,977	9,060	0
India	20,974	10,852	14,609	15,700	21,747	17,663	40,970	67,763	65,519	81,061
Jamaica	1,997	1,601	1,837	1,478	1,324	386	1,612	100	114	1,971
Mexico	13,821	275	4,355	16,554	8,902	39,548	65,389	70,475	20,446	547
Nicaragua	7,434	689	13,851	0	7,395	2,518	6,081	35,384	33,293	36,820
Venezuela	38	47	24	23	0	0	0	0	0	0
Zimbabwe	0	600	2,600	2,450	3,350	700	550	0	0	0
Brazilian and other arabicas (subtotal)	1,961,016	1,855,326	1,568,498	2,178,588	1,586,687	2,195,964	2,640,315	2,013,883	2,196,958	2,246,578
Brazil	1,961,016	1,855,326	1,568,498	2,178,588	1,586,687	2,195,964	2,640,315	2,012,233	2,196,958	2,246,578
Paraguay	0	0	0	0	0	0	0	1,650	0	0
Robustas (subtotal)	107,352	94,518	87,562	115,659	114,589	200,348	231,385	99,653	106,519	211,616
Indonesia	0	0	0	0	0	0	0	1,500	0	0
OAMCAF	93,804	90,303	86,975	101,648	97,152	179,808	211,320	79,179	76,506	194,287
Ivory Coast	93,804	90,303	86,875	101,648	97,152	179,808	211,320	79,179	76,506	194,287
Philippines	6,674	3,747	586	1,955	3,423	15,413	13,391	8,731	9,287	7,563
Thailand	0	160	0	5	5	2	5	0	0	0
Trinidad & Tobago	6,874	308	1	12,051	14,009	5,125	6,669	10,243	20,726	9,766

Source: International Coffee Organisation, London.

World Coffee Trade 265

Table 11 6: Imports by Each Importing Member Country (60kg bags Green Bean Equivalents) of Soluble Coffee Powder from all Origins

Destination	1973	1974	1975	1976	1977	1978	1979	1980	1981	1982
Grand Total	3,581,064	3,649,413	3,209,564	3,879,539	3,436,635	3,627,933	4,925,366	4,232,754	4,872,386	4,583,627
USA	1,511,151	1,621,786	1,108,846	1,552,058	1,280,885	1,237,070	1,542,276	1,281,699	1,348,367	1,190,295
EEC (subtotal)	1,519,546	1,343,305	1,349,265	1,566,494	1,447,088	1,672,666	2,360,740	1,928,009	2,370,874	2,254,250
Belgium	71,985	77,990	75,350	85,335	85,225	77,110	107,230	98,378	145,740	155,420
Denmark	33,750	16,800	23,350	22,600	15,450	22,650	20,300	18,900	27,367	27,461
France	240,785	302,290	285,620	375,770	371,215	383,805	320,570	329,420	361,595	343,495
Germany, F.R. of	181,164	144,549	119,649	128,949	145,443	113,796	227,934	255,594	392,751	462,048
Greece	55,599	45,150	64,400	64,102	63,150	60,331	94,994	82,238	95,334	103,379
Ireland	31,715	25,799	33,814	53,606	32,152	41,752	58,191	49,992	56,607	62,727
Italy	31,781	41,537	36,847	62,812	46,553	35,807	55,636	43,910	46,436	55,570
Netherlands	131,090	139,290	62,030	74,595	112,715	100,875	122,830	84,130	130,740	105,105
United Kingdom	741,677	549,900	648,205	698,725	575,185	836,540	1,353,055	965,447	1,114,304	939,045
Other Members (subtotal)	550,367	684,322	751,453	760,987	708,862	718,197	1,022,350	1,023,046	1,153,145	1,139,082
Australia	4,794	9,736	10,823	30,453	31,683	41,519	63,430	58,871	139,776	116,378
Austria	82,070	94,585	94,400	108,585	84,625	92,890	95,615	105,655	96,595	104,180
Canada	225,734	190,084	242,214	256,286	163,650	223,904	218,514	296,338	342,600	310,558
Cyprus	3,586	3,787	2,058	3,804	3,766	5,091	4,283	6,108	8,446	8,934
Finland	14,382	19,385	17,449	21,543	18,288	19,197	16,297	16,055	13,974	15,822
Japan	60,243	187,273	205,406	194,929	245,663	176,048	418,679	316,633	339,259	364,988
New Zealand	0	67	715	0	0	0	6	114	206	313
Norway	34,024	39,428	41,362	51,621	55,848	42,731	49,383	42,480	47,719	51,700
Portugal	22,840	31,850	4,366	9,630	1,095	275	190	145	340	125
Singapore	8,706	11,710	15,748	14,717	19,548	27,268	37,162	57,551	49,758	56,416
Sweden	63,741	59,296	78,384	36,564	47,597	53,253	73,438	85,194	75,417	84,522
Switzerland	28,447	34,121	38,528	31,890	36,495	32,471	41,983	36,002	38,755	24,446
Yugoslavia	1,800	3,000	0	965	604	3,550	3,370	1,900	300	700

Source: International Coffee Organisation, London.

Table 11.7: Exports (60kg bags × 1000) of all Forms of Coffee by Exporting Member Countries to Non-member Countries

	12 Month Periods to December					
Exporting Member	1977	1978	1979	1980	1981	1982
Total	4,486	5,342	5,740	6,548	8,246	9,047*
A. Sub-total: Members entitled to a basic quota	4,393	5,284	5,646	6,485	8,187	8,896
Colombian Milds	490	616	770	1,123	1,062	1,534
Colombia	375	526	678	954	624	902
Kenya	99	60	37	53	164	366
Tanzania	16	30	55	116	274	266
Other Milds	901	917	1,237	1,182	1,974	2,815
Costa Rica	149	113	87	26	329	432
Dominican Republic	85	55	96	129	107	36
Ecuador	105	205	248	166	112	486
El Salvador	37	23	13	—	29	0
Guatemala	27	6	29	8	240	706
Honduras	1	13	0	36	79	93
India	434	415	484	740	625	608
Mexico	19	25	236	38	62	233
Nicaragua	—	4	2	28	183	88[a]
Papua New Guinea	—	0	0	3	186	117
Peru	44	58	42	8	22	16
Brazilian and Other Arabicas	1,520	1,993	2,070	2,294	2,778	1,842
Brazil	1,242	1,631	1,746	1,951	2,536	1,732
Ethiopia	278	362	324	343	242	110
Robustas	1,482	1,758	1,569	1,886	2,373	2,705
Angola	716	613	576	584	494	240
Indonesia	9	75	56	185	611	1,497
OAMCAF	(485)	(814)	(764)	(959)	(1,184)	(884)
Benin	0	0	0	0	0	0
Cameroon	0	0	5	44	48	169
Central African Republic	0	6	—	—	0	0
Congo	2	1	0	0	0	0
Gabon	0	0	0	0	0	0
Ivory Coast	340	717	632	693	816	588
Madagascar	143	90	127	222	320	127
Togo	0	0	0	0	0	0
Philippines	0	0	0	2	6	2
Uganda	247	252	161	152	38	62
Zaire	25	4	12	4	40	20
B. Sub-total: Members exempt from basic quotas	93	58	94	63	59	151
Arabicas	38	41	67	36	28	112
Bolivia	16	12	32	26	8	9
Burundi	9	19	26	9	2	35
Haiti	1	—	—	0	0	0
Jamaica	1	—	2	—	—	2
Malawi	0	0	0	0	0[b]	0
Panama	0	0	0	0	0	0

Paraguay	0	0	0	0	7[c]	10[c]
Rwanda	11	10	7	1	6	50
Venezuela	0	0	0	0	0	0
Zimbabwe	0	0	0	0	5	6
Robustas	55	17	27	27	31	39
Ghana	0	0	0	0	0	0
Guinea	15	11	16	16	6	3
Liberia	0	1	1	2	18	32
Nigeria	0	0	0	0	0	0
Sierra Leone	26	0	2	0	0	0
Sri Lanka	0	0	0	0	—	0
Thailand	0	0	0	0	0	0
Trinidad and Tobago	14	5	8	9	7	4

Note: Exports to Hong Kong, Hungary and Israel which ceased to be Members on 1 October 1982 are included in this table as exports to non-members.

* Preliminary
— Less than 500 bags
[a] Excludes the quarters July-September and October-December 1982 for which Certificates of Origin in Form X were not received; the reported exports to non-member countries were 16,377 bags in the quarter July-September 1982 and 57,981 bags in the quarter October-December 1982
[b] Malawi reported a small volume of exports to non-members in the quarter October-December 1981 but no Certificates were received
[c] No Certificates of Origin in Form X were received from Paraguay in coffee year 1981/82

Source: International Coffee Organisation, London.

dispute the arbitrators will look at, or consult their experience of, other deliveries of the same growth made as nearly as possible to the time of shipment from the same port as the shipment in dispute.

Incidentally, almost all disputes whether as to quality or not, are commonly submitted to arbitrators for decision if the parties are unable to come to an amicable settlement. All the standard contract terms used in coffee business provide for arbitration by the rules and law in an agreed centre of trade — London, New York, Hamburg or some other such place, where a nucleus of suitable experts exists.

The Brazil business nowadays is mostly on types, and usually tailor-made types for particular users, but sometimes on stocklots. However, Brazil almost invented the techniques of coffee trading and at one time an elaborate system of grades and descriptions was used in a very efficient manner before airmails existed. Cables were the only rapid communication and grades — governing only the admixture of defects with sound beans — were used plus a small series of additional phrases to establish the bean size, colour, style (or shape) of beans, crop year and especially the cup quality. The system still exists and is used, but rarely as a basis for contracts, except in Rio and Victoria, where cup is not important. So many countries have learned their business from Brazil that many grading systems adopted to regulate business elsewhere have been based on the

Table 11.8: Production, Imports, Exports and Availability (60kg bags × 1000) of all forms of Coffee in Producing Non-member Countries during 1979

	Total production (a)	Domestic consumption (a)	Exportable production (1)–(2)	Imports	Availability (3)+(4)	Declared exports	To Members	To non-members	Declared imports by importing Members	Difference between declared and derived exports to Members (6a)–(7)
	(1)	(2)	(3)	(4)	(5)	(6)	(6a)	(6b)	(7)	(8)
Total	874	1,758	–884	600	0	0	0	0	295	–
Argentina	71	671	–600	14					2	–
Belize									–	
Burma	10	10	0						2	
Kampuchea	1	4	–3						0	
Cape Verde									–	
China People's Rep.	168	0	168	7	175				25	
China Republic	60	86	–26	26	0	–	–	–	0	0
Comoro	1	0	1	0	1				–	
Cuba	223	400	–117						85	
Equatorial Guinea	3	–	3	0	3	–	–	–	2	–
Fiji	–	4	–4	4	0	–	–	–	0	
Guyana	19	19	0		0				15	–
Laos	9	9	0	–					69	
Malaysia	143	17	–34	101	67	67	66	1	35	–3
Mozambique	10	10	0			1			–	
Saõ Tome and Principe	2	1	1						21	
South Africa	5	258	–253	288	25	25			21	
Sudan	9	38	–29	29	0				0	
Suriname	1	1	0						–	

Table 11.8: continued

Vanuatu	1	—	1			2	1
Vietnam	70	35	35			38	23
Yemen Arabic Republic	50	10	40	1	41	22	9
Yemen P.D.R.	15	17	−2	14	14	12	1
Zambia	3	8	−5	5	0	—	5	−5

Note: — Less than 500 bags
 a. Excludes Belize and Burma (production estimated to be negligable).
Source: International Coffee Organisation, London.

Table 11.9: Estimated Domestic Consumption of Coffee (60kg bags × 1000) in Exporting Member Countries

Exporting Member	1971/72	1972/73	1973/74	1974/75	1975/76	1976/77	1977/78	1978/79	1979/80	1980/81	1981/82	1982/83
Total	17,737	17,552	17,209	16,899	17,165	17,380	17,812	17,697	18,665	18,725	19,321	
A. Sub-total: Members entitled to a basic quota	16,714	16,492	16,139	15,774	15,472	15,605	16,039	15,798	16,562	16,720	17,373	
Colombian Milds	1,451	1,511	1,535	1,337	1,375	1,428	1,574	1,690	1,815	1,566	1,930	
Colombia	1,410	1,470	1,498	1,296	1,333	1,383	1,509	1,638	1,750	1,478	1,850	
Kenya	22	20	20	20	21	20	43	40	50	68	60	
Tanzania	19	21	17	21	21	25	22	12	15	20	20	
Other Milds	3,840	3,987	3,878	4,138	4,069	3,900	4,035	3,932	4,132	4,166	4,318	
Costa Rica	135	179	143	165	184	186	214	220	213	200	208	220
Dominican Republic	279	277	269	305	326	354	287	261	294	300	300	
Ecuador	259	265	270	270	270	270	271	274	278	283	285	
El Salvador	160	168	171	177	180	185	190	200	200	200	200	
Guatemala	300	300	300	304	307	311	307	306	307	307	307	310
Honduras	180	182	190	182	175	75	142	150	150	180	186	190
India	631	649	650	619	600	792	871	795	917	833	867	
Mexico	1,550	1,600	1,500	1,750	1,700	1,400	1,400	1,400	1,400	1,500	1,600	1,200
Nicaragua	76	76	78	80	80	77	100	91	139	129	130	
Papua New Guinea	3	7	3	2	2		2	2	1	1	2	
Peru	267	284	304	284	245	248	251	233	233	233	233	
Unwashed Arabicas	9,701	9,253	8,986	8,634	8,428	8,714	8,652	8,283	8,679	9,112	9,167	
Brazil	8,241	7,750	7,436	7,034	6,728	6,848	6,818	6,600	7,112	7,512	7,550	1,688
Ethiopia	1,460	1,503	1,550	1,600	1,700	1,866	1,834	1,683	1,567	1,600	1,617	
Robustas	1,722	1,741	1,740	1,665	1,600	1,563	1,778	1,893	1,936	1,867	1,958	
Angola	100	71	41	37	33	28	25	24	25	26	26	
Indonesia	1,299	1,298	1,309	1,219	1,113	1,140	1,145	1,148	1,170	1,196	1,208	
OAMCAF	(203)	(220)	(225)	(225)	(272)	(198)	(378)	(446)	(461)	(454)	(506)	
Benin	0	0	0	0	1	0	1	0	0	0	0	0
Cameroon	26	25	26	27	25	25	25	33	34	38	50	50
Central African Rep.	9	10	10	11	10	30	33	43	33	25	25	25

Table 11.9: continued

Congo	0	0	0	0	0	1	2	3	9	6	6	
Gabon	0	0	0	0	1	1	1	0	1	0	0	
Ivory Coast	35	47	47	44	150	64	183	217	217	217	217	
Madagascar	133	138	142	143	83	77	133	150	167	168	208	
Togo	0	0	0	0	1	0	0	0	0	0	0	
Uganda	20	27	25	29	32	30	30	25	30	33	35	
Zaire	100	125	140	155	150	167	200	250	250	167	183	183
B. Sub-total: Members export from basic quotas Arabicas	1,023	1,060	1,070	1,125	1,693	1,775	1,773	1,899	2,103	2,005	1,948	
	965	1,011	1,017	1,097	1,115	1,081	1,093	1,191	1,317	1,279	1,263	
Bolivia	25	24	23	28	32	33	35	33	36	40	43	
Burundi	3	2	4	3	3	3	3	3	2	2	1	
Haiti	220	224	225	247	276	262	252	257	252	253	275	
Jamaica	8	10	10	12	12	6	11	6	14	11	10	65
Malawi	0	0	0	0	0	0	0	0	0	0	1	
Panama	60	61	60	61	60	50	64	64	64	66	64	
Paraguay	13	13	13	14	17	12	7	12	15	15	15	
Rwanda	1	2	2	2	5	5	2	1	1	1	1	
Venezuela	635	675	680	730	700	700	709	805	923	882	843	920
Zimbabwe					10	10	10	10	10	9	10	
Robustas	58	49	53	28	578	694	680	708	786	726	685	
Ghana	10	10	10	11	11	15	1	0	0	0	0	
Guinea	0	0	0	0	1	10	10	8	10	37	20	
Liberia	5	5	6	6	6	10	10	0	12	0	0	
Nigeria	25	15	15	0	16	0	0	10	0	0	0	
Philippines					532	506	495	505	570	476	494	
Sierra Leone	3	4	5	5	5	5	2	0	0	0	0	0
Sri Lanka											33	33
Thailand						126	140	163	171	194	127	136
Trinidad and Tobago	15	15	17	6	7	22	22	22	23	19	11	15

Source: International Coffee Organisation, London.

visual defects, and just as in Brazil, are still using the black bean as the standard unit by which other kinds of defects are evaluated.

Having established quality standards, of one kind or another, exporters have then to make offers quoting their prices to potential buyers, dealers or users. This is now almost always done by telex in plain language messages and, initially, are not usually binding offers.

Unbinding offers are normal but when a response is shown by a potential buyer, there will often be a 'firm' bid by the buyer or a 'firm' offer by the seller. A firm bid or offer is one made with a specific time limit as a rule, the unqualified acceptance of which by the opposite party creates a binding contract. A qualified acceptance is really only a counter-offer and requires final acceptance by the first party.

At the commencement of a season, a seller at origin may have little idea of market prices and it is at this time that the dealer is important to him, and in fact, often prepares the ground by keeping the exporter in touch with market developments with an occasional letter or telex even when there is no immediate prospect of business. Some origins have sufficient coffee to carry right through the year, but in the case of smaller crops the bulk of the business may be done in say, one six-months period. A country having, say, more than one million bags is likely to be making some sales during at least nine months of the year and the very big producers for twelve months.

It is nowadays rare for origins to sell coffee, except on FOB terms (Free on Board vessel) though the West African countries still cling to their traditional CIF (Cost, Insurance and Freight) terms in many cases. Shipments may be 'prompt' which is universally understood to be within thirty days, or quite often some three or four months ahead, generally but not always, within a specified two-months period — August/September for instance which means at any time from 1st August to 30 September at sellers' option, unless otherwise stated. Where shipping services are reasonably good a single month may be specified. This is particularly so in the case of Brazil because the authorities permit registrations only in stated single months.

Most producing countries where there are private exporters insist on registration of contracts by the exporters immediately they are made; this is necessary in order properly to administer the quota limitations. Contracts have largely become standardised over the years and whereas some countries and organisations still insist on their own forms of contract, they are usually reasonably close to the conditions agreed or accepted internationally for common use by all those in regular trading.

Contracts

North American buyers invariably buy on what are known as 'green' coffee terms. A series of contracts have been worked out by the Green Coffee

Trading Organisations in New York, New Orleans and San Francisco to suit their market conditions and these are universally used in this area. The most important point to note is the so-called 'FDA clause'. There exist in the USA and in Canada very strict controls over imports, and goods evidencing any sign of a threat to health are refused entry. The inspection is by the Food and Drugs Administration (hence FDA) and its powers are absolute. Coffee stopped by the authority has to be removed, and since such stoppage is a serious risk caused by factors which operate before the goods reach the control of the importer, the importers insist that this shall normally be carried by the overseas seller. The appropriate clause provides that in case of a rejection by the FDA, the buyer is entitled to recover his purchase price and the goods are returned to the ownership of the seller, who must arrange to remove them. That can be expensive, though the risk can be insured. However, the obligation to provide coffee may still exist under the contract, unless the seller has taken care to obviate the risk of replacement by the use in the contract of some such term as 'no pass, no sale'. This is a very succinct but clearly understood phrase which means that if the coffee is returned to the seller the contract is discharged from further execution.

By far the most common contract used elsewhere, and not just exclusively in Europe, is the 'European Contract for Coffee'. This has been worked out over the last thirty years by the Committee of European Coffee Associations, of which the Coffee Trade Federation (the UK Trade organisation) is a member. The feature of this is that a single fairly simple set of conditions has been set up to provide for transactions on FOB, CIF or C&F terms. To do this, it departs from the ordinary common law understanding of FOB terms and transfers the onus of booking a vessel and freight to the seller as it is in a CIF or C&F contract. Under normal FOB terms it is the buyer's task to secure freight but in the case of coffee it has been found convenient to change that. A great many other contingencies are also provided for, giving both buyers and sellers remedies in certain eventualities. Since the contract is now used almost everywhere outside North America it is a very important contract.

Both the 'Green' contract terms and the European are concerned with loss in weight during a voyage. In order to protect buyers against shipment of underdried coffee, which became very prevalent at one time in a few countries, and which can lose a great deal of weight, the sellers in the standard conditions guarantee any loss in weight in excess of 1 per cent of the shipped weight, except where bags are lost or damaged in transit.

Also, as we have said elsewhere, every coffee contract provides that disputes shall be submitted to arbitration if the parties cannot reach agreement.

It is not necessary for the full details of contracts to be repeated on paper for each transaction and the usual method is to record the main specific details — quantity, description, price, shipment and so on — together with a statement that all other conditions are in accordance with the European Contract for Coffee (as an example) and, usually, some agreement on the venue for arbitration should it be required. The European Contract terms are, in fact, a small booklet, and every trader keeps copies for reference.

One of the important variables met in coffee trading is the freight charge. Under modern conditions freights change quite often, and are in any event not simple but compounded very often by surcharges for one pretext or another. It has, therefore, become the accepted thing to base contracts where freight is involved on the existing freight operating at the time of contract. Any variation, up or down (and it is occasionally down), is for account of the buyer.

Handling and Transportation

Practical execution of most shipment contracts has to be by means of documents — the bills of lading, invoice, certificates of origin or condition, etc. These are customarily presented through banking channels, since normally banks finance the trade. After giving his instructions for payment the buyer then uses the documents to obtain his coffee. To do this he invariably employs a specialist forwarding company; if possible a specialist not only in forwarding or warehousing generally but one with some experience of coffee. In the main coffee receiving ports this is easy enough and there is usually a group of competing houses to handle coffee readily and able to recognise and deal with damage or any abnormality revealed when the coffee is landed. This is very important as damaged coffee must be segregated before it can contaminate sound coffee; and contractual rights against a carrier or exporter can only be maintained by correct procedures. Finally, coffee is warehoused or forwarded directly to a user's own store, often in the interior miles from the coast. In many cases, it is customary to allow coffee to remain in port area for a short time before forwarded, though in recent years users have become accustomed to carrying very small physical stocks — far smaller than in years gone by.

Containers have come into use for coffee in the last twenty years against some resistance from the trade which originally had worries about the condition of coffee — a living article which requires to respire — travelling through differing climatic zones in a box. By degrees, the trade has come to accept that containers are inevitable simply because the shipowners are steadily moving away from traditional stowage. But there are still some problems, particularly as to responsibilities in certain contingencies and it is safe to say that containers are not as popular with the coffee technicians as they are with those solely concerned with transport.

Prices and Terminal Markets

Unfortunately, there is no one simple method of expressing prices in the coffee trade. In the American sector, prices are now almost always in US cents per pound, but some Latin-American exporters do occasionally go back to Spanish pounds which are slightly heavier than our pounds (100 Spanish = 101.3 lb or 46 kg). Elsewhere the most common unit of price is 'per 50 kilos' but sometimes it is 'per tonne of 1,000 kilos'. Fortunately, hundredweights of 112 lb have finally dropped out of use, but there still survive in some trading centres other usages, such as 'per 10 kilos', 'per kilo', 'per 100 kilos' and even 'per ½ kilo'.

The indicator prices formulated by the International Coffee Organisation are in US cents per pound; and this form of price is also used in the Resolutions and documents dealing with quota provisions and other Agreement matters. In practice, too, the US dollar is by far the most common currency used in international trading though in domestic markets, the local currency is used.

Apart from business transacted for shipment from origin there is, of course, quite a large volume of coffee sold within consuming countries after arrival. This is known as 'spot coffee' and is generally sold on 'in store', 'ex store' or 'free on conveyance' terms. Standard contracts for these have also been evolved by the associations — including a European Spot Contract, which clearly distinguishes the above terms and their implications.

The Terminal Markets, of which the New York market for arabica and London for robusta are the most important, are important tools of the trade, especially of the dealers. There are other markets elsewhere, notably in Paris/Le Havre where there is also a robusta market.

Unfortunately, they are not at all well understood either by the public or by the press. Their primary purpose is not to serve as a means of selling and delivering physical coffee; it is to serve as a tool of the physical trade which takes place quite separately. This is not the place to embark on an exposition of the whole technique and development of these markets but a simple example of just one way in which it can serve may illustrate how a Terminal market can help a user.

Let us suppose a roaster undertakes an engagement to supply coffee at a given price for a period of up to a year ahead. Obviously he will be extremely vulnerable if, in the middle of this period, there is a serious frost disaster in Brazil causing prices to soar as they did in 1976. He can, however, buy protection in the Terminal market by buying in an appropriate forward position at the price of the day he begins the commitment. In theory he has bought a standard grade (not origin), a standard volume at an agreed price on some kind of spot condition according to the terms of the market in question. But he is not committed to waiting for the contract to mature and accepting any physical coffee which happens to be tendered. At some convenient moment in his own judgement he buys in the physical market the precise quality he needs, and sells his Terminal contracts. The clearing house will then offset his Terminal sales and purchases and remit the difference, or require payment if appropriate. In theory, at least, the movements in the Terminal market should be roughly similar to those in the physical market. Any profit on the Terminal should roughly compensate for any extra cost on the physical coffee — or vice versa. Note that the physical transactions are the reverse of those on the Terminal. That of course is the simplest form of Terminal protection and many more complex schemes have been evolved by the dealers and Terminal brokers for their own use and those of their clients.

To succeed, of course, the Terminal markets must maintain some link with the physical market; without that they would be useless. This is the reason why it is essential for the seller to the Terminal market to have the facility to tender physical coffee if he so decides, and for the buyer to have the facility to hold his position to the last day of the delivery month, if necessary, to enforce physical delivery. In practice, buyers never do insist since they can never choose what growth or grade will be tendered, but they do retain a very useful freedom in cases where physical coffee is relatively scarce to wait until an attractive price is bid by the 'short'. And it must be remembered that where there is a 'long' — a party who has bought with a contract still open — there must also be a 'short' who has sold coffee and needs either to tender physical coffee or buy back his contract to offset in the clearing system.

Trading on the Terminal involves dealing in standard volumes of agreed grades and growths on the terms laid down in the rules of the market. Although trading on the 'floor', as it is known, is between two traders (or brokers, in fact) every transaction is registered with the clearing house as a principal on one side or the other. This facilitates the reversing of the original transactions without reference to the first parties but so long as two members do agree to trade, the clearing house itself always remains square — neither long nor short — and the number of short and long contracts is always balanced. The clearing house is the essential administrative and financial link which makes the whole thing work.

In London, most of the major origins of robusta are tenderable, except Conilon (Brazilian robusta) and there is a series of grades, based on visual defects, with price differentials laid down in the rules. A would-be tenderer submits samples of his coffee for official grading and receives a certificate (assuming it passes in an acceptable grade) which he adds to the normal documents when tendering his coffee to the clearing house during the delivery month. The unit of volume is 5 tonnes and the coffee must be warehoused in one of the agreed warehouses in the UK, Holland, France or Hamburg. In New York, a wide range of arabica growths is accepted but the contract is for a volume of 17,250 kg and the coffee must be in New York or New Orleans.

There is some criticism of Terminal markets because of the supposedly sinister activities of speculators. But, actually, an element of speculative interest is very healthy not to say necessary, though admittedly there are times when, due to the amount of inflated money circulating since the OPEC revolution of 1973/74, the speculative element can overwhelm genuine trade interest. All the same, there must be some catalyst willing and able to enter the market and set it working when it has been reduced to paralysis by some such event as news of a Brazilian frost. At such times, all the trade wants to buy and is wholly occupied with seeking protection; it requires somebody from outside willing to risk making a price to get the market working again. In New York, there is a rule that restricts daily fluctuations of more than four cents per lb, but in London that is considered a disastrous ruling as it renders the market totally useless at the moment it is most needed.

Laymen often wonder who pays the losses when an operator draws his profit; they point out quite logically that somebody must have lost that amount. True, but exactly the same applies at Lloyds. Insurance claims are not settled out of thin air. The coffee market as a whole, through its many participants, finds the losses and receives the gains. The clearing house is always 'square' both as to the number of commitments and as to money since, when an operator's position shows a loss the clearing house calls on him to deposit cash or a bank guarantee to cover that loss, even if temporarily.

The London market has several times traded over 1 million 5-ton lots or contracts in a year and it has become clear that it does provide a service to its users and a system that has worked very satisfactorily, particularly as to the clearing and finance. There is a close supervision by, and liaison with, the Bank of England, and the clearing house itself is owned by a consortium of the UK banks. It operates for many other commodities as well as coffee.

The membership — full floor trading members, that is — is limited for practical reasons, but associate membership is available to any other organisation and they have their representatives on the Management Committee. The Association, however, deals solely with the Terminal Market and its operation.

All other business of the coffee community in the UK, including participation with the Ministry of Agriculture, Fisheries and Food in the UK delegation to the International Agreement, is in the hands of the Coffee Trade Federation. Membership includes all sides of the coffee business — even the warehouses who normally handle coffee imports. The CTF is a member of the European Committee of Coffee Associations and the European Roasters organisation.

Soluble Powder Exports

In addition to imports and trade in green coffee, there is also a limited business in soluble coffee powder manufactured overseas — Brazil being the chief supplier though there are others, including some importing countries. The UK and the USA are the largest importers but many other countries import modest quantities. In Europe, particularly Scandinavia, the usage of soluble coffee is not large enough to warrant manufacture locally in some countries whilst others, like Holland, have been exporters to the UK which is the largest market outside the USA. Some 85 per cent of the coffee used in the UK is in soluble form and of that, perhaps, a third is customarily imported ready made and mostly spray-dried.

Spray-dried soluble powder reflects the virtues of real coffee less faithfully than the more expensive freeze-dried product. In the UK, the main body of consumers is not discriminating enough to pay the premium for freeze-dried, whereas elsewhere in Europe, and even in the USA, the taste for coffee is older and more sophisticated. The imports are, of course, practically all in bulk which is packed for retail purposes in the country of usage.

Brazil and Colombia — Special Deals

A glance at the statistics of coffee quotas or exports (see Tables 11.2; 11.3; 11.5 and 11.7) will reveal how important are Brazil and Colombia; they are the giants of the market. Even in the lean years of 1976/77 Brazil exported 12 million bags which is incomparably more than any other producer, except Colombia.

It is therefore necessary to explain as succinctly as possible how and why this enormous sector of the world supply is not reflected in the International Agreement indicator system for controlling the quota. At present, at least, the system relies on the robusta and so-called Other Milds sectors, which are not much more than half of the world's coffee.

Under the 1962 and 1968 Agreements, the selectivity system was used which implied that a particular division of producers — perhaps, for example, the robusta countries could gain quota or lose it according to the movements of its own indicator price. This was in addition to, and independent of, the overall adjustments.

Brazil, then having a large accrued surplus, was determined to avoid losing quota but to remain competitive to ensure selling it. At that time, because the structure of the Agreement was different from that currently in operation, it was not possible to assess the quota as accurately as it is now, and it often happened that the quota was not fully sold. So a scheme was devised by the Brazilians to enable her to be competitive by reducing real net prices without running the risk of losing quota through selectivity. The nominal price was kept stable but confidential discounts were available to many markets. This system still continues although the details of the arrangements have changed and, in fact, are changed each year according to the needs of the moment. Colombia, too, has adopted a similar system.

So, at present, roasters are invited to guarantee to the IBC, the Brazilian controlling authority referred to earlier, to take an agreed volume within the calendar year and in return are given discounts from the overt controlled price. The discounts are variable and regulated by reference to the indicators of the robusta and Other Milds groups. The precise formula is varied from year to year, but it is clear that though these two groups are competitors of Brazil they are unable to defeat Brazil in price competition. If they reduce their prices it merely increases the discount; there is no escape and once Brazil has tied up sufficient roaster interest to guarantee selling her quota she cannot go wrong. If and when the market is seen to shrink during the year for some reason, as it has, it is the other half of the market which bears all the loss so far as volume is concerned. And volume is, if possible, even more important than a few cents in the price.

The discounts are not paid in cash but in the form of 'avisos' which are in effect credit notes to be used to offset invoices. In the past, at any rate, Brazil has been slow in issuing these avisos so at times the roasters have found themselves lending Brazil vast sums of money.

In Brazil's case, particularly, it should be noted that the private channels of trade are respected and private exporters actually select and export the coffee required by the contracting roasters. What is more, the business is usually negotiated through the customary agencies so that they can earn a small commission though it is virtually impossible for them to merchant Brazil coffee freely. So business is often found to be negotiated at the overt controlled price subject to the issue in due time of discount vouchers for anything up to 50 per cent of the price. The private exporters invoice at the overt price and the vouchers are used to offset against these invoices but for practical reasons never for more than the so-called Contribution Quota which is a form of export tax and part of the price control mechanism in Brazil; often 50 per cent of the total value.

The Colombian system is not quite the same and has the additional feature that a large proportion of the exports are actually made by the state organisation, the FNC, itself and confined to its own agents.

It is because the nominal controlled prices of both Brazilian and Colombian coffee were so obviously sheer fiction that it was agreed to operate the International Agreement simply on the basis of the other two groups which still retained some semblance of free markets responding to supply and demand. The curious thing is that the other producers seem to have little or no objection to the Brazilian and Colombian behaviour. Apart from the fact that the voting arrangements give these two countries enormous power, they invariably assert their leadership in looking after their own interests before that of their less fortunate fellows, and although there are sometimes grumbles, no objections are made in public. Some efforts have been made to question the so-called 'special supply agreements' in the Agreement meetings but producers have always been unanimous in ruling any mention of the subject as out of order.

As a matter of fact the quota arrangements are now, under the current Agreement terms, much more efficient than they were previously and there seems no real need for Brazil and Colombia to use the deal system. They should be able to sell all their quota in any case; and they have so far refused to agree to a resumption of selectivity even in a limited fashion.

Some consumers, point to the fact that surpluses do now exist in many areas, and a selectivity system would therefore be valuable and allow a little flexibility. At present, consumers have to drink what coffee is available and, as a whole, have little choice. Somebody has to drink all the coffee provided in the overall quota as it is now only just about adequate.

The present quota market is approximately 55 million bags; growth is very slow and in most recent years a decline in the USA has been only slightly exceeded by growth elsewhere in Europe and, more especially, in Japan. Japan, since the last war has become a 3 millions-a-year user from virtually nothing, and the UK can now use $2\frac{1}{2}$ millions which is some 2 millions more than in the thirties.

Non-quota Exports

The non-quota users were until recently recorded as using some 6 million bags annually, but since the 1980 resumption of the quota, this has risen to 10 million, according to the latest statistics (see Tables 11.7 and 11.8). Hence some of the fears as to possible diversion, though it is clear that many of the territories have very much increased their usage — the OPEC countries for example. Nor is it known for certain just how much coffee was re-sold by what are now quota countries to the non-members in the seventies when movements of coffee were not closely regulated.

Trade and Markets

Most international coffee trading is in terms of US dollars, and it follows that prices in other currencies tend to move with the dollar exchange against the currency in question. London Terminal market prices are in sterling per 1,000 kg so will therefore vary according to the dollar/sterling rate. Sometimes, the lay speculative element in the market is so strong that this does not happen for the reason that the speculator is not constantly converting actual coffee prices as does the regular trader in his daily business.

Prices constantly move, but now that the market is subject to a controlled flow, and so long as the control lasts, movements have been contained. World production can vary a great deal but is now 'normally' slightly larger than total usage. We have an accrued stock to cushion us against a disastrous year and, under the Agreement, all the world can foresee what is happening to the overall level of prices and the quota adjustments which can result. The world is carrying a surplus of 40 to 45 million bags which is six months' usage or a little more. As it happens, the nations agreed some years ago that a six-months supply was prudent. So, remembering that we could at almost any moment suffer a disastrous year, such as 1975, the surplus is not quite so tragic as some 'bears' would have us think. At the same time the burden of carrying a surplus from year to year is, for the majority of producing countries, no light burden. It is difficult to blame them for seizing every chance to sell coffee to realise usable cash even if the price is low.

Most roasters (the companies who actually roast and pack coffee in consuming areas) realise that a long period of low prices is a menace to them, as it is to the growers, since it means that the smallholders who grow most of the coffee in the world cannot afford to maintain, fertilise and replace trees. Yields can shrink within a few years and force prices higher. Roasters all say that it is the fluctuating retail prices which cause them to lose sales more than any other feature. Cheaper prices will hardly affect sales volume, but rapidly rising prices do, and did so very sharply indeed in 1977 after the soaring markets of 1976, and early 1977.

One of the fascinating features of the coffee market is the almost infinite range of growths and qualities, and the fact that the market can still find a place for them in spite of growing standardisation. We still find that coffee acceptable in one part of the world is unsaleable in another.

The standard in commerce was originally set by the arabica species. Robusta only came into wide usage much later; in fact mostly since 1940. The expansion roughly coincided with the growth of the soluble coffee powder industry, though this does not mean that robusta is mainly used for this purpose or that all soluble powder is based on robusta. A great deal of arabica is now used in soluble coffee manufacture, particularly for freeze-dried brands because this process does reproduce the original flavour much more faithfully.

Because arabica dominated commerce in the first place, robusta has come to be looked on as inferior by most buyers. It is used in many blends and sometimes as a main ingredient where price is all-important. In soluble powder manufacture, the yield from robusta is slightly larger than from arabica.

Other species, such as liberica and excelsa, are only used in very small quantities by special markets. Maragogipe is only a variant of arabica and shares its taste character. Because of its bean size it has been a speciality in a few European markets, where coffee can be sold as the roasted bean. The finest quality in the world is demanded by Finland, where per capita usage is very high indeed, and it is really a delight to drink coffee there. Germany, Switzerland and Sweden are also very discriminating with other European countries not far behind, though there are some wide differences even in different places within one country.

The USA uses a very wide range of qualities, but the retail market is so competitive that the average quality tends to be a modest level for price reasons, plus the fact that the public is not accustomed to the strength of brew favoured in Europe.

In the UK the growth of usage by a couple of million bags since 1940 is due to the popularity of soluble coffee, and almost all spray-dried at that. Ordinary roasted and ground usage is only slightly increased. Soluble coffee is also popular in the USA, Japan, Australia and other places. It is least popular in Scandinavia, perhaps.

Units in the roasting and packing trades in Europe and the USA have evolved into fewer and bigger concerns over the last 35 years and this has produced a demand for large volumes of standardised qualities, both in the green coffee stage and the retail stage. The very large concerns turn out enormous volumes and are bound to sacrifice the pinnacle of quality to achieve this. Super quality always comes in small quantities; to achieve uniformity over a volume the standard must be a little less than the 'Rolls Royce' peak, although so far as coffee is concerned the standard in many cases is still very acceptable indeed.

Individuals responsible for buying the raw product on behalf of their roasting plants are no longer able to taste and select individual small lots as they once did. They are obliged to leave that task to the exporter at origin, who bulks coffee to make up the required standard as mentioned earlier. This is leading to the development of quite close relationships between roasters and exporters but, however, not necessarily to the exclusion of the dealer link. The dealer still has an important role and is appreciated by the roaster as well as the exporter.

There is little or no sign that roasters wish to own plantations at origin; perhaps, the whole business is too complex for that to happen though it has happened in other trades in a slightly less complicated past.

Bibliography

The International Coffee Agreement 1983 — text available from the International Coffee Organisation, 22 Berners St, London W1

The Lomé Convention 1980 — text published by the European Community Publications Office, Luxembourg (with agencies in member countries)
The European Contract for Coffee, 1980 — published by the Committee of European Coffee Associations and obtainable in the UK from The Coffee Trade Federation Ltd, 9 Wapping Lane, London E1 9DA
The Green Coffee Contract — published by the New York Green Coffee Trade Association, New York

12 THE MICROSCOPIC STRUCTURE OF THE COFFEE BEAN

Eliane Dentan

Introduction

The histology of the coffee bean was established many years ago (Winton and Winton, 1939); however until now, the ultrastructure of this material has not been described. So far we have been unable to find a single printed work at the transmission electron microscope level apart from our own publication (Dentan, 1979). (We apologise to authors whose publications we may have missed in our literature search.) One reason for this lack of interest could be that the cytological and histological studies pose a number of technical problems. The thickness of the cell walls, and the close packing of storage material within the cytoplasm of the coffee bean parenchyma cells do not permit easy fixation, embedding and sectioning. Therefore we include details of how these problems were solved.

Material and Methods

Two species of ripe coffee beans were studied:

1. *C. canephora, var. robusta* (Ivory Coast)
2. *C. arabica, var. catuai amarelo* (Brazil)

Light Microscopy

The berries were picked at full ripeness and fixed immediately by immersion in neutral formaldehyde (Clark, 1973 : 13) or Baker fixative (Pearse, 1968 : 601) and kept for three to four days at 4°C. They may be kept longer if required. The berries are prepared for use by washing in tap water for one day. Embedding is difficult. Waxes are unsuitable, even under vacuum, because of loss of lipids and poor infiltration of the tissues. The recently developed less viscous resins as well as glycol methacrylates (such as JB-4 of Polysciences — see Brinn and Pickett, 1979 and Higuchi, Suga Dannenberg and Schofield, 1979) allow satisfactory embedding at 4°C without obvious loss of lipid. Thin sections (thickness 2 to 3 μm) were obtained and would be stained with aqueous dye solutions, without modification.

The alternative is to cut the berries without embedding. Frozen transverse sections (thickness 10 μm) are easily obtained if both specimen and microtome knife temperature can be controlled. For ripe coffee beans the ideal temperature was between −12.5°C and −15°C. The sections were picked up by the knife and directly laid on glass slides, on a thin layer of ovalbumin. They were stained after 48 hours. A Leitz 'Kryomat' with methanol refrigeration was used for the non-embedded specimens but a Reichert OMU3 ultramicrotome for the JB-4 embedded specimens.

Staining

All the stains used are described in detail in Clark (1973) and Pearse (1968, 1970). The specific stains employed were:

1. Neutral lipids : Oil Red O (Pearse, 1968 : 697, Clark; 1973 : 149)
2. Unsaturated lipids : UV-Schiff (Pearse, 1968 : 701)
3. Free fatty acids and glycerides : Nile Blue (Pearse, 1968 : 697)
4. Proteins : $HgCl_2$ — Bromophenol Blue (Pearse, 1968 : 607)
5. Pectin and pectinates : Ruthenium Red (Pearse, 1968 : 679)
6. Carbohydrates : Periodic Acid — Schiff (PAS) (Pearse, 1968 : 659; Clark, 1973 : 156)
7. Phenolic groups : Millon Reagent (Pearse, 1968 : 606)
8. Indolic groups : Xanthydrol (Pearse, 1968 : 616)
9. Lignin and mucilage : Safranin/Fast Green FCF (Clark, 1973 : 205)

Transmission Electron Microscopy

The freshly gathered berries were immediately depulped and the beans thinly sliced (1-2 mm). These slices were prefixed at 4°C for 2-3 hours in glutaraldehyde 6.5 per cent in phosphate buffer 0.15 M, pH 6.8. They were then cut into small blocks and fixed overnight in a fresh solution of the same fixative, also at 4°C. After rapid washing in phosphate buffer, they were post-fixed two hours in osmium tetroxide 2 per cent in phosphate buffer at 4°C. After washing, these blocks were stained overnight with uranyl acetate 0.5 per cent aq. The dehydration was performed in different baths of acetone, and the embedding in 'Spurr'/soft medium (Spurr, 1969). Each bath in the mixtures 'Spurr medium' acetone lasted at least 24 hours. The last bath in pure resin took one day, on a rotating shaker. The polymerisation lasted 12 hours at 70°C. Ultrathin sections were easily obtained with glass knives. They were double stained with uranyl acetate/lead citrate according to Reynolds (1963). The specific staining for polysaccharides was performed on gold grids with phosphotungstic acid 2 per cent in 1 M HCl, according to Rousseau and Hermier (1975). The sections were examined with a Philips EM 300 at 80 kV.

Scanning Electron Microscropy

Conventional techniques did not give good results. Lipid is lost when using the critical point drying system. Several approaches were evaluated and the best results were obtained as follows : thick frozen sections (30 to 50 µm) were prepared from prefixed beans. These sections were postfixed in osmium tetroxide 2 per cent in phosphate buffer, dehydrated in acetone and by the critical point process, and then covered with gold by sputtering. They were examined with a Cambridge S4-10.

Microscopic Structure of the Coffee Bean

We have been unable to find major microscopic differences between *C. canephora* var. *robusta* and *C. arabica* var. *catuai*. The structures described, therefore, correspond to the two species. A transverse section is shown diagrammatically in Figure 12.1. The specific form of the bean can be recognised as being folded upon itself. At the periphery, there is one single layer of pavimentous epidermal cells. The major part of the bean is formed of parenchymatous storage cells. The cells close to the epidermis are more elongated than those close to the centre. In the middle part of the section one can distinguish a layer of mucilaginous material of variable thickness, in which is embedded the embryo. Houk (1938) has demonstrated that the bulky portion of the bean is perisperm. Winton and Winton (1939) proposed that the mucilaginous material could represent the endosperm.

Figure 12.1: Diagrammatic Transverse Section of Coffee Bean.

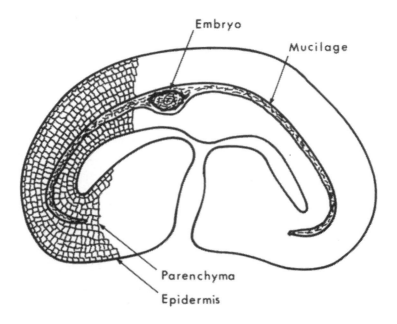

Parenchyma Storage Cells

Cytoplasm

The cytoplasm of these cells contains essentially lipids, proteins, carbohydrates and appreciable amounts of caffeine, chlorogenic acids and mineral salts (Clifford, 1975; Poisson, 1979 — see also Chapter 13).

Light Microscopy. The lipids, as shown after staining with Oil Red O, are distributed homogeneously throughout the bean (Figure 12.2) embryo included, and located close to the plasmalemma, forming a layer of variable thickness. This localisation is characteristic for oil seeds in dormancy. The UV-Schiff reaction, specific for unsaturated lipids shows their presence to be homogeneous throughout the bean (Figure 12.3). The sulphate Nile Blue staining demonstrates the distribution of free fatty acids versus glycerides. In the ripe beans examined, the amount of free fatty acids was small, and their distribution uneven, without any specific location. The middle part of the cytoplasm, free of lipids, contains chiefly proteins (Figure 12.4, $HgCl_2$ — Bromophenol Blue) or carbohydrates (Figure 12.5, PAS). The proteins are encountered mostly in the first layers of cells under the epidermis. In the robustas the cells close to the centre of the bean are often

Figure 12.2: LM (Light Microscopy) Lipids. Oil Red 0 staining. The lipids (↑) are distributed in all cells, close to the cell wall. In this and subsequent figures the scale is indicated by the horizontal bar in the right corner, the length of which is shown in μm.

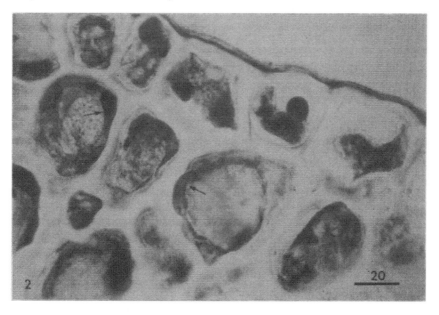

Figure 12.3: LM. Unsaturated Lipids. UV-Schiff reaction. These lipids (↑) are also distributed in all cells, close to the cell wall.

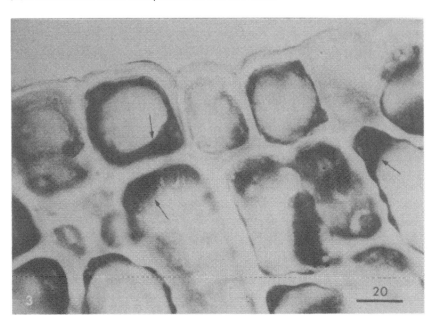

Figure 12.4: LM. Proteins. Mercuric Bromophenol Blue staining. Proteins fill the centre of the cytoplasm. Smaller 'protein bodies' (↑) are located close to the cell walls which are also faintly stained (↑↑).

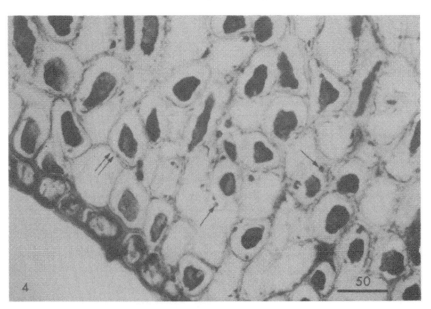

Figure 12.5: LM. Polysaccharides. Periodic acid-Schiff (PAS). Cell walls and vacuoles (↑) are stained.

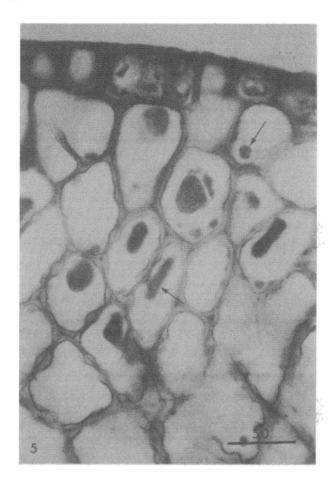

free of carbohydrates. Of course, the thick layer of mucilage, in which the embryo is embedded, is strongly stained with the PAS reaction as well as with safranin (Figure 12.6) indicating the presence of carbohydrate and lignin.

The Millon reagent is used to demonstrate phenolic groups, in the coffee bean mainly chlorogenic acids. Lignin does not show a positive reaction. A weak positive reaction was found within the cytoplasm, but the strongest reaction was seen at the periphery very close to the cell walls (Figure 12.7). The demonstration of the presence of caffeine is still a problem. All reagents used in chemical tests dissolve it, but caffeine shows a characteristic fluorescence at 360 nm. Under the fluorescence microscope it was possible to obtain the characteristic emission as small spots in the cytoplasm, close

Figure 12.6: LM. Mucilage. Safranine/Fast Green FCF. The safranine stains the mucilaginous material (↑) located in the interior of the bean.

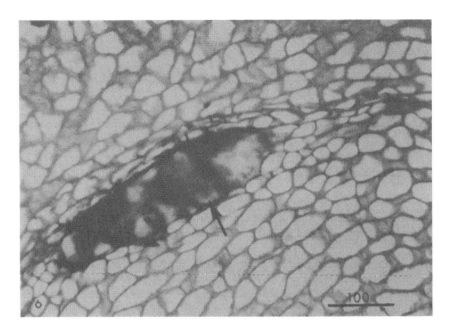

to the lipids (Figure 12.8) but, at the same wavelengths, the cell walls also gave a strong fluorescence. It is therefore impossible to detect the caffeine which is thought to be associated with chlorogenic acids adjacent to the cell walls. With the polarising microscope, it was possible to demonstrate the presence of some free caffeine in the cytoplasm. Figure 12.9 shows such small hexagonal needles. The measurement of crystallographic indices and of birefringence with the elliptical compensator according to Brace-Köhler (Jerrard, 1948) confirmed that these needles were caffeine. With the polarising microscope, it is also possible to demonstrate the presence of mineral salts such as calcium oxalate dihydrate and calcium phosphate (Figure 12.10).

Transmission Electron Microscopy. As shown in Figure 12.11, the lipids are distributed in small closely packed droplets, and covered with a fine layer of proteins. These droplets form a layer of variable thickness, in the immediate vicinity of the plasmalemma; in between, it is possible to distinguish some mitochondria, as well as small amyloplasts. Very often between the lipid droplets, small amounts of strongly osmiophilic amorphous material are also present, limited with a membrane. Since tannic substances are osmiophilic, and by analogy with the light microscopy results, we believe that this material could correspond to chlorogenic acids.

Figure 12.7: LM. Phenolic compounds. Millon reagent. Phenolic compounds are mainly found in the epidermis cell wall (↑↑), in the cytoplasm of some cells (*), as well as fine deposits along the cell walls (↑).

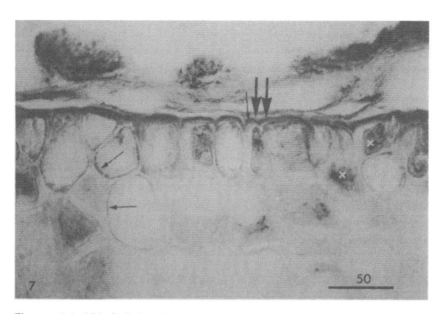

Figure 12.8: LM. Caffeine. Fluorescence. The caffeine is detected as small brilliant spots (↑) in the cytoplasm, in the vicinity of lipids. The cell walls are also fluorescent. It is therefore impossible to detect caffeine lying along the cell walls.

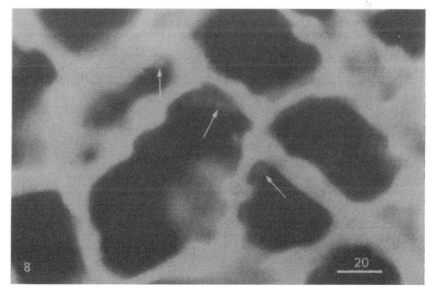

Figure 12.9: LM. Caffeine. Polarisation. Oil immersion. Small needles of caffeine are occasionally found in the cytoplasm.

Figure 12.10: LM. Polarisation. Oil immersion. In two adjacent cells are found one calcium oxalate dihydrate crystal (0), and one potassium phosphate crystal (↑).

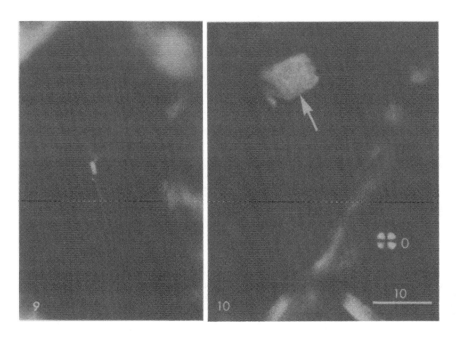

In the centre of the cytoplasm, free of lipids, are some carbohydrates, as demonstrated by specific staining with phosphotungstic acid at low pH (Figure 12.12). These carbohydrates are always included in a kind of vacuole. The amyloplasts were also stained with this technique. The greater part of the cell centre is filled with proteins (Figure 12.11), some of which are included in a type of 'protein body', limited by a membrane. These protein bodies are very often close to the lipids, or even enclosed within them. In the cells adjacent to the epidermis, we also find large vacuoles containing very osmiophilic and amorphous material, dispersed in the proteins (Figure 12.13). These vacuoles again probably represent chlorogenic acid deposits. Occasionally, we found in the lipid region some crystalline structures (Figure 12.11) which gave hexagonal diffraction patterns and could correspond to caffeine, and others which could correspond to catalase crystals.

Cell Walls

The parenchymatous cell walls of ripe coffee beans are particularly thick and do not present any intercellular space. They also exhibit characteristic

Figure 12.11: TEM (Transmission Election Microscopy). Part of parenchyma storage cell. Small lipid droplets (L) are lying on the plasmalemma. Between the lipids are starch granules (S), mitochondria (M), some crystalline structure (C) and some amorphous deposits (*). The upper part contains proteins (P). The cell wall (CW) would be crossed by some plasmodesmata.

Figure 12.12: TEM. Polysaccharides specifically stained are localised in vacuoles (↑). Starch is also stained (S). One can distinguish lipid droplets (L). The proteins are not stained (P).

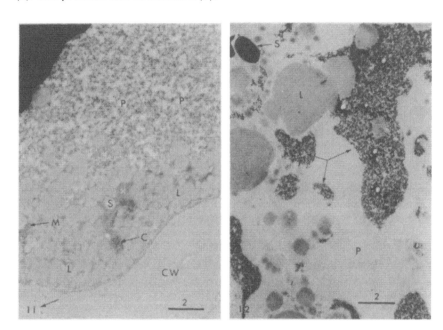

nodular structures, chiefly in the layers below the epidermal cells.

As has been described by many authors (Frey-Wissling, 1959; Frey-Wissling and Mühlethaler, 1965; Pilet, 1971; Gunning and Steer, 1975), the cell walls consist of different layers: in the centre, the middle lamella contains mostly pectic substances. On both sides, first a thin primary wall which cannot be easily differentiated from the middle lamella under the light microscope, then a thick secondary wall and lastly a tertiary wall which appears more or less warty on the interior surface. The bulk of the full-grown cell wall consists of the secondary wall.

Light Microscopy. In the coffee bean, as shown with the polarising microscope (Figure 12.14) the nodular structures are produced by the secondary walls: the middle lamella is isotropic and the weakly birefringent

Figure 12.13: TEM. Part of a parenchyma storage cell close to the epidermis. The cytoplasm contains large vacuoles (↑) full of amorphous osmiophilic material which could correspond to phenolic compounds. Small lipid droplets (L) are found at the periphery of the cell. The centre contains proteins (P) and other small vacuoles (V) full of unknown material.

Figure 12.14: LM. Cell walls. Polarisation. The cell walls present nodular structure (↑). The middle lamella, isotropic, appears black. All the brilliant regions represent the secondary cell wall which is anisotropic.

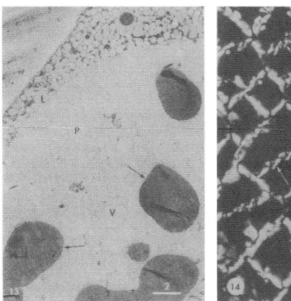

primary wall is occulted by the bright anisotropic secondary wall. All the nodes exhibit a strong birefringence. The middle lamella could be stained with Ruthenium Red, or visualised by its own fluorescence. The presence of lignin was demonstrated either with safranin staining or by its own fluorescence (Figure 12.15). We could observe that most of the nodes are partially lignified. A faint blue coloration of the cell walls after treatment with mercuric bromophenol blue confirmed that they contain a small amount of proteins. All the cell walls are, of course, strongly stained with the PAS reagent (see Figure 12.15).

Transmission Electron Microscopy. At this level, no ultrastructure can be recognised without specific treatment of thin sections of the cell walls. All constituents (cellulose, hemicellulose, lignin) consist of the same atoms

Figure 12.15: LM. Cell walls. Fluorescence. All the brilliant white regions in the cell walls correspond to the presence of lignin (↑↑). The middle lamella is fluorescent as well (↑). The proteins in the cytoplasm are also faintly fluorescent (*).

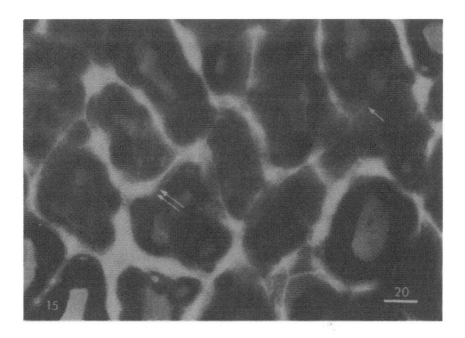

(O, C, H) and present the same contrast under the electron beam. Figure 12.16 shows a characteristic node. We can distinguish clearly the middle lamella and some plasmodesmata crossing the walls, but it is not possible to observe cellulosic fibrils. To render them visible would require recourse to the method of maceration, as described by Frey-Wissling, Mühlethaler and Wyckoff (1948). Nevertheless, if we use a specific staining for polysaccharides such as, for example, phosphotungstic acid, according to Rousseau and Hermier (1975), we obtain a better contrast (Figure 12.17). We can still distinguish the middle lamella as well as plasmodesmata which contain some stained material. We suggest that maybe the dark zone without visible structure on both sides of the middle lamella could correspond to the lignified region. Outside we can see the parallel texture of the cellulosic fibrils. Figure 12.17 shows perpendicularly-cut plasmodesmata crossing the cell wall. Very often, between the cell wall and the plasmalemma we can observe an important region, the periplasm, filled with strongly osmiophilic amorphous material (Figure 12.18). According to results obtained in light microscopy, we suggest again that these black deposits could correspond to chlorogenic acids extruded from the cytoplasm.

Figure 12.16: TEM. Characteristic nodular cell wall. One can distinguish the middle lamella (ML) and numerous plasmodesmata (D) crossing the cell wall. Small lipid droplets are lying on the plasmalemma (L). One starch granule (S).

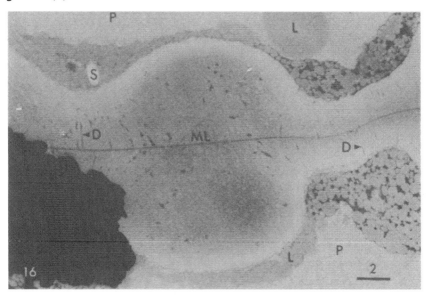

Figure 12.17: TEM. Cell wall. Specific staining for polysaccharides. On the two sides of the cell wall it is possible to distinguish cellulosic fibrils (↑↑). The cell wall is crossed by several plasmodesmata (D) containing stained material (↑). Small starch granules are also stained (S).

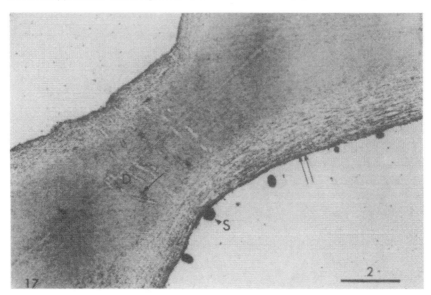

Figure 12.18: TEM. Periplasm. In the space between the cell wall (CW) and the plasmalemma (↑), strong osmiophilic amorphous deposits are visible (*), which could correspond to phenolic compounds. Some lipids (L) are lying on the plasmalemma.

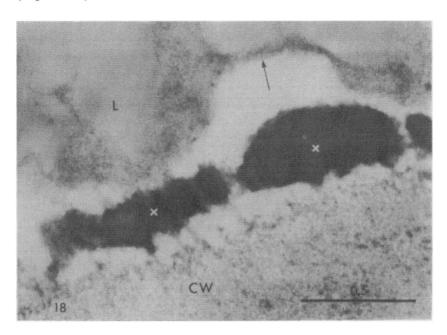

Epidermal Cells

Light Microscopy

In transverse sections, these cells present a more or less pavimental aspect. They form generally a single layer at the periphery of the bean. As for parenchyma storage cells, their cytoplasm contains lipids (Figure 12.2) proteins (Figure 12.4) carbohydrates (Figure 12.5) and a large amount of phenolic compounds (Figure 12.7).

The external cell wall had its surface well stained with Oil Red O denoting the presence of surface waxes (Figure 12.2). Under the waxes, the cutin was stained with safranin. The underlying subcuticular layer contained various deposited substances: 5-hydroxytryptamine (serotonin) derivatives, as demonstrated with the Xanthydrol reagent, were located exclusively in this region (Figure 12.19). This layer appears only during the last weeks of ripening. The Millon reagent also gave a strong positive reaction (Figure 12.7) but defined a thicker layer than could be attributed to the serotonin derivatives alone.

The different layers (waxes, cuticle, sub-cuticular layer, middle lamella) could be easily distinguished under the polarising microscope : the waxes

Figure 12.19: LM. Serotonin. Xanthydrol reaction. A strong positive reaction is visible in the epidermis cell wall (↑↑). In the cytoplasm of some cells, some tryptophan is still present (↑). Indole compounds are also detected.

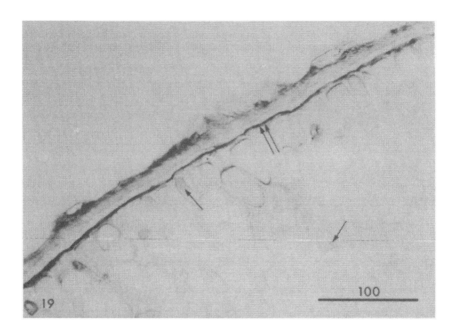

and the cuticular layer are anisotropic, the cuticle and the middle lamella are isotropic. Examination of stained sections under the polarising microscope depicts in which layers the different constituents are localised.

Electron Microscopy

The transmission electron microscope does not reveal the different layers seen in polarising microscopy. Nevertheless, at high magnification, we could see on the surface the crystallised waxes, which present a lamellar aspect (Figure 12.20) that corresponds to a smectic phase for liquid crystals, optically active under the polarising microscope. The resultant structure is called homeotropic. It is optically homogenous and shows the characteristics of a positive uniaxial crystal. It indicates that we have a phase very close to the solid crystals. These waxes formed a layer approximately 2 µm thick and they represent an effective barrier against desiccation of the beans. Under the waxes, as shown by pictures at lower magnification, was a thick area which reached and occluded the middle lamella (Figure 12.21). In this region, we could observe a number of small black deposits (see Figure 12.20) which could correspond either to phenolic compounds and/or terpenes, the two being equally osmiophilic. Examina-

Figure 12.20: TEM. Epidermis cell wall. The waxes on the surface of the bean present a characteristic myelinic structure (↑). Under the waxes are black deposits which correspond either to phenolic compounds, or to terpenes.

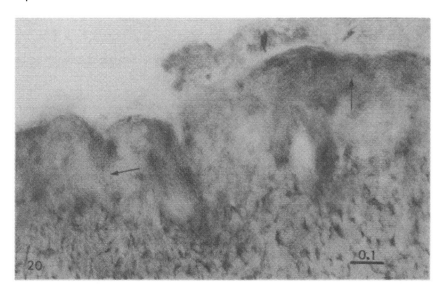

Figure 12.21: TEM. Epidermis cell wall (CW). The dark zone at the periphery comprise waxes and the cuticular layer, encrusted with various constituents. At the bottom right, there is a small part of the cytoplasm with proteins (P), lipids (L) and a starch granule (S).

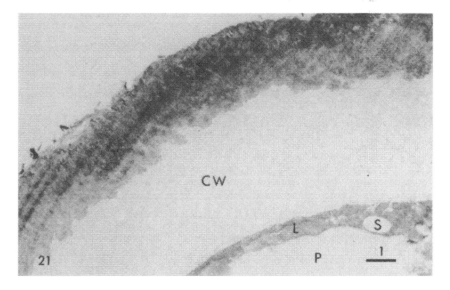

tion under the scanning electron microscope of the external surface of the epidermal cells, showed that these cells were relatively elongated and covered with folded waxes (Figure 12.22).

Embryo and Mucilage

We did not study the embryo specifically, but as we very often obtained a part of it in whole bean sections, we observed that it was made up of more or less quadrate cells whose cytoplasm contained a large amount of lipids, proteins, carbohydrates and phenolic compounds. The cell walls were thin and regular. Figure 12.23 shows a longitudinal view of the embryo, as seen under the scanning electron microscope. The mucilage in which it was embedded disappeared completely during manipulations.

The mucilage itself was well stained with safranin and PAS. It did not show any definite structure at the light microscope level. Under the transmission electron microscope, we could see that the cell walls were extremely swollen and contained much unidentified dark osmiophillic material (Figure 12.24). These osmiophilic deposits could correspond again to phenolic compounds.

Figure 12.22: SEM (Scanning electron microscopy). Epidermal cells. External view of the bean, showing the elongated epidermal cells covered with folded waxes (↑).

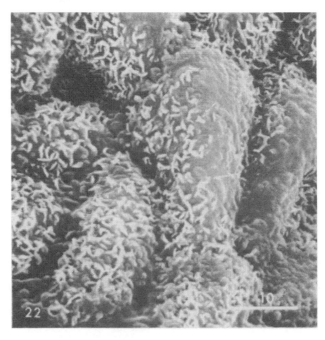

Figurre 12.23: SEM. Transverse section of coffee bean, showing the embryo with its two cotyledons (E), the parenchyma storage cells and on the outside, a small part of the silverskin (↑).

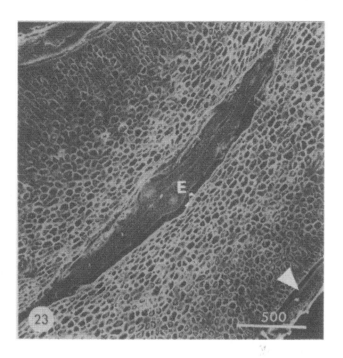

Conclusions

The bulky part of the coffee bean consists of parenchyma storage cells. These cells accumulate material in the cytoplasm as well as in the cell walls. The cell walls are thick and partially lignified. They do not present intercellular spaces, but are crossed by many plasmodesmata. They present a characteristic knotty structure. According to Frey-Wissling (1959), they contain a special hemicellulose which is not linked either to cellulose or uronic acids. This hemicellulose is metabolised during the germination process (Frey-Wissling 1959; Fahn 1974).

The epidermal cell wall, protected by a layer of crystallised waxes, contains different constituents. In the cuticular layer are found terpenes, chlorogenic acids and 5-hydroxytryptamine derivatives. Chlorogenic acids were also detected in the cytoplasm of epidermal and parenchymatous cells. The largest quantity was found in the periplasm of these cells. We know that some phenols are able to react with metals by chelation, and they can form complexes with osmium tetroxide. This characteristic explains the electronic density of phenolic complexes, such as anthocyanins or tannins. For this reason, we assume that the amorphous strongly osmiophilic deposits

Figure 12.24: TEM. Central mucilaginous region. On the left of the middle lamella (ML), a normal cell wall (CW), the cytoplasm of a parenchyma storage cell with lipids (L), proteins (P) and some osmiophilic deposits (↑). On the right part of an excessively swollen cell wall containing numerous small osmiophilic deposits (↑↑) which could correspond to phenolic compounds.

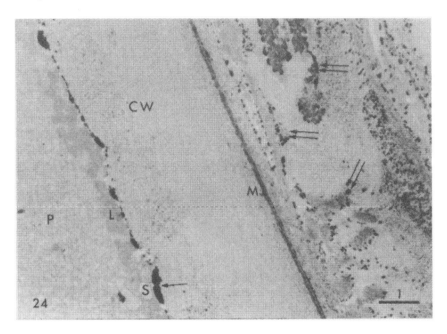

observed in periplasm and cytoplasm correspond to chlorogenic acids.

The lipids, excluding terpenes and waxes, are present in small droplets closely packed and lying on the plasmalemma. Their distribution is homogenous throughout the whole bean. The proteins, present in small quantities in the cell walls, fill the centre of the cytoplasm. They are often included in 'protein bodies' and their distribution is uniform throughout the bean. The polysaccharides are localised essentially in the cells walls (cellulose and hemicellulose), but carbohydrates are also present in the cytoplasm where they fill a kind of vacuole. In all cells (embryo excluded) small amyloplasts are observed. The central mucilage which coats the embryo is also a source of polysaccharides. This mucilage, which is water-soluble, is best observed in frozen sections.

The localisation of caffeine still presents technical problems. By fluorescence, it was possible to detect its presence within the cytoplasm of parenchymatous cells. As the primary fluorescence of the cell walls is strong at the same wavelength, it was impossible to detect caffeine which might be complexed to chlorogenic acids in the periplasm. The parenchymatous cells also contain mineral salts among which were found calcium oxalate dihydrate, calcium phosphate and potassium salts.

Acknowledgements

The author thanks Mrs M. Beaud-Weber for the photographic work, Mrs M-L. Dillmann for her valuable contribution in histological work, Mr V. Fryder for the drawing of Figure 12.1, Dr D. Farr for the reading and correcting of this manuscript, Dr E. Illy, of Illycaffè, Trieste, for interesting discussions, Dr M. Sondhal of the Agronomic Institute of Sao Paulo State at Campinas (Brazil) for providing *catuai* samples and Professor D.M. da Silva (CENA, University of Sao Paulo, at Piracicaba, Brazil).

References

Brinn, N.T. and Pickett, J.P. (1979) 'Glycol methacrylate for routine, special stains, histochemistry, enzyme histochemistry and immunohistochemistry' *Journal of Histotechnology*, 2, 125-30
Clark, G. (1973) *Staining Procedures, used by the Biological Stain Commission* 3rd edn. The Williams & Wilkins Co., Baltimore
Clifford, M.N. (1975) 'The composition of green and roasted coffee beans' *Process Biochemistry*, *10* 20-3, 29 and *10* (4) 13-16, 19
Dentan, E. (1979) 'Structure fine du grain de café vert', *8ᵉ Colloque Scientifique International sur le Café*, ASIC, Paris, 59-64
Fahn, A. (1974) *Plant Anatomy* 2nd edn. Pergamon Press, Oxford, p. 91
Frey-Wissling, A. (1959) *Die pflanzliche Zellwand* Springer-Verlag, Berlin. 140-50
Frey-Wissling A. and Mühlethaler, K. (1965) *Ultrastructural Plant Cytology*, Elsevier Publishing Company, Amsterdam pp. 275-89
Frey-Wissling A., Mühlethaler, K. and Wyckoff, R.W.G. (1948) 'Mikrofibrillenbau der pflanzlichen Zellwände', *Experientia*, 4, 475-6
Gunning, B.E.S. and Steer, M.W. (1975) *Ultrastructure and the Biology of Plant Cells*, Edward Arnold, London pp 15-32
Higuchi, S., Suga, M., Dannenberg, A.M. and Schofield, B.H. (1979) 'Histochemical demonstration of enzyme activities in plastic and paraffin embedded tissue sections', *Stain Technology*, 54, 5-12
Houk, W.G. (1938) 'Endosperm and perisperm of coffee with notes on the morphology of the ovule and seed development', *American Journal of Botany*, 25, 56-61
Jerrard, H.G. (1948) 'Optical compensators for measurements of elliptical polarisation', *Journal of the Optical Society of America*, 38, 35-9
Pearse, A.G.E. (1968) *Histochemistry. Theoretical and Applied*, Vol. I, 3rd edn. J. & A. Churchill, London
Pearse, A.G.E. (1970) *Histochemistry, Theoretical and Applied*, Vol. II, 3rd edn. J. & A. Churchill, London
Pilet, P.E. (1971) *Les Parois Cellulaires*, Doin editeur, Paris
Poisson, J.(1979) 'Aspects chimiques et biologiques de la composition du café vert', *8ᵉ Colloque Scientifique International sur le Café*, ASIC, Paris, 33-57
Reynolds, E.S. (1963) 'The use of lead citrate at high pH as an electron-opaque stain in electron microscopy', *Journal of Cell Biology*, 17, 208-12
Rousseau, M. and Hermier, J. (1975) 'Localisation en microscopie électronique des polysaccharides de la paroi, chez les bactéries en sporulation', *Journal de Microscopie et Biologie Cellulaire*, 23, 237-48

Schubert, W.J. (1973) 'Lignin', in L.P. Miller (ed.) *Phytochemistry*, Vol. III. Van Nostrand Reinhold Company, New York p. 132

Spurr, A.R. (1969) 'A low-viscosity epoxy resin embedding medium for electron microscopy', *Journal of Ultrastructure Research, 26,* 31-43

Winton, A.L. and Winton K.B. (1939) 'Seeds of the madder family Coffee'. *The Structure and Composition of Foods,* Vol. IV John Wiley and Sons Inc., New York, pp. 139-62

13 CHEMICAL AND PHYSICAL ASPECTS OF GREEN COFFEE AND COFFEE PRODUCTS

M.N. Clifford

Introduction

This chapter summarises the vast literature on the composition* of green coffee beans paying particular attention to those components which are peculiar to coffee. The corresponding data are given for roasted beans and where possible for soluble powders. Attention is focused on compositional factors that might be determinants of acceptability, and situations where the data are incomplete or contradictory with the intention of provoking thought, comment and further investigation.

Physical Properties and Water Content

Individual green beans vary in mass from below 100 mg to over 200 mg with considerable overlap between arabicas and robustas, but with some evidence of variation associated with the geographical origin. Such variation may be associated with the amount of water available to the bushes (see Cannell, Chapter 5, p. 126). The true relative density of the beans ranges from 1.15 to 1.42 (Derbesy *et al.*, 1969; Ferreira *et al.*, 1971a; Roffi *et al.*, 1971; Xabregas *et al.*, 1971; Melo, Fazuoli, Teixeira and Amorim, 1980).

At the start of the roasting process loosely bound water is driven off and some shrinkage occurs, particularly with arabicas. As evaporative cooling declines, so the bean temperature rises and an exothermic pyrolysis begins in the temperature range 140°-160°C, and leads to the formation of the well-known colour, aroma and taste of roasted coffee products. The pyrolysis peaks between 190° and 210°C with enthalpies of 230 to 375 J/g, and charring begins at about 230°C if it is not arrested. Acceptable dry matter losses range from some 3 per cent for a very pale roast to some 14 per cent for a very dark roast: the corresponding figures for total roasting loss (dry matter and water) are some 10 per cent and 25 per cent respectively. A large quantity of carbon dioxide is produced; its expansion generates internal pressures in the range 5.5 to 8.0 atmospheres and accounts for the swelling of the bean by some 170 - 230 per cent during commercial roasts, its partial escape for the loss of dry matter (Fobé, Nery and Tango, 1967; Natarajan, Balachandran and Shivashankar, 1969; Quijano-Rico *et al.*,

---

*All compositional data are given on a dry mass basis (dmb).

1975; Radtke, 1975; Raemy, 1981; Raemy and Lambelet, 1982).

Although arabicas and robustas ultimately swell to a similar extent, data recorded by Dalla Rosa, Lerici, Riva and Fini (1980) suggest these two species behave differently early in the roasting process, e.g.:

robusta 16 per cent volume increase at 172°C, 20 min and 1.65 per cent pyrolysis loss
arabica 16 per cent volume increase at 188°C, 30 min and 4.87 per cent pyrolysis loss
robusta 32 per cent volume increase at 210°C, 10 min and 2.32 per cent pyrolysis loss
arabica 2.9 per cent volume increase at 210°C, 10 min and 2.52 per cent pyrolysis loss.

These data imply that arabicas and robustas differ in one or more of the following: vapour (CO_2) production; vapour (CO_2) retention; vapour (CO_2) expansion; resistance of the cell wall complex to tensile stress. These points are discussed further on pages 319 and 357.

Water sorption isotherms or related data are given in Diaz, Gomez, Felsner and Fritsch (1973), Hayakawa, Matas and Hwang (1978) and Stirling (1980). According to Diaz *et al.*, raw beans contain 1 per cent of strongly bound water, a further 4 per cent of weakly bound water and a further 5 per cent of water with low mobility. Under these circumstances (some 10 per cent water content) raw beans have an a_w between 0.5 and 0.6 and are stable for up to 1 year provided the temperature does not exceed 25°C.

Data on the enthalpy of coffee extracts and water-binding properties of roasted coffee and soluble coffee powders are given by Riedel (1974), Hayakawa *et al.*, (1978), Maier (1981), Santanilla, Fritsch and Müller-Warmuth (1981). Roasted beans and soluble coffee powders are normally stored at water contents not exceeding 4 per cent which, at 20°C, correspond to water activities in the range 0.10 to 0.30. Such low values normally favour oxidative changes (Fennema, 1976) but roasted coffee products are protected by high contents of radical-accepting phenolic compounds and inert atmospheres produced during roasting (carbon dioxide) or during packaging (vacuum or nitrogen). For further discussion of deterioration during storage and packaging see page 350 and Chapter 14 respectively.

A knowledge of the water content of green coffee and roasted coffee products is required because: (1) it influences the water activity and stability during storage; (2) it provides a monitor of coffee extract drying processes; (3) there are often limits set by national and international legislation (see Clarke, 1979).

The analysis of water content is not simple. The difficulties, which are not peculiar to coffee and which are discussed fully in general texts on food analysis, arise because different methods of analysis give different emphasis to free water, bound water, other volatile substances, dry matter lost by pyrolysis and other constituents that interfere in chemical determinations.

With reference to green coffee and roasted coffee products there have been comparative studies of methods based upon oven drying at atmospheric pressure, vacuum oven drying, azeotropic distillation, Karl Fischer titration, GC, NMR, dielectric properties and near infra-red spectrometry (Haevecker, 1969; Paardekooper, Driesen and Cornelissen, 1969; Chassevent and Dalger, 1971; Lopes et al., 1971; Diaz et al., 1973; Gabriel-Jurgens, 1975).

The standard methods for green coffee based upon Guilbot and d'Ornano (1964) are the reference method (BS 5752: Part 1, 1979), which involves a very lengthy 48°C, 2kPa vacuum dehydration using phosphorus pentoxide (IS 1446-1978) as desiccant, and the routine method (IS 1447-1978), which is an oven heating method (130°C) in two stages with grinding of the green beans between the stages. According to Clarke (personal communication) this routine method has not been generally popular with coffee processors and an overnight (16 hours) mass loss at 105°C from 5 g of properly sampled, whole, green beans has been proposed as an alternative (IS 6673-1983). For soluble coffee powder a 16 hour vacuum oven method (70°C, about 50 mm Hg absolute pressure) is used (BS 5752, Part 6, 1983; IS 3726-1983). This method has recently been incorporated by the EEC into the Coffee Products Directive Legislation.

Ash and Minerals

The ash content and the contents of individual minerals are of interest for several reasons. Trace elements are of importance in plant nutrition (see Chapter 6). Quijano-Rico and Spettel (1973) have speculated that the higher copper levels in robustas may be linked with their greater disease resistance. Transition elements could act as catalysts during roasting though which, if any, catalyse desirable reactions is not known. Clarke and Walker (1974, 1975) and Maier (1980) have discussed the limitations of using ash, sulphated ash or potassium contents of soluble coffee powders as indicators of the yield from the originating green coffee. Within the EEC this is controlled by legislation and should not exceed 1 kg of finished soluble product from 2.3 kg of green beans (EEC Directive 77 436; SI 1420: 1978). Such methods are satisfactory for process control where the green bean ash/potassium content is known. However, green beans are variable and if the particular data are not available, and general values are used, such analyses can only indicate the yield of soluble powder to within ± 10 per cent of the true value. Similarly the ashed mineral content of the green bean varies but generally is near to 4 per cent with the potassium content (expressed as the element) forming some 40 per cent of the oxide ash. Wet processed beans have lower contents than dry processed. For further information, including the contents of other minerals, see Ferreira *et al.* (1971a,b,c), Waenke, Palme, Spettel and Jagoutz (1973), Clarke and Walker (1974) and Quijano-Rico and Spettel (1975).

There are no significant changes in the minerals during roasting and the majority of elements are extracted easily during domestic brewing and commercial extraction. Values for the ash content of commercial soluble coffee powders range from 6.6 to 12.9 per cent (Villanua, Carballido and Baena, 1971; Angelucci *et al.*, 1973; Vol'per, Il'enko-Petrovaskaya, Lazarev and Solov'eva 1973; Trugo and MacRae, 1981). Fragoso, Ferreira and Peralta (1971) give some data for coffee brews.

Non-Protein Nitrogen

Purine Nitrogen

Caffeine (1,3,7-trimethylxanthine) is the major purine in green coffee. Smaller quantities of theobromine (3,7-dimethylxanthine) and traces of theophylline (1,3-dimethylxanthine) have been recorded by some analysts (see Poisson, 1977) but not all (Sontag and Kral, 1980). Traces of several methyluric acids have been found in *C. liberica* and *C. exceloides* but not the major commercial species (Citroreksoko, Petermann, Wanner and Baumann, 1977). There have been studies on caffeine biosynthesis in coffee (Baumann and Wanner, 1977; Raju, Ratageri, Venkataramanan and Gopal, 1981); the breeding of low caffeine coffees has been discussed in Chapter 3. Caffeine is said to occur in coffee pulp. In the seed it occurs free within the cytoplasm and bound to the cell wall (Zuluaga-Vasco, Bonilla and Quijano-Rico, 1975; Pfrunder, Wanner, Frischknecht and Baumann, 1980). The bound caffeine may be associated with the chlorogenic acids which form a π-electron complex (Horman and Viani, 1971) and this would be consistent with the location suggested for these compounds by microscopical studies (see Dentan, Chapter 12:301).

There are many analytical methods for the determination of caffeine in coffee and coffee products. Early methods were based primarily upon UV spectrometry but more recent methods generally employ a chromatographic separation to avoid interference particularly when analysing roasted beans and soluble powders. Both ISO (Vitzthum, 1974) and Association of Official Analytical Chemists (AOAC) (1975) have standard methods. There are many rapid, often automated, methods for routine industrial purposes (e.g. monitoring decaffeination) and many have been compared with the standard methods and in general found to give very similar results (e.g. Lopes, 1971; Vitzthum, Barthels and Kwasny, 1974; Quijano-Rico, Acero, Morales and Piedrahita, 1977; Sloman, 1980). To ensure that bound caffeine is determined the samples should be extracted by boiling in water for 30 minutes. Some data for green beans are summarised in Table 13.1. With occasional exceptions (see Carelli, Lopes and Monaco, 1974) the caffeine content parallels the chlorogenic acids content, with robustas higher than arabicas. The virtual absence of caffeine from some wild species is notable. The influence of wet and dry processing upon caffeine content has been studied; wet processing after removal of the parchment (not a commercial practice) leads within two days to a 30 to 40 per cent reduction by leaching (Chassevent et al., 1969; Vincent et al., 1977).

Green bean dewaxing processes reduce the caffeine content (Gal, Windemann and Baumgartner, 1976) and decaffeination virtually eliminates it. Clarke has described such processes in Chapter 14. There are legal maxima for residual caffeine, typically 0.08 to 0.1 per cent for roasted beans and 0.3 per cent for soluble powders. European legislation for solvent residues has been summarized by Clarke (1979). Roasting volatilises some 10 per cent of the caffeine originally present, but because the total weight loss may be greater the caffeine content may appear to rise (e.g. see Merrit and Proctor, 1959; Fobé et al., 1967). Commercial extraction transfers the caffeine almost completely to the soluble coffee powder; typical data are summarised in Table 13.2.

Domestic brewing similarly transfers the caffeine rapidly to the brew. It has been reported that the caffeine content of a cup of coffee (undecaffeinated) may vary from 40 mg to at least 160 mg, partly because of variations in green bean caffeine contents but particularly because of variations in cup size and brew strength (Burg, 1975; Bunker and McWilliams, 1979; Ndjouenkeu, Clo and Voilley, 1981; Ulrich, 1982; Blauch and Tarka, 1983). It is thus unsound to use an estimate of 'cups consumed' as a measure of caffeine intake. Contrary to popular belief it is now recognised

Table 13 1: The Caffeine Content of Green Coffee Beans

Reference	Material Analysed	Caffeine content % dmb Range	Mean ± SD
Charrier (1975)	130 strains of arabica	0.58-1.70	—
	680 strains of robusta	1.16-3.27	—
	several Paracoffea	absent	—
Chassevent et al. (1973a)	? Mascarocoffea	trace[b]	—
Merrit and Proctor (1959)	7 South American arabicas	1.07-1.29	1.17±0.08[a]
Menchu et al. (1967)	39 Guatemalan arabicas	0.53-0.79	0.63±0.06[a]
Roffi et al. (1971)	8 Angolan arabicas	1.21-1.45	1.32±0.07[a]
Ferreira et al. (1971a)	34 Timor hybrids	—	1.15±0.18
Wurziger et al. (1977b)	7 Ivory Coast arabustas	1.47-1.83	1.72±0.12[a]
Roffi et al. (1971)	26 Angolan robustas	2.18-2.72	2.42±0.13[a]

Note: a. Recalculated from published data.
 b. not exceeding 45 µg/kg.

Table 13 2: The Caffeine Content of Soluble Coffee Powders

Reference	Material Analysed	Caffeine content % dmb Range	Mean ± SD
Villanua et al. (1971)	14 bought in Spain	3.84-6.64	5.47±0.75[a]
Vol'per et al. (1973)	— bought in Russia	4.5-5.1	—
Angelucci et al. (1973)	15 made in Brazil — arabica	1.63-3.79	2.68±0.68[a]
	1 made in Brazil — robusta	4.64	—
Sloman (1980)	5 bought in USA	1.83-3.49	2.94±0.48[a]
Smith (1981)	3 coffee-chicory powders bought in UK	2.16-2.33	—
Trugo et al. (1983)	12 bought in UK	2.83-4.83	3.83±0.56[a]

Note: a. Recalculated from published data.

that caffeine has a very limited role in beverage bitterness (see page 353). Other physiological effects are discussed by Bättig in Chapter 15.

Trigonelline

Trigonelline, the N-methylbetaine of pyridine-3-carboxylic acid, is found in all commercial and some wild species. Quantitative data are summarised in Table 13.3. Green bean processing, dewaxing and decaffeination procedures have little effect upon the trigonelline content (Merrit and Proctor, 1959; Chassevent et al., 1969; Ferreira et al., 1971a; Baltes, Petersen and Degner, 1973), but roasting causes progressive destruction. The degradation products include the vitamins nicotinic acid (niacin) and nicotinamide and a range of aroma volatiles which include pyridines and pyrroles (Hughes and Smith, 1949; Merrit and Proctor, 1959; Viani and Horman, 1974, 1975). Soluble coffee powders contain some 1 to 1.7 per cent trigonelline (Trugo, MacRae and Dick, 1983). Kwasny (1978) used the progressive conversion of trigonelline into nicotinic acid and nicotinamide as the basis of a method of retrospectively determining the colour of the roast, not only for whole beans but also for soluble coffee powders.

Table 13 3: The Trigonelline Content of Green Coffee Beans and Soluble Coffee Powder

Reference	Material Analysed	Trigonelline content % dmb Range	Mean ± SD
Merrit and Proctor (1959)	7 South American arabicas	0.97-1.15	1.08±0.08[a]
Ferreira et al.. (1971a)	34 Angolan arabicas	—	0.71±0.12
	34 Angolan robustas	—	0.88±0.12
	34 Timor hybrids	—	0.62±0.15
Trugo et al.. (1983)	13 soluble coffee powders bought in the UK	0.94-1.69	1.36±0.22[a]

Note: a. Recalculated from published data.

Free Amines and Amino Acids

Some quantitative data for the free amino acid contents are summarised in Table 13.4. Hassan (1970) and Tressl, Holzer and Kamperschroer (1982) have suggested that there might be significant species differences (robustas some 50 per cent higher than arabicas), and some success has been claimed in using the amino acids profile as an index of botanical and even geographical origin (Campos and Rodrigues, 1971; Pereira and Pereira, 1971). During storage of the green bean, particularly at elevated temperatures, it has been reported that there are changes due in part to proteolysis (e.g. increases in alanine, isoleucine and tyrosine) and losses of free amino acids in non-enzymic browning reactions. Such losses were accompanied by losses in reducing sugars, production of a brown discolouration and reduction in beverage quality (Pokorny, Con, Smidrkalova and Janicek, 1975). Such disruptions of the original amino acids profiles would limit the value of the profiles as indicators of origin, and may explain why the data in Table 13.4 show little difference between arabicas and robustas. A study of free amines (Amorim, Basso, Crocomo and Teixeira, 1977b) found

Table 13 4: The Free Amino Acids Content of Green Coffee Beans

Amino Acid	Content % dmb[a] arabicas	robustas
Total	0.37-2.40	0.66-2.88
Strecker-active amino acids		
Glycine	up to 0.03	0.02-0.03
Alanine	0.05-0.24	0.09-0.11
Valine	up to 0.08	0.02-0.03
Isoleucine	up to 0.04	0.01-0.02
Leucine	up to 0.03	0.02-0.03
Phenylalanine	0.01 to 0.08	0.03-0.04
Tyrosine	up to 0.04	0.01-0.02
Others		
Aspartic acid	0.05-0.33	0.07-0.09
Asparagine	0.05-0.30	0.08-0.09
Glutamic acid	0.11-0.49	0.07-0.09
Threonine	up to 0.006	up to 0.01
Serine	0.03-0.19	0.04-0.05
Proline	0.03-0.14	0.04-0.05
Aminobutyric acid	0.03-0.33	0.10-0.11
Lysine	up to 0.07	0.01-0.02
Histidine	up to 0.04	up to 0.006
3-methylhistidine	up to 0.005	up to 0.006
Tryptophan	0.01-0.02	0.04-0.05
Arginine	up to 0.04	0.02-0.03
Cysteine	trace	
Methionine	trace	

Note: a. Data compiled from Walter and Weidemann (1969), Pokorny et al. (1974, 1975) and Tressl et al. (1982).

putrescine (38 to 54 μg/kg), spermidine (14 to 20 μg/kg) and spermine (8 to 10 μg/kg).

Roasting causes extensive, if not complete, destruction of all these amino compounds which supply much, if not all, of the nitrogen which is incorporated into the heterocyclic aroma volatiles (Hassan, 1970; Pokorny, Con, Bulantova and Janicek, 1974; Amorim et al., 1977b). It has been shown (Hassan, 1970; Tressl et al., 1982) that the amino acid profile influences the yield of certain volatiles during roasting and thus the character and perhaps acceptability of the beverage (see page 348).

Proteins

Kjeldahl Nitrogen Determinations and Amino Acid Profiles

Crude protein contents calculated from total nitrogen contents must be corrected for caffeine and ideally also for trigonelline nitrogen. If such corrections are made there does not seem to be any significant difference between the protein contents of arabicas and robustas (see Table 13.5), or any significant effect that can be attributed to the method of green bean processing (Menchu and Ibarra, 1967; Ferreira et al., 1971a; Roffi et al., 1971; Xabregas et al., 1971). Walkowski (1981) has reported that the content of soluble protein declines during storage of the green bean — this could be associated with polyphenol oxidase activity (see page 315). Roasting probably causes only a slight loss of protein nitrogen, i.e. as volatiles, but certainly causes significant chemical transformation. The hydrolysis and amino acid analysis of protein-rich fractions isolated from roasted beans has shown significant destruction of arginine, cysteine/cystine, lysine, serine and threonine which is in keeping with the reactive nature of their side chains (Thaler and Gaigl, 1963; Feldman, Ryder and Kung, 1969; Roffi et al., 1971; Lopes, 1974; Pokorny et al., 1974). Trugo and MacRae (1981) have reported that soluble coffee powders contain some 2 to 3.5 per cent total nitrogen of which some 40 to 50 per cent can be accounted for by caffeine and trigonelline. Of the remaining nitrogen only 6 to 12 per cent was insoluble in 5 per cent trichloroacetic acid corresponding to some 5.5 to 12 per cent 'protein' in the soluble powder, if the traditional conversion factor of 6.25 is used. This 'protein' has not been characterised.

Extraction and Characterisation

According to Thaler (1975) approximately one third of green bean protein is associated with the cell wall arabinogalactan. According to Eskin (1979) such cell wall proteins have high hydroxyproline contents but this amino acid does not seem to have been recorded in coffee protein hydrolysates. Much of the remaining protein is free in the cytoplasm but some of the important enzymes, e.g. polyphenol oxidase are bound to membranes.

Table 13.5: The Protein Content of Green Coffee Beans

Reference	Material Analysed	Crude Protein			
		KN × 6.25[a]		(KN-CN) 6.25[b]	
		Range	Mean ± SD	Range	Mean ± SD
Menchu et al. (1967)	59 Guatemalan arabicas	11.79-15.00	13.63 ± 0.62[c]	11.16-14.37[c]	12.15 ± 0.64[c]
Roffi et al. (1971)	8 Angolan arabicas	—	—	10.31-11.75	11.07 ± 0.42[c]
Xabregas et al. (1971)	9 Angolan arabicas				
	dry processed	13.63-14.68	14.19 ± 0.33[c]	—	11.50 ± 0.34[c]
	wet processed	14.20-15.46	14.64 ± 0.58	—	11.81 ± 0.59[c]
Ferreira et al. (1971a)	34 Angolan arabicas	—	—	11.01-12.15	11.70 ± 0.51[c]
	34 Timor hybrids	—	—	11.21-12.22	11.68 ± 0.40[c]
	34 Angolan robustas	—	—	11.20-13.05	12.72 ± 0.51[c]
Roffi et al. (1971)	30 Angolan robustas	—	—	10.00-12.05	11.25 ± 0.70[c]

Note: a. KN — Kjeldahl Nitrogen
  b. CN — Caffeine Nitrogen
  c. Recalculated from published data

Polyacrylamide gel electrophoresis (PAGE) of albumins and what appear to be albumin-globulin-glutelin mixtures from arabicas and robustas has given results of chemotaxonomic interest (Centi-Grossi, Tassi-Micco and Silano, 1969; Payne, Oliveira and Fairbrothers, 1973; Bade and Stegemann, 1982). Amorim *et al.*, (1974c) have observed quantitative differences in globulins and glutelins that may be linked with the quality of some Brazilian arabicas. Further PAGE studies have demonstrated that the poorest grades of Brazilian coffee have a greater content of proteins in the 20,000 to 26,000 dalton range and near 9,000 daltons (Amorim and Josephson, 1975). Agar gel electrophoresis (at pH 7.0) has indicated that there are also significant differences in net charge: proteins from the best quality green coffees migrated towards the anode whereas proteins from other grades migrated, at varying rates, towards the cathode. These observations were explained as follows. Gross disruption (i.e. grinding) of a good quality green bean (i.e. having high polyphenol oxidase activity) mixes active enzyme and substrate (CQA). Quinones are produced and react with nucleophiles (e.g. lysine ε-NH_2 and cysteine-SH) in the proteins altering the protein extractability and electrophoretic mobility. Grinding poorer quality beans with low polyphenol oxidase activity produces less quinone, less derivatisation and less change in extractability and electrophoretic mobility (Amorim *et al.*, 1974c, 1975a; Liberato, Byers, Dennick and Castagnoli, 1981).

Enzymes

Coffee beans, in common with all plant material, contain a large range of enzymes. Many of these, apparently including β-glycosidases, proteases and lipases, have not been closely studied in coffee. However by analogy with other seeds such enzymes could be active during green bean processing and storage and might well produce changes, such as the production of various aglycones, free amino acids and free fatty acids (see pages 311, 326 and 338 respectively), that could influence the beverage quality. The following coffee bean enzymes have been studied in more detail.

(1) α-galactosidase (EC 3a.1.22) during the soaking and germination of seeds (Courtois and Le Dizet, 1963; Shadaksharaswamy and Ramachandra, 1968a);
(2) malate dehydrogenase (EC 1.1.1.37) as a chemotaxonomic marker (Payne *et al.*, 1973);
(3) acid phosphatase (EC 3.1.3.2) (Berndt and Meier-Cabell, 1975);
(4) laccase/*p*-diphenol oxidase (EC 1.10.3.2) (Rotenberg and Iachan, 1972), peroxidase (EC 1.11.1.7) (Amorim *et al.*, 1977a) and more extensively polyphenol oxidase/tyrosinase/catechol oxidase (EC 1.14.18.1) as indicators of green bean quality.

In vivo polyphenol oxidase (PPO) is bound to membranes and is activated only on release. It has been found in coffee pulp, in the outer layers of the bean and in the central region of the bean. It has been demonstrated that PPO activity in the green bean is a good predictor of the beverage quality judged organoleptically (Amorim and Silva, 1968; Sanint and Valencia, 1970; Amorim *et al.*, 1974c, 1976). Although arabicas and robustas contain the same five isozymes, the specific activity of PPO is a function of the species, the variety and seed maturity. Freshly harvested robustas and freshly harvested immature beans have the greater specific activities (Valencia, 1972; Griffin and Stonier, 1975; Oliveira, Amorim, Silva and Teixeira, 1976; Solov'eva, 1979).

Many factors that were known to be associated with reduced beverage quality have since been shown to be associated with low PPO activity or to lead to lower PPO activity during storage. These include cultural factors, method of green bean processing and time and conditions of storage (Sanint and Valencia, 1970; Valencia 1972; Oliveira *et al.*, 1976, 1979b; Solov'eva, 1979; Melo *et al.*, 1980). However, certain factors, such as the use of insecticides, can reduce beverage quality without reducing PPO activity (Oliveira, Silva, Teixeira and Amorim, 1979a).

It is thought that the sequence of events leading to low PPO activity is membrane damage, enzyme activation, the oxidation of CQA to quinones and enzyme inhibition by the quinones. The reduction in beverage quality is associated with the cause of the original membrane damage, but not necessarily with the action of the PPO *per se*.

Carbohydrates in Green Coffee Beans

Low Molecular Mass Carbohydrates

Low molecular mass carbohydrates are soluble in water and/or 80 per cent ethanol. They include the well-known mono-, di- and oligosaccharides. Some quantitative data are summarised in Tables 13.6 and 13.7. Sucrose is the major sugar of green coffee beans. It is found at a higher level in arabicas than robustas, the latter apparently having a higher content of reducing sugars. Traces of raffinose, stachyose, ribose, mannose and galactose have been reported. The monosaccharides probably form via enzymic hydrolysis during storage of the green bean or during extraction prior to analysis (Courtois, Percheron and Glomaud, 1963; Glomaud, Percheron and Courtois, 1965; Shadaksharaswamy and Ramachandra, 1968a; Hassan, 1970; Lerici, Lercker, Minguzzi and Matassa, 1980a; Clifford and Griffiths, 1982; Tressl *et al.*, 1982).

Amorim *et al.*, (1974a, 1975b) concluded that there was no direct relationship between the sugar content of green Brazilian arabicas and beverage quality. Pokorny *et al.*, (1975) observed losses of reducing sugars

Table 13.6: Free Mono and Disaccharides in Green and Roasted Coffee Beans and Chicory Root

Reference	Material analysed	Sucrose	Sugar Content % dmb Glucose	Fructose	Galactose	Mannose
	UNROASTED					
Tressl et al. (1982)	2 arabicas	8.2-8.3	0.03-0.04	0.02-0.03	0.04	
Hassan (1970)	3 arabicas[a]	6.2-8.6				
Barbiroli (1965)	9 arabicas	5.1-8.5	up to 1.2	up to 0.3		
Feldman et al. (1969)	2 arabicas	4.6-5.5				
Tressl et al. (1982)	2 robustas	3.3-4.1	0.16-0.18	0.19-0.21	0.07-0.08	
Hassan (1970)	3 robustas[b]	2.2-4.0				
Barbiroli (1965)	1 robusta	6.6	0.5	nd		
Sannai et al. (1982)	1 chicory root		2.68	12.44		
	ROASTED					
Tressl et al. (1982)	2 arabica-robusta blends	0.01	<0.01	nd	0.01-0.015	up to 0.04
Barbiroli (1965)	6 arabicas	0.9-1.7	nd	nd	nd	nd
Kröplien (1971)	2 chicory root		1.21-4.10	2.44-7.71		
Pazola and Cieslak (1979)	3 chicory root			8.9-13.7		
Wight and Van Niekerk (1983)	4 chicory root	1.29-1.71	0.77-0.96	2.05-2.31		

nd = not detected
Notes: a. total reducing sugars 0.16-0.41
b. total reducing sugars 0.45-0.71

during the storage of green beans at elevated temperature and associated the loss with discoloration and reduced beverage quality. Hassan (1970) and Tressl et al., (1982) reported that the profile of volatiles produced on roasting is in part determined by the ratio in the green bean of sugars to amino acids (see page 348) and that differences in this ratio may explain in part the organoleptic differences between arabicas and robustas.

Air-dried chicory root is characterised by a high fructose content, fig by a high reducing sugars content and cereals by a low sugar content (Kent, 1966; Whiting, 1970; Sannai, Fujimori and Kato, 1982).

Polysaccharides

Robustas contain some 38 to 48 per cent of polysaccharides; arabicas possibly more, but data confusing (Thaler and Arneth, 1969a; Thaler 1970; 1975). The acid hydrolysis of green coffee beans from which the low molecular mass and non-saccharidic constituents have been removed (ether followed by water, or ether followed by 70 per cent acetone and 80 per cent ethanol) and chromatographic examination of the hydrolysates have established the presence of mannose, galactose, glucose, arabinose, xylose, rhamnose and galacturonic acid in the polysaccharides.

The major reserve polysaccharide (β-mannan) encrusts the cell-wall complex which typically consists of cellulose microfibrils embedded in a continuous phase of lignin, pectin and hemicelluloses. These polymers are in close association, physically entangled and held by secondary forces, and possibly primary covalent linkages. There are certainly covalent linkages between a cell-wall arabinogalactan and a hydroxyproline-rich cell wall protein (Whistler and Richards, 1970 — for review see Eskin, 1979). Fractionation is not possible without physical disruption and scission of some covalent bonds, and the yield and nature of the material obtained is a function of the method employed. Following removal of low molecular mass constituents the first polysaccharide fraction is normally extracted with hot water. Pictet and Moreau (1969) have shown that the yield, molecular mass and identity of the constitutent sugars in the water-soluble fraction are a function of the extraction temperature. Raising the extraction temperature above 130°C extracted significantly greater amounts of araban and galactan. Little glucan was extracted even at 180°C. Pectins may be removed from the water-insoluble residue using ammonium oxalate, dilute hydrochloric acid or EDTA. Yield and composition depend upon the solvent. The pectin-free insoluble residue is treated with acid sodium chlorite to degrade lignins and rupture polysaccharide-protein linkages. This solubilises a polysaccharide fraction (Aufschluss Polysaccharid - Thaler) and leaves an insoluble holocellulose. This holocellulose consists of the cell wall complex and includes the encrusting reserve mannan. Thaler has not fractionated further but some workers remove hemicelluloses with alkali to leave a final residue of α-cellulose, rich in cellulose but containing other polysaccharides. A quite different method gives values for crude cellulose, the insoluble material remaining after sequential extraction of the ground bean with hot acid and hot alkali. Crude cellulose consists of cellulose, lignins, cell wall proteins and possibly other polysaccharides.

Celluloses are linear β1–4 homoglucans which vary in chain length and crystallinity. Wolfrom and Patin (1964) reported some 5 per cent of cellulose in a green bean. Assuming that all glucan is cellulose, then green arabicas and robustas could contain a maximum of between 4 and 9 per cent (Wolfrom, Plunkett and Laver, 1960; Thaler and Arneth, 1969a,b; Thaler, 1970). These figures are consistent with crude cellulose contents in the range 10 to 12.5 per cent (Ferreira *et al.*, 1971a; Lerici *et al.*, 1980a).

Hemicellulose is the dominant component of the cell wall complex (Whistler and Richards, 1970) and it occurs in the primary and the secondary cell wall. The term hemicellulose embraces a large number of heteroglycans. The composition of those isolated from coffee is a function of the extraction procedure. Wolfrom *et al.*, (1960) recovered some 5 per cent hemicellulose from a green arabica using 10 per cent KOH. Hydrolysis of this fraction released rhamnose, xylose, arabinose, galactose and a trace of mannose. Subsequent extraction with 18 per cent NaOH gave a 5 per cent yield of a β1–4 mannan containing only 2 per cent galactose. This would appear to be derived from the reserve polysaccharide (see page 319). Proteolysis of the residual cell wall complex released on arabinogalactan (approximately two arabinose to five galactose residues in 8.5 per cent yield) that *in vivo* was probably bound to cell wall proteins, probably acylated with aromatic acids and probably accounted for all the green bean araban. Once released the polysaccharide was slightly water-soluble and probably corresponds to Thaler's hot water-soluble fraction and degradation polysaccharides which form some 9 to 13 per cent of green arabicas but only 6 to 8 per cent of green robustas (Thaler and Arneth, 1968; Thaler, 1970). These arabinogalactan hemicelluloses are thought to have a β1–3 galactopyranose backbone bearing some short galactose side chains, with labile arabinofuranose residues linked α1–3 to the backbone and the side-chain galactose (Wolfrom and Patin, 1965) and would thus appear to be a Type II arabinogalactan (see Aspinall, 1981). Glomaud *et al.*, (1965) have isolated a hot water-soluble fraction containing arabinose (3 parts), galactose (6 parts) and galacturonic acid (1 part). This fraction seems to have a composition intermediate between typical hemicellulose and pectic substances.

Pectic substances are thought to be a mixture and/or a chemical combination of three polysaccharides: (1) a linear β1–4 galactan; (2) a short-chain araban containing arabinofuranose units, some linked α1–5 others α1–3; and (3) the major component, a high-molecular-mass polysaccharide consisting of blocks of α1–4 galacturonan joined by α1–2 rhamnopyranose. The uronic acid is esterified with methanol to a variable extent.

Pectins form between 1 and 4 per cent of the typical cell wall complex (Whistler and Richards, 1970) and are localised largely in the primary cell wall and adjacent middle lamella. Yields range from 0.8 to 3 per cent for green arabicas and robustas (Wolfrom et al., 1960; Thaler and Arneth, 1968; Shadaksharaswamy and Ramachandra, 1968b; Amorim et al., 1974a). With the exception of the fraction isolated by Glomaud et al., (see page 318) there do not seem to be any quantitative data for galacturonic acid content although many workers noted its presence in the material they were analysing.

Green coffee beans contain very little starch. Pictet and Moreau (1969) found only 0.48 per cent of hot-water-soluble (up to 180°C) glucan in green beans. The major reserve polysaccharide is a water-insoluble crystalline galactomannan which occurs outside the plasmalemma, encrusts the cellulose microfibrils and which at some time may also have a structural role. Galactomannans of various galactose-mannose ratios are well known as storage polysaccharides in Leguminosae and Palmae but much less studied in other botanical families (Courtois, 1979; Meier and Reid, 1982). The reserve polysaccharide of green arabica coffee beans contains within its structure a $\beta 1-4$ mannan with some 2 per cent of galactose attached via $\alpha 1-6$ single unit side chains. The average chain length was 45 units giving a mass near 7,300 daltons (Wolfrom, Laver and Patin, 1961). The quantitatively greater alkali-insoluble portion of this mannan has yet to be characterised. An estimate of the content of this reserve may be obtained from the mannose content of Thaler's holocellulose (Thaler and Arneth, 1968).

Such calculations suggest 25 to 30 per cent mannan in green arabicas and 19 to 22 per cent in green robustas (Thaler and Arneth, 1968; Thaler, 1970). It is possible that much, if not all of the galactose in the holocellulose is associated with the reserve polysaccharide. If this assumption is accepted provisionally, two interesting observations emerge — the contents of reserve polysaccharide become of a similar order (29 to 36 per cent) in both species, but the mannose-galactose ratio is near 4 to 1 for arabicas compared to 2 to 1 for robustas. By analogy with the known properties of other seed galactomannans one would expect the galactomannan in robustas to be less crystalline than that in arabicas if it does have a greater side chain galactose content. The lower crystallinity occurs because the more numerous side chains would limit hydrogen bonding by interfering with chain packing. However, the magnitude of this effect would depend not only on the number of branches but also whether they occurred in blocks or in a more uniform distribution (see Dea and Morrison, 1975); unfortunately this information is not available. If there are such differences in galactomannan structure and crystallinity, they could explain why robustas swell more readily under milder conditions of roasting (see page 306) and why robustas deliver more carbohydrate than arabicas during commercial extraction, possibly despite a lower content (see page 317).

Carbohydrates in Roasted Coffee and Coffee Products

Total Carbohydrates

Carbohydrates are the major constituent of roasted coffee and the extent to which they are solubilised during commercial extraction is an important factor in determining the yield of soluble coffee powder. These carbohydrates are complex and poorly characterised but much changed physically and chemically from those in the green bean. In particular there is no longer such a clear difference in molecular size between the dialysable/alcohol-soluble low molecular mass fraction and the non-dialysable/alcohol-insoluble polysaccharide fraction. This section summarises those data which have been obtained without any attempt at such a subdivision.

For a given pyrolysis loss arabicas deliver their maximum yield of total carbohydrate, representing some 17 per cent of the roasted bean, at a 44 per cent yield of solubles. Under similar conditions a robusta delivers carbohydrate equivalent to 15 per cent of the roasted bean, but this increases progressively to a maximum of 21 per cent at a total solubles yield of 58 per cent, i.e. robustas have the potential to deliver more carbohydrate possibly despite a smaller content in the green bean (Thaler, 1979), a feature of considerable importance to soluble coffee producers. This may reflect a higher galactose content in the galactomannan (see page 319).

Kazi (1979) and Pazola and Cieslak (1979) have examined the carbohydrates of coffee extenders. The carbohydrates of roasted barley and roasted rye are primarily dextrins (α1–4 glucans) derived from starch, whereas those from roasted dandelion or chicory root are primarily fructans derived from inulin. The fructose and/or fructan content of coffee-fig and coffee-chicory mixtures has been used to monitor the extender content (Kazi, 1979). Total ketose sugars after hydrolysis were reported as 47 to 49 per cent for roasted chicory, 69 to 72 per cent for spray-dried chicory powder, 21 to 24 per cent for roasted fig and 0.8 to 1.9 per cent for soluble coffee powders. These values for soluble coffee powder are significantly higher than the chromatographic data summarised in Table 13.7. While this implies that the resorcinol reagent (like the phloroglucinol reagent — see Clifford, 1974) is not completely specific for free and bound fructose, it does not necessarily imply that such a method is unsuitable for routine monitoring of chicory or fig levels.

Thaler's scheme for the fractionation of the water-soluble carbohydrates of roasted coffee is summarised in Figure 13.1 and is referred to in the sections which follow.

Table 13.7: Free Monosaccharides in Soluble Coffee Powders (% dmb)

Reference	Material analysed	Glucose	Fructose	Galactose	Mannose	Arabinose	Total
Trugo and MacRae (1982)	7 bought in UK	up to 0.48	up to 0.36[c]	up to 1.08	up to 0.16	0.34-1.55	0.86-2.90
Angelucci et al. 1973	16 made in Brazil						10.18-14.67
Kröplien (1971)	11	0.03-0.09	up to 0.03	0.20-0.93[a]	0.12-1.05[b]	0.43-2.48	1.05-4.24
Pictet (1975)	15 made in USA Germany and Switzerland						2.74-6.70

Notes: a. including up to 0.03% glucose
b. including up to 0.3% fructose
c. up to 0.36 for six samples with 1.04% for the seventh — one wonders whether this much higher level is indicative of the incorporation of chicory.

Figure 13.1: Fractionation Scheme for the Water-soluble Carbohydrates of Roasted Coffee

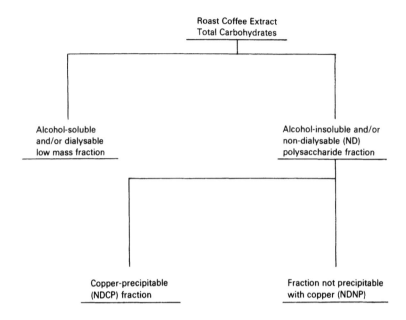

Low Molecular Mass Carbohydrates

Sugars are lost rapidly on roasting and may be completely destroyed. The occasional detection of traces of galactose, mannose, rhamnose and arabinose suggests that polysaccharide hydrolysis can occur during roasting and commercial extraction (Kröplien, 1971; Lerici *et al.*, 1980a; Tressl *et al.*, 1982; Trugo and MacRae, 1982). Thaler has reported (Thaler, 1974; Asante and Thaler, 1975a) that the yield of dialysable carbohydrate (mass below 10,000 daltons, see Figure 13.1) is a function of the roast colour and extraction rate. The yield from a roasted robusta (18.7 per cent total roasting loss) rose progressively to 11.84 per cent at an extraction rate of 58 per cent. The yield from a similarly roasted arabica rose progressively to 8.41 per cent at 43.6 per cent and then declined to 5.45 per cent at 53.2 per cent. The reason for this decline is not clear but may indicate that these fragments were converted during extraction into a form that was no longer detectable by the enzymic analysis employed. This dialysable carbohydrate contained substantial amounts of mannose and galactose, less arabinose and little glucose but was not characterised further.

An increase in reducing power in lightly roasted coffee (Pokorny *et al.*, 1974) should not necessarily be attributed to the formation of reducing sugars but rather to reducing substances. Sugar analyses of roasted coffee products based upon reducing power, before or after hydrolysis of non-reducing sugars, consistently indicate much higher sugar contents than do chromatographic methods. For example Angelucci *et al.*, (1973) reported 5.24 to 13.67 per cent reducing sugars in Brazilian soluble coffee powders, whereas Trugo and MacRae (1982) reported 0.44 to 2.03 per cent for total monosaccharides in soluble coffee powders bought in the UK. Some but not all of this apparent discrepancy can be attributed to reducing oligosaccharides, which Pictet (1975) reported at some 5 to 6 per cent in total. The smaller mass fraction of these oligosaccharides, 4 to 5 per cent yield with an apparent degree of polymerisation (DP) up to 5, had a high (45-49 per cent) arabinose content, giving some 2-2.5 per cent arabinose. This probably corresponds to the pentosans (2.11-4.15 per cent) reported by Angelucci *et al.* and the arabinogalactans reported by Wolfrom and Anderson (1967) and Thaler (1979) which had arabinose contents equivalent to some 2-3.5 per cent of the soluble coffee powder. The remaining oligosaccharide (1-1.5 per cent) has been reported to be a galactomannan with various but significant mannose contents (40-95 per cent) and an apparent DP in the range 15 to 25 (Wolfrom and Anderson, 1967; Pictet, 1975).

Polysaccharides

The polysaccharides are the most important aroma-binding agent in the beverage (Maier, 1975; Maier and Krause, 1977) and they also contribute to the viscosity or mouthfeel (see page 354).

Light microscopy has shown no obvious disruption of the polysaccharide-rich cell wall complex (Thaler, 1975). Even exhaustive extraction of a whole roasted bean leaves the cell wall apparently intact. However, some changes have occurred since a variable amount (7-28 per cent) of the glycosidic material extracted from roasted coffee no longer dissolves in 72 per cent sulphuric acid. It has been suggested that this is because of the formation of intra-residue (2-3 or 3-4) or inter-residue anhydrides during roasting. The extracted material is brown and incorporates nitrogen (Wolfrom *et al.*, 1960; Thaler, 1975).

The relative stability of the glycans increases in the sequence araban, galactan, mannan and glucan. Part of the polymeric material is fragmented probably via carbonium ions, β-eliminations and retro-aldolisations (Baltes, 1975). A further part is solubilised but not so extensively degraded. For example, Thaler has reported that between 6 and 30 per cent of the glucan, between 50 and 60 per cent of the mannan and up to 75 per cent of the galactan in the green bean holocellulose is made water-extractable during roasting (Thaler and Arneth, 1969a,b; Asante and Thaler, 1975b).

The yield of non-dialysable (ND) polysaccharide (see Figure 13.1) rose progressively with the total solubles yield for both arabica and robusta. The galactose and mannose contents were usually of a similar order and together represented more than 90 per cent of the ND polysaccharide.

The composition of the non-dialysable, copper-precipitatable (NDCP) polysaccharide depended on the species and the roasting process. There was never more than 1 per cent glucan (usually less than 0.5 per cent). Mannan dominated forming 94 per cent of the NDCP from normally roasted arabicas but only 88 per cent in similarly roasted robustas. The balance was galactan. Galactose content declined in darker roasts as the content of non-hydrolysable glycosidic material increased. For a given solubles yield, darker roasts delivered more NDCP polysaccharide which represented between 1.6 and 3.5 per cent of the green bean, 1.8 per cent and 4.4 per cent of the roasted bean, 6.6. and 12.7 per cent of the total extractable solids from either species (Ara and Thaler, 1976).

However, for a given pyrolysis loss arabicas and robustas behaved differently when the extraction rate was increased progressively. At low solubles yield from an arabica the NDCP polysaccharide formed some 80 per cent of the total ND polysaccharide. Raising the solubles yield increased the delivery of ND polysaccharide by raising the delivery of non-dialysable non-precipitatable (NDNP) polysaccharide. In contrast a robusta delivered an essentially constant amount of NDNP but an increasing amount of NDCP (Ara and Thaler, 1977; Thaler, 1979). The nature of the NDNP material was not discussed but presumably it did not contain mannose in the form and/or quantity required for complex formation with alkaline copper.

Summary

It is not yet possible to give a precise account of the carbohydrates of coffee extract or beverage. It is the author's opinion that these water-soluble carbohydrates are derived from the green bean arabinogalactan, galactomannan and pectic substances. No significant information is available for this last group.

Surviving arabinose seems to be found mainly as small arabinogalactan oligomers (forming perhaps 5 per cent of soluble coffee powders), or in the free form (perhaps up to 1 per cent). Most of the surviving mannose seems to be in large mannose-rich, non-dialysable molecules (forming perhaps 5-15 per cent of soluble powders) whereas most of the surviving galactose (forming perhaps 5-20 per cent of soluble powders) occurs as dialysable galactomannan fragments containing some 30-50 per cent galactose. The balance occurs in the previously mentioned arabinogalactan oligomers and in the free form (up to 1 per cent of soluble powders). Many, if not all, of these soluble carbohydrates contain nitrogen and peculiar anhydrosugar residues.

Table 13.8: The Crude Lipid Content of Green Coffee Beans

References	Material examined	Crude Lipid % dmb Range	Mean ± SD
Xabregas et al. (1971)	18 Angolan arabicas	15.5-16.4	16.01±0.27[a]
Wurziger (1977) Wurziger et al. (1977b)	5 arabicas	13.2-14.4	13.47±0.47[a]
Kulaba and Robins (1981)	20 Kenyan arabicas	15.30-18.59	17.35±0.94[a]
Wurziger et al. (1977b)	7 Ivory Coast arabustas	10.4-12.8	11.53±0.70[a]
Xabregas et al. (1971)	45 Angolan robustas:		
	18 Amboin province	10.2-11.2	10.66±0.28[a]
	18 Ambriz province	8.8-9.7	9.28±0.26[a]
	9 Cazengo province	10.1-11.2	10.41±0.32[a]
Wurziger (1977) Wurziger et al. (1977b)	8 robustas	7.9-12.3	9.79±1.29[a]
Chassevent et al. (1973a)	7 Mascarocoffea	12.6-30.0	—

Note: a. Recalculated from published data.

Lipids

The terms crude and total lipid refer to all material extracted by a specified, usually non-polar solvent, and may include non-lipid substances such as caffeine. The yield of crude lipid is a function not only of the composition of the bean but also the conditions of extraction particularly particle size and surface area, choice of solvent, duration of extraction, and whether caffeine is excluded (Janicek and Pokorny, 1970; Folstar, Pilnik, de Heus and Van der Plas, 1975a; Folstar, 1976). Folstar has recommended a particle size not exceeding 0.5 mm and a 6 h extraction with petroleum ether (b.p. 40°-60°C). Some earlier data are misleading because the extraction had not been optimised: the data summarised in Table 13.8 are thought to be reliable. Arabicas have a higher crude lipid content than robustas, and some data suggest that within a species the crude lipid content depends upon the location at which the bean is grown (Menchu and Ibarra, 1967; Xabregas et al., 1971).

The crude lipid includes the cuticular wax which coats the coffee bean (see Dentan, Chapter 12). Folstar and colleagues have extracted the wax (yield 0.2-0.3 per cent) from whole green beans using boiling chlorinated hydrocarbons. The main constituents are petroleum ether — insoluble hydroxytryptamides (C-5-HT), which are found in green beans in the range 40-120 mg/100 g. The C_{20} and C_{22} amides are dominant; minor constituents of the wax include the C_{18}, C_{24}, WD-hydroxy C_{20} and W-hydroxy C_{22} fatty acid amides. The C-5-HT are thought to be antioxidants, and it has been suggested that premature removal of the wax leads to a fall in bean quality during storage. Polishing, dewaxing and decaffeination remove some 50-90 per cent of the wax and roasting destroys some 40-60 per cent of that remaining yielding a range of volatiles including indoles and cresol (Wurziger and Harms, 1969; Viani and Horman, 1975; Amorim *et al.*, 1977a; Wurziger, 1977; Wurziger, Drews and Bundesen, 1977a; Hunziker and Miserez, 1979; Folstar *et al.*, 1979, 1980). According to Van der Stegen (1979) coffee brew may contain up to 2.3 mg C-5-HT per litre when the beverage is prepared by percolation of untreated beans. The C-5-HT could not be detected in beverages prepared by a filtration method, or by percolating dewaxed beans.

The bulk of the crude lipid is a typical seed oil which is distributed throughout the cytoplasm of the internal tissues (see Dentan, Chapter 12). Some 80-90 per cent of this oil consists of triglycerides with unsaturated C_{18} fatty acids located preferentially at position 2 (Folstar, Pilnik, de Heus and Van der Plas, 1975b). Arabicas differ from robustas only in terms of yield, but some wild species may be more distinct in terms of fatty acid profiles (Chassevent *et al.*, 1973a). Undesirable oxidative and hydrolytic rancidity have been observed during storage of green beans (Janicek and Pokorny, 1970; Walkowski, 1981) but normal roasting seems to have little chemical effect upon even the polyunsaturated fatty acids (Carisano and Garibaldi, 1964; Fobé *et al.*, 1967; Lopes, 1974). During roasting the oil is expelled to the surface imparting a characteristic appearance to dark roasts, and trapping part of the escaping aroma volatiles. Transfer of this layer to the roaster surface probably accounts for the 1-2 per cent loss of crude lipid from dark roasts. Such expelled oil can be used for aromatising soluble coffee powders (see Clarke, Chapter 14). This exudate must not be permitted to overheat since charring, or even combustion, may occur with concomitant formation of methylene radicals and polynuclear aromatic hydrocarbons (PNAH). Crude lipid normally forms less than 1 per cent of soluble powders, but contains oxidation susceptible volatiles and unsaturated fatty acids (Angelucci *et al.*, 1973; Simonova and Solov'eva, 1980). Such oxidation can occur during storage and cause a fall in quality.

Recent analyses indicate that unsaponifiable lipids form some 10-18 per cent of the seed oil, corresponding to some 2-3 per cent of green arabicas. Many older data are unreliable either because the seed oil was not fully extracted, or because the unsaponifiables were not fully recovered from the saponified oil (Folstar, 1976; Kulaba and Robins, 1981). Coffee wax has a smaller content (~ 2 per cent) of unsaponifiables but the composition is similar with small quantities of hydrocarbons, tocopherols, pigments and phospholipid, and some 20 per cent sterols; esterified diterpene alcohols form the balance. Gibson (1971) suggested that these diterpenes may be the precursors of the endosperm pigments (see page 342).

The major diterpene alcohols, which are probably unique to coffee beans, occur preponderantly as esters of saturated fatty acids, particularly palmitic, behenic and arachidic (Kaufmann and Hamsager, 1962; Folstar *et al.*, 1975b). These diterpenes contain a fused furan ring which renders them very sensitive to acid, heat and light; this instability made characterisation difficult. The structures for cafestol and kahweol shown in Figure 13.2 are taken respectively from Scott et al., (1962) and Lam, Sparnins and Wattenberg (1982). Other structures which have been proposed for kahweol have differed in the number and location of the double bonds. Kahweol is found in arabicas (0.7-1.1 per cent of the green bean — Kulaba and Robins, 1981) and arabustas but not robustas. There are no recent quantitative data for cafestol content; Kaufmann and Hamsager (1962) suggest that it is the major diterpene even in arabicas.

Quantitatively minor diterpenes include 19-hydroxy-(−)kaur-16(17)-ene and 16,17-dihydroxy-9(11)-kaurene-18-oic acid (Obermann and Spiteller, 1975; Wahlberg, Enzell and Rowe, 1975). Structures are shown in Figure 13.2. Nagasampagi, Rowe, Simpson and Goad (1971) suggested that abiet-15-en-13β,19-diol was present but the identification was shown to be incorrect by Wahlberg et al., (1975).

The diterpene content declines during storage; they are partially removed during dewaxing and are extensively destroyed during roasting although sufficient kahweol survives for arabicas to be distinguished from

Figure 13.2: Free and Esterified Diterpenes

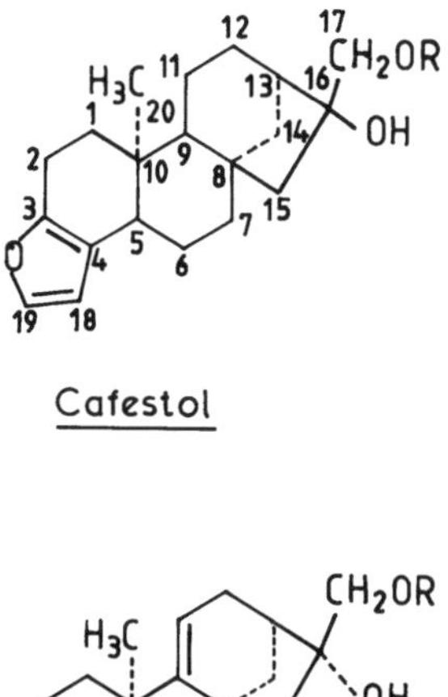

Cafestol

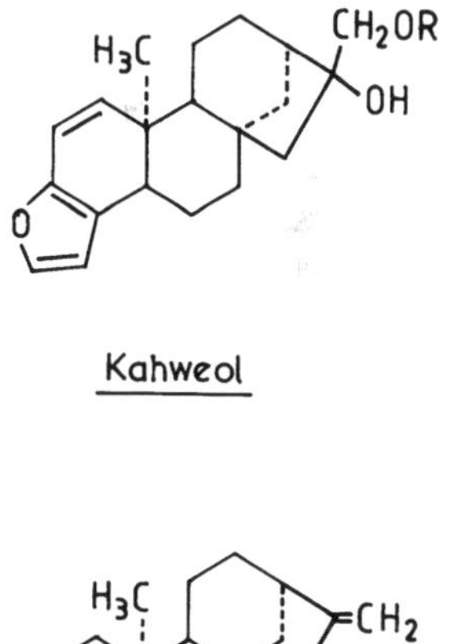

Kahweol

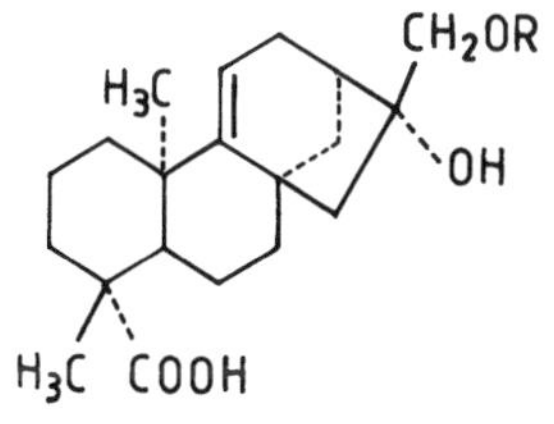

16,17-dihydroxy-9(11)-kaurene-18-oic acid

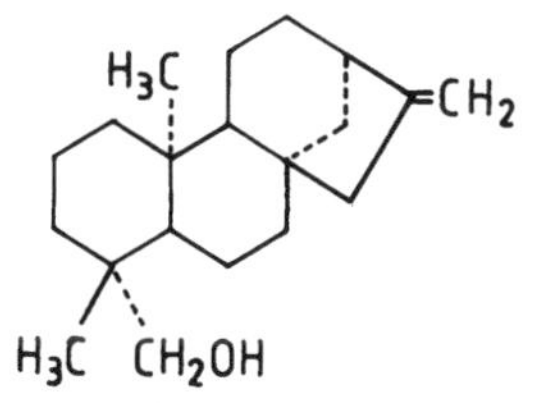

19-hydroxy-(−)kaur-16(17)ene

Note: R − H or fatty acid ester. Structures as reported by Scott et al., (1962); Obermann and Spiteller (1975); Wahlberg et al., (1975); Lam et al., (1982).

robustas by a spot test (Wurziger and Purazrang, 1972; Wurziger, 1977; König, Rahn and Vetter, 1980). Gautschi *et al.*, (1967a) have suggested that diterpene degradation may be the origin of volatile terpenes, naphthalenes and quinolines.

The sterols, unlike the diterpenes, are typical of seed oils and consist of sitosterols (some 43-57 per cent), stigmasterols (some 20-28 per cent) and campesterols (some 15-19 per cent) (Alcaide *et al.*, 1971; Nagasampagi *et al.*, 1971; Tiscornia, Centi-Grossi, Tassi-Micco and Evangelisti, 1979). Tiscornia has suggested that it may be possible to identify the geographical origin of an arabica by its content of the quantitatively minor sterols, particularly Δ^5-avenasterol, Δ^7-stigmasterol and Δ^7-avenasterol. Tocopherols (α, β and γ) together form some 400-600 µg/g of oil or some 5-10 mg/100 g of green bean (Folstar, Van der Plas, Pilnik and de Heus, 1977). Folstar (1976) also found small quantities of many saturated hydrocarbons, with C_{29} dominant in coffee wax and C_{18}, C_{20} and C_{28} dominant in wax-free oil.

Diterpene glycosides

In contrast to the free diterpenes and diterpene esters the diterpene glycosides are soluble in water and methanol. The presence of small quantities of some aglycones in commercial green coffee beans suggest that these compounds can be attacked by β-glycosidases during green bean storage (Maier, 1981). The structures of some compounds in this group have not yet been established. The most studied are cafamarine of *C. buxifolia* and mascaroside of *C. vianneyi* (see Figure 13.3). These bitter glycosides are of chemotaxonomic interest because they seem to replace caffeine in the seeds of *Mascarocoffea* (de Rostolan, 1971; Chassevent, Gerwig and Bouharmont, 1973b; Hamonnière, Ducrouix, Pascard-Billy and Poisson, 1975). Smaller quantities of similar compounds, including atractyloside, occur in the commercially important species, with much higher levels in arabicas (0.07-0.11 per cent) than in robustas and arabustas (up to 0.007 per cent). At least one of these is a cafestol derivative and resembles mascaroside. Structures are shown in Figure 13.3. Although progressively degraded during roasting the residue passes into soluble coffee powders and beverage and might contribute to brew bitterness (Ludwig, Obermann and Spiteller, 1974; Obermann and Spiteller, 1975; Maier and Wewetzer, 1978; Maier, 1981; Maier and Mätzel, 1982).

Quinic Acid and its Derivatives

Quinic Acid and its Derivatives in Green Beans

Quinic acid occurs free and in esterified form. Free quinic acid is commonly found at levels in the range 0.4-0.7 per cent, but may be as high as

Figure 13.3: Diterpene Glycosides

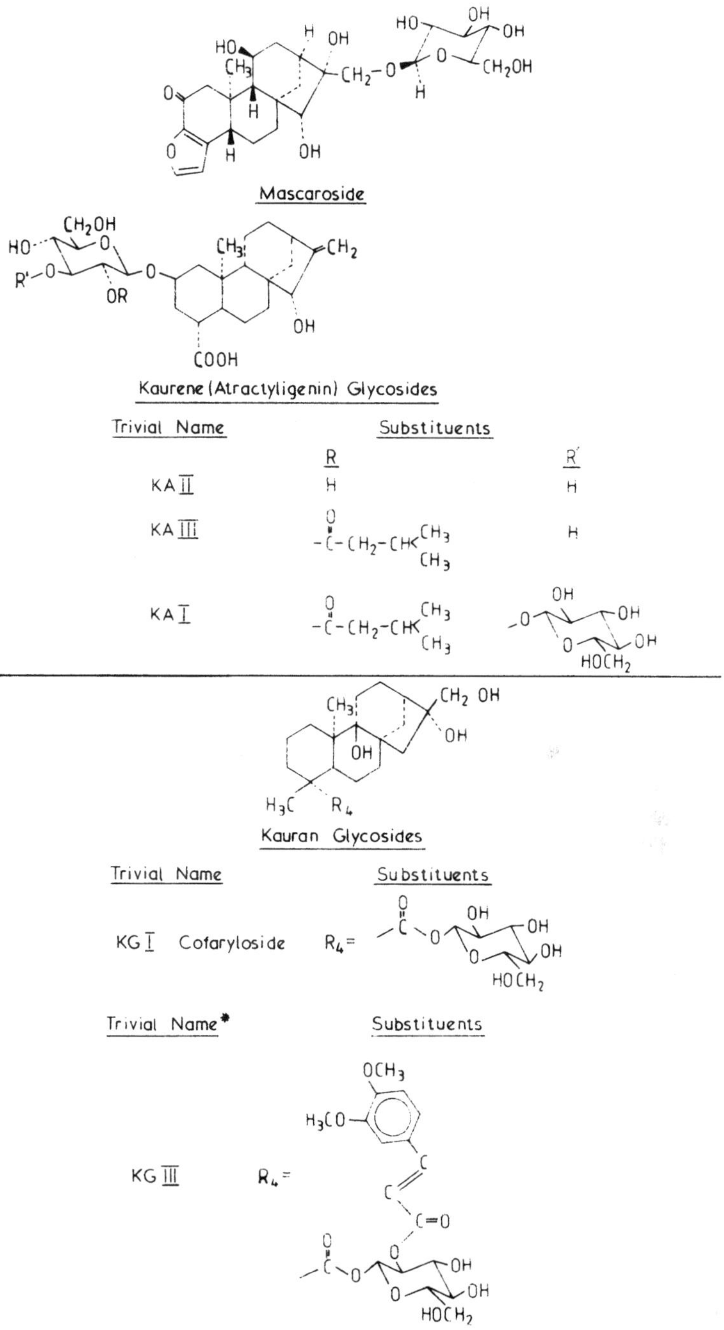

Mascaroside

Kaurene (Atractyligenin) Glycosides

Trivial Name	Substituents	
	R	R'
KA II	H	H
KA III	$-\overset{O}{C}-CH_2-CH\!\!<\!\!\overset{CH_3}{CH_3}$	H
KA I	$-\overset{O}{C}-CH_2-CH\!\!<\!\!\overset{CH_3}{CH_3}$	(sugar with OH, OH, OH, HOCH$_2$)

Kauran Glycosides

Trivial Name	Substituents
KG I Cofaryloside	$R_4 = -\overset{O}{C}-O-$(sugar with OH, OH, OH, HOCH$_2$)

Trivial Name*	Substituents
KG III	$R_4 =$ (feruloyl-sugar ester: OCH$_3$, H$_3$CO-aryl-C=C-C(=O)-O-sugar-OH,OH,HOCH$_2$, with $-\overset{O}{C}-O-$)

Note: *As used by Maier (1981). Structures as reported by Hammonière et al., (1975); Maier and Wewetzer (1978); Maier (1981).

330 *Chemical and Physical Aspects*

1.5 per cent in discoloured green beans (Ohiokpehai, 1982). The esters, known as cholorgenic acids (CGA) form a quantitatively important fraction of green and roasted coffee beans, soluble coffee powders and coffee brews. The CGA may be subdivided into groups usually of three isomers each on the basis of the number and identity of the acylating residues (see Figure 13.4): (a) the caffeoylquinic acids (CQA); (b) the dicaffeoylquinic acids (diCQA); (c) the feruloylquinic acids (FQA); (d) the *p*-coumaroylquinic acids (CoQA); and (e) several caffeoylferuloylquinic acids (CFQA).

Green coffee beans contain a range of as yet unknown but possibly related substances which are equivalent to some 5 per cent of the total CGA content of arabicas and some 1.0 per cent in the case of robustas (Van der Stegen and Van Duijn, 1980; Ohiokpehai, 1982). These unknowns could include phenolic glucosides (Amorim *et al.*, 1974b; Moll

Figure 13.4: Coffee Bean Chlorogenic Acids

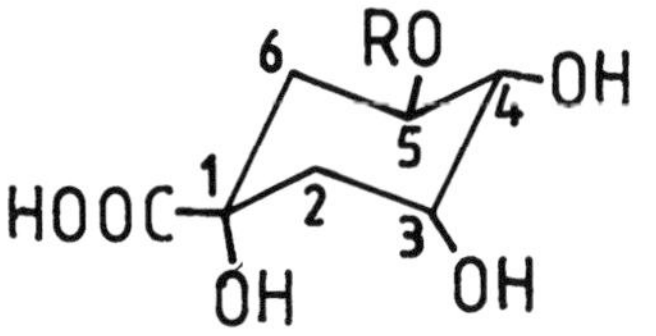

Quinic Acid and CGA*
Favoured Chair Conformation

R = H = Free Quinic Acid
R = caffeoyl = 5-CQA
R = feruloyl = 5-FQA
R = p-coumaroyl = 5-CoQA

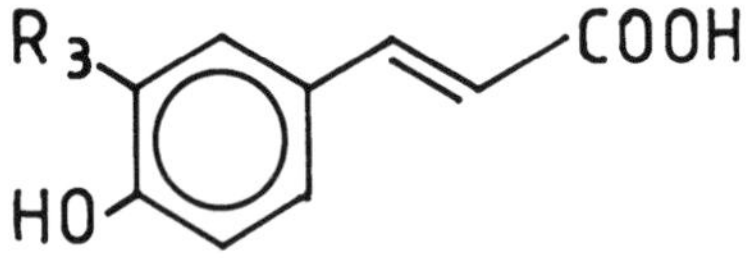

Cinnamic Acids occurring in CGA

R_3	
OH	caffeic acid
OCH_3	ferulic acid
H	p-coumaric acid

Note: *Numbering of carbon atoms as recommended by UPAC (1976). Acylation of coffee bean CGA also involves positions 3 and 4.

and Pictet, 1980) and esters of glucose instead of quinic acid (Clifford and Griffiths, 1982).

According to Sondheimer (1964) chlorogenic acid was first reported in coffee by Robiquet and Boutron in 1837. Characterisation of this fraction has occupied many years during which the CGA accumulated a very complex and confusing trivial nomenclature. This nomenclature is outlined, and so far as is possible, explained in Table 13.9. More recently for nomenclature purposes quinic acid and its esters have been treated as cyclitols (IUPAC, 1976).

The naturally occurring isomer of quinic acid is 1L-1(OH),3,4/5-tetra-hydroxycyclohexane carboxylic acid (see Figure 13.4). In the preferred configuration the carboxyl, and hydroxyls 4 and 5 are equatorial, with hydroxyls 1 and 3 axial. Few authors as yet comply with these IUPAC recommendations common practices are to place hydroxyl 5 above the ring with hydroxyls 1,3 and 4 below and/or to number in the reverse direction. *The latest IUPAC recommendations are used in this book and previously published material has been converted to correspond.* In coffee beans quinic acid is not esterified at position 1, but such compounds are known elsewhere (see Table 13.9).

There have been several studies of CGA biosynthesis. The enzymes responsible for the ultimate step in the synthesis of trans 5-CQA, trans 5-FQA and trans 5-CoQA have been isolated from a range of plants including coffee cell tissue culture (see Gross, 1981). The monoesters acylated at position 5 are dominant in green coffee beans but species are known where a 3-ester, 4-ester or diester are dominant (e.g. see Ruveda, Deulofeu and Galmarini, 1964; Whiting and Coggins, 1975; Martino, Debenedetti and Coussio, 1979; Möller and Herrmann, 1983). At present the origin of these related compounds remains uncertain, but where they dominate they are probably not artefacts. In coffee it should be noted that the 3- and 4-monoesters (but not the diesters) could form by acyl migration of the dominant 5-ester. Such migration is more likely in aqueous than in alcoholic solvents (Haslam, Makinson, Naumann and Cunningham, 1964; Hanson, 1965; Nichiforesco, 1970; see page 338).

In green coffee beans the 5-isomers account for some 70-85 per cent of the CQA and 80-85 per cent of the FQA. The 4-isomers equal or slightly exceed the 3-isomers. In green arabicas 3,5-diCQA>4,5-diCQA>3,4-diCQA but in green robustas and arabustas 4,5-diCQA is dominant with the 3,4-diCQA equal to or slightly exceeding 3,5-diCQA (Van der Stegen and Van Duijn, 1980; König and Stürm, 1982; Maier and Grimsehl, 1982a, b). Some typical data are shown in Table 13.10. Ohiokpehai (1982) has studied the accumulation of chlorogenic acids as the seeds mature. The patterns were complex and varied between species and cultivars. Chicory root contains several CQA including diCQA, and chicory leaves contain caffeoyltartaric (caftaric acid) and dicaffeoyltartaric

Table 13.9: The Trivial Nomenclature of the Chlorogenic Acids

Trivial Name	Introduced	Current Interpretation	Notes
Band 510	Sondheimer (1958)	4-CQA Scarpati and Esposito (1963)	
Chlorogenic acid	Payen (1846)	CGA General term including all quinic acid esters	
Cryptochlorogenic acid	?	4-CQA Scarpati and Esposito (1963)	
Cynarin(e) (1965)	Panizzi and Scarpati (1954)	1,5-diCQA Panizzi and Scarpati (1965)	Apparently restricted to Artichoke (*Cynara scolymus*). Labile and converted easily into 1,3-diCQA during extraction. Originally assigned 1,4-diCQA
Hauschild's substance	Hauschild (1935)	3-CQA lactone Ruveda et al. (1964)	Can be formed during extraction
Isochlorogenic acid	Barnes et al. (1950)	Coffee bean di CQA. Scarpati and Guiso (1964) and CFQA Corse et al. (1965)	Originally Barnes et al. (1950) suggested 3-CQA and its lactone, later Uritani and Miyano (1955) suggested 4-CQA
Isochlorogenic acid a, b and c	Scarpati and Guiso (1964) Corse et al. (1965) Maier et al. (1982a)	Refer to individual components of the isochlorogenic acid	The assignments by the two groups are not in complete agreement
n-chlorogenic acid		5-CQA	Prefix used to distinguish from the all embracing general term (see above)
Neochlorogenic acid	Corse (1953)	3-CQA Scarpati and Esposito (1963)	
Pseudochlorogenic acid	Uritani and Migano (1955)	Not used recently	A poorly defined mixture of CQA and diCQA (Hermann 1967)

Table 13.10: The Contents of Individual Chlorogenic Acids in Green Coffee Beans

Reference	Samples Analysed	CQA			diCQA			FQA		
		3	4	5	3,4	3,5	4,5	3	4	5
Arabicas										
Van der Stegen et al. (1980)	2	0.3-0.5	0.5-0.7	3.4-4.1	0.2-0.4	0.3-0.4	0.3-0.4	0.01	0.04	0.2-0.3
Maier et al. (1982b)	7	0.4-0.8	—[a]	4.6-5.6	0.1-0.3	0.3-0.6	0.2-0.3	—[a]	—[a]	—[a]
Robustas										
Van der Stegen et al. (1980)	2	0.6-0.7	0.9	4.4-5.0	0.5-0.6	0.4-0.6	0.6-0.8	0.07	0.1	0.9-1.0
Maier et al. (1982b)	4	0.7-1.1	—[a]	5.9-6.5	0.5-0.7	0.4-0.8	0.6-1.0	—[a]	—[a]	—[a]

Note: a. No value reported.

(chicoric) acid, but whether these also occur in the roots is uncertain. (Scarpati and Oriente, 1958; Scarpati and d'Amico, 1960; Paulet and Mialoundama, 1976).

According to Dentan (see Chapter 12) the chlorogenic acids are found on the bean surface in association with the cuticular wax and in the cytoplasm adjacent to the cell walls of the endosperm parenchyma. It is not known whether the CGA at these two locations differ in composition. The cell wall chlorogenic acid may be associated with caffeine possibly in a 1:1 or 2:1 molar complex of the type studied by Horman and Viani (1971). There is some evidence that CQA may occur in a polymerised or complexed form, possibly with protein, both in the pulp (mass near 1,000 daltons) and in the bean (mass near 5,000 daltons). These substances are not well characterised but are said to be inhibitors of indole acetic acid oxidase. The pulp phenols might include condensed or hydrolysable astringent tannins (Griffin and Stonier, 1975; Zuluaga-Vasco and Tabacchi, 1980).

Since characterisation of the CGA began in 1846 many methods of analysis have been developed. The more widely used methods are categorised in Table 13.11 which includes notes on specificity and precision. Further information can be found elsewhere (Clifford and Wight, 1976; Clifford and Staniforth, 1977; Ohiokpehai, 1982). Provided that allowances are made for the idiosyncrasies of the analytical method employed it is clear that robustas generally have a significantly higher CGA content than arabicas, some 7-10 per cent and 5-7.5 per cent respectively. Further data are given in Table 13.12. The various hybrids are intermediate. The non-commercial species fall into two groups. Other *Coffea*, with the exception of *C. salvatrix*, resemble the commercial species. The *Mascarocoffea* and *C. salvatrix* have low CGA and low caffeine contents (Chassevent *et al.*, 1973b; Carelli *et al.*, 1974; Maier and Grimsehl, 1982b).

It will be seen from the data in Tables 13.11 and 13.12 that many studies involving CGA have used analytical methods that do not distinguish individual subgroups, and which in some cases lack precision. The conclusions so obtained may be valid but those which should be treated with caution until such time as they are confirmed, include:

(1) variations in fertiliser treatment have no effect upon CGA content (Chassevent *et al.*, 1973b);
(2) cultivars grown at different locations do not differ in CGA content (Vilar and Ferreira, 1973);
(3) cultivars differing in disease resistance do not differ in CGA content (Carelli *et al.*, 1974);
(4) on a commercial scale variations in the method of green bean processing do not affect the CGA content, although laboratory scale 'wet processing' for 48 hours caused a 40 per cent reduction in CGA (Chassevent *et al.*, 1969; Vilar and Ferreira, 1973; Vincent *et al.*, 1977).

Table 13.11: Some Notes on Methods of Chlorogenic Acids Analysis

Basis of Method	Idiosyncrasies	Specimen References	Coefficient of Variation
1. Spectroscopy at 320-330nm — no purification of extract	Detects other UV-absorbing substances particularly in roasted beans. If 5-CQA used as standard the diCQA are over estimated by 15%	Weiss (1953) Rubach (1969) Carelli et al. (1974)	up to 1.84%
2. As method 1 but with purification of extract	Less interferences than method 1 - otherwise similar	Weiss (1957) Lehmann and Hahn (1967) Kwasny (1975)	up to 4.77% up to 6% 10.2%
3. Methods for 1,2-dihydroxy phenols, e.g. borate or molybdate reagents	Detects some roasting degradation products. If 5-CQA used as standard the diCQA are overestimated by 37%	Sloman and Panio (1969) Kwasny (1975) Clifford and Staniforth (1977) Ohiokpehai (1982)	0.3% 4.1% 8.7% 1.5%
4. Method for 1,2-dihydroxy-phenols and mono-methyl ethers	Similar to 3 but measure FQA also	Clifford and Staniforth (1977) Ohiokpehai (1982)	4.6% 5.1%
5. Method for bound quinic acid — e.g. thiobarbituric acid	Complex, multi-step method — little interference. If 5-CQA used as standard the di CQA are under estimated by 31%	Mesnard and Devaux (1964) Ohiokpehai (1982)	— 4.1%
6. Chromatographic methods mainly LC	Informative, time consuming, require scarce standards (see 1) regarding UV detection Some silylating agents cause acyl migration	Ohiokpehai (1982) Möller and Herrmann (1982)	1.2%
7. Use of oxidising agents to detect reducing substances, e.g. permanganate, iodine, phosphomolybdate reagents.	Detects other reducing agents	Moores et al. (1948) Pokorny et al. (1972) Amorim et al. (1974b) Kulaba (1981)	up to 2.14% — 2.18%[b]
8. Methods using protein precipitation	Primarily detects diCQA but the interactions are complex and might include other (non-phenolic) substances	Pokorny et al. (1972) Zuluago-Vasco and Tabacchi (1980) Ohiokpehai, (1982) Clifford and Ohiokpehai (1983)	5.87% — —

Note: a. Additional references will be found in Table 13.12.
b. Folin Ciocalteu and Turnbulls reagents.

Table 13.12: The Chlorogenic Acids Content of Green Coffee Beans

Type of method[a] and reference	Range reported (%dmb) and number of samples analysed				
	Arabicas	Robustas	Interspecific Hybrids	Other *Coffea*	*Mascarocoffea*
Type 1					
Merrit and Procter (1959)	6.86-8.05 (7)				
Carelli *et al.* (1974)	6.63-8.20 (10)	10.30(1)	9.46[b](1)	5.67-9.82[c](7)	
				2.70[c](1)	
Type 2					
Ferreira *et al.* (1971a)	6.90-8.02	10.84-11.70			
Roffi *et al.* (1971)	6.67-7.11 (8)	8.80-11.20 (32)	7.86-9.60[d]		
Hausermann and Brandenburger (1961)	7.71-7.41 (2)	9.02 (1)			
Chassevent *et al.* (1973a)	7.10 (1)	9.20-9.60			0.1-1.20 (21)[c]
Type 3					
Hausermann and Brandenburger (1961)	6.32-6.98 (2)	7.92 (1)			
Clifford *et al.* (1976, 1977)	5.57-6.17 (3)	6.90-8.18 (3)	6.46 (1)[e]		
Rees and Theaker (1977)	7.20-7.60 (3)	8.50-9.10 (4)	8.30 (1)[e]		
Kulaba (1981)	5.80-6.40 (20)				
Ohiokpehai (1982)	5.80-6.70 (5)	7.40-8.30(3)	6.60 (1)[e]		
Type 4					
Clifford *et al.* (1976, 1977)	5.78-6.45 (3)	7.80-9.36 (3)	7.06 (1)[e]		
Rees and Theaker (1977)	7.90-8.40 (3)	11.70-12.30 (4)	10.10 (1)[e]		
Ohiokpehai (1982)	5.90-6.80 (5)	7.80-9.50 (3)	7.00 (1)[e]		
Type 5					
Ohiokpehai (1982)	4.10-5.90 (3)	6.20-6.29 (3)	6.10 (1)[e]		
Clifford (unpublished)	4.07-6.20 (3)	7.39 (1)		4.38 (1)[c]	

Table 13.12: continued

Type 6			
Rees and Theaker (1977)	6.80-7.60 (4)	8.10-10.10 (7)	8.60 (1)[e]
Van der Stegen and Van Duijn (1980)	5.53-6.53 (2)	8.96-10.03 (2)	8.39 (1)[e]
Maier and Grimsehl (1982a)	6.05-7.15 (7)	8.01-9.58 (4)	7.33 (1)[e]
			6.70-7.54 (4)[c]
Type 7			
Kulaba (1981)	6.30-7.60 (20)[f]		
	7.20-9.00 (20)[g]		

Note: a. See Table 13.11
b. Tetraploid arabica-robusta hybrid
c. Number of species
d. Timor hybrid
e. Arabusta
f. Folin Ciocalteu reagent
g. Turnbulls Reagent

It has been observed that old, and particularly discoloured green beans have not only lower extractable CGA contents, but lower polyphenol oxidase activity and lower CQA content relative to the diCQA content (Walkowski, 1981; Ohiokpehai, Brumen and Clifford, 1982; Maier, personal communication, see also pages 315 and 343). Such observations are consistent with polyphenol oxidase attacking CQA and producing insoluble brown pigments, and with diCQA not being a substrate (*in vitro*). The oxidation/polymerisation products have not been characterised although there are some model system data for the oxidation of catechol (Forsyth and Quesnel, 1957; Forsyth, Quesnel and Roberts, 1960). Northmore (1967) suggested that coffee bean endosperm pigments may be CGA oxidation products. (See page 342 for further discussion of this hypothesis).

It has been suggested that astringent beverage is associated with green beans having either a high CGA content or a high diCQA content relative to the CQA content. The mechanism of the astringent sensation is discussed on page 354.

There are some data which suggest that green beans contain phenolic glycosides, one of which might be scopolin (Thier *et al.*, 1968; Amorim *et al.*, 1974b; Moll and Pictet, 1980).* According to Amorim these are hydrolysed when the beans are attacked by moulds, thus accounting for the lower content of such substances in poor quality Brazilian arabicas. Gibson and Butty (1975) have linked the white fluorescence of stinkers with caffeic acid released from CGA by hydrolysis, but such glycosides are an alternative source.

During coffee bean dewaxing and decaffeination there is some degradation of CGA yielding 4-vinylcatechol and 4-vinylguaiacol (Gal *et al.*, 1976; Rahn, Meyer and König *et al.*, 1979). There is considerable acyl migration during dewaxing which raises the content of the 3- and 4-monoesters at the expense of the 5-monoester (Van der Stegen and Van Duijn, 1980; König and Stürm, 1982).

Quinic Acid and its Derivatives in Coffee Products

Roasting of coffee beans causes progressive and ultimately extensive loss of extractable CGA. Ohiokpehai (1982) reported that the relative rates of destruction of CQA, diCQA and FQA (as groups) were 1, 0.56 and 0.09 in a commercially roasted Tanzanian arabica; Pictet and Rehacek (1982) have made similar observations. Within these groups the isomers acylated at position 3 and/or 4 are somewhat more stable than those acylated at position 5. Maier considers 3-CQA to be the most stable and 3,5-diCQA

---

*In a later publication (Pictet and Rehacek, 1982) these putative glycosides appear to have been redesignated FQA.

Table 13.13: The Chlorogenic Acids Content of Roasted Coffee Beans

Bean	Reference	Pyrolysis Loss (%)	Method of analysis[a] and CGA content (% dmb)						
			CQA	HPLC[b] diCQA	FQA	Total	Periodate[c]	Molybdate[d]	TBA[e]
Arabicas									
Kenya	Rees and Theaker (1977)	6.6	2.0	0.2	0.4	2.6			
		8.4	1.0	nd	0.2	1.2			
Colombia	Van der Stegen and Van Duijn (1980)	—	1.89	0.25	0.5[f]	2.64			
Tanzania	Ohiokpehai (1982)	7	2.11	0.59	0.18	2.88	4.7	4.4	4.8
		10	1.37	0.48	0.18	2.03	3.9	3.3	4.1
		11	1.19	0.46	0.18	1.83	3.6	3.1	3.7
Robustas									
Zaire	Rees and Theaker (1977)	4.9	1.6	0.4	0.6	2.6			
		8.9	0.3	0.2	0.1	0.6			

Note: a. See Table 13.11
b. Type 6
c. Type 4
d. Type 3
e. Type 5
f. Authors reported peaks contaminated

the least stable of the caffeic acid-containing CGA. Many other data including some from model systems are consistent with these observations (Corse, Layton and Patterson 1970; Clifford, 1972; Van der Stegen and Van Duijn, 1980; König and Stürm, 1982; Maier and Grimsehl, 1982b). The progressive loss of CGA, particularly when compared with the content of a constituent such as caffeine which is essentially stable to roasting, has been used to monitor the roast severity (Pekkorinen and Porka, 1963; Pictet and Rehacek, 1982). Selected data for the CGA contents of roasted coffee beans are shown in Table 13.13.

Depending upon the roast colour, up to 6 g of CGA may be lost from 100 g of green arabicas and up to 9 g from green robustas. The fate of this CGA has not been studied comprehensively but only some 50 per cent of the CGA destroyed during medium roasts (4-5 g) has been found in the brown pigments or accounted for as free phenols or free quinic acid. Although some of the balance may have been lost in the roaster gases (Clifford, 1972; Tressl, Kossa, Renner and Koppler, 1976; Tressl, Grünewald, Koppler and Silwar, 1978a; Tressl, Bahri, Koppler and Jensen, 1978b; Tressl et al., 1982) it would seem that a significant amount must remain in an uncharacterised form.

Quinic acid is one of the products formed from CGA during roasting, but the yield is not sufficient to account for all the extractable CGA that has been destroyed (Blanc, 1977; Nakabayashi and Kojima, 1980; Maier and Grimsehl, 1982b; Ohiokpehai, 1982). Quinic acids may epimerise and/or lactonise under conditions similar to those encountered in coffee roasting. Epimerisation, particularly at position 4 is favoured when the hydroxyl is esterified. At least one quinide (quinic acid lactone) and one CGA lactone have been detected in extracts of roasted coffee (Wolinsky, Novak and Vasileff, 1964; Corse and Lundin, 1970; Anderson, 1972; König and Stürm, 1982).

Model system studies supported by analyses of roasted coffee suggest that the major products from quinic acid are catechol (although this can

Table 13.14: Chlorogenic Acid Content in Soluble Coffee Powders

Reference	Material analysed	Method of analysis[a]	Total chlorogenic acids (% dmb)
Villanua et al. (1971)	14 bought in Spain	Type 1[a]	8.64-12.02
Vilar and Ferreira (1971)	4 bought in Portugal	Type 1[a]	8.52-11.50
Angelucci et al. (1973)	16 made in Brazil		
	15 arabicas	Type 1[a]	5.23-8-35
	1 robusta		8.02
Vol'per et al. (1973)	Russian	—	5.30-7.40
Pictet (1975)	15 Various	Type 3[a]	3.20-5.20

Note: a. See Table 13.11 for comparison of methods of analysis

also form from the caffeic acid residue), quinol, pyrogallol and 1,2,4-trihydroxybenzene. Caffeic acid yields primarily 4-methyl, 4-ethyl and 4-vinylcatechols and 3,4-dihydroxycinnamaldehyde. The other cinnamic acids yield the corresponding products in similar percentage yields except that benzaldehydes replace the cinnamaldehyde. The quantities found in roasted coffee reflect the relative levels of the precursors in the green bean (Pypker and Brouwer, 1969; Clifford, 1972; Tressl et al., 1976, 1978a, b, 1982). Some acyl migration or degradation may occur during commercial extraction or domestic brewing but there are no published data. Table 13.14 summarises data for soluble coffee powders, some of which are clearly inflated by interference from other products of roasting.

Other Acids

This section deals with those acids not previously discussed (for free amino acids see page 311; for free fatty acids see page 328; for quinic acid and its derivatives see page 328). Published data suggest that sound green beans contain citric acid (0.5 per cent), malic acid (0.5 per cent), oxalic acid (0.2 per cent) and tartaric acid (0.4 per cent) but there has been no comparison of arabicas with robustas (Mabrouk and Deatherage, 1956; Lentner and Deatherage, 1959; Nakabayashi and Kojima, 1980). Northmore (1969) reported elevated acetic acid content in stinkers.

Blanc (1977) commented that the corresponding data for roasted coffee and soluble coffee powders are extremely variable, reflecting not only differences in the material analysed, but also differences in the analytical methods employed. During roasting there is continuous production and simultaneous volatilisation and degradation of acids. The yield of acids, as judged by the beverage pH value and titratable acidity, appears to be a function of the origin of the green beans, the method of green bean processing, the method of roasting, roast colour and method of brewing. Typically the beverage acidity is greater for roasted arabicas compared to roasted robustas, for wet processed beans compared to dry processed, for medium roasts compared to dark roasts and for pressure roasting compared to normal roasting (Sivetz, 1972). The production of a less acid beverage from dark roasts has been attributed to the destruction of CGA which is characteristic of dark roasting (Lentner and Deatherage, 1959; Kung, McNaught and Yeransian, 1967; Feldman et al., 1969), and more recently to the binding of acids by the bean matrix (Nakabayashi, 1978). The greater acidity of pressure-roasted beans has been attributed to greater retention of volatile acids (Sivetz, 1973). Blanc (1977) reported between 1.4 and 4.8 per cent of non-phenolic acids in soluble coffee powders. Feldman et al., (1969) reported that such acids made the least contribution (milliequivalents) to beverage organic acidity: volatile (C_1 to C_{10}) acids made a greater and phenolic acids the greatest contribution. Maier (personal communication) has commented that phosphoric acid ($pK_1 = 1.96$) is potentially the most important contributor. The organoleptic aspects of acidity are discussed further on page 352.

Seed and Fruit Pigments

Immature fruits contain chlorophyll and have considerable photosynthetic capacity (see Cannell, Chapter 5). At harvest maturity most cultivars are red due to anthocyanins (Poisson, 1977). The pigments of the yellow-fruited cultivars have not been reported. The silverskin may be pigmented with carotenoids and chlorophylls in which case it is intensely green. Reddish (foxy) silverskins contain only carotenoids (Gibson, 1971; Ohiokpehai, 1982).

Although unroasted beans are commonly referred to as green (and this usage is accepted in British and International standards (see BS 5752: Part 4, 1980) wet-processed arabicas may be blue-green, green, yellow, brown or black in order of declining quality. Good quality robustas are yellow turning brown or black on deterioration. The colour of the endosperm develops during green bean processing but the nature of the pigments is not established. Chlorophyll is not present. On the basis of alkaline model system studies Northmore (1967) suggested that these pigments were oxidation products of chlorogenic acids or magnesium chlorogenate. Gibson (1971) discounted Northmore's hypothesis with the argument that green beans were not alkaline and suggested cafestol and kahweol as precursors thus providing a plausible explanation for the different colouration of arabicas and robustas. Wurziger and Harms (1969) have suggested the involvement of wax C-5-HT although these occur only on the surface and the pigments occur throughout the endosperm. Recently Pierpoint (1982) has shown that blue and yellow pigments which resemble allagochrome (Habermann, 1973) can be formed by interaction of CQA-quinone and the ε-amino group of lysine. Moreover in an earlier publication Habermann (Garrick and Habermann, 1962) reported that the leaves of *C. arabica* were capable of producing allagochrome. These observations suggest that PPO* and CQA might be precursors of the endosperm pigments. This is an attractive hypothesis because:

(1) CQA and PPO occur throughout the bean (see Dentan, Chapter 12);
(2) the desirable blue and yellow pigments could be analogues of those

---

*In view of Northmore's hypothesis it is interesting to note that PPO generates an alkaline environment adjacent to its substrate-binding site (see Singleton, 1972).

reported by Pierpoint: the desirable green pigment could be a mixture of both;

(3) the undesirable brown and black endosperm pigments could be due to elevated PPO activity following cell damage: the relatively low CQA content in such discoloured beans (Ohiokpehai *et al.*, 1982; see also page 338) would be consistent;

(4) the different pigmentation of arabicas and robustas could be due to species differences in PPO activity/isozyme spectrum (see page 315) and/or to differences in CQA:diCQA ratio (see page 338).

However, it would be premature to discount the possible involvement of the C-5-HT and/or the diterpenes, or the suggestion (Pokorny *et al.*, 1974) that the brown pigments are formed by sugar-amino acid interactions during storage.

Volatile Constituents in Green Coffee Beans

Some 180 volatile substances have been identified in green coffee beans. The characteristic odour has been attributed to a group of methoxy pyrazines: aliphatic hydrocarbons, carbonyls, acids, alcohols and thiols; furans, pyrroles, pyridines and quinolines; phenols, aromatic amines and carbonyls have also been detected. Arabustas are similar to robustas; arabicas are distinguished by a large content of terpenes and fewer aromatic compounds (Merritt, Robertson and McAdoo, 1969; Vitzthum, Werkhoff and Ablanque, 1975; Gutmann, Werkhoff, Barthels and Vitzthum, 1977).

Meritt *et al.*, (1969) have suggested that the range of volatiles may depend upon the geographical origin; certainly the range is affected by green bean processing and its monitoring by GC has been proposed as a method of judging green bean quality (Guyot, Cros and Vincent, 1982). The odour of defective green arabicas has been linked variously with acetaldehyde; acetic acid; carbonyls; several alcohols and esters (Northmore, 1969; Rodriquez, Frank and Yamamoto, 1969; Barel, Challot, Hahn and Vincent, 1973; Gibson, 1973a,b; Amorim *et al.*, 1976; Cros, Guyot and Vincent, 1979).

Sensory Aspects of Coffee Products

When a consumer is presented with a roasted coffee product it is the visual and olfactory senses which are stimulated first; if these sensations are acceptable beverage may be prepared and judged also upon taste and mouthfeel. Odour, taste and mouthfeel together constitute flavour, and if even one of these attributes is not to the consumers' liking, the beverage

344 Chemical and Physical Aspects

may be rejected. Odour is associated with volatile compounds. Colour, taste and mouthfeel are associated primarily with non-volatile pigments and sapins. However, compounds of low volatility are known to be important in coffee odour (Gianturco and Giammarino, 1965; Radtke, Springer and Mohr, 1966a, b) and the boundary between volatile odour compounds and non-volatile sapins is not sharp. For convenience volatile compounds will be defined here as those compounds which without derivatisation have been studied by GC.

The Pigments and Colour of Coffee Products

The pigments of roasted coffee and soluble coffee powders are probably heterogeneous with respect to molecular size and structure. They can be divided conveniently into dialysable and non-dialysable fractions but model system studies have suggested that pigments produced from small precursors become darker as condensation progresses (i.e. mass increases) (Motai, 1976), and thus the dialysable and non-dialysable pigments may be very similar.

The colour of coffee products has been studied by reflectance spectroscopy. The reflectance properties of particulate products depend upon the particle size distribution since small particles appear paler than large (Smith and White, 1965; Paardekooper *et al.*, 1969; Pictet and Rehacek, 1982). Clarke (1965) reported that the particle size distribution obtained by grinding roasted coffee beans is a function, *inter alia*, of the severity of the roast-

Figure 13.5: Low Molecular Mass Pigments Isolated from Model Systems

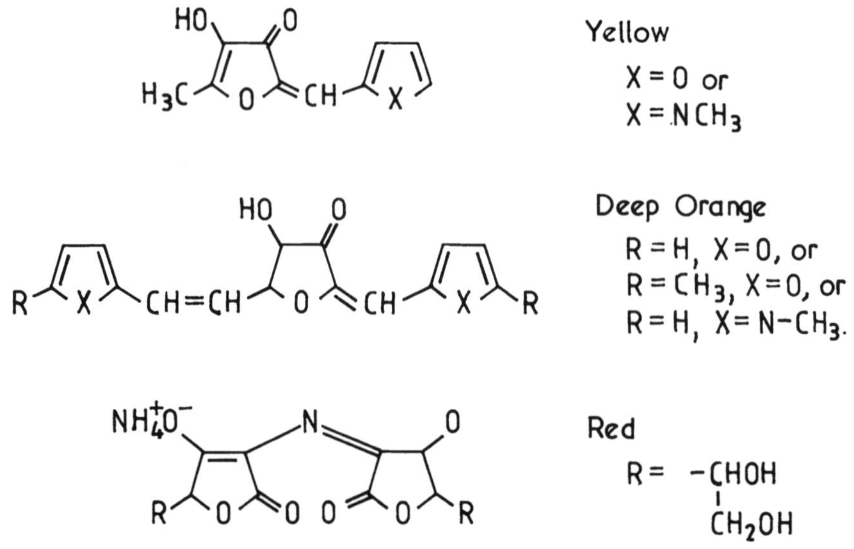

Yellow
$X = O$ or
$X = NCH_3$

Deep Orange
$R = H$, $X = O$, or
$R = CH_3$, $X = O$, or
$R = H$, $X = N-CH_3$.

Red
$R = -\underset{|}{C}HOH$
CH_2OH

Source: Nursten (1981); O'Reilly (1982).

ing process, darker roasts giving a greater number of small particles. Accordingly reflectance measurements should be carried out on specific size fractions.

Recent studies suggest that at the start of the roasting process the bean surface darkens more rapidly than the interior, but that at the end of commercial roasting the interior may darken more rapidly than the surface. Rapid darkening of the interior begins earlier in arabicas than robustas (Lerici, Dalla Rosa, Magnanini and Fini, 1980b; Dalla Rosa et al., 1980). Unfortunately, it is not clear whether these workers have allowed for the effects that variations in particle size distribution have upon the reflectance properties after grinding. However Pictet and Rehacek (1982) have confirmed that arabicas and robustas behave differently on roasting.

Pangborn (1982) has studied the visual properties of coffee beverage. Colour intensity and turbidity increased significantly with increasing temperature of extraction (despite constant extracted solids content), with increased contact time and with increased holding time (at 95°C) after filtration. Colour intensities, turbidity and especially the level of sediment varied with the type of equipment used for brewing: colour intensity was much lower when hard water was used for brewing.

Neither domestic brewing nor commercial extraction remove all the pigment: according to Thaler (1975) the residue may be released by acid sodium chlorite. The low mass pigments of roasted coffee products have not been studied and very little is known even about such pigments produced in model systems. Recent reviews (Nursten, 1981; O'Reilly, 1982) have suggested the following classes of compound may be involved: (1) conjugated polycondensates of furans; (2) similar condensates incorporating furans and pyrroles or pyridones; (3) 1,4-quinones derived from aliphatic vicinal diones via double aldol condensations, and further derivatives thereof. The structure of some coloured compounds isolated from model systems are shown in Figure 13.5.

Olsson, Pernemalm, Popoff and Theander (1977) have commented that 1,2- and 1,4-dihydroxyphenols could be important precursors of pigments in model systems. Farr-Jones and Clifford (1983) have demonstrated facile colour formation from the interaction of aryl aldehydes, 1,2-dihydroxyphenols and nitrogen or sulphur-containing nucleophiles. Such precursors are plentiful in roasting coffee and one may anticipate with confidence interactions of the types summarised here.

Several groups have suggested that the mean mass of the larger water-soluble pigments falls in the range 5,000 to over 20,000 daltons. Yields of 12 to 15 per cent have been recovered even from light roasts (Streuli, 1962; Aurich, Hofmann, Klöcking and Mücke, 1967; Maier, Diemair and Ganssmann, 1968; Clifford, 1972; Maier and Büttle, 1973). Trugo and MacRae (1982) have suggested that the range may extend to 250,000 daltons.

Table 13.15: Low Molecular Mass Phenols in Coffee Pigment Hydrolysates

3-hydroxybenzoic acid[a]
4-hydroxybenzoic acid[a]
3-methoxy-4-hydroxybenzoic acid[b]
3,4-dihydroxybenzoic acid[b]
3,5-dihdroxybenzoic acid[a]
3,4,5-trihydroxybenzoic acid[b]
3,4-dihydroxybenzaldehyde[b,c]
3,4-dihydroxycinnamic acid[a,b]
3-methoxy-4-hydroxycinnamic acid[a,b]
3,4,5-trihydroxybenzene[b,c]

Note: a. Klöcking et al. (1971)
b. Clifford (1972)
c. Found in coffee volatiles by Tressl et al. (1978a,b, 1982)

Fractionation has yielded polypeptide-phenol and polypeptide-carbohydrate complexes. Hydrolysis has released a typical range of amino acids and either mannose and galactose or a range of phenols. Aurich *et al.*, (1967) suggested that between two-thirds and three-quarters of the pigment nitrogen was present in a non-peptide form, possibly inter- or intrachain amides (e.g. see Bjarnason and Carpenter, 1970) or larger mass analogues of the nitrogen-containing polycondensates discussed earlier (see page 345). The range of phenols released is shown in Table 13.15. The hydrolysis systems employed would have degraded any intact CGA (Clifford, 1972) and thus the phenols detected may be artefacts. However, pyrogallol and 3,4-dihydroxybenzaldehyde have been found free in roasted coffee (Tressl *et al.*, 1978a,b, 1982) and at least these two phenols may have been produced during roasting and incorporated into the pigment.

The Aroma of Coffee Products

A major impetus to the study of coffee aroma has been the commercial awareness of its importance in determining consumer acceptance of coffee products, particularly stored soluble coffee powders. Coffee aroma is unique: the complexities of its composition and formation remain a challenge to the analyst even after at least a century of investigation.

According to Walter and Weidemann (1969) the first study of coffee aroma was published in 1880 by Bernheimer. By 1929 nearly 50 volatiles had been identified — now (late 1983) over 700 are known. Useful listings are given by Walter and Weidemann (1969), Weidemann and Mohr (1970), Vitzthum (1975) and Maier (1981). A brief classification is given in Table 13.16. Carbon dioxide is quantitatively the major volatile; the organic constituents form some 500-600 mg/kg of the roasted bean (Tressl *et al.*, 1982). The yield and profile of volatiles depends upon the botanical origin of the green bean and the roast colour. Arabicas and robustas have been distinguished by computer analysis of 47 selected (but uncharacterised) volatiles, and even 1 per cent robusta detected in arabica-robusta mixtures by its greater content (up to × 20) of sulphur-containing volatiles (Gianturco and Giammarino, 1965; Biggers, Hilton and Gianturco, 1969; Pypker and Brouwer, 1969; Nurok, Anderson and Zlatkis, 1978; Pictet and Rehacek 1982).

The routes and mechanisms, by which these many volatiles are produced, are of considerable interest and great complexity; there are recent reviews by Baltes (1975), Nursten (1981) and Vernin and Vernin (1982). The pathways that have been proposed are based upon theoretical considerations supported by model system studies and may (by considerable simplification) be subdivided artificially into the following broad groups:

(1) Pathways involving degradation or conversion of a green bean component into a structurally related compound(s) such that the origin seems fairly obvious, e.g. CGA to phenols; trigonelline to pyridines and pyrroles; C-5-HT to indoles; diterpenes and sterols to alicyclic compounds and acyclic terpenes; Strecker-active amino acids to carbonyls.

(2) Pathways, usually involving several steps, which are sufficiently complex to make the precursor-product relationship less obvious:
    (a) degradative, e.g. fragmentation of carbohydrates to yield saturated and unsaturated hydroxycarbonyl compounds, or of proteins to yield N or S-containing species;
    (b) synthetic, e.g. interaction of electrophilic carbonyl compounds with nucleophiles (oxygen, nitrogen or sulphur-containing) to yield heterocyclic products — many of the reactants being derived via the degradative pathways. These complex synthetic reactions may provide alternative, albeit less obvious, routes to some of the products obtained by simple degradations (Group 1 above).

Table 13.16: Roasted Coffee Volatiles

Class of Compound	Number	Comment
Aliphatic	156	Includes 54 carbonyls
Alicyclic	26	
Aromatic	79	Includes 45 phenols
Furans	103	possibly unique 34[a]
Thiophenes	28	possibly unique 11[a]
Pyrroles	63	possibly unique 3[a]
Pyrazines	70	possibly unique 4[a]
Pyridines	9	
Thiazoles	28	possibly unique 8[a]
Oxazoles	28	possibly unique 19[a]

Note: a. According to Vernin and Vernin (1982) these volatiles had been found only in roasted coffee.

It would be a mistake to assume that a green bean with a greater content of a precursor will *necessarily* produce a roasted bean with a greater content of the associated product(s). The immediate product may react further, or be lost in roaster gases; alternatively the two beans may differ physically (pH value, temperature and pressure) during roasting, thus modifying the chemical reactions.

Hassan (1970) reported a greater total carbonyl content, as yield of dinitrophenylhydrazones, in the aroma volatiles from a freshly roasted arabica (0.16 per cent) compared to a freshly roasted robusta (0.12 per cent) and a commercial soluble coffee powder (0.05 per cent). The arabica-robusta difference was linked to the greater destruction of glucose, fructose and sucrose, both in absolute terms and relative to the destruction of amino acids.

Recently, more extensive (semi)-quantitative analyses of the aroma complex have been reported; the data provided by Tressl's group (Tressl *et al.*, 1978a, b, 1981b) are summarised in Table 13.17. These data, and particularly those which have been linked with a defined green bean composition (Tressl *et al.*, 1982) illustrate and in general support the points discussed above. For example, greater contents of N-alkylpyrroles and pyrazines in roasted robustas compared to roasted arabicas were linked by Tressl with greater contents of Strecker-active amino acids in green robustas. Similarly the greater content of FQA in green robustas can be linked with greater contents of structurally related guaiacols in roasted robustas. Although the greater contents of catechol and 4-methylcatechol in roasted robustas can be linked with greater destruction of caffeic acid and quinic acid residues (from CGA), such a simple precursor-product relationship cannot explain the greater contents of several other structurally related phenols in arabicas (see Table 13.17). Such di- and trihydroxyphenols are not particularly volatile (much less so than the guaiacols) and so are not likely to have been lost from the robustas in the roaster gases. To rationalise these data one must consider the possibility of different physical constraints modifying the yields — either reducing the yield in robustas or raising it in arabicas possibly from carbohydrate as reported by Olsson *et al.*, (1977) for certain model systems.

Many investigators have sought an individual compound(s) that possesses the *full* aroma of freshly roasted coffee. Table 13.18 lists several coffee aroma volatiles said to have a *coffee-like* odour when pure. Structures are shown in Figure 13.6. Furyl-2-methanethiol is of particular interest, since depending upon concentration its odour may be reminiscent of freshly roasted coffee or of stale coffee (see also page 350).

Possibly these components are important in coffee aroma but to obtain a clearer indication of their contribution (aroma value) requires a knowledge of the concentrations which are likely to be present and their threshold values in the coffee product. In practice a bland dearomatised coffee extract may be used since this will take account of synergistic or antagonistic interactions. The synergistic lowering of odour threshold values for some phenols has been demonstrated by Maga (1978). Adsorption, e.g. of

Table 13.17: Coffee Aroma — Summary of Quantitative Data (mg/kg)

Compound	Arabica	Arabusta	Robusta	Soluble powder	Threshold in water
Phenol	13.0	9.5	17.0		0.050
2-methylphenol	1.2	0.7	1.1		0.065
4-methylphenol	1.3	0.3	1.0		0.001
3-methylphenol	0.7	1.0	1.2		0.068
3-ethylphenol		0.4			
4-vinylphenol	0.2	0.2	0.2		
Guaiacol	2.7	3.9	8.4		0.021
4-ethylguaiacol	0.3	1.2	5.6		
4-vinylguaiacol	9.5	18.4	19.5		0.005
Vanillin	5.2	4.4	5.0		
Catechol	80	95	120		
4-methylcatechol	16	10	13		
Quinol	40	25	30		
4-ethylcatechol	37	20	80		
4-vinylcatechol	25	15	25		
Pyrogallol	45	25	35		
1,2,4-trihydroxybenzene	20	6	13		
3,4-dihydroxcinnamaldehyde	10	5	12		
3,4-dihydroxybenzaldehyde	20	8	9		
Furfuryl alcohol	300	150	520		
2-Furoic acid	80	50	55		
5-Hydroxymethylfurfural	35	30	10	70-200	
Furaneol	50	35	25		0.1
Ethylfuraneol	8	4	2		
Cyclotene	40	17	26	70-110	
Isomaltol	8	2	2		
Maltol	39	20	45	60-120	
5-Hydroxymaltol	15	13	6		
5-Hydroxy-5,6-dihydromaltol	13	12	10		
Furyl-2-methanethiol	1.10		2.00	3.90	5×10^{-6}
Furyl-2-methyl methyl sulphide	1.10		2.20	2.80	
Furyl-2-methyl methyl disulphide	0.12		0.65	0.60	
5-Methylfuran-2-methanethiol	0.19		0.11	0.01	5×10^{-5}
5-Methylfuran-2-methyl methyl sulphide	0.09		0.06	0.04	
5-Methylfuran-2-methyl methyl disulphide	0.03		0.02	0.015	
Furyl-2-methyl ethyl sulphide	+		0.01		
Difurylsulphide	0.06		0.13		
2-Methyl-3-(methylthio)-furan	+				5×10^{-5}
2-Methyl-3-(methyldithio)-furan	+				1×10^{-5}
Kahweofuran	1.16		0.85	0.60	0.005

volatiles by polysaccharides (Maier, 1975), and chemical interactions, e.g. between basic volatiles and non-volatile acids, both lower the headspace concentrations of volatiles and thus effectively raise the thresholds. In practice no volatile functions in isolation.

According to Tressl the most important aroma impact compounds are found in an analytical fraction which is rich in alkyl furans and sulphur-containing volatiles. At present many of the heterocyclic volatiles in coffee are thought to be unique to this commodity (Vernin, 1982; for summary see Table 13.16) and this fact suggests that such compounds might contribute significantly to the unique nature of its aroma.

Staling

Staling refers to the deterioration in taste and odour which occurs when roasted coffee and coffee products are stored prior to the preparation of beverage by the consumer. This phenomenon has been studied extensively since it determines the commercial shelf life. According to Arackal and Lehmann (1979) the staling of freshly roasted coffee, 'becomes noticeable after 10 to 11 days and leads to a clearly recognisable stale taste and smell after 6 to 8 weeks and to rancidity after 4 to 5 months'. Unfortunately these workers did not specify the storage conditions. Contact with oxygen is particularly damaging but Radtke has shown that the rate of oxygen uptake/deterioration is a function of: the partial pressure of residual oxygen content; water content; temperature; and is approximately twice as fast in roast and ground beans compared to whole roast beans. According to Radtke roast and ground coffee has a commercial storage life of six months at 23°C provided the residual oxygen content does not exceed 1 per cent and the water content does not exceed 4 per cent. An increase in

Table 13.18: Possible Coffee Aroma Impact Compounds

Compound	Odour threshold in water (μg/kg)	Description
Furyl-2-methanethiol[a]	5	At 10-500μg/kg-freshly roasted coffee At 1-10mg/kg-staled coffee, sulphury note
Kahweofuran[a]	5	At 10-100μg/kg-cooked meat with slight coffee note
N-furyl-2-methylpyrrole[b]		Stale coffee
2-ethylfuran[c]		Powerful, burnt, sweet, coffee-like
N-ethyl-2-formyl pyrrole[c]		Burnt, roasted coffee
Thiobutyrolactone[c]		Green, burnt coffee
3-Methyl-2-acetylthiophene[c]		Coffee-like

Note: a. Tressl *et al.* (1981a)
 b. Tressl *et al.* (1979)
 c. Vernin (1982)

any one of these variables significantly reduces the storage life (Heiss, Radtke and Robinson, 1977; Radtke, 1979).

The entrapped carbon dioxide is instrumental in achieving this storage life. Normally an oxidation-sensitive commodity such as whole roast beans with a water activity near 0.20 would be expected to stale very rapidly. Grinding liberates a substantial part of the protective carbon dioxide and this explains why roast and ground coffee is less stable.

During storage many volatiles (including several furans, thiophenes, aliphatic carbonyls, and vinylguaiacol) decline in concentration. These losses are attributed to oxidation; certain other volatiles increase in concentration, possibly as a result of oxidation, but possibly also as a result of other interactions (Heiss *et al.*, 1977; Tressl, Grünewald, Kamperschroer and Silwar, 1979, 1981a; Simonova and Solov'eva, 1980; Radtke-Granzer and Piringer, 1981; Tressl *et al.*, 1981b). Of particular interest in this respect are *N*-furyl-2-methylpyrrole, which when pure has an odour of stale coffee, and furyl-2-methanethiol which at low concentrations has an odour of freshly roasted coffee *but* at higher levels an odour of stale coffee (see Table 13.18).

Commercially it is important to be able to determine the useful remaining storage life, and this requires a measure of the extent to which staling

Figure 13.6: Possible Coffee Aroma Impact Compounds

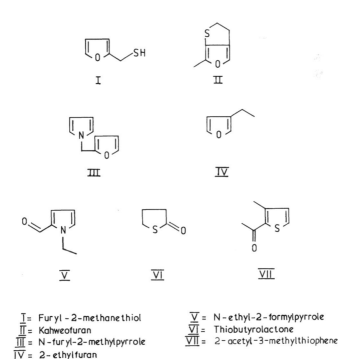

I = Furyl-2-methanethiol
II = Kahweofuran
III = N-furyl-2-methylpyrrole
IV = 2-ethylfuran
V = N-ethyl-2-formylpyrrole
VI = Thiobutyrolactone
VII = 2-acetyl-3-methylthiophene

Table 13.19: Terms Developed by Judges to Describe Taste/Flavour of Coffee

Term	Description
Sweet	Resembles taste of sucrose; perceived mainly on tip of tongue
Salty	Resembles taste of salt; perceived mainly on anterior sides of tongue
Sour	Resembles taste of acid; perceived mainly on posterior sides of tongue
Bitter	Resembles taste of caffeine or quinine; perceived mainly at base of tongue
Balanced	Well-blended flavour, with no unduly prominent flavour characteristics
Flat	Dull, weak, flavourless, lacking in characteristic coffee flavour
Stale	Old, unfresh, but non-rancid (distinguished from flat).
Rancid	Old, oxidised-oil flavour
Astringent	Harsh, drying, puckery sensation; resembles sensation from unripe persimmon
Metallic	Taste associated with iron, copper, brass, or a tarnished silver-plated spoon
Burnt	Overcooked, caramelised, scorched flavours

Source: Reprinted from Pangborn (1982) with permission.

has occurred. Vitzthum and Werkhoff (1979) have discussed the use of several aroma index values for this purpose, in particular the ratio of 2-methylfuran to butan-2-one. This ratio correlates closely with the sensory rating of roast and ground coffee and declines steadily when such coffee is stored in contact with air. The original value is influenced by the species and geographical origin of the green beans, the roast colour and the fineness of grinding (Reymond, Chavan and Egli, 1962; Arackal and Lehmann, 1979; Kwasny and Werkhoff, 1979). Radtke-Granzer and Piringer (1981) have developed this approach by using GC-MS to obtain absolute values for the contents of 2-methylpropanal, 3-methylbutanal, diacetyl and 2-methylfuran.

The Taste of Coffee Products

Coffee products are graded by trained and experienced tasters, who have an extensive vocabulary to describe the desirable and undesirable attributes of the beverage. Table 13.19 is taken from Pangborn (1982) and defines 11 attributes that she has investigated.

Many less well-defined terms also are found in the literature, particularly to describe off-tastes. For example bricky, cereal, chemical, common, earthy, grassy, green, harsh, mocca, musty, onion, oxidised, papery, pungent, unclean and woody (Ferreira *et al.*, 1971a; Van Roekel, 1975; Pictet and Vuataz, 1977; Kulaba, 1978).

Acidity and Sourness. Commonly the layman considers acidity and sourness to be synonymous and even Pangborn (see Table 13.19) does not distinguish them. Traditionally coffee graders consider acidity a desirable attribute whereas sourness is undesirable and indicative of faulty green bean processing (Northmore, 1969). The sour note is associated with a mixture of acids, alcohols and esters produced by microbial fermentation. Acidity is associated with protons and thus proton donors. The sensation produced by an inorganic acid is largely a function of its degree of ionisation. Maier (1983) considers phosphoric acid (pK_1~2) to be the major source of coffee beverage acidity. Organic acids in the beverage are generally weaker and their sensory contribution is modified by a response to their anions.

According to Sivetz (1973) wet-processed, freshly harvested, high grown arabicas produce the most acid beverage. Dry-processed robustas are least acid and dry-processed arabicas are intermediate for a given roast colour. Beverage acidity is influenced also by the method and severity of roasting (see page 341), method of extraction or brewing and brew water alkalinity. Acidity is considered a particularly important characteristic of medium roasts, and high acidity gives a better quality and more intense aroma to the beverage, possibly by modifying the relative concentrations of acidic and basic volatiles in the headspace (Sivetz, 1973; Pictet and Vuataz, 1977; Moll and Pictet, 1980; Voilley, Sauvageot, Simatos and Wojcik, 1981; Pangborn, 1982). There is general agreement that pH 4.9 to pH 5.2 is the preferred range for coffee beverage (Sivetz, 1972; Pangborn, 1982). According to Sivetz, beverage prepared from roasted arabicas falls in this range when subjected to the optimum roast but values up to pH 6.0 were found when lightly roasted. At their optimum robustas were less acid, pH 5.0-5.8. American soluble coffee powders yielded beverages with values in the range pH 4.7-5.2.

Bitterness. Bitterness is a primary taste for which there are specific receptors on the tongue. According to Belitz *et al.*, (1983) a compound need contain only one electrophilic or nucleophilic polar group and a hydrophobic group or region in order to evoke a bitter response. Such structural features occur in many classes of molecule. An element of bitterness is expected and even desirable in coffee beverage, but can become objectionable in dark roasts and at high extraction rates. Bitterness has commonly been attributed to caffeine which would normally exceed its taste threshold of 0.3-1.0 mmol/l (60-200 µg/100 ml), but even decaffeinated beverage has pronounced bitterness and more recently it has been reported that caffeine never accounts for more than 10 per cent of beverage bitterness (Belitz, 1975; Voilley, Sauvageot and Durand, 1977; Panghorn, 1982).

Belitz (1975) has stated that trigonelline is not a significant contributor to beverage bitterness and has suggested that heterocyclic compounds may be responsible. Perceived bitterness may be masked at elevated viscosity, and by astringent substances (Lea and Arnold, 1978; Pangborn, Gibbs and Tassan, 1978).

Astringency. Astringency is not a primary taste, and has been little studied. Many astringent molecules are bitter and these sensations may be confused. However, Lea and Arnold (1978) established that the two sensations can be distinguished and consistently quantified by experienced taste panel members. Astringency may also be distinguished by objective protein-precipitation methods (see Clifford and Ohiokpehai, 1983); and the ability to precipitate salivary proteins and glycoproteins is probably the fundamental characteristic of an astringent compound. Most of the compounds known to be astringent possess at least two 1,2-dihydroxyphenyl residues and the relative astringency increases exponentially as the number of such residues increases. Compounds that bind to and raise the net charge of the astringent phenol — salivary protein complex may impede its precipitation and thus weaken the astringent sensation. The CQA are known, and caffeic acid is likely, to have this effect (Whiting and Coggins, 1975; Ohiokpehai *et al.*, 1982). Subject to certain constraints of pH value and protein pI value the interaction is probably non-specific for the protein and such molecules may also bind to the tongue and soft palate. Such non-specific binding could block primary taste receptors and such binding might explain the observation that astringent substances can reduce perceived bitterness (Lea and Arnold, 1978), yet permit the astringency to be detected as a dominant note. However, this bitterness-masking effect might also be a result of elevated viscosity in the buccal cavity, a point discussed on page 353.

It has been suggested that diCQA are the astringent components of coffee brew. They have a taste threshold in coffee brew between 0.05 and 0.1 mg/ml and at 1 mg/ml impart a 'lingering metallic bitter taste' in the presence of 0.45 mg/ml CQA (Ohiokpehai *et al.*, 1982; Clifford and Ohiokpehai, 1983). Various investigations have linked astringent beverage with the use of green beans from immature fruit, discoloured green beans; darker roasts and arabicas rather than robustas. Such beans are known to have either greater diCQA contents, or more diCQA relative to CQA (see page 338). (Ferreira *et al.*, 1971a; Amorim *et al.*, 1974b; Ohiokpehai *et al.*, 1982; Pictet and Rehacek, 1982; Clifford and Ohiokpehai, 1983).

Body. Body is synonymous with mouthfeel and viscosity. Smith (1983) has said that there is no simple relationship between beverage viscosity measured instrumentally and body judged subjectively. In view of the dilute nature of coffee brew (instrumental) viscosity is presumably associated with macromolecules. It has been reported that body in wines, ciders, legumes and pulses has been associated with low levels of astringent phenols (Arnold and Noble, 1978; Lea and Arnold, 1978; Ariga and Asao, 1981). An interaction in the buccal cavity between, e.g. CQA, diCQA and salivary proteins, could contribute to the body in the mouth without affecting instrumental viscosity measured on the brew. Such an interaction could account for the discrepancy reported by Smith; substances which raise viscosity have been reported to mask bitterness (see page 353) and such elevation, as well as the blocking of bitterness receptor sites mentioned previously, may explain the masking of perceived bitterness by astringent compounds.

Chemical Aspects of Coffee Quality Assurance

Caplan (1978) has defined quality as 'fitness for a purpose'. Blanchfield (1980) prefers the definition 'a multicomponent measure of the extent to which the units of a product, which a seller is willing and able to offer at a price, consistently meet the requirements and expectations of the group of buyers willing to buy that product at that price'. Cost and consistency of performance have a considerable influence upon the consumer and thus are important to the processor who has to compromise, perhaps for example between a particularly delectable odour and taste associated with a particularly costly arabica and a more mundane odour and taste associated with a cheaper robusta which delivers a higher yield of soluble coffee powder.

For the chemist there are three particular interesting aspects to quality.

(1) *What substances in the product/beverage determine (a) consumer acceptance or preference and (b) rejection?* Before answering this question one must note that preference is notoriously idiosyncratic and one must think in terms of a family of acceptable products/beverages which individuals will prefer to a lesser or greater extent. In attempting to answer this question one can point to the desirability of a few µg/kg of the fresh coffee aroma impact compounds detected by Tressl and colleagues (see page 348 and Table 13.18) and to the desirability of a beverage pH value in the range pH 4.9-5.2 (see page 353). Equally one may refer to the need to avoid certain defective notes in the beverage such as sourness (associated with defective green beans, see page 352), acid, burnt or bitter notes (often associated with dark roasts, see page 353), phenolic, astringent, metallic notes (associated with irregular phenols contents, see pages 338 and 354), stale and rancid notes (associated with incorrect or unduly long storage, see page 326). This is far from a complete answer.

(2) *What are the precursors in the green bean and the reaction sequences leading to these desirable or undesirable substances?* Since the first question cannot be answered completely, clearly the second cannot be. One may seek an answer to this question by comparing the physicobiochemical systems which constitute the major green beans of commerce, and which for a given roast colour yield organoleptically distinct beverages. The

following generalised descriptions (Clarke, personal communication) apply to typical light or medium roasts:

wet processed arabicas — aromatic with a fine acidity and some astringency;
dry processed arabicas — less aromatic and less acid but with greater body;
dry processed robustas — more neutral aroma, more cereal-like, with body and some harsh, bitter and astringent notes, least acid.

Table 13.20 summarises the chemical composition of arabicas and robustas. However, this summary is based upon limited data and may not perfectly represent normal commercial beans. The lack of data comparing wet and dry processed beans is a particular deficiency.

Only arabicas contain kahweol but there is no evidence to link its presence directly with odour or taste in the beverage. There is a clear difference in caffeine content but in view of the relatively small contribution which this compound makes to bitterness (see page 353) it is unlikely that caffeine can be a major contributor to the organoleptic differences. Robustas have a higher CGA content, particularly FQA and diCQA which respectively could be responsible for smokier aromas and more astringent tastes (see pages 348, 354). Arabicas have a greater content of thermally-labile carbohydrates (sugars and araban). Robustas may have a greater

Table 13.20: A Comparison of Chemical Composition of Green Arabicas and Robustas

Component	Typical Content (% dmb)	
	Arabicas	Robustas
Kahweol	0.7-1.1	not detected
Caffeine	0.6-1.5	2.2-2.7
CQA	5.2-6.4	5.5-7.2
diCQA	0.7-1.0	1.4-2.5
FQA	0.3-0.5	0.5-1.5
Sucrose and reducing sugars	5.3-9.3	3.7-7.1
Total free amino acids	0.4-2.4	0.8-0.9
including Strecker-active	0.1-0.5	0.2-0.3
Araban	9-13	6-8
Reserve mannan	25-30	19-22
Reserve galactan	4-6	10-14
Other polysaccharides	8-10	8-10
Triglyceride	10-14	8-10
Protein	~12	~12
Trigonelline	~1	~1
Other lipids	~2	~2
Other acids	~2	~2
Ash	~4	~4
Total	90-114	86-107

content of free and particularly Strecker-active amino acids, but not all data are consistent (see Table 13.20 and pages 311 and 318). This requires clarification since Hassan and Tressl and colleagues have linked such differences in amino acid contents with differences in the composition and character of the aroma.

Hassan (1970) considered that the molar ratio of reactive constituents was more important than the absolute contents or absolute destruction during roasting. He demonstrated that for total roasting losses of some 17 to 19 per cent arabicas lost some 35 moles of simple sugars for every mole of free α-NH_2-nitrogen (effectively free amino acids) compared to some 10 moles of simple sugars in similarly roasted robustas. Tressl's data (Tressl *et al.*, 1982) are consistent with Hassan's.

Arabicas have a higher polysaccharide content, although less is delivered to the beverage (see page 00). Beverage polysaccharide is an important aroma binding agent (Maier, 1975; Maier and Krause, 1977) but it is not clear whether qualitative or quantitative differences in polysaccharide content can influence aroma character. It has been suggested that the galactomannan of arabicas is more stable during roasting than that of robustas (see page 319) and this might influence volatiles formation chemically (i.e. by lesser or greater degradation) and/or physically by influencing the internal pressure during roasting (see page 306).

Arabicas have a distinctly higher triglyceride content but in view of its relative stability during roasting (see page 326) it is difficult to see how this might make a significant and direct chemical contribution, although it may provide a non-polar environment and thus it may exert an indirect effect upon the reactions that occur during roasting. As the triglyceride melts and is expelled during roasting it could act as a physical barrier to the loss of volatile reactants, aroma constituents and internal pressure thus in arabicas reinforcing the effect proposed for the galactomannan.

To summarise one can suggest that a significant part of the arabica-robusta difference can be attributed to differences in reaction conditions, particularly internal pressure and differences in reactant ratios particularly thermally-labile carbohydrates, amino nitrogen, FQA and diCQA. It has been suggested that the content of at least some of these substances at harvest might vary within species (e.g. see pages 311 and 318). The relative effects of wet and dry processing have not been studied in sufficient detail to permit useful comment, but it has been reported that the ratio of sugars/amino acid and the CQA/diCQA ratio change during storage and discolouration (see pages 318 and 333) and such changes may contribute to the reduced beverage quality associated with beans which have deteriorated during storage.

Amorim and colleagues (summarised in Amorim *et al.*, 1973, 1975b) examined four grades of Brazilian green arabicas and by multiple linear regression analysis were able to explain almost 70 per cent of the variations in (professionally judged) organoleptic quality by variations in green bean composition. Unfortunately many of the chosen methods of chemical analysis showed significant inter-method correlation (e.g. total nitrogen, NaCl-soluble proteins, NaOH-soluble proteins, phosphate buffer-soluble proteins and total CGA, water-soluble phenols, methanol-soluble phenols, hydrolysable phenols). These inter-method correlations led to uncertainty in estimating regression coefficients and this may have been a major reason why no convincing relationships were demonstrated between green bean composition and beverage quality. However, this comment is made with the benefit of hindsight: Amorim's group were limited to the information and methods available in the late 1960s and their extensive investigations remain a landmark in the study of coffee quality. It remains to be seen whether a similar study using modern methods of analysis will lead to the identification of more precise relationships.

(3) *What methods may be used to monitor green bean quality, the quality of process intermediates and the quality of the final product?* The most widely used test of green bean quality is a visual assessment followed by trial roasting, brewing and evaluation of the beverage by experienced tasters. The polyphenol oxidase activity (see page 315) and GC analysis of green bean headspace volatiles (see page 343) are slower methods for obtaining similar information. The GC method might also differentiate arabicas from robustas. Other than visual recognition the only truly practical, i.e. rapid, test of botanical origin seems to be Wurziger's colour test based upon the presence of kahweol (see page 327). CGA subgroup analysis and possibly polyacrylamide gel electrophoresis of extracted proteins are feasible but slow by comparison (see pages 314 and 333), amino acid profiles are slow and problematical (see page 333). However, CGA subgroup analysis or sterol analysis, while slow, appear to offer some possibility of monitoring geographical origin (see page 328). The use and efficiency of dewaxing procedures may be monitored by measuring residual C-5-HT (see page 327), monitoring the levels of 4-vinylcatechol and 4-vinylguaiacol, or possibly by measuring acyl migration within the CGA (see page 338). Similarly decaffeination processes are monitored by measuring the residual caffeine (see page 309) and checking for solvent residues, e.g. see Van Rillaer, Janssens and Beernaert (1982).

The roasting process is controlled primarily by standardising charge size and moisture content, and the time and temperature of the operation. Water content can be estimated rapidly on green bean and products by several methods, particularly near-infra red spectroscopy (see page 307). The severity of the roasting process may be estimated by measuring the dry matter loss, the destruction of trigonelline (see page 310), the destruction of chlorogenic acid (see page 340), by monitoring the ratio of 2-methylfuran to butan-2-one (see page 352) or by reflectance measurements applied to the whole or the ground roast bean (see page 344).

Soluble coffee production requires control of water hardness and pH value (see page 353) and a check on green bean yield probably by potassium or ash measurements (see page 307) or possibly by a relative density/soluble solids estimation on the liquor prior to concentration and drying.

The presence of coffee substitutes in roast and ground products may be detected by microscopy, or by carbohydrates analysis; this second approach is essential for liquid extracts and soluble powders (see page 320). Deterioration of coffee products during storage and distribution may be monitored subjectively, or by aroma index and headspace oxygen analyses (see page 351).

In closing, one must recognise that, despite over 100 years of exponentially accelerating investigations, not even one of the questions posed (see pages 355-9) can be answered conclusively. However, progress has been made and it would seem that further significant progress is within reach provided that the breadth of factors contributing to quality are appreciated and accommodated in future investigations. In this respect it is essential that the precise origins (botanical, geographical, agricultural, processing and storage) of the green beans are known and that the chemical and physical characteristics before and during roasting are determined using well-defined methods, and these results correlated with the organoleptic and compositional characteristics of the roast bean and beverage.

References

Alcaide, A., Devys, M., Barbier, M., Kaufmann, H.P. and Sen Gupta, A.K. (1971) 'Triterpenes and sterols of coffee oil,' *Phytochemistry*, 10, 209-10

Amorim, H.V., Basso, L.C., Crocomo, O.J. and Teixeira, A.A. (1977b) 'Polyamines in green and roasted coffee,' *Journal of Agricultural and Food Chemistry*, 25, 957-8

Amorim, H.V., Cruz, A.R., St. Angelo, A.J., Dias, R.M., Melo, M., Teixeira, A.A., Gutierrez, L.E. and Ory, R.L. (1977a) 'Biochemical, physical, and organoleptical changes during raw coffee quality deterioration,' in *8th International Colloquium on the Chemistry of Coffee*, ASIC, Paris, pp. 183-6

Amorim, H.V., Cruz, V.F., Teixeira, A.A. and Malavolta, E. (1975b) 'Chemistry of Brazilian green coffee and the quality of the beverage. V Multiple linear regression analysis', *Turrialba*, 25, 25-8

Amorim, H.V. and Josephson, R.V. (1975) 'Water-soluble protein and non-protein components of Brazilian green coffee beans', *Journal of Food Science*, 40, 1179-84

Amorim, H.V., Legendre, M.G., Amorim, V.L., St. Angelo, A.J. and Ory, R.L. (1976) 'Chemistry of Brazilian green coffee and the quality of the beverage. VII. Total carbonyls, activity of polyphenol oxidase and hydroperoxides', *Turrialba*, 26, 193-5

Amorim, H.V., Malavolta, E., Teixeira, A.A., Cruz, V.F., Melo, M., Guercio, M.A., Fossa, E., Breviglieri, O., Ferrari, S.E. and Silva, D.M. (1973) 'Relationship between some organic compounds of Brazilian green coffee with the quality of the beverage', in *6th International Colloquium on the Chemistry of Coffee*, ASIC, Paris, pp. 113-27

Amorim, H.V. and Silva, D.M. (1968) 'The relationship between the polyphenol oxidase activity of coffee beans and the quality of the beverage', *Nature, 219*, 381-2

Amorim, H.V., Teixeira, A.A., Breviglieri, O., Cruz, V.F. and Malavolta, E. (1974a) Chemistry of Brazilian green coffee and the quality of the beverage. I Carbohydrates', *Turrialba, 24*, 214-6

Amorim, H.V., Teixeira, A.A., Guercio, M.A., Cruz, V.F. and Malavolta, E. (1974b) 'Chemistry of Brazilian green coffee and the quality of the beverage. II Phenolic compounds', *Turrialba, 24*, 217-21

Amorim, H.V., Teixeira, A.A., Melo, M., Cruz, V.F. and Malavolta, E. (1974c) 'Chemistry of Brazilian green coffee and the quality of the beverage. III Soluble proteins', *Turrialba, 24*, 304-8

Amorim, H.V., Teixeira, A.A., Melo, M., Cruz, V.F. and Malavolta, E. (1975a) 'Chemistry of Brazilian green coffee and the quality of the beverage. IV. Electrophoresis of proteins in agar gel and its interaction with chlorogenic acids, *Turrialba, 25*, 18-24

Anderson, L. (1972) 'The Cyclitols' in W. Pigman and D. Horton (eds.), *The Carbohydrates, Chemistry and Biochemistry*, Academic Press, London, pp. 520-79

Angelucci, E., Yokomizo, Y., de Moraes, R.M., de Campos, R.B., Miya, E.E. and Figueireido, I.B. (1973) 'Chemical and sensory evaluation of the main Brazilian instant coffees', in *6th International Colloquium on the Chemistry of Coffee*, ASIC, Paris, pp. 178-83

Ara, V. and Thaler, H. (1976) 'Studies of coffee and coffee substitutes. XVIII. Dependence on the quality and composition of a high polymer galactomannan on the coffee species and the degree of roasting', *Zeitschrift für Lebensmittel Untersuchung und Forschung, 161*, 143-50

Ara, V. and Thaler, H. (1977) 'Studies of coffee and coffee substitutes. XIX. Dependence of the quality of a highly polymeric galactomannan on the degree of extraction of coffee extracts', *Zeitschrift für Lebensmittel Untersuchung und Forschung, 164*, 8-10

Arackal, T. and Lehmann, G. (1979) 'Measurement of the quotient 2-methylfuran/2-butanone for unground coffee during storage out of contact with air', *Chemie, Mikrobiologie, Technologie der Lebensmittel, 6*, 43-7

Ariga, T. and Asao, Y. (1981) 'Isolation, identification and organoleptic astringency of dimeric proanthrocyanidins occurring in Azuki beans', *Agricultural and Biological Chemistry, 45*, 2709-12

Arnold, R.A. and Noble, A.C. (1978) 'Bitterness and astringency of grape seed phenolics in a model wine solution', *American Journal of Enology and Viticulture, 32*, 5-13

Asante, M. and Thaler, H. (1975a) 'Investigations on coffee and coffee substitutes. XVI. Polysaccharides in extracts of a robusta coffee', *Chemie, Mikrobiologie, Technologie der Lebensmittel, 4*, 110-6

Asante, M. and Thaler, H. (1975b) 'Investigations on coffee and coffee substitutes; XVII. Behaviour of polysaccharide-complexes of robusta coffee during roasting, *Zeitschrift für Lebensmittel Untersuchung und Forschung, 159*, 93-6

Aspinall, G.O. (1981) 'Constitution of plant cell wall polysaccharides', in W. Tanner and F.A. Loewus (eds.) *Encyclopedia of Plant Physiology, New Series, Vol. 13B Extracellular carbohydrates*, Springer-Verlag, Berlin, pp. 3-8

Association of Official Analytical Chemists (1975) *Official Methods of Analysis, 12th Edn*, AOAC, Washington, p. 247

Aurich, R., Hofmann, R., Klöcking, R. and Mücke, D. (1967) 'Stoffe vom Huminasauretyp in Röstkaffee-Extrakten. II. Aminosäuregehalt von Kaffee-Huminsäuren', *Zeitschrift für Lebensmittel Untersuchung und Forschung, 135*, 59-64

Bade, H. and Stegemann, H. (1982) 'Protein patterns of coffee beans. Characterization by one- and two-dimensional electrophoresis', *Journal of Agronomy and Crop Science, 151*, 89-98

Baltes, W. (1975) 'Vorstufen und Entstehung von Farbe und Geschmack des Kaffees', in *7th International Colloquium on the Chemistry of Coffee*, ASIC, Paris, pp. 91-108

Baltes, W., Petersen, H. and Degner, C (1973) 'Determination of trigonelline and caffeine in coffee by reflectance spectrophotometry', *Zeitschrift für Lebensmittel Untersuchung und Forschung, 152*, 145-6

Barbiroli, G. (1965) 'Il contenuto in zuccheri nei caffè verdi e tostati', *Rassegna Chimica, 17*, 261-3

Barel, M., Challot, F., Hahn, D. and Vincent, J.-C. (1973) 'Analyse chromatographique des espaces de tête des fèves défectueuses', in *6th International Colloquium on the Chemistry of Coffee*, ASIC, Paris, pp. 95-101

Barnes, H.M., Feldman, J.R. and White W.V. (1950) 'Isochlorogenic Acid. Isolation from coffee and structure studies, *Journal of the American Chemical Society*, 72, 4178-82

Baumann, T.W. and Wanner, H. (1977) 'Tracer studies on biosynthesis of caffeine in *'Coffea arabica'*, in *8th International Colloquium on the Chemistry of Coffee*, ASIC, Paris, pp. 135-8

Belitz, H.D. (1975) 'Geschmacksaktive Substanzen im Kaffee', in *7th International Colloquium on the Chemistry of Coffee*, ASIC, Paris, pp. 243-52

Belitz, H.D., Chen, W., Jugel, H., Stempfl, H., Treleano, R. and Wieser, H. (1983) 'QSAR of bitter tasting compounds', *Chemistry and Industry*, *1983*, 23-6

Berndt, W. and Meier-Cabell, E. (1975), 'Eigenschaften einiger, Proteine sowie der säuren Phosphatase aus Rohkaffee', in *7th International Colloquium on the Chemistry of Coffee*, ASIC, Paris, pp. 225-32

Biggers, R.E., Hilton, J.J. and Gianturco, M.A. (1969) 'Differentiation between *Coffea arabica*, and *Coffea robusta* by computer revaluation of gas chromatographic profiles, *Journal of Chromatographic Science*, 7, 453-72

Bjarnason, J. and Carpenter, K.J. (1970) 'Mechanism of heat damage in proteins. 2. Chemical changes in pure proteins', *British Journal of Nutrition*, 24, 313-29

Blanc, M. (1977) 'Les acides carboxyliques du café; mise au point et résultats de différentes déterminations', in *8th International Colloquium on the Chemistry of Coffee*, ASIC, Paris, pp. 73-8

Blanchfield, D. (1980) 'Philosophy of Food Control', in P.O. Dennis, J.R. Blanchfield and A.G. Ward (eds.), *Food Control in Action*, Applied Science, London, pp. 1-14

Blauch, J.L. and Tarka, S.M. (1983) 'HPLC determination of caffeine and theobromine in coffee, tea and instant hot cocoa mixtures', *Journal of Food Science*, 48, 745-7,750

British Standards Institution (1979) 'Methods of test for coffee and coffee products. Part 1. Green coffee: Determination of moisture content' (Basic Reference Method), ISO, 1446, 1978

British Standards Institution (1980) 'Methods of test for coffee and coffee products. Part 4. Green coffee: Olfactory and visual examination and determination of foreign matter and defects', ISO, 4149, 1980

British Standards Institution (1983) 'Methods of test for coffee and coffee products. Part 6. Instant Coffee: Determination of loss in mass at 70°C under reduced pressure', ISO, 3726, 1983

Bunker, M.L. and McWilliams, M. (1979) 'Caffeine content of common beverages', *Journal of the American Dietetic Association*, 74, 28-31

Burg, A.W. (1975) 'Effects of caffeine on the human system', *Tea and Coffee Trade Journal*, *147* (1), 42-48, 88

Campos, L.S. and Rodrigues, J.M.L. (1971) 'Aplicação dos métodas cromatográficos à análise dos aminoácidos livres dos cafés verdes', in *5th International Colloquium on the Chemistry of Coffee*, ASIC, Paris, pp. 91-6

Caplan, R.H. (1978) *A practical approach to Quality Control* 3rd Edn, Business Books Ltd, London

Carelli, M.L.C., Lopes, C.R. and Monaco, L.C. (1974) 'Chlorogenic acid content in species of *Coffea* and selections of *C. arabica*', *Turrialba*, 24, 398-401

Carisano, A. and Garibaldi, L. (1964) 'Gas chromatographic examination of the fatty acids of coffee oil', *Journal of the Science of Food and Agriculture*, 15, 619-22

Centi-Grossi, M., Tassi-Micco, C. and Silano, V. (1969) 'Albumin-fractionation of green coffee seed varieties by acrylamide gel-electrophoresis, *Phytochemistry*, 8, 1749-51

Charrier, A. (1975) 'Variation de la teneur en caféine chez les caféiers', in *7th International Colloquium on the Chemistry of Coffee*, ASIC, Paris, pp. 295-302

Chassevent, F. and Dalger, G. (1971) 'Détermination de la teneur en eau des extraits de cafés en poudre soluble par spectrophotométrie dans le proche, infrarouge, la méthode de Karl Fischer et les méthodes du étuvage', in *5th International Colloquium on the Chemistry of Coffee*, ASIC, Paris, pp. 162-74

Chassevent, F., Dalger, G., Gerwig, S. and Vincent, J-C. (1973a) 'Contribution à l'étude des

Mascarocoffea', in *6th International Colloquium on the Chemistry of Coffee*, ASIC, Paris, pp. 147-54

Chassevent, F., Gerwig S. and Bouharmont, M. (1973b) 'Influence éventuelle de diverses fumures sur les teneurs en acides chlorogéniques et en cafeine de grains de caféiers cultivés', in *6th International Colloquium on the Chemistry of Coffee*, ASIC, Paris, pp. 57-60

Chassevent, F., Vincent, J.-C., Hahn, D., Pougneaud, S. and Wilbaux, R. (1969) 'Etude de relations éventuelles gustatives ou chimiques en function de la préparation du café Robusta au stade primaire', in *4th International Colloquium on the Chemistry of Coffee*, ASIC, Paris, pp. 179-85

Citroreksoko, P.S., Petermann, J., Wanner, H. and Baumann, T.W. (1977) 'Detection of trace amounts of methylated uric acids in crude caffeine from different sources', in *8th International Colloquium on the Chemistry of Coffee*, ASIC, Paris, pp. 143-6

Clarke, R.J. (1965) 'Roasted coffee grinding techniques', *Food Processing and Marketing*, 26, 9-14

Clarke, R.J. (1979) 'International coffee standardisation and legislation', *Food Chemistry*, 4, 81-96

Clarke, R.J. and Walker, L.J. (1974) 'Potassium and other mineral contents of green, roasted and instant coffees', *Journal of the Science of Food and Agriculture*, 25, 1389-404

Clarke, R.J. and Walker, L.J. (1975) 'The inter-relationship of potassium contents of green, roasted and instant coffees', in *7th International Colloquium on the Chemistry of Coffee*, ASIC, Paris, pp. 159-64

Clifford, M.N. (1972) *The Phenolic Compounds of Green and Roasted coffee Beans*, PhD Thesis, University of Strathclyde, Glasgow

Clifford, M.N. (1974) 'Specificity of acidic phloroglucinol reagents', *Journal of Chromatography*, 94, 321-4

Clifford, M.N. and Griffiths, T.I. (1982) 'The analysis of sugars in immature green coffee beans'. Unpublished data.

Clifford, M.N. and Ohiokpehai, O. (1983) 'Coffee astringency', *Analytical Proceedings*, 20, 83-6

Clifford, M.N. and Staniforth, P.S. (1977) 'A critical comparison of six spectrophotometric methods for measuring chlorogenic acids in green coffee beans', in *8th International Colloquium on the Chemistry of Coffee*, ASIC, Paris, pp. 109-14

Clifford, M.N. and Wight, J. (1976) 'The measurement of feruloylquinic acids and caffeoylquinic acids in coffee beans. Development of the technique and its preliminary application to green coffee beans', *Journal of the Science of Food and Agriculture*, 27, 73-84

Corse, J. (1953) 'A new isomer of chlorogenic acid from peaches', *Nature*, 172, 771-2

Corse, J., Layton, L.L. and Patterson, D.C. (1970) 'Isolation of chlorogenic acids from roasted coffee', *Journal of the Science of Food and Agriculture*, 21, 164-8

Corse, J. and Lundin, R.E. (1970) 'Diastereomers of quinic acid, chemical and NMR studies', *Journal of Organic Chemistry*, 35, 1904-9

Corse, J., Lundin, R.E. and Waiss, A.C. (1965) 'Identification of several components of isochlorogenic acid', *Phytochemistry*, 4, 527-9

Courtois, J.E. (1979) 'Les galactomannanes des legumineuses dans l'alimentation', *Annales de la Nutrition et l'Alimentation*, 33, 189-98

Courtois, J.E. and Le Dizet, P. (1963) 'Recherches sur les galactomannanes. IV. Action de quelques preparations enzymatiques sur les galactomannanes de Trèfle et Gleditschia', *Bulletin de la Société de Chimie Biologique*, 45,731-41

Courtois, J.E., Percheron, F. and Glomaud, J.C. (1963) 'Recherches préliminaires sur les oligo et polysaccharides du café vert (*Coffea canephora* var. *robusta*)', in *1st International Colloquium on the Chemistry of Coffee*, ASIC, Paris, pp. 231-6

Cros, E., Guyot, B. and Vincent, J.-C (1979) 'Profil chromatographique de la fraction volatile du café, *Cafe, Cacao, Thé*, 23, 193-201

de Rostolan, J. (1971) 'Composition et caractéristiques chimiques de *Coffea* sauvages de Madagascar. VI. Recherche et dosage de la cafamarine et d'autres substances voisines dans quelques caféiers sauvages ou cultivés, in *5th International Colloquium on the Chemistry of Coffee*, ASIC, Paris, pp. 149-53

Dalla Rosa, M., Lerici, C.R., Riva, M. and Fini, P. (1980) 'Coffee processing: chemical, physical and technological aspects. V. Evaluation of some physical characteristics of coffee during heat treatment at constant temperature, *Industrie delle Bevande*, No. 50, 466-72

Dea, I.C.M. and Morrison, A. (1975) 'Chemistry and interactions of seed galactomannans', *Advances in Carbohydrate Chemistry and Biochemistry*, 31, 241-312

Derbesy, M., Venot, C., Santos, A.-C., Benfredj, A. and Busson, F. (1969) 'Cafés verts de l'Angola', *Qualitas Plantarum et Materiae Vegetablis*, 18, 367-75

Diaz, J.S. Gomez, G.S., Felsner, G. and Fritsch, G. (1973) 'Estudio de la señal ancha resonancia magnética nuclear (Resonancia de Protones) en muestras de café colombiano preparado en diferentes condiciones', in *6th International Colloquium on the Chemistry of Coffee*, ASIC, Paris, pp. 172-7

EEC Directive 77/436 (1977) 'Council Directive of 27th June, 1977 on the approximation of the laws of the Member States relating to coffee extracts and chicory extracts', *Official Journal of the European Communities*, 20

Eskin, N.A.M. (1979) *Plant Pigments, Flavors and Textures: The Chemistry and Biochemistry of Selected Compounds*, Academic Press, New York, pp. 176-83

Farr-Jones, S.L. and Clifford, M.N. (1983) Unpublished data

Feldman, J.R. Ryder, W.S. and Kung, J.T. (1969) 'The importance of non-volatile compounds to the flavour of coffee', *Journal of Agricultural and Food Chemistry*, 17, 733-9

Fennema, O.R. (1976) 'Water and Ice', in O.R. Fennema (ed.), *Principles of Food Science, Part 1, Food Chemistry*, Marcel Dekker Inc., New York, p. 32

Ferreira, L.A.B., Vilar, H., Fragoso, M.A.C., Aguiar, M.C., Cruz, M.J.R. and Goncalves, M.M. (1971a) 'Subsidos para a caracterizaçao do grao de café do 'hibrido de Timor', in *5th International Colloquium on the Chemistry of Coffee*, ASIC, Paris, pp. 128-47

Ferreira, L.A.B., Oliveira, E.F., Vilar, H. and Aguiar, M.C. (1971b) 'Contribuições para a identificação da genuidade do café, in *5th International Colloquium on the Chemistry of Coffee*, ASIC, Paris, pp. 79-84

Ferreira, L.A.B., Fragoso, M.A.C., Peralta, M.F., Silva M.C.C. and Rebelo, M.C. (1971c) 'Constituintes minerais dos cafés de Angola', in *5th International Colloquium on the Chemistry of Coffee*, ASIC, Paris, pp. 51-62

Fobé, L.A., Nery, J.P. and Tango, J.S. (1967) 'Influence of the roasting degree on the chemical composition of coffee', in *3rd International Colloquium on the Chemistry of Coffee*, ASIC, Paris, pp. 389-97

Folstar, P. (1976) *The Composition of Wax and Oil in Green Coffee Beans*, Centre for Agricultural Publishing and Documentation, Wageningen

Folstar, P., Pilnik, W., de Heus, J.G. and van der Plas, H.C. (1975a) 'On the analysis of oil in green coffee beans', *Mitteilungen aus dem Gebiete der Lebensmittel Untersuchung und Hygiene*, 66, 502-6

Folstar, P., Pilnik, W., de Heus, J.G. and van der Plas, H.C. (1975b) 'The composition of fatty acids in coffee oil and wax', *Lebensmittel Wissenschaft und Technologie*, 8, 286-8

Folstar, P., van der Plas, H.C., Pilnik, W. and de Heus, J.G. (1977) 'Tocopherols in the unsaponifiable matter of coffee bean oil', *Journal of Agricultural and Food Chemistry*, 25, 283-5

Folstar, P., van der Plas, H.C., Pilnik, W. Schols, H.A. and Melger, P. (1979) 'Liquid chromatographic analysis of Nβ-alkanoyl-hydroxy-tryptamide (C-5-HT) in green coffee beans', *Journal of Agricultural and Food Chemistry*, 27, 12-15

Folstar, P., Schols, H.A., van der Plas, H.C., Pilnik, W., Landheer, C.A. and van Veldhuizen, A. (1980) 'New tryptamine derivatives isolated from wax of green coffee beans', *Journal of Agricultural and Food Chemistry*, 28, 872-4

Forsyth, W.G.C. and Quesnel, V.C. (1957) 'Intermediates in the enzymic oxidation of catechol', *Biochimica et Biophysica Acta*, 25, 155-60

Forsyth, W.G.C., Quesnel, V.C. and Roberts, J.B. (1960) 'Diphenylenedioxide-2, 3-quinone. An intermediate in the enzymic oxidation of catechol', *Biochimica et Biophysica Acta*, 37, 322-6

Fragoso, M.A.C., Ferreira, L.A.B. and Peralta, M.F. (1971) 'Metais pesadas em cafés, in *5th International Colloquium on the Chemistry of Coffee*, ASIC, Paris, pp. 70-8

Gabriel-Jurgens, H.O. (1975) 'Zur Bestimmung der Feuchtigkeit in Rohkaffee. Ergebnisse aus DNA-Ringversuchen', in *7th International Colloquium on the Chemistry of Coffee*, ASIC, Paris, pp. 279-86

Gal, S., Windermann, P. and Baumgartner, E. (1976) 'Untersuchungen über die Vorgänge beim Dämpfen von Kaffeebohnen', *Chimia, 30*, 68-71

Garrick, L.S. and Habermann, H.M. (1962) 'Distribution of allagochrome in vascular plants', *American Journal of Botany, 49*, 1078-88

Gautschi, F., Winter, M., Flament, I., Willhalm, B. and Stoll, M. (1967b) 'New developments coffee aroma, a survey of coffee aroma', in *3rd International Colloquium on the Chemistry of Coffee*, ASIC, Paris, pp. 67-76

Cautschi, F., Winter, M., Flament, I., Willhalm, B., and Stoll, M. (1967b) 'New developments in coffee aroma research', *Journal of Agricultural and Food Chemistry, 15*, 15-23

Gianturco, M.A. and Giammarino, A.S. (1965) 'Considerations on the study of the aromatic constituents of roasted coffee', in *2nd International Colloquium on the Chemistry of Coffee*, ASIC, Paris, pp. 169-82

Gibson, A. (1971) 'Photochemical aspects of drying East African arabica coffees', in *5th International Colloquium on the Chemistry of Coffee*, ASIC, Paris, pp. 246-58

Gibson, A. (1973a) 'East African mild arabica coffee. Quality characteristics associated with green coffee volatiles. Part I, Dimethyl sulphide', in *6th International Colloquium on the Chemistry of Coffee*, ASIC, Paris, pp. 319-24

Gibson, A. (1973b) 'East African mild Arabica coffee. Quality characteristics associated with green coffee volatiles. Part II, Solai flavour', in *6th International Colloquium on the Chemistry of Coffee*, ASIC, Paris, pp. 325-31

Gibson, A. and Butty, M. (1975) 'Overfermented coffee beans ('stinkers'). A method for their detection and elimination', in *7th International Colloquium on the Chemistry of Coffee*, ASIC, Paris, pp. 141-52

Glomaud, J.C., Percheron, F. and Courtois, J.E. (1965) 'Teneurs comparées en oligosaccharides de quelques variétés de café vert. Etude préliminaire des polysaccharides extractibles par l'eau, in *2nd International Colloquium on the Chemistry of Coffee*, ASIC, Paris, pp. 39-43

Griffin, B. and Stonier, T. (1975) 'Studies on auxin protectors. XII. Auxin protectors and polyphenol oxidase in developing coffee fruit', *Physiologia Plantarum, 33*, 157-60

Gross, G.G. (1981) 'Phenolic acids', in E.E. Conn (ed.), *The Biochemistry of Plants*, Vol. 7, Academic Press, London, pp. 301-16

Guilbot, A. and d'Ornano, M. (1964) 'Méthode de référence fondamentale et méthodes pratiques de determination de la teneur en eau du café vert', *Café, Cacao, Thé, 8*, 293-300

Gutmann, W., Werkhoff, P., Barthels, M. and Vitzthum, O.G. (1977) 'Vergleich der Headspace. Aromaprofile von Arabusta-Kaffee mit Arabica — und Robusta-Sorten', in *8th International Colloquium on the Chemistry of Coffee*, ASIC, Paris, pp. 153-62

Guyot, B., Cros, E. and Vincent, J.-C. (1982) 'Caractérisation et identification, des composés de la fraction volatile d'un café vert Arabica sain et d'un café vert Arabica puant', in *10th International Colloquium on the Chemistry of Coffee*, ASIC, Paris, pp. 253-70

Habermann, H.M. (1973) 'Distribution patterns of allagochrome, chlorogenic acid and capacity for secondary synthesis of allagochrome', *Botanical Gazette, 134*, 221-32

Haevecker, U. (1969) 'Wasserbestimmung in Kaffee-Extrakt Resultate eines Ringuersuchs mit verschiedenen Methoden', in *4th International Colloquium on the Chemistry of Coffee*, ASIC, Paris, pp. 160-5

Hamonnière, M., Ducrouix, A., Pascard-Billy, C. and Poisson, J. (1975) 'Structure du mascaroside, principe amer de *Coffea vianneyi*', in *7th International Colloquium on the Chemistry of Coffee*, ASIC, Paris, pp. 201-4

Hanson, K.R. (1965) 'Chlorogenic acid biosynthesis. Chemical synthesis and properties of the mono-o-cinnamoylquinic acids', *Biochemistry, 4*, 2719-35

Haslam, E., Makinson, G.K., Naumann, M.O. and Cunningham, J. (1964) 'Synthesis and properties of some hydroxycinnamoyl esters of quinic acid, *Journal of the Chemical Society, 1964*, 2137-46

Hassan, M.U. (1970) *The Precursors of Coffee Aroma*, PhD Thesis, University of Strathclyde, Glasgow

Hauschild, W. (1935) 'Untersuchung über die Bestandteile das Mate', *Mitteilungen aus dem Gebiete der Lebensmittel Untersuchung und Hygiene, 26*, 329-50

Hausermann, M. and Brandenberger, H. (1961) 'Uber phenolische Pflanzeninhaltsstoffee. IV. Spektrophotometrische Bestimmungsmethode für Chlorogensäuren und ihre Anwendung auf pflanzliche Extrakte', *Zeitschrift für Lebensmittel Untersuchung and Forschung, 115,* 516-27
Hayakawa, K.-I., Matas, J. and Hwang, M.P. (1978) 'Moisture sorption isotherms of coffee products', *Journal of Food Science, 43,* 1026-7
Heiss, R., Radtke, R. and Robinson, L. (1977) 'Packaging and marketing of roasted coffee', in *8th International Colloquium on the Chemistry of Coffee,* ASIC, Paris, pp. 163-74
Hermann, K. (1967) 'Uber Hydroxy-zimtsäuren und ihre Bedeutung in Lebensmittein', *Zeitschrift für Lebensmittel Untersuchung und Forschung, 133,* 158-78
Horman, I. and Viani, R. (1971) 'The caffeine-chlorogenate complex of coffee. An NMR study', in *5th International Colloquium on the Chemistry of Coffee,* ASIC, Paris, pp. 102-11
Hughes, E.B. and Smith, R.F. (1949) 'The volatile constituents of roasted coffee', *Journal of the Society of Chemistry and Industry, 68,* 322-7
Hunziker, H.R. and Miserez, A. (1979) 'Bestimmung der 5-Hydroxytryptamide in Kaffee mittels Hochdruck-Flüssigkeitschromatographie', *Mitteilungen aus dem Gebiete der Lebensmittel Untersuchung und Hygiene, 70,* 142-52
International Standard No. 1447 (1978) 'Green coffee. Determination of moisture content (Routine method)'
International Standard No. 3726 (1983) 'Instant coffee. Determination of loss in mass at 70°C under reduced pressure'
International Standard No. 6673 (1983) 'Green coffee. Determination of loss in mass at 105°C'
IUPAC (1976) 'Nomenclature of cyclitols', *Biochemical Journal, 153,* 23-31
Janicek, G. and Pokorny, J. (1970) 'Veränderungen der Kaffeelipide während der Lagerung von Kaffeebohnen', *Zeitschrift für Lebensmittel Untersuchung und Forschung 144,* 189-91
Kaufmann, H.P. and Hamsagar, R.S. (1962) 'Zur Kenntnis der Lipoide der Kaffeebohne. I. Uber Fettsäure-Ester der Cafestols', *Fette-Seifen-Anstrichmittel, 64,* 206-13
Kazi, T. (1979) 'Coffee mixtures — Spectrophtometric method for the determination of coffee content', *Food Chemistry, 4,* 73-80
Kent, N.L. (1966) *Technology of Cereals,* Pergamon Press, Oxford, pp. 41-2
Klöcking, R., Hofmann, R. and Mücke, D. (1971) 'Stoffe vom Huminsäuretyp in Röstkaffee-Extrakten. III. Nachweis von Phenolkörpern in Kaffeehuminsäure-Hydrolysaten', *Zeitschrift für Lebensmittel Untersuchung und Forschung, 146,* 79-83
König, W.A., Rahn, W. and Vetter, R. (1980) 'Identifizierung und Quantifizierung emetisch wirksamer Bestandteile in Röstkaffee', in *9th International Colloquium on the Chemistry of Coffee,* ASIC, Paris, pp. 145-51
König, W.A. and Stürm, R. (1982) 'Gas chromatography and mass spectrometry as an aid for the investigation of high boiling coffee constituents', in *10th International Colloquium on the Chemistry of Coffee,* ASIC, Paris, pp. 271-8
Kröplien, U. (1971) 'Monosaccharides in coffees and coffee substitutes', in *5th International Colloquium on the Chemistry of Coffee,* ASIC, Paris, pp. 217-23
Kulaba, G.W. (1978) *Chemical Aspects of The Quality of Kenya Mild Arabica Coffee,* MSc Thesis, University of Nairobi, Kenya
Kulaba, G.W. (1981) Personal Communication
Kulaba, G.W. and Robins, P.A. (1981) 'Kahweol content and quality in Kenya mild arabica coffee', *Kenya Journal of Science and Technology (A), 2,* 63-72
Kung, J.T., McNaught, R.P. and Yeransian, J.A. (1967) 'Determining volatile acids in coffee beverages by NMR and gas chromatography', *Journal of Food Science, 32,* 455-8
Kwasny, H. (1975) 'Zur Bestimmung von Chlorogensäuren in Kaffee-Extrakt Ergebrisse aus DNA-Ringversuchen, in *7th International Colloquium on the Chemistry of Coffee,* ASIC, Paris, pp. 303-10
Kwasny, H. (1978) 'Die Bestimmung des Röstgrades von Bohnenkaffee im daraus hergestellten Kaffee-Extrakt', *Lebensmittelchemie und Gerichtliche Chemie, 32,* 36-8
Kwasny, H. and Werkhoff, P. (1979) 'Abhängigkeit der Aromaindex-Kennzahl M/B von verschiedenen Parametem', *Chemie, Mikrobiologie, Technologie der Lebensmittel, 6* (1), 31-2

Lam, L.K.T., Sparnins, V.L. and Wattenberg, L.W. (1982) 'Isolation and identification of kahweol palmitate and cafestol palmitate as active constituents of green coffee beans that enhance glutathione S-transferase activity in the mouse', *Cancer Research*, 42, 1193-8

Lea, A.G.H. and Arnold, G.M. (1978) 'The phenolics of ciders: Bitterness and astringency', *Journal of the Science of Food and Agriculture*, 29, 478-83

Lehmann, G. and Hahn, H.G. (1967) 'Uber die Bestimmung der Chlorogensäure und des Trigonellins', in *3rd International Colloquium on the Chemistry of Coffee*, ASIC, Paris, pp. 115-20

Lentner, C. and Deatherage, F.E. (1959) 'Organic acids in coffee in relation to the degree of roast', *Food Research*, 24, 483-92

Lerici, C.R., Dalla-Rosa, M., Magnanini, E. and Fini, P. (1980b) 'Coffee processing: chemical, physical and technological aspects. IV. Evolution of some physical characteristics during the roasting process', *Industrie delle Bevande*, 9, 375-81

Lerici, C.R., Lercker, G., Minguzzi, A. and Matassa, P. (1980a) 'Coffee processing: chemical, physical and technological aspects. III. Roasting effects on chemical composition of some coffee blends', *Industrie delle Bevande*, 9, 232-8

Liberato, D.J. Byers, V.S. Dennick, R.G. and Castagnoli, N. (1981) 'Regiospecific attack of nitrogen and sulphur nucleophiles on quinones derived from poison oak/ivy catechols (Urushiols) and analogues as models for urushiol-protein conjugate formation', *Journal of Medicinal Chemistry*, 24, 28-33

Lopes, M.H. (1971) 'Teor em cafeína de cafés espontâneos de Moçambique', in *5th International Colloquium on the Chemistry of Coffee*, ASIC, Paris, pp. 63-9

Lopes, M.H. (1974) 'Chemical and technological characteristics of Mozambique Racemosa coffees', *Cafe, Cacao, Thé*, 18, 263-76

Lopes, F., Santos, A.-C., Matos, N., Martins, L., Teixeira, A.A. and Mexia, J.T. (1971) 'Determinação do teor em água do café verde por métodos expeditos', in *5th International Colloquium on the Chemistry of Coffee*, ASIC, Paris, pp. 154-61

Ludwig, H., Obermann, H. and Spiteller, G. (1974) 'Atractyligenin — ein wesentlicher Bestandteil gerösteter Kaffeebohnen', *Chemische Berichte*, 107, 2409-11

Mabrouk, R.F. and Deatherage, F.E. (1956) 'Organic acids in brewed coffee', *Food Technology*, 10, 194-7

Maga, J.A. (1978) 'Simple phenol and phenolic compounds in food flavour', *CRC Critical Reviews in Food Science and Nutrition*, 10, 323-72

Maier, H.G. (1975) 'Bindung von Aromastoffen an die Matrix von Pulverkaffee', in *7th International Colloquium on the Chemistry of Coffee*, ASIC, Paris, pp. 211-20

Maier, H.G. (1980) 'Zur Bestimmung der Extraktionsausbeute von Kaffee-Extrakt', in *9th International Colloquium on the Chemistry of Coffee*, ASIC, Paris, pp. 113-24

Maier, H.G. (1981) *Kaffee*, Paul Parey, Berlin and Hamburg

Maier, H.G. and Buttle, H. (1973) 'Zur Isolierung und Charakterisierung der braunen Kaffeeröststoffe II', *Zeitschrift für Lebensmittel Untersuchung und Forschung*, 150, 331-4

Maier, H.G., Diemair, W. and Ganssmann, J. (1968) 'Zur Isolierung und Charakterisierung der braunen Kaffeeröststoffe', *Zeitschrift für Lebensmittel Untersuchung und Forschung*, 137, 282-92

Maier, H.G. and Grimsehl, A. (1982a) 'Die Säuren des Kaffees II. Chlorogensäuren im Rohkaffee', *Kaffee und Tee Markt*, 32, 3-5

Maier, H.G. and Grimsehl, A. (1982b) 'Die Säuren des Kaffees III. Chlorogensäuren im Röstkaffee', *Kaffee und Tee Markt*, 32, 3-6

Maier, H.G. and Krause, H.G. (1977) 'Zur Bindung flüchtiger Aromastoffe and Pulverkaffee I. Bindung kleiner Mengen', *Kaffee und Tee Markt*, 27, 3-6

Maier, H.G. and Mätzel, U. (1982) 'Atractyligenin und seine Glykoside im Kaffee', in *10th International Colloquium on the Chemistry of Coffee*, ASIC, Paris, pp. 247-52

Maier, H.G. and Wewetzer, H. (1978) 'Bestimmung von Diterpen-Glykosiden im Bohnenkaffee', *Zeitschrift für Lebensmittel Untersuchung und Forschung*, 167, 105-7

Martino, V.S., Debenedetti, S.L. and Coussio, J.D. (1979) 'Caffeoylquinic acids from *Pterocaulon virgatum* and *Pluchea sagittalis*', *Phytochemistry*, 18, 2052

Meier, H. and Reid, J.S.G. (1982) 'Reserve polysaccharides other than starch in higher plants', in F.A. Loewus and W. Tanner (eds.), *Encylopedia of Plant Physiology, New Series*, Vol. 13A. *Intracellular carbohydrates*, Springer Verlag, Berlin, pp. 418-71

Melo, M., Fazuoli, L.C. Teixeira, A.A. and Amorim. H.V. (1980) 'Chemical, physical and organoleptic alterations on storage of coffee beans', *Ciencia e Cultura* (Sao Paulo), *32*, 468-71

Menchu, J.F. and Ibarra, E. (1967) 'The chemical composition and the quality of Guatemalan coffee', in *3rd International Colloquium on the Chemistry of Coffee*, ASIC, Paris, pp. 146-54

Merrit, M.C. and Proctor, B.E. (1959) 'Effect of temperature during roasting cycle on selected components of whole bean coffee', *Food Research, 24*, 672-80

Merritt, C., Robertson, D.H. and McAdoo, D. (1969) 'The relationship of volatile compounds in the roasted coffee bean to their precursors', in *4th International Colloquium on the Chemistry of Coffee*, ASIC, Paris, pp. 144-8

Mesnard, P. and Devaux, G. (1964) 'Nouvelle methode de dosage colorimetrique de l'acid quinique et de ses esters naturels', *Bulletin de la Société Chimique de France*. Part 1, 43-7

Moll, H.R. and Pictet, G.A. (1980) 'La chromatographie liquide haute performance appliquée à certains constituants specifiques du café, in *9th International Colloquium on the Chemistry of Coffee*, ASIC, Paris, pp. 87-98

Möller, B. and Herrmann, K. (1982) 'Analysis of quinic acid esters of hydroxycinnamic acids in plant material by capillary gas chromatography and high-performance liquid chromatography', *Journal of Chromatography, 241*, 371-9

Möller, B. and Herrmann, K. (1983) 'Quinic acid esters of hydroxycinnamic acids in stone and pome fruit', *Phytochemistry, 22*, 477-81

Moores, R.G., McDermott, D.L. and Wood, T.R. (1948) 'Determination of chlorogenic acid in coffee', *Analytical Chemistry, 20*, 620-4

Motai, H. (1976) 'Viscosity of melanoidins formed by oxidative browning, *Agricultural and Biological Chemistry', 40*, 1-7

Nagasampagi, B.A., Rowe, J.W., Simpson, R. and Goad, L.J. (1971) 'Sterols of coffee', *Phytochemistry, 10*, 1101-7

Nakabayashi, T. (1978) 'Changes of organic acids and pH roast of coffee', *Journal of the Japanese Society of Food Science and Technology, 25*, 142-6

Nakabayashi, T. and Kojima, Y. (1980) 'Changes in the quinic acid contents of coffee beans during the roasting process', *Nippon Shokuhin Kogyo Gakkaishi, 27*, 108-11 cited in *Chemical Abstracts*, 1980, *93*, 44318 j

Natarajan, C.P., Balachandran, A. and Shivashankar, S. (1969) 'Triage components in coffee', in *4th International Colloquium on the Chemistry of Coffee*, ASIC, Paris, pp. 92-6

Ndjouenkeu, R., Clo, G. and Voilley, A. (1981) 'Effect of coffee-brew extraction parameters on the concentration of methylxanthines measured by HPLC', *Sciences des Aliments, 1*, 365-75

Nichiforesco, E. (1979) 'Sur la composition des dérivés cafeylquiniques des feuilles d'artichaut (*Cynara scolymus*)', *Plantes medicinales et phytotherapie, 4*, 56-62

Northmore, J.M. (1967) 'Raw bean colours and the quality of Kenya arabica coffee', in *3rd International Colloquium on the Chemistry of Coffee*, ASIC, Paris, pp. 405-14

Northmore, J.M. (1969) '"Over-fermented" beans and "stinkers" as defectives of arabica coffee', in *4th International Colloquium on the Chemistry of Coffee*, ASIC, Paris, pp. 47-54

Nurok, D., Anderson, J.W. and Zlatkis, A. (1978) 'Profiles of sulphur-containing compounds obtained from arabica and robusta coffees by capillary column gas chromatography', *Chromatographia, 11*, 188-92

Nursten, H.E. (1981) 'Recent development in studies of the Maillard Reaction', *Food Chemistry, 6*, 263-77

Obermann, H. and Spiteller, G. (1975) '16,17-Dihydroxy-9(11)-kauren-18-säure-ein Bestandteil des Röstkaffees', *Chemische Berichte, 108*, 1093-100

Ohiokpehai, O. (1982) *Chlorogenic acid content of green coffee beans*, PhD Thesis, University of Surrey

Ohiokpehai, O., Brumen, G. and Clifford, M.N. (1982) 'The chlorogenic acids content of some peculiar green coffee beans and the implications for beverage quality', in *10th International Colloquium on the Chemistry of Coffee*, ASIC, Paris, pp. 177-86

Oliveira, J.C., Amorim, H.V., Silva, D.M. and Teixeira, A.A. (1976) 'Polyphenol oxidase enzymatic activity of four species of coffee beans during storage', *Cientifica*, 4, 114-9

Oliveira, J.C., Silva, D.M. Teixeira, A.A. and Amorim, H.V. (1979a) 'Effects of the application of insecticides to control coffee and tea borers on the polyphenol oxidase activity and the beverage quality of coffee', *Cientifica*, 7, 221-4

Oliveira, J.C., Amorim, H.V., Silva, D.M. and Teixeira, A.A. (1979b) 'Effects of the origin, pulping types and storage of coffee on the polyphenol oxidase activity and beverage quality', *Cientifica*, 79-84

Olsson, K., Pernemalm, P., Popoff, T. and Theander, O. (1977) 'Formation of aromatic compounds from carbohydrates. V. Reaction of D-glucose and methylamine in slightly acidic, aqueous solution', *Acta Chemica Scandinavica B*, 31, 469-74

O'Reilly, R. (1982) *The nature of the chemical groupings responsible for the colour of the products of the Maillard Reaction.* PhD Thesis, Department of Food Science, University of Reading

Paardekooper, E.J.C., Driesen, J. and Cornelissen, J. (1969) 'Some simple analytical methods in coffee processing', in *4th International Colloquium on the Chemistry of Coffee*, ASIC, Paris, pp. 131-9

Pangborn, R.M. (1982) 'Influence of water composition, extraction procedures and holding time and temperature on quality of coffee beverages', *Lebensmittel Wissenschaft und Technologie*, 15, 161-8

Pangborn, R.M., Gibbs, Z.M. and Tassan, C. (1978) 'Effect of hydrocolloids on apparent viscosity and sensory properties of selected beverages', *Journal of Texture Studies*, 9, 415-36

Panizzi, L. and Scarpati, M.L. (1954) 'Isolamento e costituzione del principo attivo del carciofo', *Gazzetta Chimica Italiana*, 84, 792-805

Panizzi, L. and Scarpati, M.L. (1965) 'Sugli acidi 1,4-ed 1,5-dicaffeilchinici', *Gazzetta Chimica Italiana*, 95, 71-82

Paulet, P. and Mialoundama, F. (1976) 'Changes in content of chlorogenic and isochlorogenic acids during vernalisation of the roots of *Cichorium intybus* in connection with their ability to bloom *in vitro*', *Soviet Plant Physiology*, 23, 466-9

Payen, A. (1846) Cited in Reichstein, T. and Staudinger, H. (1955). 'The aroma of coffee', *Perfumery and Essential Oil Record*, 46, 86-8

Payne, R.C. Oliveira, A.R. and Fairbrothers, D.E. (1973) 'Disc electrophoretic investigation of *Coffea arabica* and *coffea canephora*. General protein and malate dehydrogenase of mature seeds', *Biochemical Systematics*, 1, 59-61

Pazola, Z. and Cieslak, J. (1979) 'Changes in carbohydrates during the production of coffee substitute extracts, especially in the roasting process', *Food Chemistry*, 4, 41-52

Pekkorinen, L. and Porka, E. (1963) 'Effect of the degree of roasting of coffee on the absorbance of the extract and total solids therein', *Zeitschrift für Lebensmittel Untersuchung und Forschung*, 120, 20-5

Pereira, A. and Pereira, M.M. (1971) 'Acidos aminados de cafés', in *5th International Colloquium on the Chemistry of Coffee*, ASIC, Paris, pp. 85-96

Pfrunder, R., Wanner, H., Frischknecht, P.M. and Baumann, T.W. (1980) 'An attempt to localise caffeine in the cell by its washout kinetics', in *9th International Colloquium on the Chemistry of Coffee*, ASIC, Paris, pp. 169-76

Pictet, G.A. (1975) 'Les hydrates de carbone du café instantané — un procède nouveau pour leur fractionnement et leur évaluation quantitative', in *7th International Colloquium on the Chemistry of Coffee*, ASIC, Paris, pp. 189-200

Pictet, G. and Moreau, A. (1969) 'Les glucides du café vert, leur solubilisation à l'eau et leur évaluation quantitative', in *4th International Colloquium on the Chemistry of Coffee*, ASIC, Paris, pp. 75-84

Pictet, G. and Rehacek, J. (1982) 'Contrôles analytiques du degré de torrefaction', in *10th International Colloquium on the Chemistry of Coffee*, ASIC, Paris, pp. 219-34

Pictet, G. and Vuataz, L. (1977) 'Etude des techniques d'infusion', in *8th International Colloquium on the Chemistry of Coffee*, ASIC, Paris, pp. 261-70

Pierpoint, W.S. (1982) 'A class of blue quinone-protein coupling products: The allagochromes?', *Phytochemistry*, 21, 91-5

Poisson, J. (1977) 'Aspects chimiques et biologiques de la composition du café vert', in *8th International Colloquium on the Chemistry of Coffee*, ASIC, Paris, pp. 33-58

Pokorny, J., Con, N.-H., Bulantova, H. and Janicek, G. (1974) 'Veränderungen von Aminosäuren und reduzierenden Zuckern während des Röstens von Kaffee', *Nahrung*, *18*, 799-805
Pokorny, J., Con, N.-H. and Janicek, G. (1972) 'Determination of chlorogenic acids in green coffee', *Scientific Papers of the Institute of Chemical Technology*, Prague E33.
Pokorny, J., Con, N.-H., Smidrkalova, E. and Janicek, G. (1975) 'Non-enzymic browning XII. Maillard reactions in green coffee beans on storage', *Zeitschrift für Lebensmittel Untersuchung und Forschung*, *158*, 87-92
Pypker, J. and Brouwer, H. (1969) 'Headspace analysis of the less volatile constituents of coffee', in *4th International Colloquium on the Chemistry of Coffee*, ASIC, Paris, pp. 122-30
Quijano-Rico, M. and Spettel, B. (1973) 'Análisis instrumental por activación con neutrone térmicos de algunas muestras de café', in *6th International Colloquium on the Chemistry of Coffee*, ASIC, Paris, pp.102-5
Quijano-Rico, M. and Spettel, B. (1975) 'Determinatión del contenido en vorios elementos en muestras de cafés de diferentes variedades', in *7th International Colloquium on the Chemistry of Coffee*, ASIC, Paris, pp. 165-73
Quijano-Rico, M., Bautista, E., Chaparro, F., Zamudio, V., Ortiz, P. and van Helden, J. (1975) 'Estudio par análisis térmico diferencial-espectrometria de masas de muestras de café verde', in *7th International Colloquium on the Chemistry of Coffee*, ASIC, Paris, pp. 125-32
Quijano-Rico, M., Acero, F.M.T., Morales, E. and Piedrahita, C. (1977) 'Algunos métodos para el estudio del contenido del contenido en cafeina de muestras variadas en diferentes condiciones', in *8th International Colloquium on the Chemistry of Coffee*, ASIC, Paris, pp. 125-34
Radtke, R. (1975) 'Das problem der CO_2-Desorption von Röstkaffee unter dem Gesichtspunkt einer neuen Packstoffentwicklung', *Kaffee und Tee Markt*, *25*, 323-33
Radtke, R. (1979) 'Information about oxygen consumption of roast coffee and its influence on the sensorial evaluated quality of coffee beverage', *Chemie, Mikrobiologie, Technologie der Lebensmittel*, *6*, 36-42
Radtke, R., Springer, R. and Mohr, W. (1966a) 'Coffee aromas and their changes during storage of roasted coffee. II. Importance of aroma fraction of the middle volatile', *Zeitschrift Für Lebensmittel Untersuchung und Forschung*, *128*, 321-33
Radtke, R., Mohr, W. and Springer, R. (1966b) 'Coffee aroma alterations during storage. Less volatile aromatic compounds and volatile acids', *Zeitschrift für Lebensmittel Untersuchung and forschung*, *129*, 349-59
Radtke-Granze, R. and Piringer, O.G. (1981) 'Zur Problematik der Qualitätsbeurteilung von Röstkaffee durch quantitative Spurenanalyse flüchtiger Aromakomponenten', *Deutsche Lebensmittel Rundschau*, *77*, 203-10
Raemy. A. (1981) 'Differential-thermal-analysis and heat-flow calorimetry of coffee and chicory products', *Thermochimica Acta*, *43*, 229-36
Raemy, A. and Lambelet, P. (1982) 'A calorimetric study of self-heating in coffee and chicory', *Journal of Food Technology*, *17*, 451-60
Rahn, W., Meyer, H.W. and König, W.A. (1979) 'Effects of KVW-process on the composition of phenolic components of green and treated coffee', *Zeitschrift für Lebensmittel Untersuchung und Forschung*, *169*, 346-9
Raju, K., Ratageri, M.C., Venkataramanan, D. and Gopal, N.H. (1981) 'Quantitative changes of caffeine and total nitrogen in the reproductive parts of *arabica* and *robusta* coffee', *Journal of Coffee Research*, *11*, 70-5
Rees, D.I. and Theaker, P.D. (1977) 'HPLC of chlorogenic acid isomers in coffee', in *8th International Colloquium on the Chemistry of Coffee*, ASIC, Paris, pp. 79-84
Reymond, D., Chavan, F. and Egli, R.H. (1962) 'Changes in roasted coffee aroma induced by staling', in J.M. Leitch (ed.) *First International Congress of Food Science and Technology*. Vol. III. *Quality Analysis and Composition of Foods*, Gordon and Breach Science Publishers, London, pp. 595-602
Riedel, L. (1974) 'Calorimetric measurements on the system coffee extract/water'. *Chemie, Mikrobiologie, Technologie der Lebensmittel*, *3*, 108-12
Robiquet, P.J. and Boutron, C. (1837) Cited in Sondheimer, E. (1964) 'Chlorogenic acids

and related depsides', *Botanical Reviews*, 30, 667-712

Rodriguez, D.B., Frank, H.A. and Yamamoto, H.Y. (1969) 'Acetaldehyde as a possible indicator of spoilage in green kona (Hawaiian) coffee', *Journal of the Science of Food and Agriculture*, 20, 15-17

Roffi, J., Santos, A.-C., Mexia, J.T., Busson, F. and Maigrot, M. (1971) 'Cafés verts et torréfies de l'Angola. Etude chimique', in *5th International Colloquium on the Chemistry of Coffee*, ASIC, Paris, pp. 179-200

Rotenberg, B. and Iachan, A. (1972) 'Contribution to the enzymatic study of the green coffee bean', *Revista Brasileira de Tecnologia*, 3, 155-9

Rubach, K. (1969) *Beitrag zur Analytik der Hydroxyzimtsäure-ester des Kaffees*. Dissertation, Technische, Universität, Berlin

Ruveda, E.A., Deulofeu, V. and Galmarini, O.L. (1964) 'The lactone of neochlorogenic acid: 5-*O*-caffeoylquinide', *Chemistry and Industry*, 1964, 239-40

Sanint, B. and Valencia, A. (1970) 'Actividad enzimatica en el grano de café en relacion con la calidad de la bebida. I. Duracion de la fermentacion', *Cenicafé*, 21, 59-71

Sannai, A., Fujimori, T. and Kato, K. (1982) 'Studies on flavour compounds of roasted chicory root', *Agricultural and Biological Chemistry*, 46, 429-33

Santanilla, J.D., Fritsch, G. and Müller-Warmuth, W. (1981) 'NMR investigation of the internal movement of binding of water in green and roasted coffee beans', *Zeitschrift für Lebensmittel Untersuchung und Forschung*, 172, 173-7

Scarpati, M.L. and d'Amico, A. (1960) 'Isolamento dell acido monocaffeiltartarico della cicoria (*Cichorium intybus*)', *La Ricerca Scientifica*, 30, 1746-8

Scarpati, M.L. and Esposito, P. (1963) 'Neochlorogenic acid and Band 510', *Tetrahedron Lett*, 1147-50

Scarpati, M.L. and Guiso, M. (1964) 'Structure of the three dicaffeoylquinic acids of coffee', *Tetrahedron Lett*, 2851-3

Scarpati, M.L. and Oriente, G. (1958) 'Chicoric acid (dicaffeyltartaric acid): its isolation from chicory (*Cichorium intybus*) and synthesis', *Tetrahedron*, 4, 43-8

Scott, A.I., Sim, G.A., Ferguson, G., Young, P.W. and McCapra, F. (1962) 'Stereochemistry of the diterpenoids. Absolute configuration of cafestol', *Journal of the American Chemical Society*, 84, 3197-9

Shadaksharaswamy, M. and Ramachandra, G. (1968a) 'Changes in the oligosaccharides and the alpha galactosidase of coffee seeds during soaking and germination', *Phytochemistry*, 7, 715-9

Shadaksharaswamy, M. and Ramachandra, G. (1968b) 'The acid soluble polysaccharides of coffee seeds', *Current Science*, 37, 583-4

Simonova, V.N. and Solov'eva, T. (1980) 'Effect of packaging method on the quality of instant coffee during storage', *Ernährung*, 4, 77-8

Singleton, V.L. (1972) 'Common plant phenols other than anthocyanins, contribution to coloration and discoloration' in C.O. Chester (ed.) *The Chemistry of Plant Pigments; Advances in Food Research Supplement* 3, Academic Press, London, pp. 143-91

Sivetz, M. (1972) 'How acidity affects coffee flavour', *Food Technology, Champaign*, 26, 70-7

Sivetz, M. (1973) 'Comparison of changes in roasted coffee beans in pressurised oxygen free vs. atmospheric roasters', in *6th International Colloquium on the Chemistry of Coffee*, ASIC, Paris, pp. 199-221

Sloman, K.G. (1980) 'Caffeine methodology: semi-automatic method for the determination of caffeine in green and processed coffees', in *9th International Colloquium on the Chemistry of Coffee*, ASIC, Paris, pp. 159-68

Sloman K.G. and Panio, M. (1969) 'Determination of chlorogenic acid in coffee', Technicon International Congress, 1969, Chicago, Illinois, *Advances in Automated Analysis*, 2, 83-5

Smith, A. (1983) 'Sensory aspects of coffee quality' presented at a Society of Chemical Industry meeting, London.

Smith, R.F. and White, G.W. (1965) 'The measurement of the colour of ground coffee and instant coffee by means of the Gardner automatic colour difference meter' in *2nd International Colloquium on the Chemistry of Coffee*, ASIC, Paris, pp. 207-11

Smith, R.M. (1981) 'Determination of 5-hydroxymethylfirfural and caffeine in coffee and chicory extracts by high performance liquid chromatography', *Food Chemistry*, 6, 41-5

Solov'eva, T.Ya (1979) 'Comparative study of polyphenol oxidase and peroxidase activities of different commercial varieties of green coffee during storage', in E.N. Lazarev (ed.), *Problems of Quality and Biological Value of Food Products*, Leningradski Institu Sovetskoi Torgorli im. F. Engel'se Leningrad, USSR, pp. 27-34, cited in *Food Science and Technology Abstracts* (1982) *14*, 11-1813

Sondheimer, E. (1958) 'On the distribution of caffeic acid and the chlorogenic acid isomers in plants', *Archives of Biochemistry and Biophysics*, *74*, 131-8

Sontag, G. and Kral, K. (1980) 'Determination of caffeine, theobromine and theophylline in tea, coffee, cacao and beverages by HPLC', *Mikrochimia Acta*, *11*, 39-52

Statutory Instrument 1978 No. 1420 (1979) *The Coffee and Coffee Products Regulations 1978*, HMSO, London

Stirling, H. (1980) 'Storage research on Kenya arabica coffee', in *9th International Colloquium on the Chemistry of Coffee*, ASIC, Paris, pp. 189-200

Streuli, H. (1962) 'Fraktionierung von Farb- und Geschmacksstoffen des Röstkaffees mittels Sephadex G25', *Chimia*, *16*, 371-2

Thaler, H. (1970) 'Studies of coffees and coffee substitutes XIV. Polysaccharides of the raw beans of *Coffee canephora var. robusta*', *Zeitschrift für Lebensmittel Untersuchung und Forschung*, *143*, 342-8

Thaler, H. (1974) 'Investigations on coffee and coffee substitutes. XV. Polysaccharides in extracts of arabica coffee', *Chemie, Mikrobiologie, Technologie der Lebensmittel*, *3*, 1-7

Thaler, H. (1975) 'Makromolekulare Strukturen in Kaffee', in *7th International Colloquium on the Chemistry of Coffee*, ASIC, Paris, pp. 175-88

Thaler, H. (1979) 'The chemistry of coffee extraction in relation to polysaccharides', *Food Chemistry*, *4*, 13-22

Thaler, H. and Arneth, W. (1968) 'Untersuchungen an Kaffee und Kaffee-Ersatz. XI. Polysaccharide der grünen Bohren von *Coffee arabica*', *Zeitschrift für Lebensmittel Untersuchung und Forschung*, *138*, 26-35

Thaler, H. and Arneth, W. (1969a) 'Untersuchungen an Kaffee und Kaffee-Ersatz. XIII. Verhalten der Polysaccharid-Komplexe des rohen Kaffees beim Rösten', *Zeitschrift für Lebensmittel Untersuchung und Forschung*, *140*, 101-9

Thaler, H. and Arneth, W. (1969b) 'Veränderungen hochpolymerer Kohlenhydrate beim Rösten von Arabica-Kaffee', in *4th International Colloquium of Arabica Coffee*, ASIC, Paris, pp. 174-8

Thaler, H. and Gaigl, R. (1963) 'Untersuchungen an Kaffee und Kaffee-Ersatz. VIII. Das Verhalten der Stickstoffsubstanzen beim Rösten von Kaffee', *Zeitschrift für Lebensmittel Untersuchung und Forschung*, *120*, 357-63

Thier, H.-P., Bricout, J., Viani, R., Reymond, D. and Egli, R.H. (1968) 'Scopolein, ein Bestandteil des Rohkaffees', *Zeitschrift für Lebensmittel Untersuchung und Forschung*, *137*, 1-4

Tiscornia, E., Centi-Grossi, M., Tassi-Micco, C. and Evangelisti, F. (1979) 'The sterol fraction of coffee oil (*Coffea arabica* L.)', *Rivisti Italiana Della Sostanze Grasse*, *56*, 283-92

Tressl, R., Kossa, T., Renner, R. and Koppler, H. (1976) 'Gas chromatogaphic-mass spectrometric investigations on the formation of phenolic and aromatic hydrocarbons in food', *Zeitschrift für Lebensmittel Untersuchung und Forschung*, *162*, 123-30

Tressl, R., Grünewald, K.G., Koppler, H. and Silwar, R. (1978a) 'Phenols in roasted coffees of different varieties', *Zeitschrift für Lebensmittel Untersuchung und Forschung*, *167*, 108-10

Tressl, R., Bahri, D., Koppler, H and Jensen, A. (1978b) 'Diphenols and caramel compounds in roasted coffees of different varieties', *Zeitschrift für Lebensmittel Untersuchung und Forschung*, *167*, 111-4

Tressl, R., Grünewald, K.G., Kamperschroer, H. and Silwar, R. (1979) 'Behaviour of some minor volatile aroma components during coffee staling', *Chemie, Mikrobiologie, Technologie der Lebensmittel*, *6*, 52-7

Tressl, R., Grünewald, K.G., Kamperschroer, H. and Silwar, R. (1981a) 'Formation of pyrroles and aroma-contributing sulphur compounds in malt and roasted coffee', *Progress in Food and Nutrition Science*, *5*, 71-9

Tressl, R., Grünewald, K.G. and Silwar, R. (1981b) 'Gas chromatographic-mass spectrometric study of N-alkyl and N-furfurylpyrroles in roasted coffee', *Chemie, Mikrobiologie, Technologie der Lebensmittel*, 7, 28-32

Tressl, R., Holzer, M. and Kamperschroer, H. (1982) 'Bildung von Aromastoffen in Röstkaffee in Abhängigkeit vom Gehalt an Aminosäuren und Reduzierenden Zuckern', in *10th International Colloquium on the Chemistry of Coffee*, ASIC, Paris, pp. 279-92

Trugo, L.G. and Macrae, R. (1981) Personal communication

Trugo, L.G. and Macrae, R. (1982) 'The determination of carbohydrates in coffee products using HPLC', in *10th International Colloquium on the Chemistry of Coffee*, ASIC, Paris, pp. 187-92

Trugo, L.G., Macrae, R. and Dick, J. (1983) 'Determination of purine alkaloids and trigonelline in instant coffee and other beverages using high performance liquid chromatography', *Journal of the Science of Food and Agriculture*, 34, 300-6

Ulrich, R. (1982) 'Neue Aspekte der Kaffee-Physiologie', in *10th International Colloquium on the Chemistry of Coffee*, ASIC, Paris, pp. 293-308

Uritani, I. and Miyano, M. (1955) 'Derivatives of caffeic acid in sweet potato attacked by black rot', *Nature*, 175, 812-4

Valencia, A.G. (1972) 'Actividad enzimatica en el grano de café en relacion con la calidad de la bebida de café', *Cenicafé*, 23, 3-18

Van Rillaer, W., Janssens, G. and Beernaert, H. (1982) 'Gas chromatographic determination of residual solvents in decaffeinated coffee', *Zeitschrift für Lebensmittel Untersuchung und Forschung*, 175, 413-5

Van Roekel, J. (1975) 'Sensory analysis of coffee and coffee-related products', in *7th International Colloquium on the Chemistry of Coffee*, ASIC, Paris, pp. 259-64

Van der Stegen, G.H.D. (1979) 'The effect of dewaxing of green coffee on the coffee brew', *Food Chemistry*, 4, 23-31

Van der Stegen, G.H.D. and Van Duijn, J. (1980) 'Analysis of chlorogenic acids in coffee', in *9th International Colloquium on the Chemistry of Coffee*, ASIC, Paris, pp. 107-12

Vernin, G. and Vernin, G. (1982) 'Heterocyclic aroma compounds in foods: Occurrence and organoleptic properties' in G. Vernin (ed.) *Chemistry of Heterocyclic Compounds in Flavours and Aromas*, Ellis Horwood, Chichester, pp. 72-150

Viani, R. and Horman, I. (1974) 'Thermal behaviour of trigonelline', *Journal of Food Science*, 39, 1216-17

Viani, R. and Horman, I. (1975) 'Determination of trigonelline in coffee', in *7th International Colloquium on the Chemistry of Coffee*, ASIC, Paris, pp. 273-8

Vilar, H. and Ferreira, L.A.B. (1971) 'How chlorogenic acid may contribute to the quantative evaluation of coffee in soluble coffees mixed with substitutes', in *5th International Colloquium on the Chemistry of Coffee*, ASIC, Paris, pp.120-7

Vilar, H. and Ferreira, L.A.B. (1973) 'Acide chlorogénique dans des variétés régionales d'Angola', in *6th International Colloquium on the Chemistry of Coffee*, ASIC, Paris, pp. 135-41

Villanua, L., Carballido, A. and Baena, L. (1971) 'Determinaciones analiticas en extractos solubles de café', *Anales de Bromatologia*, 23, 259-84

Vincent, J.-C., Guernot, M.C., Perriot, J.J., Hahn, J. and Gueule, D. (1977) 'Influence de différents traitements technologiques sur les caractéristiques chimiques et organoleptiques des cafés Robusta et Arabusta', in *8th International Colloquium on the Chemistry of Coffee*, ASIC, Paris, pp. 271-84

Vitzthum, O. (1974) 'International Standards Organisation Document ISO/TC34/SC8/W62 (Ser. 91) 233E of 1974-06-20

Vitzthum, O.G. (1975) *Chemie und Bearbeitung des Kaffees*, Springer-Verlag, Berlin

Vitzthum, O.G. and Werkhoff, P. (1979) 'Messbare Aromaveränderungen bei Bohmenkaffee in Sauerstoffdurchlässiger Verpackung', *Chemie, Mikrobiologie, Technologie der Lebensmittel*, 6, 25-30

Vitzthum, O.G., Barthels, M. and Kwasny, H. (1974) 'Schnelle gaschromatographische. Coffein-bestimmung in coffeinhaltigem und coffeinfreiem Kaffee mit dem Stickstoffdetektor', *Zeitschrift für Lebensmittel Untersuchung und Forschung*, 154, 135-40

Vitzthum, O.G., Werkhoff, P. and Ablanque, E. 1975) 'Flüchtige Inhaltsstoffee des Rohkaffees', in *7th International Colloquium on the Chemistry of Coffee*, ASIC, Paris, pp. 115-23

Voilley, A., Sauvageot, F. and Durand, D. (1977) 'Influence sur l'amertume d'un café boisson, de quelques paramètres d'extraction', in *8th International Colloquium on the Chemistry of Coffee*, ASIC, Paris, pp. 251-60
Voilley, A., Sauvageot, F., Simatos, D. and Wojcik, G. (1981) 'Influence of some processing conditions on the quality of coffee brew', *Journal for Food Processing and Preservation*, 5(3), 135-43
Vol'per, I.N., Il'enko-Petrovskaya, T.P., Lazarev, E.N. and Solov'eva T.Ya. (1973) 'Konservnaya i Ovoshchesushil naya Promyskolennost', 4, 10-11, cited in *Food Science & Technology Abstracts*, (1974) 6, 4H467
Waenke, H., Palme, H., Spettel, B. and Jagoutz, E. (1973) 'Moderne Methoden der Elementenanalyse und ihre mögliche Anwending an Proben biologischen Ursprungs', in *6th International Colloquium on the Chemistry of Coffee*, ASIC, Paris, pp. 89-94
Wahlberg, J., Enzell, C.R. and Rowe, J.W. (1975) 'Ent-16-kauren-19-ol from coffee', *Phytochemistry*, 14, 1677
Walkowski, A. (1981) 'Changes in factors determining coffee bean quality during storage', *Lebensmittelindustrie*, 28, 75-6
Walter, W., and Weidemann, H.-L (1969) 'Verbindungen der Kaffeearomas', *Zeitschrift für Ernährungswissenschaft*, 9, 123-47
Weidemann, H.-L. and Mohr, W. (1970) 'Uber die Spezifität des Röstkaffeearomas', *Lebensmittel Wissenschaft und Technologie*, 3, 23-32
Weiss, L.C. (1953) 'Report on chlorogenic acid in coffee', *Journal of the Association of Official Agricultural Chemists*, 36, 663-70
Weiss, L.C. (1957) 'Report on chlorogenic acid in coffee', *Journal of the Association of Official Agricultural Chemists*, 40, 350-4
Whistler, R.L. and Richards, E.L. (1970) 'Hemicelluloses' in W. Pigman and D. Horton (eds.) *The Carbohydrates Chemistry/Biochemistry*, Vol. IIA, 2nd edn, Academic Press, New York, pp. 447-69
Whiting, G.C. (1970) 'Sugars' in A.C. Hulme, *The Biochemistry of Fruits and Their Products*, Vol. 1, Academic Press, London, pp. 3-5
Whiting, G.C. and Coggins, R.A. (1975) 'Estimation of the monomeric phenolics of ciders', *Journal of the Science of Food and Agriculture*, 26, 1833-8
Wight, W.A. and van Niekerk, P.J. (1983) 'Determination of reducing sugars, sucrose and inulin in chicory root by high performance liquid chromatography', *Journal of Agricultural and Food Chemistry*, 31, 282-5
Wolfrom, M.L. and Anderson, L.E. (1967) 'Polysaccharides from instant coffee powder', *Journal of Agricultural and Food Chemistry*, 15, 685-7
Wolfrom, M.L., Laver, M.L. and Patin, D.L. (1961) 'Carbohydrates of the coffee bean. II. Isolation and characterisation of a mannan', *Journal of Organic Chemistry*, 26, 4533-5
Wolfrom, M.L. and Patin, D.L. (1964) 'Isolation and characterization of cellulose in the coffee bean', *Journal of Agricultural and Food Chemistry*, 12, 376-7
Wolfrom, M.L. and Patin, D.L. (1965) 'Carbohydrates of the coffee bean. IV. An arabinogalactan', *Journal of Organic Chemistry*, 30, 4060-3
Wolfrom, M.L., Plunkett, R.A. and Laver, M.L. (1960) 'Carbohydrates of the coffee bean', *Journal of Agricultural and Food Chemistry*, 8, 58-65
Wolinsky, J., Novak, R. and Vasileff, R. (1964) 'A stereoscopic synthesis of (±)-quinic acid', *Journal of Organic Chemistry*, 29, 3596-9
Woodman, J.S., Giddey, A. and Egli, R.H. (1967) 'Carboxylic acids of brewed coffee', in *3rd International Colloquium on the Chemistry of coffee*, ASIC, Paris, pp. 137-43
Wurziger, J. (1977) 'Diterpene in Kaffeeölen zur Beurteilung von Rohkaffee nach Art und Bearbeitung', *Fette-Seifen-Anstrichmittel*, 79, 334-9
Wurziger, J., Drews, R. and Bundesen, G. (1977a) 'Uber Arabusta- Roh- und Röstkaffee', in *8th International Colloquium on the Chemistry of Coffee*, ASIC, Paris, pp. 101-8
Wurziger, J. and Harms, U. (1969) 'Carbonsäure-hydroxytryptamide in rohen und gerösteten Kaffeebohnen', in *4th International Colloquium on the Chemistry of Coffee*, ASIC, Paris, pp. 85-91
Wurziger, J. and Purazrang, H. (1972) 'Ein einfacher Nachweis von Robusta-Kaffeebohnen', *Deutsche Lebensmittel-Rundschau*, 68, 117-8
Wurziger, J., Capot, J. and Vincent, J.-C. (1977b) 'Uber untersuchungen *an Arabusta*-Rohkaffees', *Kaffee und Tee Markt*, 27, 3-8

Xabregas, J., Gomes, V., Santos, A.-C., Nogueira, C., Concalves, A. and Mexia, J.T. (1971) 'Análise de algumas características des cafés verdes de Angola', in *5th International Colloquium on the Chemistry of Coffee*, ASIC, Paris, pp. 27-44

Zuluaga-Vasco, J., Bonilla, C. and Quijano-Rico, M. (1975) 'Contribución al estudio y utilización de la pulpa de café, in *7th International Colloquium on the Chemistry of Coffee*, ASIC, Paris, pp. 233-42

Zuluaga-Vasco, J. and Tabacchi, R. (1980) 'Contribution à l'étude de la composition chimique de la pulpe de café, in *9th International Colloquium on the Chemistry of Coffee*, ASIC, Paris, pp. 335-44

14 THE TECHNOLOGY OF CONVERTING GREEN COFFEE INTO THE BEVERAGE

R.J. Clarke

Introduction

Though infusions of green coffee, and fermented drinks made from the whole fruit, have been prepared at different times in history, only after roasting the green coffee beans is the beverage as we know it brewed and consumed today. The large-scale roasting of green coffee has been established for some considerable time, leading to the sale of roasted whole beans in suitable packages, in which form the product is reasonably stable. The advent of vacuum packaging machinery allowed coffee roasters to make their product in the pre-ground form which then has considerable stability to deterioration on storage. The housewife has the responsibility for brewing the roasted coffee with water, to obtain her final desired product in the cup, and since now she has a variety of appliances at her disposal, different grinds of coffee from fine to coarse, suiting the appliance, have also to be offered. The roaster must also offer different coffee blends and degrees of roast. The housewife will however finally have to dispose of the spent grounds.

As a further development in convenience, it was no longer necessary to dispose of the grounds after the introduction of soluble or instant coffee. Furthermore, this product was much more flavour stable in opened packages. Whilst historically, particularly in the USA, there have been numerous soluble coffee ventures, really large-scale manufacture for the retail market took place after World War II. Since then the technology of manufacture has become increasingly sophisticated, and with it, a corresponding increase in the quality of instant coffees available. Liquid coffee extracts have also been offered from time to time, including one long-selling product in the UK. Without additions of such substances as sugar, however, they are not sufficiently stable at ambient temperatures, though attempts have also been made to market deep-frozen products.

The literature on the science and technology of roasted coffee, and particularly instant coffee manufacture, is relatively sparse excepting that of patents, and of publications of a trade journal character, which is extremely large. The two volumes by Foote and Sivetz (1963), and its single volume update by Sivetz and Desrosier (1979), served as a prime source of information, not least to patentees, but have been criticised for some controversial and non-scientific elements. The proceedings of the biennial

International Colloquia on Coffee published by the Association Scientifique Internationale du Café (Paris) are also a fruitful source of information. The Noyes Data Corporation in their food technology reviews provide useful US patent reviews, though non-critically assessed, and not comprehensive; Pintauro (1975) has collated review No. 28 on Coffee Solubilization.

The following description of current technology, within the short compass available, guides the scientific reader to issues, problems and solutions devised by others. The patents quoted are usually only a fraction of those published on any particular subject, with some preference being given to British patent numbers.

Selection and Pre-Preparation of Green Coffee

Green coffee imported from the various producing countries of the world is the basic raw material for the manufacture of roasted and of soluble coffee. There is, however, also substantial manufacture, which takes place in the producing countries themselves (e.g. Brazil), and therefore uses mainly indigenous green coffee. The preparation of green coffee for export has already been fully described in Chapter 10 with a description of the types and grades that are available to importers.

The selection of green coffees for roast (and ground) coffee manufacture is primarily a matter of blending though many roasters also market individual coffees, designated by country of origin (e.g. 100 per cent Columbian, etc.), but all designed for consumer taste requirements. There are various principles of blending, but with considerable skill, experience and art required in practice, which have been typically described (Davids, 1980) together with close consideration of cost factors in the final product. Various combinations of 'Mild' coffees may be used, but with Brazils (dry processed arabicas) generally included, characterised by their 'blendability' and lower cost. Speciality considerations may also apply, by the selection of maragogipes, peaberries, large sized beans, and Mocca coffees, by smaller roasters. National and regional tastes are important in selection, so that, for example, the cheaper robusta green coffees, usually darkly roasted, are very popular in France and Italy. It is fair to say that the Swedish and German manufacturers offer the highest proportion of high quality coffees for their markets, despite the fact that in Germany, there are considerable extra governmental taxes on coffee. Similar principles of blending green coffee apply in the manufacture of instant coffee, again with the important considerations of cost factors, and of quality factors determined by the overall process used, which are of increasing sophistication as subsequently described. The older processes of drum or simple spray drying would not always justify the use of better qualities rather than poorer, (judged initially by liquoring tests on the brewed roast and ground coffee). Robusta coffees are widely used in blends for instant coffee, dependent on market brand factors of cost and consumer taste requirement in different countries.

After selection, the pre-preparation of the green coffee for subsequent conversion into roasted or soluble coffee is very little. Green coffee is delivered in bags (usually of 60 kg content weight, but see Chapter 11, p. 251). It is important that such coffee should be properly warehoused in the factory, in respect of environmental conditions, indeed at all times through transportation from the overseas source to usage. Many of these factors are well described in a National Coffee Association (USA) booklet (1979), and Guidelines being developed by the ISO Coffee Subcommittee. Commercial green coffee may well however contain fractional percentage weights of foreign matter, such as dirt and stones, tending to settle to the bottom of the bags. As in large scale manufacture, storage silos are used immediately prior to blending and roasting, it is usual to tip the contents of the bags onto conveyors, leading to cleaning and de-stoning equipment, which are typically described by Foote and Sivetz (1963: 187-93).

There is however, a distinct form of pre-preparation required in the manufacture of decaffeinated roasted or soluble coffees. The extraction of caffeine to a very low level (less than 0.1 per cent dry basis, which will give less than 0.3 per cent in the resultant instant coffee) will either be carried out by companies specialising in these techniques (e.g. Coffex AG), or in the premises of the roasted or instant coffee manufacturers. Decaffeination was first practised as early as 1905 by the Hag Company in Bremen, Germany who took out a patent on their process. The process of decaffeinating green coffee is essentially one of solvent extraction for caffeine, followed by removal of caffeine and re-use of the solvent. Organic solvents are effective, provided the green coffee is first well moistened with water, typical solvents are methylene chloride and ethyl acetate. Residual solvent in the coffee is removed to very low levels (ppm) by steaming/drying the extracted beans. Other processes are used, e.g. water extraction only followed by organic solvent contact of the extracting water to recover the caffeine, and re-use, first described in US patent 2,309,092 (1943); and more recently the use of supercritical carbon dioxide as the solvent, first patented in 1971 (GP 2,005,293). Solvents will not be equally or totally selective for caffeine; some aroma precursors of the green coffee may be extracted to a greater or lesser extent, though with good decaffeination practice, the brew difference between decaffeinated and non-decaffeinated coffees of the same type will be small. The technology is again complex, with a large patent literature, developing the various processes outlined. Direct decaffeination of roasted coffee or coffee extracts is much less usual in practice, though a number of patents also cover this technique. Detailed descriptions of decaffeination processes are also available in the general literature by Foote and Sivetz (1963: 207-215 Vol. 2), Katz (1980) and Rothmeyr (1980: 82-6).

A more limited form of pre-treatment is that used by a number of German manufacturers, by solvent washing or steaming and other processes to remove the outer coffee 'wax', 5-hydroxytryptamide, and/or other substances claimed responsible for gastric upsets by some consumers (see Chapter 13).

Roasting

Process Factors

Roasting is a time-temperature dependent process, whereby chemical changes are induced in the green coffee, with a loss of dry mass primarily as gaseous carbon dioxide and other volatile products of the pyrolysis. About half of the total carbon dioxide generated will be retained within the roasted coffee, together with a major proportion of the important volatile flavour substances. Roasting is normally carried out under atmospheric pressure with hot air and combustion gases as the primary heating agent, though heat may also be provided by contact with hot metal surfaces, solely or supplementarily. After the initial removal of moisture, roasting proper starts at a green bean temperature of about 200°C, after which through exothermic reactions, escalation of effect readily occurs requiring considerable control of the process for a given degree of roast. These exothermic reactions have been studied recently in the laboratory by differential thermal analysis methods (Quijano Rico, 1975: Raemy and Lambelet 1982). Other methods of applying heat have been used in limited practice.

Degree of roast plays a large part in determining the flavour characteristics of extracts subsequently brewed from the roasted coffee, whatever the blend. The degree of roast is qualitatively assessed from colour, for example, simple categorisation as a light, medium or dark roast. Various sub-divisions have been devised, such as French or Italian roasts; but in large-scale commercial practice, a final quantitative assessment is made by colour reflectance readings of the roasted coffee surface, shortly after roasting but after grinding in a clearly specified manner. There are various commercial instruments available for reporting this colour in arbitrary units, of the prepared sample (Smith, 1963, Sivetz and Desrosier, 1979). Roast colour will also be broadly correlated with percentage loss of coffee matter, expressed on a dry basis; so that, a light roast will show about 3-5 per cent loss, medium 5-8 per cent, dark 8-14 per cent together with the moisture (variable) that the green beans will have contained. More specific correlations can be developed for particular roasting methods and blends.

Clearly, chemical composition, in respect of both volatile and involatile substances will be strongly determined by the degree of roast. These changes have been studied in detail by many workers and summarised by Clifford (see Chapter 13) and Maier (1981). Both chlorogenic acid(s) and trigonelline are destroyed, according to the degree of roast, and may indeed be used as a measure of it (Kwasny, 1978 see also Chapter 13). Of commercial interest also, are any differences in extractable soluble solids, whether by home brewing, laboratory (including exhaustive) extraction or in instant coffee manufacture, though reliable published information is hard to come by (Kroplien, 1963; Wilbaux 1967).

In roasting, times of heating range in practice from a few minutes to about 30 minutes; beyond that time is generally regarded as undesirable for flavour. For any given degree of roast, the time is determined by conventional heat transfer factors, i.e. the relative movement of the beans-air (velocities), bean-bean and bean-contact surfaces together with any internal radiative heat effects. Air temperatures may range from 540°C in older roasters, to 430°-480°C and to 370°C or lower, often reflecting differences in air velocities used. These considerations in design are especially discussed in detail by Foote and Sivetz (1963: 203-39).

The physical changes in the coffee beans during roasting are also technically important. Expansion of the beans takes place, including a 'popping' phase, leading to considerably decreased density, as a function of degree of roast but also of the speed of roasting. This decrease of density is followed through into the ground roasted coffee, as a decrease of bulk density. This feature is an important determinant in deciding subsequently, the size of packages, particularly cans. Green coffees from different sources and age start with different bean densities, and furthermore roast somewhat differently, for example, 'harder' beans roast more slowly to a given colour. Though a matter of taste opinion, optimum flavour is developed by different green coffees at different roast colours, so that whereas single roasting of a green coffee blend is general, it is often recommended that different coffees are roasted separately and subsequently blended. Uniformity of roast colour is generally desired, though this may be frustrated by individual differences in beans, e.g. immature beans giving rise to 'pales'; control of roasting avoiding individual charred beans is important.

The roasting process is followed by immediate cooling, in which the addition of water by a few per cent plays a part. This addition of water, however, enables much greater uniformity of particle size in subsequent commercial grinding. A further by-product of roasting is the variable production of 'chaff', which is the 'roasted' silverskin released when present from the surface of the bean into the conveying air. Robusta coffees tend to have more than arabica; a further small quantity left in the cleft or centrefold of the beans will be released on air-veying and grinding. The overall amounts are small, but arrangements are made to separate, usually by a cyclone, from the discharging gases.

Process Equipment

Large-scale roasting plant, or roasters, are now largely of US or German manufacture, with several companies being old-established from the nineteenth century. The most generally used principle is that of the horizontal rotating drum, for tumbling the green coffee beans in a current of hot air, either batch or continuously operated. Once through air flow was general, but hot air re-circulation is now more usually adopted. The former is, however, generally associatd with further combustion (after-burners) to minimise atmospheric pollution by the discharged gases, so that re-circulation of these gases is a logical extension for fuel economy. It may still be necessary to have an 'after burner' or catalytic system to lower the emission of organic carbon compounds, (Anon 1975).

The horizontal drum principle is typified by the Probat-Werke roasters (series GO) in a standard 4-bag roaster (240 kg batch) with a roasting time of some 10-12 minutes, and turn-round of 15-18 min. Hot air is provided by direct fuel oil/gas combustion, with air intake controlled by a damper, which enters the back-end of the double-walled solid drum, and leaves to a stack. Probat in the RR series, adopt rc-circulation of the gases through the main burder. Barth, also of Germany, manufacture batch drum roasters (Tornado models).

The drum principle is also used in the 'Thermalo' roasters of the original Jabez Burns Company (US) or Neotec in Germany. These roasters, are either batch, with a perforated drum, to allow cross hot air flow also, or continuous, first introduced in 1940. In batch roasters, the cessation of roasting is determined apart from the control of the temperature programme applied, by the addition of quench water and then discharge into a cooling tray with up-draft of cool air. With continuous roasters, there is a cooling section of the rotating drum, separated by a heat lock, through which the roasted beans pass. Gas velocities in the roasting section are high, the roasting times down to five minutes or so, with consequent high throughputs (up to about 3,000 kg/hour). Details of practical operation of these roasters are described by Foote and Sivetz (1963: 210-220).

Other mechanical principles have been adopted, in particular the Gothot roaster (Rapido and Rapido Nova) which uses a fixed vertical vessel but with rotating paddles to assist heat transfer from the hot air to the green coffee, again providing short roasting times (i.e. about 6 min in the Rapido-Nova). Discharge takes place into a second vessel, with cooling air. A recirculation system for the roaster gases can be readily incorporated. In recent years, Probat have marketed a batch roaster, on yet another principle, the Radial-Turbo Roaster (RZ series) in which a rotating bowl is used for intimate contact of hot gases and beans. Discharge is arranged after a short roasting time from the periphery of the bowl. Again a recirculation system is incorporated. The logical extension of the use of high velocity air is in fully 'fluidised bed' roasting, where various designs and performance have been described (Sivetz, 1975; Vincent, 1977). For example, a fully commercial unit is the Wolverine Jet Zone Roaster.

The Smithern roaster differs from all others, in that roasting takes place under pressure (50-150 psig) with a closed circulation of nitrogen gas (USP 3,615,688 (1971) and 3,825,221 (1975)). Closed pressure systems are generally reported to increase brew acidity of such roasted beans. A study has been made on its performance (Sivetz, 1973) together with a report of other pressure roasting systems.

Grinding

Process Factors

Grinding of roasted coffee, whether by the manufacturer or the housewife, is necessary to allow sufficiently rapid infusion or percolation of its soluble content by 'boiling' or hot water. Clearly the finer the particle size, the more rapid will be this rate of infusion, but in practice, the method of brewing and equipment used will dictate the degree of grinding which is desirable. This is equally true in the process of grinding roasted coffee, subsequently to be extracted in large-scale plant for instant coffee manufacture. Degree of grind is usually assessed from result of screen or sieve analyses, using a number of different mesh sizes.

It is not possible to grind coffee to a single uniform particle size, a distribution of particle size around an average is to be expected. This distribution is best expressed graphically in a plot of cumulative oversize by weight for each screen used on a probability scale versus the corresponding mesh size (aperture, in microns) on a linear scale. The average particle size may be assessed as the aperture at which there is 50 per cent cumulative oversize. The slope of the straight line plot (except at the extremes) indicates the degree of uniformity of particle size, which may be quantitatively assessed as a coefficient of variation (CV). The CV is in fact, the percentage ratio of the aperture for 16 per cent over-size minus aperture for 50 per cent retention divided by aperture for 50 per cent retention (or average). The screens chosen in a nest, usually not more than 5, should be chosen from a $\sqrt{2}$ series of screen sizes, now internationally standardised (British Standard 410, 1976).

In general, single stage grinding (i.e. once through a single grinding device) will give a relatively high CV figure, and show a high proportion of fine particles. Multi-stage grinders (e.g. with a series of grinding rolls, progressively reducing particle size) will be found to give lower CV figures, typically ± 30 per cent for roasted coffee. Similarly, closed circuit grinding with return of oversize to a single grinding device will decrease CV figures. Uniformity of grinding is also dependent upon the condition or brittleness of the roasted coffee, its moisture content and degree of roast. For this reason, grindability is improved by increasing moisture content of roasted coffee by so-called quenching in the roaster. Really fine grinding (i.e. an average of about 50 μm or less, average particle size) is best achieved by cryogenic grinding (i.e. at sub-zero temperatures by use of solid CO_2 or liquid nitrogen), which can have other advantages (BP 2,006,603B (1982)).

Really fine grinding, however, is not required in roast and ground coffee for brewing or large-scale percolation. Practical grinds for home-brewing purposes may be divided qualitatively by the terms 'coarse', 'medium', 'fine' or 'very fine', or according to the brewing device for which they are intended, e.g. 'jug', 'percolator', 'drip-pot', 'filter' or 'espresso'. Recommended standards characterising degree of grind by screen analyses (often in terms of percentage oversize at two screen sizes only, which is adequate) have been published from time to time in various countries (Clarke, 1965; IEC 661; BS 3999 Part 8). Home percolators and 'jugs' will tend to require 'coarse' to 'medium' grinds, in order to minimise the amount of fines finding their way into the cup, typically the average size will be between 1100 and 600 μm, whilst modern filter devices (automatic) use finer grinds, typically in Europe at 500-400 μm though coarser in the USA (between 600 and 700 μm). With filter papers, the problem of sedimenting fines in the cup is negligible, but the coffee bed should not consist of such fine particles as to impede percolation.

Roasted and ground coffee for large-scale extraction generally require relatively coarse grinds, dependent upon the plant design and operation. Allowable pressure drops in long percolation beds have to be considered.

The grinding of coffee will release a further quantity of chaff. This material can be disposed of, though it does contain a substantial proportion of coffee soluble solids. Often however, it is incorporated with the ground coffee in a so-called 'normaliser', by mixing under the action of rotating blades in a trough-shaped vessel. Particularly with dark-roast coffees, this process is an advantage to absorb any exuding oil on the surface, and keep the ground coffee reasonably flowable in subsequent handling. 'Normalising' has a further function in that it is a means of controlling the bulk density of roast and ground coffee, though only within certain limits, by variation of blade speed and disposition. This is of importance in subsequent packing for retail sale, though less so in large-scale extraction. The combination of grinding, and 'normalising' can be referred to as 'granulising'. The grinding of coffee will also release its inherent CO_2 and some of its more volatile components, at the time of grinding and subsequently but at a declining rate. The release of CO_2 is again of importance in subsequent packing operations (see page 389). The release of volatile components necessitates rapid transfer of the roasted and ground coffee to extraction equipment in instant coffee manufacture.

Process Equipment

For large-scale production of ground roasted coffee at whatever grind for retail sale or subsequent instant coffee manufacture, the multi-stage grinding machine is almost essential. The best known and probably most used type is the Gump Grinder, which has been manufactured in different models for a large number of years. It is a multiple-roll grinder, with two to four sets of rolls below a feeding roll, according to degree of grind required. The cutting surfaces of these rolls are especially serrated for grinding coffee based on the original LePage patented (1905) design for each roll. Roasted coffee beans require a cutting, rather than a tearing or crushing action. The construction of this grinder, operation and important aspects of its maintenance is described by Foote and Sivetz (1963: 239-50, Vol. 1).

Extraction

Process Factors

After grinding, the large-scale extraction of roast and ground coffee is the next stage in the manufacture of instant coffee. Both soluble solids and volatile aroma and flavour compounds need to be extracted. Liquid water is the only solvent used in practice for extracting the soluble solids, which however, will also extract the volatile components. The latter, being present in quite small quantities, will generally be found solubilised in the extracting water, though some components (e.g. non-polar) will require larger operating ratios of water to coffee than others (polar) for complete extraction. Prior passage of steam through roasted and ground coffee in a bed, with collection of a volatile-containing condensate, has been recommended in numerous patents (e.g. BPs 982,521 (1965); 1,206,296 (1967); 1,424,263 (1976); 1,466,881 (1976); 1,525,808 (1978)). Organic solvents are not used for extracting volatile or soluble solids for instant coffee though there are patented processes. Since roasted coffee contains coffee oil, in which many of the volatile components are located, mechanical pressing or expelling has also been recommended as a means of separately isolating these components, prior to water extraction. Inert gases, like CO_2, whether supercritical or not, have also been described for this purpose (e.g. BP 1,106,468, (1968)).

Contact between an extractable solid foodstuff and water may be arranged by any one of three main methods. In all these methods, however, unlike single-stage home brewing, in order to obtain efficient extraction with an adequately high concentration of soluble solids in the final extract (to lessen amount of water to be subsequently handled, and removed to form a dry product), it is necessary to arrange multi-stage or continuous counter- or co-current operation. So-called 'slurry' extraction is obtained by the use of tanks or vessels for contacting, and separating devices such as centrifuges between the stages. Slurry extraction does enable the use of relatively fine grinds for the roasted coffee. The most largely used method is that of percolation batteries, deriving from the 'Shanks' battery in which the coffee is held as a bed in vessels (or vertical columns) with internal separation of liquor from one stage or column to the next. The flow of water to the solid foodstuff is countercurrent, though the draw-off of extract is intermittent. As each column is exhausted, so it is replaced in the battery by a 'fresh' column. A truly continuous countercurrent system can be established within a single horizontal cylindrically shaped vessel, by movement of the solids by screw conveyor against the flowing water. The grind of coffee in these two methods needs to be relatively coarse. Other designs are possible, but these are the main methods used in extracting coffee.

Coffee extraction has a number of features which distinguish it from other extractions of materials. Roasted coffee is a cellular substance, with matrix of carbohydrate material and oil, and can become compressible on extraction. Extraction of coffee soluble solids by water at 100°C provides a yield from the roasted coffee of up to about 30 per cent, depending upon the conditions and number of stages used. It was generally recognised however, that such extracts did not provide very satisfactory powders on drying, with problems in producing adequate flavour retention and free-flowing and 'non-tacky' powders. Corn syrup solids or other similar carbohydrate materials were therefore often added (up to about 50 per cent) to give 'soluble coffees with added carbohydrates'. It was then realised that roast coffee itself was a source of suitable carbohydrates (see Chapter 13), readily solubilisable by use of extracting water at higher temperatures up to 175°C (USP 2,324,526 (1943)); and also by use of dilute acids (USP 2,687,355 (1954)) at 100°C. The first large-scale use of such elevated temperatures developed about 1950, for which a percolation battery with columns operable under pressure (to keep water liquid at all times) was particularly suitable. In this way instant coffee made from 100 per cent coffee could be satisfactorily manufactured. The effective yield of coffee solubles was then well beyond 30 per cent, figures primarily dependent upon the feed water temperature used (and pressure rating of the system). In fact, a controlled temperature regime has to be established across a battery percolation set, the most 'spent' coffee is contacted with the feed water, whilst the 'fresh' coffee is contacted with liquor from previous stages at a temperature around 100°C. Continuous extractors have similarly to be designed, in which the required complete temperature profile is either built into a single unit or into two separate units, operating at two different temperatures (one high, the other low), with external transfer of partially extracted grounds from one unit to the other.

As a unit operation of chemical engineering, extraction is based upon the mass-transfer of diffusible components from a solid phase to a liquid phase. The fundamentals are extremely complex, especially reflected in the mathematical equations needed to model these systems, as will be noted from the various recent studies that have been made (Schwartzberg, 1976; Bruin and Spanninks, 1979; Spanninks, 1979; Besson, 1983). In the same way, the operation in practice has many elements of craft, skill and experience. In optimising their operation for extract quality, yield from coffee, concentration of extract and productivity factors, numerous variants of procedure have been described, largely in the patent literature (e.g. USP 3,655,398 (1972); BP 1,547,242 (1979).

It is probable that roasted coffee extraction is not a completely physical operation, with some chemical changes taking place, as for example, some cleavage of large molecular mass polysaccharides at the temperatures conventionally used for commercial extraction. Like roasted coffee the composition of coffee extracts is complex, (see Chapter 13), and the proportions of different soluble components extracted will differ somewhat with yield of solubles taken. Caffeine, chlorogenic acids and numerous constituents are very easily extracted, whilst the use of higher temperatures will cause increasing solubilisation of less-diffusible substances. Extract composition in respect of polysaccharide content has been experimentally studied in detail (Thaler, 1979) and does not show any substantial monosaccharide formation by hydrolysis (Kroplien, 1974). Only a negligible amount of free coffee oil enters into the extract.

Extraction of roasted coffee requires also a number of ancillary processes. Coffee extracts as drawn off, may still contain a proportion of fine suspended coffee grounds and other insoluble matter, which generally need to be removed by centrifugal filtration procedures. The finally spent grounds have to be disposed of; pressure percolation systems are particularly convenient for their discharge. The hydraulic pressure in a spent column (e.g. $200\,lb/in^2$) after its isolation from the rest of the battery is available for pushing out its contents on release to atmospheric pressure. In the course of this expansion, some 10-15 per cent of this water will be vapourised, but the spent-grounds will still contain some 80 per cent of water, which are then usually mechanically pressed to around 60 per cent w/w moisture content. The spent grounds will then appear dry and be capable of easy further handling. Coffee extract as it emerges hot from an extractor, has to be immediately cooled through heat exchangers, and kept cool before proceeding to subsequent operations.

Process Equipment

A description of typical large-scale percolation battery plants is given by Foote and Sivetz (1963: 261-319), together with the important various items of ancillary equipment that are required. Percolation batteries will consist of 5-8 columns, one of which will be out of stream to allow discharge of the spent contents and re-filling with fresh coffee. Such equipment is supplied by Niro A/S; but it is probable that most large-scale manufacturers of instant coffee will have fabricated plant to their own particular designs, and requirements. The overall time of contact of coffee with water is variable, of the order of 180 minutes or less; though the cycle time (30 min or less) of batch intermittent percolators is important in determining output per hour.

Various types of continuous extractors are available for solid-liquid contact; but the continuous extractors supplied also by Niro A/S (Kjaergaard and Andresen, 1973) are those which have been especially designed for coffee extraction. There is advantage in such extractors, in that unlike percolation batteries, substantial interstage piping and valving is not required. The substantial contact times required are similar to those of percolation batteries.

Slurry systems consisting of tanks or pressure vessels, and interstage separating devices (centrifuges) are not now widely used, except that they have a use in dealing separately with coffee fines, in secondary streams, from grinding and screening out 'unders' and from oil expelling. Their really large-scale use in a large number of stages would represent high capital expense compared with percolation batteries, for similar performance.

Drying Coffee Extracts

Drying Processes

Whereas early ventures into instant coffee employed drum-drying, spray-drying quickly established itself on account of favourable appearance and easy control of bulk density of finished powder that can be effected. Conventional spray-drying however, causes high losses of the flavour volatile substances from the original extract. In the 1950s and early 1960s, freeze-drying equipment became available and suitable for large-scale use on coffee extracts, providing greater retention of flavour and other product advantages. Even so, freeze-drying should be operated to produce granules, and not fine powder. The increasing consumer preference for granules then spurred the marketing of so-called agglomerated instant coffee, that is, granules formed by adhesion of spray-dried particles by water or steam-wetting techniques, followed by finish-drying. There was a corresponding development in milk powders, for 'instant milks'. Such granules have essentially the flavour characteristics of the original spray-dried powders, which due to further technical developments in this process, can, and usually have markedly improved flavour retention over the older products.

Coffee extraction techniques as previously described, lead to extracts with solubles concentrations of less than about 25 per cent, and often much less. Direct spray-drying at these concentrations will lead to substantial flavour loss. The effect of various variables (including concentration) in spray drying on the retention of different volatile substances, of different relative volatilities and other relevant physical characteristics have been studied in depth by Professor Thijssen and colleagues from 1965 onwards, as reported in numerous papers (see for example, Bomben, Bruin and Thijssen, 1973: 65-74). Other workers have also studied these variables, including effects on particle morphology (Karel, 1973; King, 1980).

Whilst freeze-drying can be conducted effectively at these concentrations, higher concentrations are again advantageous, especially since the cost of removing water by freeze-drying is high compared with other methods. Volatile retention is also determined by various operational factors including the rate of freezing prior to the drying itself (Bomben *et al.*, 1973; Karel, 1973). The characteristics of the freezing curve are also important for freeze drying (e.g. see enthalpy-concentration diagram for coffee extract, Riedel, 1974) and process conditions must be chosen to ensure sublimation of the ice. Operating vacuums of less than 1 mm Hg are required.

Increased solubles concentration as with other foodstuffs liquids can be achieved by two main methods, evaporation and freeze-concentration, both well-studied scientifically and technically, especially well covered in a 1971 Symposium on Foodstuffs Evaporation held in London (Clarke, 1971; Gray, 1971; Shinn, 1971; Wiegand, 1971). Evaporation, however, is also responsible for loss of volatile substances, according to the percentage water evaporated (Bomben *et al.*, 1973; 19-25) especially significant with coffee extracts. Due to the viscosity of coffee extracts, however, final concentrations possible are limited depending upon the type of evaporator. Wiped film evaporators are capable of achieving concentrations up to 70 per cent. When evaporation is used, two stages may be employed, a preliminary stripping and condensation of required volatiles, followed by bulk evaporation and condensate discarded. The desired condensate, together with other flavour distillates from other sources in the soluble coffee process, can then be added to the bulk evaporate for feed to the drier, whether spray or freeze. Numerous variants of this kind of technology are clearly possible, well-testified to by the considerable patent literature, (typically USP 3,244,530 (1966); BP 1,265,206 (1972); BP 1,563,230 (1980)) though evaporation may also be used merely as a means of increasing solubles concentration for economy reasons. Freeze concentration is an alternative to evaporation, though probably much less used, but with the advantage that flavour volatiles are retained within the concentrate (Bomben *et al.*, 1973: 45-50). Extract viscosity factors are again limiting, to around 40 per cent for coffee extracts. Reverse osmosis is probably even less used, except for concentrating very dilute extract streams.

Whilst improved spray and freeze-drying techniques have led to greatly improved flavour retention, some coffee volatile substances (of low molecular mass and high volatility) especially those responsible for characteristic coffee headspace aroma, are very difficult to retain, so that direct addition of coffee oil to the dried powder is recommended (e.g. BP 1,525,808 (1978)), thus by-passing drying systems.

Spray Drying Equipment

Spray drying of coffee extracts demands finished powders for direct marketing of a desired colour, and appearance and moisture content, but a bulk density (either assessed free flow or packed) such that a typical-sized teaspoon dispenses about 2 g or less. The latter requirement in turn demands particle sizes and distribution, averaging about 300 μm which is not possible to obtain in the type of drier conventionally used in the production of fine milk powders, that is, spray driers of wide diameter and relatively low height, fitted with spinning disc 'atomisers'. In general, therefore spray driers for coffee extract are relatively narrow in relation to height (or drying path), into which 'atomisers', or spraying devices, providing small spray angles, especially centrifugal nozzles, need to be fitted. Such driers are manufactured by a number of companies such as Niro A/S, Stork-Bowen and Anhydro A/S, but it is probably true to say that most large instant coffee manufacturers will provide themselves with driers and 'atomisers' of their own specific design. There are a number of other features of general importance; the supply of hot air (usually direct combustion gases from refined fuel oils or gas) and its entry arrangements into the drier (usually co-current or radial); the discharge of dried powder from the bottom of the tower, and the separation (either internal or external) of the dustier fractions from the main product, together with a final collection of very fine powder from the outlet air, otherwise a pollutant to the atmosphere. A good detailed description of spray driers is provided by Masters (1972), though not really covering important points of operation for coffee extracts, for which the other references should be consulted. Any subsequent or simultaneous process of agglomeration is carried out in equipment, of which a wide variety of designs and operable plants are available for coffee, and other foodstuffs generally (Jensen, 1975). The patent literature is again replete with methods (Pintauro, 1975: 177-90).

Freeze-Driers

Small-scale and laboratory freeze driers became widely available in the 1950s and used in the pharmaceutical industry. From about 1960, large freeze driers became available, from such companies as Leybold, Atlas and Stokes. Numerous experimental marketings of various foodstuffs both solid and liquid were initiated. Most of these ventures eventually lapsed, due to the high capital and running costs of freeze-drying, though its application to coffee extracts prospered, by the quality and appearance advantages offered. Freeze driers, in the batch mode, consist of a cylindrical chamber, carrying trays of the deeply frozen product resting on hollow shelves, through which heating water can be passed. The chamber is kept however under a high vacuum, so that the ice in the product sublimes off, and the water vapour allowed to condense on suitably positioned condensers. The frozen extract may be in thin slabs, and after being freeze-dried, removed from the chamber and ground to a powder. The preferred type of procedure, outlined in a number of patents (Pintauro, 1975: 134-54) is, however, to grind the frozen slabs into still frozen granules, which are then loaded onto the trays. The under-sized particles are recycled by a number of different methods. In this way, the required sized granules are ready made. It is important that either slabs or granules remain frozen during drying, determined by the vacuum used and the temperature programme. The time of drying is often long, 8 hours or more dependent upon equipment design, including that of the trays. A further commercial variant is the foaming of the extract, immediately prior to freezing, especially when high concentration extracts are being dried, also described in patent and other literature (Pintauro, 1975: 139-54; Petersen, 1977). Some driers instead of using conductive heat transfer, use radiant heat (Atlas A/S, Copenhagen). Semi-continuous freeze driers also became available, e.g. Leybold tunnel driers. These use an intermittent feed into, and discharge of product from a compartmented tunnel of chambers with interlocks, operating at different temperature regimes as required. Truly continuous freeze driers have also been made (Lurgi, GmbH).

The overall economics of operation of concentration-drying systems has been described in a number of papers (e.g. Van Pelt, 1977).

Packing Roast Coffee and Soluble Coffee

Roast Coffee

Both roasted whole beans, and roasted and ground coffee contain large amounts of gas, primarily carbon dioxide, immediately after manufacture, which is only slowly released, with time. This phenomenon is of importance in the packing of, and the packaging required for these products, presenting problems which have been overcome in a variety of ways. The initial content of carbon dioxide in roasted beans depends upon the type of green coffee (primarily whether arabica or robusta), and importantly on the degree of roast. This content can be measured (by collection to 'infinite' time), and is of the order of 2-6 ml CO_2 (at NTP)/g of roast whole bean. On grinding, dependent upon grind size, a large proportion is, however, immediately released, but a high amount of gas may still remain especially with coarse grinds. The remaining 1-3 ml is then slowly released to the atmosphere in an exponential manner, and many hours may pass before the amount reduces to say 5 per cent of its original. The rate of release will also be dependent on grind size, temperature and external pressure. It can be readily seen therefore that attempts to pack a roast and ground coffee (or whole beans) before sufficient CO_2 has been allowed to escape, in a package of limited total take-up volume for gas may well cause the package (even cans) to burst eventually, due to excess internal pressure. The conventional solution to this problem, is to vacuum pack, i.e. to maintain some initial vacuum within the pack. this solution, however, still requires close consideration of the vacuum to be applied, the size of the package in relation to weight/volume of the contents, and whether any period of holding after grinding for 'degassing' is still required. In practice, it may be found with fine grinds and the use of a high vacuum (e.g. 15 mm Hg pressure absolute), packing can be commenced very shortly after grinding. This solution has been practised for many years with metal cans; or more latterly with so-called 'hard-packs', which are laminate flexible bags, collapsing on the contents under high vacuum. There is however, less surplus volume available for take-up of released carbon dioxide, before such packs would become undesirably soft.

There are, however, other considerations; roasted coffee is very sensitive to effect on quality by oxygen; for example, some 70 ml of oxygen pick up per pound of coffee is enough to cause the roasted coffee to become stale. More scientific investigations (Heiss and Radtke, 1977) have suggested a maximum in-pack level of 0.12 mg oxygen/g of coffee. Certainly the aim is to have an in-pack oxygen content of less than 0.5 per cent for a shelf-stability of up to one year. The use of a very high vacuum helps to achieve this, but the conditions surrounding the holding of roasted and ground coffee for many hours before packing can allow the absorption of oxygen, which may not be readily released in the vacuum packing operation. A lower vacuum can be used in packaging, together with a pre-inert gas flushing stage, in order to reduce the oxygen level. All these matters have been closely examined at the Packaging Institute of the Technical University of Munich in a series of papers (Radtke, 1973, 1975, 1982; Heiss and Radtke, 1977). A further solution to the problem of potential deterioration during de-gassing has been offered by a process of controlled vacuum degassing (BP 1,200,635 (1967))); whilst another uses a simple non-return valve, the Goglio valve, for the release of gas in the package itself (Heiss and Radtke, 1977).

The stability of roast and ground coffee in packages is also dependent upon its initial moisture content (usually recommended to be not more than 4 per cent). The package itself should be impermeable to the ingress of further moisture, and oxygen, and not allow egress of volatile aroma compounds. These conditions are of course, readily possible with metal cans; but with flexible packages, careful selection is needed for the package material. The types of packing machinery in general use are described by Sivetz and Desrosier (1979: 279-314).

Instant Coffee

The packing of instant coffee presents fewer problems. Instant coffee in Europe and the USA is now generally packed in glass jars, with a sealed diaphragm at the mouth of the jar and closing lid, the pack is then impermeable to moisture ingress. With instant coffee processed to have a high proportion of retained aroma/flavour volatiles such as freeze-dried, it is also necessary to pack these coffees in an atmosphere of inert gas, either carbon dioxide or nitrogen or mixtures of both, so that residual oxygen content is as low as possible (less than 4 per cent). The operation of filling machinery for large-scale manufactured instant coffee, is necessarily high speed (Foote and Sivetz, 1963: 548-81, Vol. 1).

References

Anon. (1975) 'Clearing the air', *Coffee International 2*, 36-93
Besson, A. (1983) 'Mathematical model of leaching' in C. Canterelli and C. Peri (eds.) *Progress in Food Engineering*, Forster-Verlag A.G., Küsnacht, Switzerland, pp. 147-56
Bomben, J.L., Bruin, S. and Thijssen, H.A.C. (1973) 'Aroma recovery and retention in concentration and drying of foods' in *Advances in Food Research*, Avi, Westport, *20*, 2-111
Bruin, S. and Spanninks, J.A.M. (1979) 'Mathematical simulation of the performance of solid-liquid extractors', *Chemical Engineering Science*, *34*, 199-215
British Patents
(1965) 982,521 Nestlé Products Ltd, 'Aromatisation of Coffee Products'
(1967) 1,200,635 Kenco Coffee Company Ltd, 'Conditioning Roasted Coffee'
(1967) 1,206,296 General Foods Corporation, 'Production of Soluble Coffee'
(1968) 1,106,206 Nestlé Products Ltd, 'Vegetable Extracts'
(1968) 1,106,468 Nestlé Products Ltd, 'Vegetable Extracts'
(1972) 1,265,206 Proctor and Gamble Company, 'Preparation of Instant Coffee'
(1976) 1,424,263 Proctor and Gamble Company 'Stable Concentrated Flavourful Aromatic Product'
(1976) 1,466,881 General Foods Ltd, 'Continuous Production of Aromatic Distillates under Vacuum'
(1978) 1,525, 808 Société des Produits Nestlés SA, 'Aromatisation of Coffee Products'
(1979) 1,547,242 Société des Produits Nestlés SA, 'Coffee Extraction Process'
(1980) 1,563,230 General Foods Ltd, 'Production of Soluble Coffee'
(1982) 2,006,603B General Foods Ltd, 'Coffee Product and Process'
British Standard 410 (1976) 'Test Screens', British Standards Institution, Park Street, London W.1
British Standard 3999, Part 8 'Electric Coffee Percolators'
Clarke, R.J. (1965) 'Roasted coffee grinding techniques', *Food Processing and Marketing*, pp. 9-12
Clarke, R.J. (1971) 'Evaporation of heat sensitive liquids', *Journal of Applied Chemistry and Biotechnology*, *21*, 349-50
Clarke, R. J. (1976) 'Food engineering and coffee', *Chemistry and Industry*, 17 April, pp. 362-5
Davids, K. (1980) *The Coffee Book*, Whittet Books, Weybridge, England, pp 21-59
Foote, H.E. and Sivetz, M. (1963) *Coffee Technology*, Vols. 1 and 2, Avi. Westport, Conn., USA

German Patents
(1971) 2,005,293 Studien Gesellschaft Köhle, Mülheim 'Verfahren zur Entkoffeinierung von Rohkaffee'
(1973) 2,212,281 Hag, Bremen, 'Verfahren zur Entkoffierung von Rohkaffee'
Gray, R.M. (1971) 'The plate evaporator', *Journal of Applied Chemistry and Biotechnology*, 21, 359-62
Heiss, R. and Radtke, R. (1977) 'Packaging and marketing of roasted coffee' (in English), *8th International Colloquium on the Chemistry of Coffee*, ASIC, Paris, pp 163-74
IEC Standard 661 (1980) 'Methods for measuring performance of household coffee makers', *International Electrotechnical Commission*, Geneva
Jensen, J.D. (1975) 'Recent advances in agglomerating, instantizing and spray drying', *Food Technology*, 29, 60-70
Karel, M. (1973) 'Fundamentals of dehydration processes' in A. Spicer (ed.) *Advances in Pre-concentration and Dehydration of Foods*, Applied Science Publishers, London, pp. 45-94
Katz, S. (1980) 'Decaffeination of coffee', *9th International Colloquium on the Chemistry of Coffee*, ASIC, Paris, pp. 295-301
King, F. Judson (1980) 'Chemical engineering in coffee technology' *9th International Colloquium on the Chemistry of Coffee*, ASIC, Paris, pp. 237-50
Kjaergaard, O.G. and Andresen, E., (1973) 'Preparation of coffee extracts by continuous extraction', in *6th International Colloquium on the Chemistry of Coffee*, ASIC, Paris, pp. 234-9
Kroplien, U. (1963) *Green and Roasted Coffee Tests*, Gordian, Hamburg
Kroplien, U. (1974) 'Monosaccharides in coffee extraction' *Journal of Food Science*, 22, 110-16
Kwasny, H. (1978) 'Determination of the degree of roasting of coffee beans in coffee extracts', *Lebensmittel und Gerichtliche Chemie*, 32, 36-8
Maier, H.G. (1981) *Kaffee*, Parey, Hamburg
Masters, K. (1972) *Spray Drying*, Leonard Hill, London
NCA Booklet (1979) '*Health and safety in the importation of green coffee*', National Coffee Association of America, 120 Wall Street, New York
Petersen, E.E. (1977) 'Five steps of the freeze-dry process', *Tea and Coffee Trade Journal*, (April), 14-16
Pintauro, N. (1975) 'Freeze drying processes' in *Coffee Solubilization*, Noyes Data Corporation, Park Ridge, NJ, pp. 126-36
Quijano Rico, M. (1975) 'Study by differential thermal analysis — mass spectrophotometry of green coffee samples' (in Spanish), *7th International Colloquium on the Chemistry of Coffee*, ASIC, Paris, pp. 125-32
Radtke, R. (1973) 'General view on the actual state of coffee packing' (in German), *6th International Colloquium on the Chemistry of Coffee*, ASIC, Paris, pp. 188-98
Radtke, R. (1975) 'Das Problem der CO_2 — Desorption von Rostkaffee unter dem Gesichtspunkt einer neuen Packstoffentwicklung' in *7th International Colloquium on the Chemistry of Coffee*, ASIC, Paris, pp. 323-34
Radtke, R. (1982) 'Advances in quality evaluation of roasted coffee' (in German), *10th International Colloquium on the Chemistry of Coffee*, ASIC, Paris, pp. 81-98
Raemy, A. and Lambelet, P. (1982) 'A calorimetric study of self-heating in coffee and chicory', *Journal of Food Technology*, 17, 451-60
Riedel, L. (1974) 'Calorimetric investigations on the system, coffee-water', *Chemisches Mikrobiologie Technologie Lebensmittel*, 4, 108-12
Rothmeyr, W.W. (1980) 'Food process engineering' in *Food Technology in the 1980's*, The Royal Society, London, pp. 82-6
Schwartzberg, H.G. (1976) 'Continuous counter-current extraction in the food industry', *Chemical Engineering Progress*, 76, 67-85
Shinn, B.E. (1971) 'The centritherm evaporator in the food industry', *Journal of Applied Chemistry and Biotechnology*, 21, 366-71
Sivetz, M. (1973) 'Comparison of changes in roasted coffee beans in pressurized oxygen-free versus atmospheric roasters', *6th International Colloquium on the Chemistry of Coffee*, ASIC, Paris, pp. 199-221

Sivetz, M. (1975) 'Fluid bed drying and roasting of coffee beans, *7th International Colloquium on the Chemistry of Coffee*, ASIC, Paris, pp. 359-66

Sivetz, M. and Desrosier N.W. (1979) *Coffee Technology*, AVI, Westport, Connecticut

Smith, R.F. (1963) 'Measurement of colour of ground coffee by the Gardner meter, *2nd International Colloquium on the Chemistry of Coffee*, ASIC, Paris, pp. 207-11

Spanninks, J.A.M. (1979) *'Design procedures for solid-liquid extractors'*, PhD. Thesis. Agricultural University of Wageningen

Thaler, H. (1979) 'Chemistry of coffee extraction in relation to polysaccharides', *Food Chemistry*, 4, 13-22

United States Patents

(1943) 2,309,092 General Foods Corporation, 'Water Decaffeination'

(1943) 2,324,526 Inredeco Inc. (Nestle-Morgenthaler). 'Manufacture of Soluble Dry Extracts'

(1954) 2,687,355 National Research Corporation Cam. Mass, 'Acid Hydrolysis'

(1966) 3,244,530 General Foods Corporation, 'Recovering Aromatics from Coffee Extracts'

(1971) 3,615,688 Smithern Industries Ltd, 'Coffee Roaster'

(1972) 3,655,398 General Foods Corporation, 'Process for Manufacture of Coffee Extract'

(1975) 3,825,221 Smithern Industries Ltd, 'Coffee Roaster'

Van Pelt, W.H. (1977) 'Concentration of coffee extracts to high products concentration', *8th International Colloquium on the Chemistry of Coffee*, ASIC, Paris, pp. 211-16

Vincent, J-C (1977) Roasting of coffee in a fluidized bed' (in French), *8th International Colloquium on the Chemistry of Coffee*, ASIC, Paris, pp. 217-26

Wiegand, J. (1971) 'Falling film evaporation in the food industry', *Journal of Applied Chemistry and Biotechnology*, *21*, 351-8

Wilbaux, R. (1967) 'Report on collaborative investigations relative to the methods of determination of soluble extract in roasted coffee', *3rd International Colloquium on the Chemistry of Coffee*, ASIC, Paris, pp. 77-83

15 THE PHYSIOLOGICAL EFFECTS OF COFFEE CONSUMPTION

K. Bättig

Introduction

Coffee is a widely and extensively consumed recreational beverage. Its origins and spread across the world have been described by Smith (see Chapter 1), and he has drawn attention to the opposition it aroused in the eyes of political and religious authorities. Despite such opposition the coffee drinking habit survived and still flourishes, helped no doubt by claims for benefits for health, mood and mental stimulation.

Since the beginning of this century a vast amount of scientific literature has been devoted to attempts to define objectively the particular actions and underlying mechanisms of this beverage. However, actual understanding is still scarce. The measurable effects are in most cases modest and subtle, often positive and beneficial, and any effects tend to be similar over a relatively wide dosage range.

Whatever effects are attributed specifically to the consumption of coffee, one must assume that they are associated with compounds which in dietary terms are found only, or almost only, in coffee beverage. Immediately this focuses attention upon caffeine, the chlorogenic acids, possibly the sparingly water-soluble diterpenes kahweol and cafestol and possibly the ill-defined and numerous miscellaneous products of roasting (see Chapter 13).

Of these substances scientific and public attention has been focused upon caffeine because of its known pharmacological actions. The diterpenes and chlorogenic acids by comparison have been little studied, and the miscellaneous products of roasting not at all, unless one considers studies which used whole coffee beverage, and then specific effects cannot be distinguished or assigned to specific components.

Individual green coffees, individual roasted coffees and individual soluble coffee powders may vary considerably in their detailed composition. These factors, complicated by personal preferences regarding brew strength, cup size and frequency of consumption, to say nothing of variations in what else may be consumed simultaneously and vagaries of memory when retrospectively reporting are a major limitation in performing and interpreting epidemiological studies, especially retrospective studies.

Consumption, Absorption and Elimination

The most important facts about consumption, composition, absorption and elimination of the different substances contained in coffee have been reviewed in great detail in a volume by Eichler (1976). An updated and comprehensive account of composition is given in Chapter 13 and accordingly only a brief résumé is given here of the content in coffee brew of those constituents which have been examined for physiological activity.

Composition

Caffeine, which is the main physiologically active ingredient (Kaplan, Holmes and Sapeika, 1974), is classed as an alkaloid because it occurs in plants, because it forms salts with acids and because of its pharmacological actions. Depending on the techniques of preparation, a cup of coffee contains about 80 mg caffeine with a range from about half this figure for a 'weak' coffee to about double this amount for a typical espresso. Although the caffeine content of tea leaves is in the order of 2 per cent and more, coffee beverages usually contain more caffeine than tea beverages, the reason being that more coffee than tea is used for an ordinary serving. Soft drinks such as Coca-Cola contain only about 30-50 mg caffeine per bottle (360 ml), and chocolate contains only a few mg caffeine (per 100 g). Whereas coffee contains mainly caffeine, tea also contains theophylline, and in cocoa nuts, theobromine constitutes the main active ingredient. Typical coffee brews contain some 30-120 mg total chlorogenic acids per 100 ml with CQA dominant and diCQA exceeding FQA. There would, however, appear to be some ill-defined roasting products (see Chapter 13). At present there are no data for the kahweol or cafestol content of coffee brew.

Consumption

The consumption of coffee varies widely across different cultures and countries. A reliable estimation of *per capita* consumption per year is difficult due to many different factors. A recent study (Roggenkamp, 1982) based on gliding averages for periods of three years for imports of raw coffee classified a series of countries into four groups. Low consumption (less than 3.5 kg) was obtained for all eastern European socialist countries, for Canada, North America, and for Portugal, Spain and Great Britain among the western European countries. Medium levels of consumption (3.5-7 kg) were obtained for the USA and the European countries, Italy, Austria, France and the Federal Republic of Germany. Heavy consumption (7-10.5 kg) was obtained for Switzerland, Norway and the three Benelux countries. Extreme consumption was noted for the three Scandinavian countries, Denmark, Sweden and Finland. These figures are well in line with data presented earlier for the USA and the different European countries (Pan American Coffee Bureau 1972). There are also great differences between countries with respect to the proportion of coffee consumed as soluble preparations (Butler, 1983). Great Britain, Ireland and Australia rank at the top with more than 70 per cent consumed as solubles (the raw/ soluble conversion factor can be estimated at about 2.5), followed by Japan (50 per cent) and Canada (39 per cent). Figures around 20 per cent were calculated for the USA, Switzerland and Spain, whereas the proportion was lower for all other countries, reaching the lowest value for Italy with only 1 per cent. Even more difficult is the estimation of the average consumption of caffeine *per se*. For the United States, which appears within the list of coffee consumption in the middle range, the average caffeine consumption has been estimated at about 200 mg/day, with 90 per cent of this amount being obtained from coffee.

Coffee consumption also varies considerably between individuals, as indicated by the usually large variances obtained in questionnaire self-reports. However, little is known about possible factors underlying such differences. Equivocal results have been obtained even for such general aspects as sex, time of day and age (Gilbert, 1976). More consistently, coffee consumption was found to be higher in cigarette smokers (Thomas, 1973). Some studies have also reported increased coffee consumption for periods of increased psychic stress (Conway, Vickers, Ward and Rahe, 1981; Verner and Krupka, 1982).

Absorption, Metabolism and Elimination

Caffeine is readily and rapidly absorbed after both enteral and parenteral application. It is rapidly distributed throughout all tissues as a function of their water content. In man, caffeine appears in all tissue fluids about 5 minutes after application, and peak blood plasma levels are reached after about 20 to 30 minutes (Marks and Kelly, 1973). The substance also penetrates easily into the different organ systems including the brain, gonadal tissues, placenta, fetal tissue and maternal milk (Berlin, 1981: Warszawski and Gorodischer, 1981).

In man, caffeine is almost entirely metabolised; only about 1 per cent is excreted via the urine in its non-metabolised form. The main metabolites in man belong to the classes of demethylated xanthines, uric acids or uracil derivations of caffeine (Bonati *et al.*, 1982; Callahan *et al.*, 1982; Tang-Liu, Williams and Riegelman, 1983). The half-life of caffeine in human plasma has been reported to be about three to six hours, but varies widely between individuals. A considerable amount of data is available by now for the caffeine plasma levels reached with normal coffee consumption. Newly developed methods for determining the saliva concentration of methylxanthines are also of great help in assessing plasma levels by a non-invasive procedure (Khanna *et al.*, 1982). These figures are important for comparisons with animal experiments. For single bolus intakes in the order of 200 to 300 mg caffeine, plasma levels reach values between 5 and 10 mg/1 plasma (Axelrod and Reichenthal, 1953). With such a dose distributed over several servings per day, the plasma levels can be expected to be in the range below 5 mg/1.

For comparisons with animal experiments it should be remembered that dosages indicated as mg/kg body weight neglect the important impact of the differences between species in metabolic rate, which is dependent on surface area and amounts for different species including man, mouse and rat to about 1,000 kcal/m^2 (Bell, Davidson and Scarborough, 1953). Dose equivalents calculated on the basis of body weight corrected for body surface (= 'metabolic weight') are therefore considerably lower than dose equivalents based solely on body weight. Calculated on the basis of body weight alone, a factor of roughly 3 to 4 seems to be appropriate for comparisons between man and the rat. This would fit similarly for toxic and psychotropic stimulant effects. The lethal dose for 50 per cent of a group (LD_{50}) for the rat has been reported to be 150 mg/kg and the lowest fatal dose in humans to be 57 mg/kg (Boyd, Dolman, Knight and Sheppard, 1965; Peters, 1967). On the other hand, doses between 250 and 500 mg have been reported reliably to produce psychostimulant effects in humans (ca 4-7 mg/kg) as opposed to a range of 8-20 mg/kg in the rat, as reviewed in more detail below. However, taking into account a correction for metabolic weight, it has been estimated that lethal doses for rats, cats and humans are in about the same order (Peters, 1967).

Several different conditions have been found to alter the time course of elimination considerably. Whereas caffeine clearance is relatively slow in patients with alcoholic liver disease (Statland and Demas, 1980), it is accelerated up to more than twofold in heavy smokers (Kotake *et al.*, 1982). In women, the use of contraceptives has been reported to double the half-life (Patwardhan, Desmond, Johnson and Schenker, 1980), as is also the case for the second and third trimesters of pregnancy, a finding suggested to be due to a decrease in biotransformation caused by the altered hormonal milieu (Aldridge, Bailey and Neims, 1981; Knutti, Rothweiler and Schlatter, 1982). A roughly tenfold increase in the half-life has been observed in neonates, with gradual normalisation by the age of about 6 months (Aranda, Collinge, Zinman and Watters, 1979; Parsons and Neims, 1981; Gorodischer and Karplus, 1982). The slow clearance in neonates has been ascribed to the fact that neonates are deficient in N-demethylation, which causes them to excrete caffeine in the urine mainly as the unchanged compound (Horning, Butler, Nowlin and Hill, 1975).

Coffee and Extracaffeinic Substances

Up to now most of the scientific literature on 'coffee' has dealt with caffeine. Coffee, however, is more than just savoury caffeine, as it contains literally hundreds of other substances (Eichler, 1976). Furthermore, the exact composition varies considerably with different species of the plant, with different origins and with different treatments. Besides proteins, polysaccharides, monosaccharides and a multitude of other substances, one finds in coffee beverages notably a rather high amount (relative to other dietary sources) of chlorogenic acids and associated conversion products (see Chapter 13), which is usually several times that of caffeine and may reach amounts of up to 10 per cent of the dry mass of green coffee beans. Chlorogenic acids have therefore attracted more scientific interest than many other extracaffeinic constituents of coffee. This group of compounds is complex and includes at least 13 isomers (Clifford and Staniforth, 1979). The older literature concerning these substances is characterised by a rather confusing nomenclature and only more recent classifications (Clifford, 1979) provide the clarity needed for future detailed research into the effects of the different substances belonging to this class.

Eichler (1976), in his review on coffee, mentioned studies showing that chlorogenic acids somewhat delay the clearance of caffeine, that they accelerate the resorption of caffeine from the intestine and that some compounds of this class form complex bonds with caffeine. Furthermore, chlorogenic acids were described as stimulating intestinal motility (Czok and Lang, 1961a). Whether this class of substances has any central effects or not has been the subject of only a few studies. Czok and Lang (1961b) reported a decrease in the thresholds for convulsive electric shocks, whereas Krönig and Künkel (1973) found no evidence for stimulating effects on the electroencephalogram (EEG) with coffee substances other than caffeine. Possible important biological effects have, however, recently been suggested for kahweol palmitate and to a lesser extent for cafestol palmitate (see Chapter 13). Lam, Sparnins and Wattenberg (1982), using extracts of these substances from green coffee beans and in adding 20 per cent green coffee to the diet of mice, found a potent enhancement of glutathione S-transferase, an enzyme which constitutes a major detoxification system.

Extracaffeinic substances certainly also play an important role in taste quality, which in turn may influence spontaneous consumption. Ohiokpehai, Brumen and Clifford (1982) have presented analytical data suggesting that ratios of different chlorogenic acid isomers play a role in coffee astringency and thereby coffee quality. Astringency (see also Chapter 13) occurs through precipitation of salivary proteins or glycoproteins, thus removing their lubricating action. Low astringency was found to be correlated with acceptable mouth sensations, whereas higher levels have been blamed for less acceptable sensations. The taste and smell of coffee are important aspects in any research on behavioural effects for several reasons. As will be seen in later sections, many studies failed to obtain clear dose-effect relations, and several studies also suggest that cognitive factors may be important in mediating the effects of drinking coffee, since these can be affected by the information given to the subjects with respect to presence, absence or dosage of caffeine. If caffeine is given by means of coffee, one can therefore expect a mixture of true pharmacological effects, of effects elicited by taste, smell and situational conditions and of effects based on mechanisms of learning and conditioning.

A number of studies controlled for this aspect by comparing effects of decaffeinated coffee with and without the addition of caffeine. Several studies suggested that the presence of caffeine cannot be detected by taste. Hall, Bartoshuk, Cain and Stevens (1975) have provided data indicating that for the majority of people the threshold for detection of caffeine is at the rather low level of 65 µg/ml, and it is widely accepted that caffeine accounts for no more than 10 per cent of coffee beverage bitterness.

Apart from the implications of non-caffeinic substances on behavioural and physiological functions, their significance in chronic toxicology is even less clear. The multitude of different substances makes difficult any analytical or direct approach in studying single substances and the task becomes even more complex in view of the interactions to be expected between single substances.

Cellular Basis of Caffeine Action

Three main hypotheses for explaining the basic cellular actions of xanthines have been discussed so far. They concern xanthine effects on the translocation of intracellular calcium, on increasing the accumulation of cyclic adenosine 3',5'-monophosphate (cyclic AMP) due to inhibition of phosphodiesterase and, finally, on the competitive blocking of adenosine receptors. The first two hypotheses are hampered by the fact that, in general, for these mechanisms dose ranges far above the therapeutic range are needed.

Studies on the effects of xanthines on intracellular translocation of calcium have concentrated mainly on the effects of caffeine on skeletal muscle. The substance increases the twitch response of the isolated frog sartorius muscle preparation. It is believed that this effect is due to increased release of calcium from the terminal cisternae of the sarcoplasmic reticulum. This effect has, however, been shown to occur at relatively moderate dose levels only under particular experimental circumstances (Katz, Repke and Hasselbach, 1977). Normally, the effect can be observed only at dose ranges which far exceed the therapeutic range. The exact mechanism is not yet known, and besides a direct action on the release mechanism (Ito, Osa and Kuriyama, 1973) the possibilities of a decreased calcium re-uptake (Ohba, 1973) or a direct action on the sarcoplasmic reticulum (Thorpe and Seeman, 1971) have also been proposed. Furthermore, it also seems likely that caffeine facilitates transmitter release by increasing calcium release at neuromuscular junctions (Onodera, 1963).

The phosphodiesterase hypothesis implies that caffeine might indirectly increase the concentration of cyclic AMP, which plays an important role in the process of synaptic transmission, particularly in the case of adrenergic synapses. In early studies cyclic AMP was found to play an important role in liver and muscle glycogenolysis (Rall, Sutherland and Berthet, 1957). Later, caffeine was found to facilitate the glycogenolytic effect of epinephrine (Berthet, Sutherland and Rall, 1957), and this effect was found to be correlated with the inhibition of cyclic nucleotide phosphodiesterase (EC 3.1.4.17) and the consequent accumulation of cyclic AMP (Sutherland and Rall, 1958). Cyclic AMP also serves as a second messenger in the mediation of the effects of several other hormones (Sutherland, Robison and Butcher, 1968). However, the effect of caffeine on cyclic AMP is not only considerably weaker than that of theophylline, but it also remains modest in magnitude even at concentrations well above the therapeutic range (Beavo et al., 1970). Thus, it remains questionable as to whether and in what functions phosphodiesterase inhibition may play a role in the mediation of caffeinic effects.

The adenosine hypothesis, which was originated by the observation that methylxanthines competitively block adenosine binding at specific receptor sites of a wide variety of cells (Sattin and Rall, 1970; Baer and Drummond, 1979), has received considerably more experimental support over the last years than both the phosphodiesterase and calcium hypotheses. Recently it has been demonstrated that metabolically stable analogues of adenosine are, even at very low dose levels, highly potent depressants of overt motor behaviour and that across different xanthines the behavioural stimulating action increases with increasing binding affinity to labelled adenosine receptors (Snyder et al., 1981). Although with such compounds motor depression is seen even with minimal doses, with increasing dose levels animals remain alert and responsive to nociceptive stimulation, and even a several thousandfold dose is not lethal. This can be taken as indicative that the behavioural effects are due to central actions rather than to systemic effects such as hypotension. This notion was further supported by the finding that a xanthine analogue not penetrating the blood-brain barrier failed to antagonise the behavioural depression induced by adenosine analogues.

Great interest has therefore been concentrated over the last years on the role of central adenosine receptors. Adenosine inhibits neuronal firing rather consistently (Phillis and Wu, 1981), and the methylxanthines are the most specific antagonists for this action. Although histochemical methods for localising adenosine receptor sites are not yet available, the technique of autoradiographic marking has produced some important first indications towards a mapping of the brain for adenosine receptors (Goodman and Snyder, 1982). According to this study, low densities of adenosine receptors were detected in white matter and hypothalamic nuclei, intermediate densities in certain layers of neocortex, in the striatum and in certain thalamic nuclei, whereas the highest densities were localised in molecular layers of the cerebellum, hippocampus and dentate gyrus, and in some thalamic nuclei. However, there is still a long way to go towards a detailed understanding of the functional interactions between xanthines and the adenosine systems. As recently reviewed by Daly, Bruns and Snyder (1981), the situation is complicated by the existence of different types of adenosine receptors which are located at the outer membrane surface or inside the cells and which may be either inhibitory or stimulatory in nature, depending upon their differential sensitivity to different adenosine analogues and drug interactions and by their existence in a wide variety of cell types. Nevertheless, there is good evidence that adenosine appears to act in a modulatory fashion to control adenylcyclase by promoting the formation of cyclic AMP, which in turn is involved in different second messenger functions and the phosphorylation of proteins. It also seems likely now that adenosine acts presynaptically as an inhibitor of the release of such different transmitters as acetylcholine, noradrenaline and γ-aminobutyric acid (GABA). Taking into consideration that for the neurobehavioural functions studied so far the xanthines, including caffeine, act as adenosine antagonists already at dose levels well within the therapeutic range, it seems highly probable that the adenosine hypothesis will attract increased scientific interest in the near future (Daly, 1983).

Effects on the Vegetative System

All xanthines, including caffeine, affect the functioning of the vegetative system. The mechanisms of action are, however, rather complex, and the direction of action often varies from one subsystem to another.

Effects on Smooth Muscles

Although caffeine, to a lesser extent than theophylline and theobromine, exerts a direct relaxant effect on the smooth muscles of the bronchioles, biliary tract, gastrointestinal tract and portions of the vascular system, the mechanism of action is rather complex. It was proposed that the variation encountered for this effect across the smooth muscles of different organ systems may be due to the relative importance of electromechanically coupled contractions caused by the action potential frequency in spontaneously contracting muscles, such as in the intestine, and the unimportance of this phenomenon in other muscles, such as those of the pulmonary arteries (Somlyo and Somlyo, 1967). The predominant effect of caffeine seems to be a decrease in or elimination of action potentials (Imai and Takeda, 1967; Somlyo and Somlyo, 1967), and this effect tends to follow the pattern of agonistic actions on beta-receptors. However, in some cases biphasic effects with initial contraction followed by relaxation have also been reported, particularly for higher doses of the drug (Sunano and Miyazaki, 1973).

With respect to different organ systems, relaxation has been reported quite unequivocally for the uterus, but the therapeutic value of xanthines, as for instance in the case of imminent abortion (Crainicianu, 1969), is doubtful. For the gall bladder, an antagonistic effect has been shown for the xanthines, particularly theophylline. On the other hand, coffee consumption is followed by a cholecystokinetic effect which may be due, at least in part, to constituents of the beverage other than caffeine, such as products of the roasting process (Keiner and Weder, 1967). The effects on the intestine seem to vary from species to species and also for different parts of the tract. Whereas coffee beverages can cause increases in peristalsis or even diarrhoea (Kretschmer, 1936), the effects of caffeine are less consistent and weaker than those of other xanthines (Mitznegg, 1976). Perhaps the most important effect of xanthines on smooth muscles consists in the relaxation of the bronchial muscles, which becomes particularly apparent for spasmic conditions, such as with asthma or after histamine. However, in this respect theophylline and theobromine are again much more potent than caffeine (Hume and Rhys-Jones, 1961). Since stimulation of respiration by xanthines relies not only on bronchodilation but also on a stimulation of the medullary response to carbon dioxide and on an increased pulmonary blood flow due to vasodilation, the net sum of all these actions has to be considered. The minute-volume of respiration was found to be increased after 300-500 mg caffeine (Starr et al., 1937), and even smaller doses were found to antagonise the depressing effects of opioids (Bellville et al., 1962).

Effects on Secretion

The effect of xanthines on exocrine glandular systems is mainly to increase secretion. For caffeine such effects have received particular attention in the case of gastric secretion. Both acid and pepsin production increase after moderate doses of caffeine in man (Debas, Cohen, Holubitsky and Harrison, 1971). It has also been known for a long time that substances in coffee other than caffeine are also very important for increases in gastric secretion. Cohen and Booth (1975) have shown that the effects of decaffeinated coffee are only slightly smaller than those of normal coffee and that each is about twice as potent as the equivalent of pure caffeine (Cohen, Debas, Holubitsky and Harrison, 1971). In medical practice these findings find their counterpart in the common recommendation for ulcer patients and patients with gastrointestinal diseases in general to abstain from coffee. However, although this secretogenic action of coffee is well known, the exact mechanism and its possible relation to the role of cyclic

AMP in the regulation of secretion remains poorly understood (Jacobson and Thompson, 1976).

Caffeine, as well as other methylxanthines, also has a stimulating effect on the endocrine secretion of different glands. The best known effect in this respect is probably the increase in circulating catecholamines and renin (Robertson et al., 1978). Stimulation of insulin and parathyroid hormone has been reported for theophylline only (Bowser, Hargis, Henderson and Williams, 1975; Somers, Devis, Van Obberghen and Malaisse, 1976), and this for dosages which exceed the therapeutic range. Inhibition of secretion was observed for the release of histamine from mast cells; this effect has been related to the anti-inflammatory action of caffeine in various experimental model systems (Vinegar et al., 1976). As caffeine in moderate doses also potentiates the anti-inflammatory effect of aspirin and phenylbutazone, this interaction could be considered as a rationale for the frequent combinations of these drugs in medical use. For subjective pain, however, the addition of 130 mg caffeine was recently found to have no effect on pain reduction induced by acetaminophen (Winter, Appleby, Ciccone and Pigeon, 1983).

Metabolism

Caffeine at moderate doses (3-9 mg/kg) produces a slight but measurable increase in the metabolic rate in the order of about 10 per cent (Means, Aub and DuBois, 1917; Haldi, Bachman, Ensor and Wynn, 1941). This effect reaches its maximum between one and three hours after application and can also be observed in habitual coffee drinkers. During the early phase of this period elimination of carbon dioxide was found to be increased, followed by a compensatory reduction for the later phase (Haldi et al., 1941).

Several effects on the intermediary metabolism have also been described, but the evidence is less consistent. Although caffeine is widely believed to raise blood glucose levels, several more recent studies (Avogaro, Capri, Pais and Cazzolato, 1973; Daubresse et al., 1973; Oberman et al., 1975) have found only little or no effect. A lipolytic effect by producing increased plasma levels of free fatty acids has been evidenced in several investigations (Avogaro et al., 1973; Daubresse et al., 1973; Portugal-Alvarez et al., 1973), but the magnitude of this effect seems to vary considerably across subjects. It was found to be greater in subjects with normal weight than in obese subjects (Oberman et al., 1975) and in younger than in older subjects (Avogaro et al., 1973). Furthermore, it was also found that the lipolytic effect of caffeine can be blocked by insulin (Portugal-Alvarez et al., 1973) and by food intake (Studlar, 1973). The possibility that caffeine might be atherogenic, however, has been discounted by Avogaro et al., (1973). In the rabbit no atherogenic effects were found even after prolonged chronic treatment with relatively high doses of coffee or caffeine (Czochra-Lysanowicz, Gorski and Kedra, 1959).

Diuretic Action

Although all methylxanthines possess diuretic action, caffeine is the least powerful substance among them (Armitage, Boswood and Large, 1961). It was noticed quite early that the diuretic action of caffeine is rapid, of short duration and dependent on the amount of simultaneous fluid intake (Lie, 1930) and that tolerance to this develops rather rapidly (Kihara, 1928). Increases in renal blood flow seem to be a major contributing factor in caffeine diuresis (Fülgraff, 1969).

Effects on the Cardiovascular System

Although the cardiovascular system belongs to the vegetative system, the circulatory effects of xanthines, particularly theophylline and caffeine, are so prominent that they merit more detailed consideration. Furthermore, the circulatory effects of these substances are complex and mediated in part through antagonistic actions. The main actions appear by now to be the direct effects on the heart and the vascular tissues and the indirect effects through increased release of catecholamines and possibly the renin angiotensin system (Robertson et al., 1978; Burghardt et al., 1982), whereas effects on the vagal and vasomotor centres of the brain stem remain less well documented.

Effects on the Heart

A number of studies have shown that caffeine increases both the force and the rate of contraction in isolated mammalian preparations (de Gubareff and Sleator, 1965). Heart-lung experiments have further shown that this effect is not an indirect consequence of coronary dilation (Smith and Jensen, 1946). It has been reported for caffeine that the duration of the action potentials increased in isolated atria at low frequencies of stimulation, resembling catecholamine action, whereas at high frequencies of stimulation the rise in the action potentials was faster and the duration shorter, resembling the action of calcium (Gualtierotti, 1955a, 1955b; Shibata and Hollander, 1967).

Several mechanisms may underlie such stimulant effects of caffeine and other xanthines on the heart muscle tissue. The administration of 250 mg caffeine in humans has been seen to increase plasma adrenaline and noradrenaline by about 100 per cent and 50 per cent, respectively (Robertson et al., 1978). However, with theophylline it was observed that the increase in contractive force of cat papillary muscles was not abolished but only decreased with beta-antagonists, suggesting that the catecholamine hypothesis can explain this stimulating action only in part (Marcus, Skelton, Grauer and Epstein, 1972). Another frequently discussed hypothesis suggests that the xanthines inhibit phosphodiesterase, leading to increased levels of cyclic AMP, which in turn is followed by increased glycogenolysis and a rise in glucose-6-phosphate levels (Ellis, 1956, 1959; Belford and Feinleib, 1962). The validity of this hypothesis might be put in question, however, by the facts that caffeine is in this respect considerably less potent than other xanthines and that even maximal doses of caffeine were found to inhibit phosphodiesterase by only a few per cent (Beavo et al., 1970). The potency of caffeine to release calcium from the cisternae of the sarcoplasmic reticulum suggests another mechanism of action. This effect has been demonstrated for caffeine with low therapeutic doses in skeletal muscle preparations (Katz et al., 1977), and similar effects in skeletal and cardiac muscle tissue have also been shown (Blinks, Olson, Jewell and Braveny, 1972).

Effects on Blood Vessels

Therapeutic doses of caffeine characteristically produce a decline in peripheral resistance which is generally modest, independent of arterial blood pressure and short lasting (Ogilvie, Fernandez and Winsberg, 1977). As this effect is modest, however, it remains of little value in the treatment of peripheral vascular disease. In animal preparations as well as with the techniques of cardiac catheterisation in patients, the xanthines have particularly been observed to increase coronary blood flow. However, this increase hardly contributes to an increase in the oxygen supply to the cardiac muscle and may, at least in part, also be an indirect consequence of the simultaneous increases in heart work. In contrast to the dilating action of xanthines, including that of caffeine on peripheral blood vessels, these substances are known to increase cerebrovascular resistance (Moyer et al., 1952). This effect is believed to be at the base of the clinically observed relief for migraine and other types of headache caused by cerebrovascular distention.

Effects on the Medullary Centres

Caffeine, which is generally less potent in its effects on cardiac functions, coronary dilation, smooth muscle relaxation and diuresis than theophylline and theobromine, is more active than these drugs in stimulating the medullary respiratory, vasomotor and vagal centres. However, also these effects are rather modest, and the stimulation of the respiratory functions may become apparent only in cases where these centres are depressed by barbiturates, opioids or other drugs. Electrophysiological studies have shown increases in firing rates of neurones in the brain stem reticular formation of the rat after administration of 1-2 mg/kg caffeine (Foote et al., 1978). It has frequently been assumed that such stimulant effects of the drug on the brain stem also include sufficient vagal stimulation to exert a slowing effect on heart rate and thus to counteract the direct stimulant effect of the drug on heart rate. However, direct evidence for such a hypothesis remains poorly documented (Rall, 1980).

Combined Effects

In view of the partly antagonistic effects of xanthines on the different subsystems it is no surprise that the net effects of caffeine were found to be modest or absent in most of the experimental studies. With dosages over a relatively wide range, not only slight brachycardia but also tachycardia or no effects have been reported. With excessive doses, however, tachycardia becomes the prominent response to the drug. Cardia arrhythmias have also been reported for persons who use caffeine beverages excessively.

A recent series of studies (Robertson *et al.*, 1978; Robertson *et al.*, 1981;) seems to indicate that the development of tolerance might also constitute an important factor in the cardiovascular effects of caffeine. They found in subjects who did not consume coffee and abstained from other caffeine-containing preparations a net increase in both systolic and diastolic blood pressure with heart rate first increasing and then decreasing. These effects, as well as the accompanying increases in plasma renin and norepinephrine, disappeared gradually after the initial dose of 250 mg had been given repeatedly over a week. In line with this observation a recent study by Freestone and Ramsay (1982) observed in patients with mild hypertension a lowering of blood pressure if they had previously abstained over night from cigarette smoking and coffee drinking. A moderate but significant increase in the order of about 10 mmHg was obtained for a short period of about 15 minutes when they were allowed to smoke two cigarettes and for about two hours when they were given 200 mg caffeine, either alone or in combination with smoking. On the other hand, pulse rate increased significantly only after smoking.

Effects on the Central Nervous System

Surprisingly few studies so far have been concerned with the effects of coffee and caffeine on the central nervous system. In healthy male subjects Gibbs and Maltby (1943) observed a shift in EEG power toward the fast side (high frequency, low amplitude) of the total spectrum, and this effect was very similar to that produced by active behavioural attention. Goldstein, Murphree and Pfeiffer (1963) made a similar observation after administering 250 mg of caffeine and showed further that the effect was similar to that obtained with other stimulant drugs. Opposite effects with respect to the different frequency bands of the EEG, usually seen after alcohol, were prevented by caffeine in a study by Czarnecka and

Gruszczyński (1978). Cloutte, Glass and Butler (1977) remarked a decrease in alpha prevalence, and Künkel (1976), using 200 mg of caffeine, observed in addition a decrease in theta power, an increase in alpha dominant frequency and a decrease in theta dominant frequency. He further observed that effects of coffee and caffeine were indistinguishable, but different from the effects of the two placebos, warm water and decaffeinated coffee. Only a few researchers have also investigated effects on stimulus elicited changes in the EEG. Ashton, Millman, Telford and Thompson (1974) found CNV, the contingent negative variation which occurs in the EEG prior to an expected stimulus, to increase significantly after ingestion of 300 mg caffeine. On the other hand, Spilker and Callaway (1969) and more recently Elkins *et al.*, (1981) did not find any effects of caffeine on stimulus-evoked responses of the EEG. Several animal studies suggest that the caffeine-induced activation of the EEG might be due to an increase in cortical activation through the reticular formation. Using *encéphale isolé* preparations Schallek and Kuehn (1959) found that in the cat 10 mg/kg caffeine enhanced the cortical response to stimulation of the reticular formation consisting of an increase in the dominant EEG frequency. Similar results with this method have also been obtained by Jouvet, Benoit, Marsallon and Courjon (1957) in the cat and by White (1963) in the rat. In the waking dog Hori and Yoshii (1963) found with 10 mg/kg both a general EEG activation and an enhancement of its evoked response to click stimuli. Goldberg, Horvath and Meares (1974) obtained similar effects in the waking rabbit in response to light flashes. More direct evidence for an action of caffeine on the reticular formation was obtained by Hirsh, Forde and Pinzone (1974) and Forde and Hirsh (1976), who found caffeine at low doses to have direct effects on single neurone activity in the reticular formation, and using a similar technique, Chou, Forde and Hirsh (1980) reported that the arousing effect of caffeine might in part be related to the suppressive effect of the drug on the medial thalamic nuclei system.

Several studies involving transmitter research suggest that caffeine stimulation might, at least in part, be due to a dopaminergic stimulation, since caffeine increased circling in mice after unilateral destruction of dopaminergic neurones (Von Voigtlander and Moore, 1973; Fuxe and Ungerstedt, 1974 Fuxe *et al.*, 1975; Watanabe, Watanabe, Hagino and Ikeda, 1978). Consistent with these reports, it was also found that caffeine increases the turnover rate of the catecholamines, noradrenaline and dopamine, in the brain (Berkowitz, Tarver and Spector, 1970; Waldeck, 1971), and Andén and Jackson (1975) observed that in the rat caffeine potentiated the stimulation of locomotor activity induced by local injection of dopamine into the nucleus accumbens. Thus, in summary, these experiments present good evidence that caffeine enhances cortical arousal and that this effect is mediated by the ascending activating systems of the brain.

In comparison to the studies on brain functions, hardly any studies have been carried out for clarifying the possible effects of the drug on the spinal cord. Sant'Ambrogio, Frazier and Boyarsky (1962), using electrophysiological methods, found that even at high dosages caffeine had no effect on monosynaptic reflexes in the isolated spinal cord of the cat, whereas a low dose of 5 mg/kg facilitated polysynaptic reflexes.

Spontaneous Activity

Up to now spontaneous activity has been widely studied in animals, but only exceptionally in man, as for instance recently by Rapoport et al., (1981), who found that caffeine (3 mg and 10 mg/kg) increased activity in children, whereas such increases reached significance in adults only for the subgroups of high coffee consumers and only for the higher dose level of 10 mg/kg. Although the scarcity of human data is understandable in terms of the available methodology, it is regrettable, as animal work in this field impresses by the relatively high degree of consistency of the results, in spite of different species and different testing situations having been used. Although most studies employed mice or rats, the list of tested animals also includes dogs (Kusanagi, Fujii and Inada, 1974), gerbils (Pettijohn, 1979), race horses (Fujii et al., 1972) and even hornets and bees (Ishay and Paniry, 1979), all of which were shown to increase spontaneous activity at similar dose levels.

Activity cages using different methods of recording were probably used most frequently (Siegmund and Wolf, 1952; Boissier and Simon, 1965, 1967; Nieschulz, 1968; Fog, 1969; Cox, 1970b; Gupta, Dandiya and Gupta, 1971; Hach and Heim, 1971; Kallman and Isaac, 1975; Hughes and Greig, 1976). The effect of caffeine was generally found to be biphasic with stimulation being obtained at dose levels ranging from a few up to about 25 mg/kg and depression of performance at higher levels. The maximal effective dose varied somewhat from study to study and amounted to levels between 5 and 20 mg/kg in most of these studies for both mice and rats, but such methods measure only overall motor output. With the method of direct observation of the animals in open fields it is possible to separate out different aspects of activity, such as non-specific ambulation, ambulation directed toward novelty, rearing and preening. Whereas non-specific ambulation was also increased in this type of experimentation (Gupta et al., 1971; Hughes and Greig, 1976), rearing remained unaffected (Hughes and Greig, 1976) or was increased only at a higher dose level (Gupta et al., 1971). Exploration, as evaluated by head dipping into holes in the side walls (Nieschulz, 1968) or the floor (Boissier and Simon, 1965, 1967), was found to increase within similar dose ranges, but the effect seemed to be less pronounced than that on non-specific spontaneous motility (Nieschulz, 1968).

Several studies suggest that caffeine-induced stimulation is greater when compared with control performance levels which are low due to fatigue, habituation or low motivational levels than when compared with high levels of performance. This was seen with treadmill performance in animals that were previously exhausted (Takagi, Saito, Lee and Hayashi, 1972) or in food-rewarded alley-running performance where the drug particularly improved speed in the starting section, increasing it to levels similar to those seen under control conditions only for the goal section of the alley (Wanner and Bättig, 1965). In a recent study exploratory activity was compared for tunnel mazes differing in the complexity of the alley configurations (Rosenberg, Martin, Oettinger and Bättig, 1983). With more complex alley configurations the intrasession activity decline was smaller than with a less complex alley configuration. Caffeine mainly delayed this habituation and appeared, thus, to be more effective with the simple than the complex configuration.

In summary, it appears that caffeine enhances different types of spontaneous activity in different animal species, at similar dose levels and without inducing pathological disorders such as stereotypies (Fog, 1969), which are seen with many other stimulants, particularly amphetamine and related drugs.

Effects on Motor Performance

Quite early Weiss and Laties (1962) came to the conclusion that caffeine 'prolong[s] the amount of time during which an individual can perform physically exhausting work'. Both the possible mechanisms underlying such an effect and the degree of generality of this conclusion merit closer consideration.

Mechanism of Action

Caffeine tends to contract striate skeletal muscles. This effect appears, however, only at dosages which far exceed the levels reached by the range of habitual consumption. A minimal concentration of 400 mg/ml plasma has been reported to be required for causing muscle contraction (Lin and Bittar, 1974), whereas about 6 μg/ml plasma caffeine is an average concentration found after a cup of strong coffee (Marks and Kelly, 1973). Spontaneous contractions as well as the increases in twitch tension which already appear at considerably lower doses can be obtained both after systemic application and in isolated denervated muscle preparation. The increases in twitch tension go along with a lowering of the stimulation threshold but without changes in the resting membrane potential of the action potential (Foulks, Perry and Sanders, 1971a, b). Even at concentrations below those producing increases in twitch tension caffeine increases both the cellular influx and efflux of calcium (Feinstein, 1963), but the effect of caffeine is more closely correlated with the level of intracellular calcium concentration than with the rate of cellular influx and efflux (Zett, 1966). Release of calcium from the vesicles of the sarcoplasmic reticulum constitutes an important step in the actin-myosin contraction response, and it is supposed that intracellular mobilisation of calcium may represent a major mechanism of caffeine action (Orentlicher and Ornstein, 1971). However, a direct action of caffeine on the neuromuscular end plate by facilitating acetylcholine release has also been verified (Hofmann, 1969) as well as an increase in muscular oxygen consumption (Novotny and Vyskocil, 1966) and in glycogen-glucose transformation (Danforth and Helmreich, 1964). Since the increase in the metabolic rate as well as increases in blood flow (Hirchie et al., 1971) can be observed at substantially lower dose levels than any direct action on the muscle system, it becomes likely that any effects on physical performance might be due to such indirect rather than to direct actions of the drug.

Motor Performance

In animal studies the effect of caffeine differed for different methods of measuring physical performance. Improvements were found in mice performing in a treadmill (Marriott, 1968; Takagi et al., 1972), but not in mice or rats swimming to exhaustion (Spengler, 1957; Jacob and Michaud, 1961). This latter, negative result might be explained by the fact that swimming endurance depends not only on physical capacity but also on complex motivational interaction, since in these tests rescue from the water serves as reward for early 'giving up.'

In humans caffeine might improve performances which are subject to fatigue. Foltz, Ivy and Barborka (1942, 1943) required the subjects to work on bicycle ergometers until exhaustion. This performance was then repeated after a 10-minute rest period during which 500 mg caffeine sodium benzoate was administered. Performance during this second period was significantly improved by the medication. Several studies with negative results might have been biased by the fact that the effect of training was not controlled (Spengler, 1957; Perkins and Williams, 1975). Costill, Dalsky and Fink (1978), who studied the effects on performance in well-trained competitive cyclists, found with 330 mg caffeine a prolongation of endurance by about 20 per cent. Athletic performance in short events was studied by Eichler, Klein and Stephan (1949) in 80 subjects using the double-blind technique. In comparison to preapplication performance they found improvements after 250 mg caffeine in 54 per cent of the subjects for the long jump, in 60 per cent for the shot put and in 80 per cent for the 100 m sprint, whereas with decaffeinated coffee they observed impairments in 83 per cent for long jump, in 70 per cent for the shot put and in 72 per cent for the 100 m sprint. Another study by Ambrozi and Birkmayer

(1970) measured acceleration for boxing punches in 45 subjects and found acceleration to be significantly improved after 250 mg caffeine.

Measurements of performance of single groups of muscles by the ergograph method revealed equivocal results (Merton, 1954), as did measurements of the frequency of tapping, which has often been used to determine the maximum speed of voluntary motor discharge (Gilliland and Nelson, 1939; Dahme, Lienert and Malorny, 1972; Clubley, Henson, Peck and Riddington, 1977). One possible reason for these contradictory results could be seen in the well-documented fact that caffeine, even at average dose levels, tends to increase tremor and to impair hand steadiness (reviewed by Calhoun, 1971). Impairments due to decreased hand steadiness were found for tasks such as needle threading or hitting a small target with a stylus. Such effects may appear rather late, several hours after caffeine intake, and in several studies they were preceded by a transient state of improvement or no effect (Horst, Buxton and Robinson, 1934a; Horst, Robinson, Jenkins and Bao, 1934b; Goldstein, Kaizer and Warren, 1965a).

In summary, the effect of caffeine on motor work output appear, thus, to be more complex than one might expect. Predominantly one finds more reports of beneficial than of negative or no effects. This might, at least in part, be due to the fact that the wide variety of tasks used in such studies depend to different degrees on motivational states, central nervous motor effects, energy utilisation, training, tremor and hand steadiness, which are all, and in part differentially, affected by the drug.

Human Psychomotor and Routine Performance

A great number of studies have investigated the effects of coffee or caffeine on a series of reaction time tasks differing both in response and stimulus modality. No effects were observed by Lehmann and Csank (1957), von Klebelsberg and Mostbeck (1963), Lovingood, Blyth, Peacock and Lindsay (1967) and Dahme et al., (1972), although these studies used widely differing dosages in the range from 100-700 mg caffeine. Wenzel and Rutledge (1962) even found an inverse dose-effect relation with greater improvements after 100 mg than after 200 or 300 mg caffeine for both simple and complex reaction tasks. An improvement with 300 mg caffeine was reported by White et al., (1980) and with 200 mg by Krueger, Zülch and Gandorfer (1979) and Smith, Tong and Leigh (1977). Some studies suggest that cognitive expectancies due to conditioning and/or tolerance and habituation may be responsible for such diverging effects. Along this line, similar effects for coffee and decaffeinated coffee were found by Knowles (1963) and differing effects for subjects with continued coffee consumption and subjects who abstained from coffee were reported by Eddy and Downs (1928).

Inconsistent reports are also encountered in the literature on simple and complex routine tasks. For selective letter cancellation tests as simple tasks, Feierabend and Bättig (1982) found no improvement in average performance with average doses, but they observed less fluctuation in performance. With the high dose of 750 mg Fröberg et al., (1969) found improvements in different cancellation tasks, whereas File, Bond and Lister (1982) recently obtained no effects with 450 mg. Performance in simple arithmetic tasks was found to be unaffected by Lovingood et al., (1967), but to be improved by Gilliland and Nelson (1939) and by Nash (1962). In such tasks the maintenance of vigilance might well be the crucial element for possible improvements, as suggested by an early study by Barmack (1940) which found no improvement for the initial period of testing but a net beneficial effect for later phases of continued testing. The same explanation may also hold for the improvement seen by Elkins et al., (1981) with a continuous performance test.

Complex psychomotor functions such as tasks related to car driving or typewriting seem to respond to caffeine in a similar manner. Baker and Theologus (1972) and Regina, Smith, Keiper and McKelvey (1974) found improvements in simulated car driving tests for signal detection and for acceleration and deceleration performance. Strasser and Müller-Limmroth (1971) noted an improvement, in that caffeine prevented performance decreases due to fatigue. Improvements were also noted with airplane simulators by Adler, Burkhardt, Ivy and Atkinson (1950), Seashore and Ivy (1953) and Hauty and Payne (1955).

Most of these tests have in common that they are usually administered for prolonged periods, which under normal conditions frequently involves gradual declines in performance. Maintenance of performance and vigilance is, however, also an attribute of personality and typically difficult for highly extroverted subjects. A particularly pronounced beneficial effect of caffeine in a prolonged vigilance task has been shown by Keister and McLaughlin (1972).

It is difficult to see in any type of animal performance a basis for comparison with typical human performance tests, except perhaps experiments using the different methods of operant bar pressing behaviour. Such studies have quite unequivocally revealed stimulant effects of caffeine with dose ranges that also increase different types of spontaneous activity. Increased bar pressing for food was reported for the rat by Skinner and Heron (1937) and by Solyom, Enesco and Beaulieu (1968) and for the pigeon by Blough (1957). Particularly impressive increases above control levels were reported by Webb and Levine (1978) when the animals were differentially rewarded for long intervals between successive bar presses. As a recent study by Valdes, McGuire and Annau (1982) demonstrates, such effects are also independent of the type of motivation. In this study bar pressing also increased if the animals were given water reward or intracranial self-stimulation rather than food reward, as used in most other studies.

Learning and Intellectual Performance

In animals, particularly in rodents, improvements in learning performance have been demonstrated in numerous studies. Such improvements have been seen for different task modalities, such as shock-motivated avoidance (Alpern and Jackson, 1978), discriminative lever pressing or maze learning (Cooper, Potts, Morse and Black, 1969; Kulkarni, 1972; Burov and Borisenko, 1975). The effective dose range in these studies tended to be smaller than that seen for increases in different activities and was reported to be in the order between about 1 and 10 mg/kg, although improvements were occasionally noted with doses as high as 30 mg/kg (Burov and Borisenko, 1975). There may be some evidence that the effect is due to increased perceptive stimulus under the influence of the drug, as suggested by Cox (1970a), who found an increase in spontaneous alternation, or by Hearst (1964), who found a relative decrease in stimulus generalisation. Whether such effects might also be due to a particular action of the drug on memory processes or not remains, however, more controversial. The substance was found to counteract the amnestic effects induced by pharmacological inhibition of protein synthesis (Flexner and Flexner, 1975; Flood *et al.*, 1978) or by electroconvulsive shocks (Dall'Olio, Gandolfi and Montanaro, 1978), but on the other hand, caffeine was also reported to impair long-term memory (Alpern and Jackson, 1978; Izquierdo, Costas, Justel and Rabiller, 1979). One possible explanation for some negative results (Geller, Hartmann and Blum, 1971) might be seen not only with the particular task modalities but also with the different learning ability between strains of the same species (Castellano, 1977). The interesting question as to whether chronic treatment may produce dependency or withdrawal symptoms has been touched by Vitiello and Woods (1977), who reported that chronically pretreated rats avoid flavoured solutions indicating the absence of caffeine.

Relatively few studies on learning in man have been carried out so far, and the results are in part rather anecdotal and in part controversial. Early, in 1892, Kraeplin reported acceleration in forming associations after drinking tea or coffee, whereas Hull, in 1935, found 325 mg (5 grains) of caffeine citrate had no effects on verbal rote learning, and Cattell (1930) even reported an impairment of associative and intellectual functions after 400 mg caffeine. On the other hand, two more recent studies both suggest improvement after caffeine intake. In the study by File *et al.*, (1982), caffeine in the dose range 125-500 mg improved digit symbol substitution, and in the study by Mitchell, Ross and Hurst (1974) a marginally significant increase in speed in a spaced sequential memory task was reported. Both tasks involve intellectual speed rather than intellectual power, and many of the differences between different studies suggest that caffeine might increase intellectual speed rather than power. With a mental maze learning task requiring primarily intellectual power Feierabend and Bättig (1982) recently found no effect for 300 mg caffeine. With an embedded figure task Broverman and Casagrande (1982) observed an improvement with caffeine only after the task had become routinised through previous practice. Similarly, Lienert and Huber (1966) found improvements across different intelligence test subscales predominantly for types of performance which depend on intellectual speed.

Sleep

Motor Activity

In popular belief, sleep disturbance belongs to the most notorious effects of coffee. Research on the objective effects of coffee and caffeine on sleep started quite early. Several of the early studies measured motor activity during sleep by means of pulleys and strings attached to the bed springs. Giddings (1934) analysed this activity in 28 school children for a total of 80,000 sleep hours. Significant increases in restlessness were found after a heavy meal or after a hot day but not after caffeine, which was tested only at the low dose level of about 40 mg. A dose of 3 mg/kg, given in three doses at two-hour intervals before going to bed, was also ineffective in a study by Marbach and Schwertz (1964), whereas Mullin, Kleitman and Cooperman (1933) found with 260-390 mg (4-6 grains) an increase in motility. Stradomsky (1970) compared sleep actograms in 90 subjects after drinking decaffeinated coffee and a coffee beverage containing 200 mg caffeine. No differences were found between placebo and decaffeinated coffee, but regular coffee prolonged sleep latency and decreased total sleep duration and subjective sleep quality with no differences between habitual coffee drinkers and coffee abstainers.

Self-reports

Self-reports on sleep after caffeine consumption were first studied systematically by Goldstein and his associates. In a first study (Goldstein, 1964) on 230 students he found decaffeinated coffee to have no effect and 150 or 200 mg caffeine either in capsules or in decaffeinated coffee to prolong sleep latency and to decrease 'soundness of sleep.' Morning headaches were increased in habitual coffee drinkers after a placebo night, and in these subjects the sleep disturbances after caffeine were less pronounced than in moderate consumers of coffee. Furthermore, the study also mentioned as an interesting phenomenon a 'reverse placebo' effect with less subjective disturbances when the subjects knew that they received caffeine than when they did not. In a subsequent study on 20 students (Goldstein, Warren and Kaizer, 1965b) the effect of 300 mg caffeine was studied in a double-blind manner, and the plasma caffeine levels were determined one, two, and three hours after drinking the test beverages. These caffeine plasma levels were similar across the subjects and failed to explain the large interindividual differences in the effects on sleep. Essentially similar results were obtained by Forrest, Bellville and Brown (1972) in middle-aged hospital patients.

EEG Sleep Patterns

As a third indicator of sleep quality, the investigation of EEG sleep patterns has become more popular in caffeine research over the last few years. Gresham, Webb and Williams (1963) investigated in seven students the effects on rapid-eye-movement (REM) sleep of 1 g/kg alcohol and of 5 mg/kg caffeine, given either alone or in combination. The EEG records were analysed for the first five hours of sleep, and alcohol was found significantly to shorten the REM dream phases of sleep, whereas caffeine failed to produce a significant effect on this measure. A more extended approach was used by Müller-Limmroth (1973), who not only performed a complete frequency analysis of the EEG, but also included recordings of the electrooculogram (EOG), electrocardiogram (ECG), electromyogram (EMG), respiration, activity and electrodermal responses in eight subjects. He also found no effects of caffeine (200 mg in decaffeinated coffee versus decaffeinated coffee) on the duration of the REM dream phases of sleep, but observed a reduction in 'deep' sleep (EEG sleep stages 3 and 4) for the first three hours of sleep, accompanied by an increase in light sleep (stages 1 and 2) for the same period.

Slightly different results were reported by Březinová (1974), who compared in six middle-aged subjects EEG-EOG sleep patterns for 300 mg caffeine in decaffeinated coffee versus decaffeinated coffee and no drink. He found no changes for REM sleep, but a roughly threefold increase in sleep latency, a decrease in total sleep time by nearly two hours and a doubling of the light sleep stages 1 and 2 and of the number of spontaneous awakenings. All effects were most pronounced for the first three hours of sleep, but there were also effects for the later parts of the night, particularly an increase in shifts to awakening. In accordance with other studies, the inter-individual variance was great and greater than the intra-individual variance over the five test nights carried out for each treatment condition.

In later experiments, the same author (Březinová, Oswald and Loudon, 1975; Březinová, 1976) compared caffeine insomnia with habitual insomnia in late middle-aged subjects and with insomnia caused by the withdrawal of hypnotic drugs. It became apparent from the detailed EEG analysis that the underlying mechanisms are different, consisting perhaps of increased stability of wakefulness in the case of caffeine and of decreased stability of deep sleep for the other conditions.

An additional aspect of EEG sleep analysis was considered by Bonnet and his colleagues (Bonnet, 1978; Bonnet and Webb, 1979; Bonnet, Webb and Barnard, 1979) by introducing awakening thresholds for auditory stimuli and measuring the latencies for going back to sleep as evidenced by EEG criteria. Caffeine (400 mg given as pill) was found to delay return to sleep for the entire night, but particularly for the first four hours of sleep. The threshold intensities needed to awaken the subjects and the subsequently measured auditory thresholds (Bonnet et al., 1979) were both decreased with caffeine.

The dose-effect relation for sleep disturbance using EEG measures was investigated by Karacan et al., (1976), who gave the subjects on different days before going to bed 1.1, 2.3 and 4.6 mg/kg caffeine either as regular coffee drinks or as pure substance in decaffeinated coffee. No differences were obtained between pure caffeine and regular coffee, and the different doses produced increasing symptoms of sleep disturbance such as prolonged sleep latency, decreased total sleep duration and decreased sleep quality. The EEG changes obtained ran more or less parallel to the results of the postsleep questionnaires (sleep latency, distribution of stages 1 to 4). In addition, caffeine produced a slight shift of REM phases to the initial part of the night and of deep sleep stages 3 and 4 to the later stages of the night, which also reached significance. Given the surprisingly well-fitted dose-response effects, the authors suggest that in normal subjects coffee may be instrumental in inducing insomnia.

Other Physiological Variables

Only a few and particularly older studies have measured physiological variables other than EEG, such as EMG, ECG, electrodermal responses, body temperature, etc. Mullin et al., (1933) reported an increase in body temperature with 260 and 390 mg (4 and 6 grains), but not with 130 mg (2 grains) of caffeine, together with a parallel increase in motor activity. Marbach and Schwertz (1964) administered to 12 subjects 3 mg/kg caffeine in the afternoon, followed by three additional doses of 1 mg/kg each at two-hour intervals and found a subjective decrease in sleep quality and a decrease in heart frequency throughout the night, but no effects on body temperature or motility. Müller-Limmroth (1973) reported an increase in EMG potentials (taken from the mouth cavity), but no effects for rectal temperature, heart rate, respiration or electrodermal changes.

Interaction with Other Drugs

Until now the interaction of caffeine with drugs has been studied only occasionally. Pentobarbital at a dose level of 100 mg was found to offset the sleep disturbing effects of 250 mg caffeine, as became evident from subjective reports on sleep latency, quality and duration (Forrest et al., 1972). Given alone, the same dose of pentobarbital produced effects opposite to those of caffeine for awakening thresholds (Bonnet and Webb, 1979; Bonnet et al., 1979). However, barbiturates have also become well known for inhibiting REM sleep phases, which seem to be unaffected by caffeine (Oswald, 1968). Alcohol at dosages inducing plasma levels of 0.5 parts per thousand or more also seems to have effects somewhat similar to those of a moderate dose of pentobarbital in depressing REM sleep (Gresham et al., 1963), decreasing motility and body temperature, improving subjective sleep quality (Mullin et al., 1933), and increasing heart rate (Marbach and Schwertz, 1964). Similar effects were also reported for meprobamate, flurazepam and nitrazepam (Schwertz and Marbach, 1965; Oswald, 1968).

Only recently have a few studies also investigated the interactions of cigarette smoking with caffeine. Johns (1974) observed only the frequently reported positive association between coffee drinking, alcohol consumption and cigarette smoking, but no effects on subjective sleep reports, except that both heavy smokers and frequent coffee drinkers tend to go to bed late. Soldatos et al., (1980) found, in addition, that smoking *per se* was accompanied by longer sleep latencies and shorter sleep duration, which became similar to values for non-smokers after a few days of smoking abstinence, whereas EEG sleep stage distribution was similar for smokers and non-smokers and failed to change upon smoking abstinence.

In summary, it appears from these studies on coffee and sleep that caffeine produces insomnia in a dose-dependent fashion. However, as opposed to many drugs, this effect is not very strong, does not affect REM sleep and is subject to high inter-individual variance.

Mood and Side Effects

People drink coffee because they like it rather than because of a conscious appreciation of an effect on performance. However, studies trying to verify some typical improvement in mood are quite controversial. Based on everyday experience one would expect that coffee would be stimulating and slightly euphoric, with increases in nervousness and irritability as undesired side effects after too high dosages. Nevertheless, no effects or only negligible or marginally significant effects on mood, such as 'being stimulated,' 'more interested' or 'more alert' were found in various studies, as for example by Flory and Gilbert (1943) with 160 mg caffeine, by Bachrach (1966) with 130 mg, by Nash (1962) with 100 mg/m^2 body surface and by Lader (1969) with 150 and 300 mg. Goldstein et al., (1965a), on the other hand, obtained more positive results. They noted in students with 150-

200 mg caffeine an increase in alertness and simultaneously a dose-related increase in nervousness. Later they reported (Goldstein and Kaizer, 1969) increased nervousness in light coffee consumers and increased euphoria in high consumers, a finding which suggests that habituation and/or tolerance might play important roles in the manifestation of mood effects.

More recent studies are also equivocal in this respect. Elkins *et al.*, (1981) reported for prepubertal boys with a high dose of 10 mg/kg only an increase in fidgetiness. Rapoport *et al.*, (1981) found in adults, but not in children, an increase in feeling faint or flushed with 10 mg/kg, but an increase in nervousness and jitteriness even at the lower dose level of 3 mg/kg, and this both in children and adults. File *et al.*, (1982) obtained with a series of different ratings for sedation and well-being no effects in the dose range from 125 to 500 mg. Leathwood and Pollet (1983) obtained with the modest dose of 100 mg caffeine under double-blind conditions significant increases in scores for wakefulness, vigour, clarity of mind, feeling full of ideas and efficiency, with the effects lasting for about three hours, and they argued that this positive result was a consequence of their easily understandable and straightforward questionnaires.

An attempt to measure objectively an attribute of emotionality was made by Cherek, Steinberg and Brauchi (1983), who tested with an operant response task the effect of caffeine on interpersonal aggression and found that after 4 mg/kg caffeine the subjects responded with reduced aggression to the provocations of a simulated counterpart. At the same time these subjects reported no increase in nervousness; they were unable to guess correctly whether they had received placebo or caffeine as treatment; and the reduction in aggressive responses was independent of caffeinic effects on general responsiveness.

In summary, it appears thus that the direction of mood effects may depend on a series of different factors, among which the techniques of assessment are probably of primary importance. Cognitive factors may also be a crucial element, since a study by Mitchell *et al.*, (1974) observed that positive effects on mood became significant only when the subjects were informed that they had received the drug.

Individual Differences

Large individual differences in response to coffee and caffeine are a major outcome of most studies. Constitution, personality, habituation and tolerance are perhaps the main candidates for explaining such differences, but so far only relatively few studies have been devoted to such aspects.

Constitutional differences between coffee drinkers and coffee abstainers could be assumed if hereditary aspects of the habit could be demonstrated. Conterio and Chiarelli (1962) reported that there was more concordance for the coffee drinking habit in monozygotic than in dizygotic twins. Abe (1968) reported for monozygotic twins a within-pair concordance for insomnia after drinking coffee, whereas Perry (1973a) found with the same method only support for a genetic factor related to attributes predisposing to alcohol drinking, but not for cigarette smoking or coffee drinking. He did, however, report significant genetic components for the magnitude of coffee consumption (Perry, 1973b). Pharmacogenetic differences in response to caffeine have been shown by Alleva, Castellano and Oliverio (1978) for different strains of inbred mice, both for onset and duration of barbiturate-induced sleep. Kuftinec and Mayer (1964) reported that mice with the hereditary obese hyperglycaemic syndrome were unusually sensitive to the toxic effects of caffeine, but not to those of the other compared substances.

Differential responses to caffeine in humans differing in personality have also been reported. Ulrich (1968) suggested that psychologically unstable persons might be more sensitive to undesired side effects of coffee than stable persons, an observation which was also made by Meyer, Walther and Walther (1983). More objective studies have been carried out for the interaction of the caffeinic response with the introversion-extroversion personality dimension. According to the theories of Eysenck, internal (cortical) arousal determines behavioural performance in a biphasic way with performance first improving up to an optimum level of arousal and then gradually deteriorating. As extroverts are characterised by suboptimum arousal levels, the stimulatory effect of caffeine could be expected to improve their performance, while introverts, with their chronic high or overarousal, could be expected to show no change or even deterioration in performance. Revelle, Humphreys, Simon and Gilliland (1980) observed such effects on mental performance for 'impulsivity' and 'sociability,' two subscales of extroversion, in interaction with daytime as an additional determinant of arousal. A similar result was obtained by Anderson and Revelle (1982) for proofreading as a task of information processing. Partially supportive results for the Eysenck hypothesis have also been obtained by Smith, Wilson and Jones (1983) by manipulating arousal with caffeine in introverts and extroverts and using electrodermal responses to auditory stimuli as the dependent variable.

Assuming that personality affects the caffeine response, one might also expect that personality could affect spontaneous coffee consumption. This issue has already been investigated, particularly for tobacco and also for alcohol consumption, both of which correlate significantly with extroversion (Eysenck, 1980). However, the amount of explained variance in such studies is mostly modest, and the rather consistent positive intercorrelations between coffee, alcohol and tobacco consumption leave open the possibility of multiple interactions. Therefore, data suggesting increased extroversion or increases in behaviours usually correlated with extroversion, such as increases in sexual behaviour as reported by Giese and

Schmidt (1968), remain for the time being more descriptive than explanative.

Another aspect of caffeine responsiveness might be related to habituation and tolerance, as discussed extensively by Gilbert (1976). The issue has been discussed again recently by Estler (1982), but the impression remains that further clarification through additional experimentation is still necessary.

Toxicity of Coffee and Caffeine

According to the literature, fatal intoxications with coffee or caffeine are rare (Dimaio and Garriott, 1974). The minimum intravenous dose reported to be fatal in man was administered erroneously and amounted to about 60 mg/kg (Jokela and Vartiainen, 1959). Fatal oral doses in man have been reported to be in the range 150-200 mg/kg (Peters, 1967). The principal forms of untoward response to non-fatal overdoses are mostly extensions of the pharmacological actions and include (a) central nervous system symptoms (restlessness, irritability, excitement, muscle twitchings, convulsions), (b) cardiovascular symptoms (tachycardia, palpitation, marked hypotension), (c) gastrointestinal symptoms (nausea, vomiting), (d) respiratory symptoms (tachypnoea) and (e) renal symptoms (albuminuria). Such reactions can be expected after the intake of at least 1 g or more of caffeine taken as a single dose, which is equivalent to 6 to 25 cups of coffee, depending on the strength of the beverage preparation. In contrast to such acute effects, the possible effects of the chronic intake of moderate doses have received more attention and interest over the last few years.

Myocardial Infarction

A discussion concerning a possible link between coffee consumption and ischaemic heart disease was initiated by an epidemiological prospective study carried out on the employees of Western Electric in Chicago (Paul *et al.*, 1963). However, the positive epidemiological association noted in this study was later shown by the same authors (Paul, MacMillan, McKean and Park, 1968) to be the indirect result of the close association between coffee drinking and cigarette smoking habits.

Two later studies (Boston Collaborative Drug Surveillance Program, 1972; Jick *et al.*, 1973) found that drinking more than 6 cups/day constituted an increased risk of myocardial infarction which could not be explained by other common risk factors such as age, obesity, smoking and hypertension. Since then, however, several other well-controlled studies have failed to reveal such associations (Dawber, Kannel and Gordon, 1974; Tibblin, Wilhelmsen and Werko, 1975; Heyden *et al.*, 1978;

Murray, Bjelke, Gibson and Schuman, 1981).

Curatolo and Robertson (1983), who have examined this controversy in more detail, suggest that the early reports on positive associations with this disease might have been biased by the inclusion of hospitalised patients as controls, by the fact that the number of controlled variables was insufficient, by effects of the disease on the retrospective subjective estimation of coffee consumption, etc. In fact, Hennekens et al., (1976), who also failed to find a positive association, did find a small but non-significant increase in the risk for consumers of more than 6 cups/day after they had eliminated from their analysis all variables that were not considered in the earlier study by Jick et al., (1973).

Mutagenic Effects and Carcinogenesis

Caffeine can induce chromosomal damage in plant and in mammalian cells *in vitro*. The effect appears, however, only at doses which are far in excess of those which could be tolerated *in vivo* without lethal consequences. In recent years it has also been shown by several studies that caffeine at nontoxic dose levels might enhance the lethal effects of various DNA-damaging agents by interfering with the DNA repair process (Timson, 1977; Roberts, 1978). On the other hand, a greater number of studies have even shown that caffeine inhibits tumour induction. Roberts (in preparation), who recently discussed these findings in greater detail, came to the conclusion that such effects could hardly have any relevance for human carcinogenesis, not only because the available evidence is inconsistent and the required dose levels are outside the range normally consumed by coffee drinkers, but also because such potentiation could occur only during a precise and short period of time immediately following the initial damage to DNA.

Epidemiological studies also do not seem to support the suggestion made by some studies that caffeine use might be associated with a cancer risk. Several studies have demonstrated associations between coffee consumption and urinary tract carcinogenesis. Curatolo and Robertson (1983), who have recently discussed this issue, observed that across 15 relevant studies the outcome was highly inconsistent, with the associations being weak, absent or present only for differing subsamples of the populations studied. Furthermore, all studies were retrospective, none of them could present a dose-response relationship and most of them did not sufficiently control for other intervening variables such as smoking.

A significant dose-effect relationship was recently postulated by MacMahon et al., (1981) for the association between coffee habits and cancer of the exocrine pancreas. The study was based on case-controlled interviews of patients with clinically and histologically proven cancer of the pancreas. However, as in the case of cancer of the urinary tract, several serious methodological shortcomings of the study, particularly in the selection of the control populations, prevent these data being taken as conclusive (Higgins, Stolley and Wynder, 1981; Curatolo and Robertson, 1983). In addition, other recent analyses prompted by the MacMahon study failed to obtain an association between coffee drinking and pancreatic cancer (Goldstein, 1982; Severson, Davis and Polissar, 1982).

Fibrocystic Breast Disease

Minton, Abou-Issa, Reiches and Roseman (1981) have presented evidence in several studies that coffee consumption might be associated with fibrocystic breast disease and that eliminating all sources of methylxanthines often resolves the symptoms of the disease, but again the methodology of these studies has been subject to serious criticism (Heyden, 1980, 1982). Furthermore, other studies have failed to find this association or to verify any clinical improvements after abstinence from methylxanthines (Lawson, Jick and Rothman, 1981; Marshall, Graham and Swanson, 1982).

Gastrointestinal Disorders

Coffee and decaffeinated coffee both stimulate acid secretion in man to a greater extent than the equivalent amount of caffeine (Cohen and Booth, 1975), and 'heart burn' is a frequently noted subjective symptom after coffee consumption. The effect of methylxanthines can be enhanced by cholinergic agonists, histamine, pentagastrin and food (Bieck, Oates, Robison and Adkins, 1973; Gabrys, Nyhus, Van Meter and Bombeck, 1973) and antagonised by atropine and cimetidine, an H_2-receptor antagonist (Cano, Isenberg and Grossman, 1976). Surprisingly, however, the question whether coffee and its constituents might play a role in the pathogenesis of ulcers, as suggested by Roth and Ivy (1946), has so far not been the subject of any systematic epidemiological studies.

Teratogenic Effects

The structural similarity of the methylxanthines to the DNA purines has been taken as suggestive of possible teratogenic effects. In animal studies excessive doses, such as 50-300 mg/kg in mice, were found to produce a slight increase in cleft palate and in retarded ossification of the supraoccipital bones (Elmazar, McElhatton and Sullivan, 1982). In general, other studies also revealed defects of the limbs, ectrodactyly and skeletal variants, particularly in the sternum (Collins, Welsh, Black and Collins, 1980; Wilson and Scott, in preparation). A considerable number of other studies obtained similar results in rats and mice (Timson, 1977). As these studies generally found mutagenic effects with extreme dosages, any direct relevance for human caffeine consumption should be discussed with caution.

Furthermore, the question remains open as to what extent such damage might be an indirect consequence of the treatments, since such excessive dosages were found to increase plasma corticosterone to levels which *per se* induce similar teratogenic damage (Elmazar, McElhatton and Sullivan, 1981). Beyond this suggestion, it should also be remembered that doses required to produce teratogenic effects are inevitably associated with maternal toxicity including reduced food intake, reduced body weight, convulsions and often death. As the malformations are mostly accompanied by haematomata, it appears further that vasoactive effects of the drug, perhaps mediated by the increases in catecholamine release, might be operational at the origin of the malformations. Within the range of habitual dose levels, however, teratogenic effects are hardly to be expected, as demonstrated by Nolen (1982). Instead of drinking-water, this author offered rats coffee beverages equivalent to 50, 25, or 12 cups daily for a period of six months and involving two gestation periods. The only observation made was an increase in the incidence of unossified sternebrae in the fetuses of dams given the highest dose in the form of regular coffee as opposed to caffeine in decaffeinated coffee or instant coffee. However, this transient delay in ossification which is frequently observed in teratogenic studies can appear also as a consequence of non-specific stress and is commonly not considered as a teratogenic response (Khera, 1981).

The evidence for teratogenic effects in humans has remained rather anecdotal or is based on relatively small sample studies (Mau and Netter, 1974; Weathersbee, Olsen and Lodge, 1977). Nevertheless, and in view of the numerous animal studies, in 1980 the US Food and Drug Administration issued a warning for pregnant women to avoid drugs and foodstuffs containing caffeine (Goyen, 1980). Since then, two large sample studies, which also controlled for other demographic and life-style variables, have found no association between coffee consumption and teratogenic effects (Linn *et al.*, 1982; Rosenberg, Mitchell, Shapiro and Slone, 1982). Further, these studies found no evidence for lower birth weights in the offspring of heavy coffee consumers. Thus, it seems at present that the effects on the outcome of pregnancies are minimal, if there are any at all (reviewed most recently by Sullivan and McElhatton, 1983).

Behavioural Teratology

Over the last years increased attention has been paid to the question of whether or not prenatal treatment with centrally active substances might affect the developing brain in such a way as to produce permanent functional changes in the offspring. Burgess and Monachello (1983) exposed sibling juvenile fish chronically between 50 and 100 days after fertilisation to caffeine solutions at the concentration of 15 mg/l, a concentration which approximates that obtained in human plasma after a few cups of coffee.

Some animals were sacrificed for histological analysis of midbrain structures immediately after the end of the exposure and others after they had been allowed to recover from the caffeine effects up to the age of 18 months. Caffeine facilitated the formation of dendritic spines and dendritic branching, and this effect was marked in the juvenile fish and still present to a moderate degree in the adult animals. A behavioural analysis of the juvenile animals revealed further (Burgess, 1982) that after termination of the chronic treatment the caffeine-treated animals exhibited an increased variability in interindividual spacing in schooling behaviour. Brain catecholamines and sleep states were studied in offspring of caffeine-treated rats by Enslen, Milon and Würzner (1980). Caffeine was mixed into the diet of the pregnant rats in varying concentrations up to about 20 mg/kg. After the offspring had grown up, they were monitored for their EEG sleep patterns. In the offspring of the caffeine-treated mothers a significant increase in the number of paradoxical sleep phases was obtained in a dose-dependent manner, together with a decrease in dopamine in the locus coeruleus.

Although such findings merit interest, only extended future research in the presently growing field of behavioural teratology will allow conclusions about the functional meaning of such findings.

Behavioural Disorders

Frequent claims are found in the literature that chronic and heavy use of substances containing caffeine may lead to nervousness, irritability, headache and anxiety states. This condition has also frequently been referred to as 'caffeinism.' However, so far only very few controlled studies on this question are available. Greden (1974) described three clinical cases in which caffeinism was misdiagnosed as an anxiety syndrome. All three cases reported in this study showed improvement in their symptoms after withdrawal of caffeine. A later study on 135 psychiatric patients (Winstead, 1976) observed that the heavy users among these patients scored significantly higher on the State-Trait Anxiety Index (STAI) than moderate users. This finding was more recently reconfirmed in another study on psychiatric inpatients (Greden, Fontaine, Lubetsky and Chamberlin, 1978). This study found higher STAI scores for subjects consuming more than 750 mg caffeine per day. These patients also scored higher on the Beck depression scale, but they reported less insomnia than moderate coffee consumers for coffee taken at bedtime. The two groups did not, however, differ in headache symptoms after caffeine withdrawal. A study carried out in 91 healthy subjects and using the Spielberger questionnaire confirmed this finding, but the correlations between anxiety and caffeine consumption were small and only marginally significant (Hire, 1978). Thus, it appears that the association between heavy caffeine consumption and anxiety or depression is still poorly described.

Summary

In most cultures coffee, along with tea, is a recreational beverage of high popularity as well as the main source of the pharmacologically active drug caffeine. A considerable amount of scientific work has been devoted to the analysis of the physiological and psychological effects, but many questions have remained nearly untouched and others will need reanalysis with more modern methods and approaches.

With respect to possible vegetative effects on health, particularly afflictions of the cardiovascular system and cancer, the last decade has witnessed not only a series of accusations, but also little support from more detailed epidemiological and experimental analysis. Some more popular beliefs about undesired side effects such as stomach irritation and nervousness and withdrawal effects such as headaches have, however, attracted much less scientific interest.

Behaviourally, coffee is a stimulant due to its caffeine content, but the effects are mostly subtle. Dose-response dependence is generally rather weak for the wide range from about 100 to 750 mg caffeine. The best-documented effect is perhaps the ability of caffeine to disrupt sleep, which, however, in view of detailed EEG analysis, could just as well be seen as an increased ability to maintain wakefulness. However, not only this 'specific' effect, but even more the mostly beneficial effects on different types of behavioural performance vary considerably among individuals, and the question as to what extent such differences might be explained by tolerance, habituation, constitution and personality is far from being settled.

Conditioning and learning processes may well constitute a critical factor in the experience of the effects of coffee, since according to many studies coffee 'works best' when the subjects are informed about the caffeine content. In this sense one might assume that stimuli such as smell, taste and the act of drinking associated with the low doses of caffeine might play an important role in the manifestation of the behaviourally stimulating actions.

Great advances have been made in recent years with respect to the metabolism and pharmacokinetics of caffeine and with respect to its interaction with central nervous adenosine receptor sites, and it can be hoped that such progress will promote the understanding of the behavioural effects of the substance. However, toward this end a considerable amount of work will have to be devoted to the many aspects of the problem, such as for instance the differential impact of the extracaffeinic substances in coffee, possible long-lasting effects of caffeine, as seen in experiments of behavioural teratology, and physiological effects on different organ systems.

Acknowledgement

The author wishes to thank Mrs B. Strehler for literature documentation and her invaluable and conscientious assistance in the preparation of the manuscript.

References

Abe, K. (1968) 'Reactions to coffee and alcohol in monozygotic twins', *Journal of Psychosomatic Research, 12*, 199-203

Adler, H.F., Burkhardt, W.L., Ivy, A.C. and Atkinson, A.J. (1950) 'Effect of various drugs on psychomotor performance at ground level and at simulated altitudes of 18,000 feet in a low pressure chamber', *Journal of Aviation Medicine, 21*, 221-36

Aldridge, A., Bailey, J. and Neims, A.H. (1981) 'The disposition of caffeine during and after pregnancy', *Seminars in Perinatology, 5*, 310-14

Alleva, E., Castellano, C. and Oliverio, A. (1978) 'Individual differences in barbiturate-induced sleeping time in the mouse', *Progress in Neuro-Psychopharmacology, 2*, 451-3

Alpern, H.P. and Jackson, S.J. (1978) 'Stimulants and Depressants: drug effects on memory' in M.A. Lipton, A. DiMascio and K.F. Killam (eds.), *Psychopharmacology: A Generation of Progress*, Raven Press, New York, pp. 663-75

Ambrozi, L. and Birkmayer, W. (1970) 'Ueber die Objektivierbarkeit von psychopharmakologischen Drogen am Beispiel von Coffein und coffeinfreiem Kaffee', *Internationale Zeitschrift für Klinische Pharmakologie, Therapie und Toxikologie, 3*, 167-73

Andén, N.-E. and Jackson, D.M. (1975) 'Locomotor activity stimulation in rats produced by dopamine in the nucleus accumbens: potentiation by caffeine', *Journal of Pharmacy and Pharmacology, 27*, 666-70

Anderson, K.J. and Revelle, W. (1982) 'Impulsivity, caffeine, and proofreading: a test of the Easterbrook Hypothesis', *Journal of Experimental Psychology: Human Perception and Performance, 8*, 614-24

Aranda, J.V., Collinge, J.M., Zinman, R. and Watters, G. (1979) 'Maturation of caffeine elimination in infancy', *Archives of Disease in Childhood, 54*, 946-9

Armitage, A.K., Boswood, J. and Large, B.J. (1961) 'Structure-activity relationships in a series of 6-thioxanthines with bronchodilator and coronary dilator properties', *British Journal of Pharmacology and Chemotherapy, 17*, 196-207

Ashton J., Millman, J.E., Telford, R. and Thompson, J.W. (1974) 'The effect of caffeine, nitrazepam and cigarette smoking on the contingent negative variation in man', *Electroencephalography and Clinical Neurophysiology, 37*, 59-71

Avogaro, P., Capri, C., Pais, M. and Cazzolato, G. (1973) 'Plasma and urine cortisol behavior and fat mobilization in man after coffee ingestion', *Israel Journal of Medical Sciences, 9*, 114-19

Axelrod, J. and Reichenthal, J. (1953) 'The fate of caffeine in man and a method for its estimation in biological material', *Journal of Pharmacology and Experimental Therapeutics, 107*, 519-23

Bachrach, H. (1966) 'Note on the psychological effects of caffeine', *Psychological Reports, 18*, 86

Baer, H.P. and Drummond, G.I (eds.) (1979) *Physiological and Regulatory Functions of Adenosine and Adenine Nucleotides*, Raven Press, New York

Baker, W.J. and Theologus, G.C. (1972) 'Effects of caffeine on visual monitoring', *Journal of Applied Psychology, 56*, 422-7

Barmack, J.E. (1940) 'The time of administration and some effects of 2 Grs. of alkaloid caffeine', *Journal of Experimental Psychology, 27*, 690-8

Beavo, J.A., Rogers, N.L., Crofford, O.B., Hardmann, J.G., Sutherland, E.W. and Newman, E.V. (1970) 'Effects of xanthine derivatives on lipolysis and on adenosine 3',5'-

monophosphate phosphodiesterase activity', *Molecular Pharmacology, 6*, 597-603

Belford, J. and Feinleib, M.R. (1962) 'The increase in glucose 6-phosphate content of the heart after the administration of inotropic catecholamines, calcium and aminophylline', *Biochemical Pharmacology, 11*, 987-94

Bell, G.H., Davidson, J.N. and Scarborough, H. (1953) *Textbook of Physiology and Biochemistry*, Livingston, Edinburgh

Bellville, J.W., Escarraga, L.A. Wallenstein, S.L., Wang, K.C., Howland, W.S. and Houde, R.W. (1962) 'Antagonism by caffeine of the respiratory effects of codeine and morphine', *Journal of Pharmacology and Experimental Therapeutics, 136*, 38-42

Berkowitz, B.A., Tarver, J.H. and Spector, S. (1970) 'Release of norepinephrine in the central nervous system by theophylline and caffeine', *European Journal of Pharmacology, 10*, 64-71

Berlin, C.M., Jr. (1981) 'Excretion of the methylxanthines in human milk', *Seminars in Perinatology, 5*, 389-94

Berthet, J., Sutherland, E.W. and Rall, T.W. (1957) 'The assay of glucagon and epinephrine with use of liver homogenates', *Journal of Biological Chemistry, 229*, 351-61

Bieck, P.R., Oates, J.A., Robison, G.A. and Adkins, R.B. (1973) 'Cyclic AMP in the regulation of gastric secretion in dogs and humans', *American Journal of Psychology, 224*, 158-64

Blinks, J.R., Olson, C.B., Jewell, B.R. and Braveny, P. (1972) 'Influence of caffeine and other methylxanthines on mechanical properties of isolated mammalian heart muscle. Evidence for a dual mechanism of action', *Circulation Research, 30*, 367-92

Blough, D.S. (1957) 'Some effects of drugs on visual discrimination in the pigeon', *Annals of the New York Academy of Sciences, 66*, 733-9

Boissier, J.R. and Simon, P. (1965) 'Action de la caféine sur la motilité spontanée de la souris', *Archives Internationales de Pharmacodynamie et de Thérapie, 158*, 212-21

Boissier, J.R. and Simon, P. (1967) 'Influence de la caféine sur le comportement en situation libre de la souris', *Archives Internationales de Pharmacodynamie et de Thérapie, 166*, 362-9

Bonati, M., Latini, R., Galletti, F., Young, J.F., Tognoni, G. and Garattini, S. (1982) 'Caffeine disposition after oral doses', *Clinical Pharmacology and Therapeutics, 32*, 98-106

Bonnet, M.H. (1978) 'The reliability of a depth of sleep measure and the effects of flurazepam, pentobarbital, and caffeine on depth of sleep' (Doctoral Dissertation, University of Florida, 1977), *Dissertation Abstracts International, 38*, 5632B (University Microfilms No. 77292226)

Bonnet, M.H. and Webb, W.B. (1979) 'The return to sleep', *Biological Psychology, 8*, 225-33

Bonnet, M.H., Webb, W.B. and Barnard, G. (1979) 'Effects of flurazepam, pentobarbital, and caffeine on arousal threshold', *Sleep, 1*, 271-9

Boston Collaborative Drug Surveillance Program (1972) 'Coffee drinking and acute myocardial infarction', *Lancet, 2*, 1278-81

Bowser, E.W., Hargis, G.K., Henderson, W.J. and Williams, G.A. (1975) 'Parathyroid hormone secretion in the rat: Effect of aminophylline', *Proceedings of the Society for Experimental Biology and Medicine, 148*, 344-6

Boyd, E.M., Dolman, M., Knight, L.M. and Sheppard, E.P. (1965) 'The chronic oral toxicity of caffeine' *Canadian Journal of Physiology and Pharmacology, 43*, 995-1007

Březinová, V. (1974) 'Effect of caffeine on sleep: EEG study in late middle age people', *British Journal of Clinical Pharmacology, 1*, 203-8

Březinová, V. (1976) 'Duration of EEG sleep stages in different types of disturbed night sleep', *Postgraduate Medical Journal, 52*(603), 34-6

Březinová, V., Oswald, I. and Loudon, J. (1975) 'Two types of insomnia: too much waking or not enough sleep', *British Journal of Psychiatry, 126*, 439-45

Broverman, D.M. and Casagrande, E. (1982) 'Effect of caffeine on performances of a perceptual-restructuring task at different stages of practice', *Psychopharmacology, 78*, 252-5

Burgess, J.W. (1982) 'Chronic exposure to caffeine during early development modifies spatial behavior in juvenile jewel fish', *Pharmacology Biochemistry and Behavior, 17*, 137-40

Burgess, J.W. and Monachello, M.P. (1983) 'Chronic exposure to caffeine during early development increases dendritic spine and branch formation in midbrain optic tectum', *Development Brain Research*, 6, 123-9

Burghardt, W., Geist, D., Grün, M., Staib, A.H. and Wernze, H. (1982) 'Does caffeine influence the sympathoadrenal-system, renin-angiotensin-aldosterone-system and blood pressure?' in F. Mantero, E.G. Biglieri and C.R.W. Edwards (eds.), *Serono Symposium No. 50, Endocrinology of Hypertension*, Academic Press, London and New York, pp. 415-21

Burov, Y.V. and S.A. Borisenko (1975) 'Effect of psychotropic drugs on capacity for mental and physical work in rats', *Bulletin of Experimental Biology and Medicine*, 79, 140-3

Butler, D. (1983) 'Optimistic for the future', *Coffee and Cocoa International*, 6, 10; 13

Calhoun, W.H. (1971) 'Central nervous system stimulants' in E. Furchtgott (ed.), *Pharmacological and Biophysical Agents and Behavior*, Academic Press, New York, pp. 181-268

Callahan, M.M., Robertson, R.S., Arnaud, M.J., Branfman, A.R., McComish, M.F. and Yesair, D.W. (1982) 'Human metabolism of [1-methyl-^{14}C]- and [2-^{14}C]caffeine after oral administration', *Drug Metabolism and Disposition*, 10, 417-23

Cano, R., Isenberg, J.I. and Grossman, M.I. (1976) 'Cimetidine inhibits caffeine-stimulated gastric acid secretion in man', *Gastroenterology*, 70, 1055-7

Castellano, C. (1977) 'Effects of pre- and post-trial caffeine administrations on simultaneous discrimination in three inbred strains of mice', *Psychopharmacology*, 51, 255-8

Cattell, R.B. (1930) 'The effects of alcohol and caffeine on intelligent and associative performance', *British Journal of Medical Psychology*, 10, 20-33

Cherek, D.R., Steinberg, J.L. and Brauchi, J.T. (1983) 'Effects of caffeine on human aggressive behavior', *Psychiatry Research*, 8, 137-45

Chou, D.T., Forde, J.H. and Hirsh, K.R. (1980) 'Unit activity in medial thalamus: comparative effects of caffeine and amphetamine', *Journal of Pharmacology and Experimental Therapeutics*, 213, 580-5

Clifford, M.N. (1979) 'Chlorogenic acids — their complex nature and routine determination in coffee beans', *Food Chemistry*, 4, 63-71

Clifford, M.N. and Staniforth, P.S. (1979) 'A critical comparison of six spectrophotometric methods for measuring chlorogenic acids in green coffee beans', *Huitième Colloque Scientifique International sur le Café, Abidjan, 1977*, Association Scientifique Internationale du Café, Paris, pp. 109-13

Cloutte, G.R., Glass, A. and Butler, S.R. (1977) 'The influence of caffeine on the prevalence of alpha rhythm', *Electroencephalography and Clinical Neurophysiology*, 43, 533

Clubley, M., Henson, T., Peck, A.W. and Riddington, C. (1977) 'Effects of caffeine and cyclizine alone and in combination on human performance and subjective ratings', *British Journal of Clinical Pharmacology*, 4, 652

Cohen, M.M., Debas, H.T., Holubitsky, I.B. and Harrison, R.C. (1971) 'Caffeine and pentagastrin stimulation of human gastric secretion', *Gastroenterology*, 61, 440-4

Cohen, S. and Booth, G.H., Jr. (1975) 'Gastric acid secretion and lower-esophageal sphincter pressure in response to coffee and caffeine', *New England Journal of Medicine*, 293, 897-9

Collins, T.F.X., Welsh, J.J., Black, T.N. and Collins, E.X. (1980) 'A comprehensive study of the teratogenic potential of caffeine in rats when given by oral intubation', Report from Division of Toxicology, Bureau of Foods, Food and Drug Administration, Washington, DC.

Conterio, F. and Chiarelli, B. (1962) 'Study of the inheritance of some daily life habits', *Heredity*, 17, 347-59

Conway, T.L., Vickers, R.R., Jr., Ward, H.W. and Rahe, R.H. (1981) 'Occupational stress and variation in cigarette, coffee, and alcohol consumption', *Journal of Health and Social Behavior*, 22, 155-65

Cooper, B.R., Potts, W.J., Morse, D.L. and Black, W.C. (1969) 'The effects of magnesium pemoline, caffeine and picrotoxin on a food reinforced discrimination task', *Psychonomic Science*, 14, 225-6

Costill, D.L., Dalsky, G.P. and Fink, W.J. (1978) 'Effects of caffeine ingestion on metabolism and exercise performance', *Medicine and Science in Sports*, 10, 155-8

Cox, T. (1970a) 'The effects of caffeine, alcohol, and previous exposure to the test situation on spontaneous alternation', *Psychopharmacologia*, 17, 83-8
Cox, T. (1970b) 'The effects of long-term administration of caffeine on growth and activity in rats', *Psychonomic Science*, 18, 285-6
Crainicianu, A. (1969) 'Pharmacodynamic action of caffeine. Inference in clinical obstetrics', *Revue Française de Gynécologie et d'Obstetrique*, 64, 415-20
Curatolo, P. and Robertson, D. (1983) 'The health consequences of caffeine', *Annals of Internal Medicine*, 98 Part 1, 641-53
Czarnecka, E. and Gruszczyński, W. (1978) 'Clinical and experimental electroencephalographic investigations of the combined effect of ethyl alcohol with caffeine, hydroxyzine and diazepam', *Electroencephalography and Clinical Neurophysiology*, 45, 7P-8P
Czochra-Lysanowicz, Z., Gorski, M. and Kedra, M. (1959) 'The effect of nicotine and caffeine on the development of arteriosclerosis in rabbits', *Annales Universitatis Mariae-Curie Sklodowska*, 14, 181-206
Czok, G. and Lang, K. (1961a) 'Chlorogensäure-Wirkungen am Magen-Darmkanal', *Arzneimittel-Forschung*, 11, 545-9
Czok, G. and Lang, K. (1961b) 'Zur erregenden Wirkung von Chlorogensäure', *Arzneimittel-Forschung*, 11, 448-50
Dahme, G., Lienert, G. and Malorny, G. (1972) 'Einflüsse von Alkohol und Kaffee auf die Psychomotorik sowie auf die subjektive Einschätzung des eigenen Befindens', *Zeitschrift für Ernährungswissenschaft*, 14, 36-46
Dall'Olio, R., Gandolfi, O. and Montanaro, N. (1978) 'Effects of pre- and post-trial caffeine administrations upon "step-down" passive avoidance behavior in rats submitted or not to electroconvulsive shock', *Pharmacological Research Communications*, 10, 851-8
Daly, J.W. (1983) 'Role of ATP and adenosine receptors in physiologic processes: summary and prospectus' in J.W. Daly, Y. Kuroda, J.W. Phillis, H. Shimizu and M. Ui (eds.), *Physiology and Pharmacology of Adenosine Derivatives*, Raven Press, New York, pp. 275-90
Daly, J.W., Bruns, R.F. and Snyder, S.H. (1981) 'Adenosine receptors in the central nervous system: relationship to the central actions of methylxanthines', *Life Sciences*, 28, 2083-97
Danforth, W.H. and Helmreich, E. (1964) 'Regulation of glycolysis in muscle. I. The conversion of phosphorylase b to phosphorylase a in frog sartorius muscle', *Journal of Biological Chemistry*, 239, 3133-8
Daubresse, J.C., Franchimont, P., Luyckx, A., Demey-Ponsart, E. and Lefèbvre, P. (1973) 'Effects of coffee and caffeine on carbohydrate metabolism, free fatty acid, insulin, growth hormone and cortisol plasma levels in man', *Acta Diabetologica Latina*, 10, 1069-84
Dawber, T.R., Kannel, W.B. and Gordon, T. (1974) 'Coffee and cardiovascular disease: observations from the Framingham study', *New England Journal of Medicine*, 291, 871-4
Debas, H.T., Cohen, M.M., Holubitsky, I.B. and Harrison, R.C. (1971) 'Caffeine-stimulated gastric acid and pepsin secretion: dose-response studies', *Scandinavian Journal of Gastroenterology*, 6, 453-7
Dimaio, V.J.M. and Garriott, J.C. (1974) 'Lethal caffeine poisoning in a child', *Forensic Science*, 3, 275-8
Eddy, N.B. and Downs, A.W. (1928) 'Tolerance and cross-tolerance in the human subject to the diuretic effect of caffeine, theobromine and theophylline', *Journal of Pharmacology and Experimental Therapeutics*, 33, 167-74
Eichler, O. (1976) *Kaffee und Coffein*, 2nd edn., Springer-Verlag, Berlin
Eichler, O., Klein, H.-W. and Stephan, H. (1949) 'Kaffeewirkung bei sportlichen Uebungen', *Naunyn-Schmiedebergs Archiv für Pharmakologie und experimentelle Pathologie*, 206, 251-7
Elkins, R.N., Rapoport, J.L., Zahn, T.P., Buchsbaum, M.S., Weingartner, H., Kopin, I.J. Langer, D. and Johnson, C. (1981) 'Acute effects of caffeine in normal prepubertal boys', *American Journal of Psychiatry*, 138, 178-83
Ellis, S. (1956) 'The metabolic effects of epinephrine and related amines', *Pharmacological Reviews*, 8, 485-565
Ellis, S. (1959) 'Relation of the biochemical effects of epinephrine to its muscular effects', *Pharmacological Reviews*, 11, 469-79

Elmazar, M.M.A., McElhatton, P.R. and Sullivan, F.M. (1981) 'Acute studies to investigate the mechanism of action of caffeine as a teratogen in mice', *Human Toxicology*, 1, 53-63

Elmazar, M.M.A., McElhatton, P.R. and Sullivan, F.M. (1982) 'Studies on the teratogenic effects of different oral preparations of caffeine in mice', *Toxicology*, 23, 57-71

Enslen, M., Milon, H. and Würzner, H.P. (1980) 'Brain catecholamines and sleep states in offspring of caffeine-treated rats', *Experientia*, 36, 1105-6

Estler, C.-J. (1982) 'Caffeine' in F. Hoffmeister and G. Stille (eds.), *Psychotropic Agents. Part III. Alcohol and Psychotomimetics, Psychotropic Effects of Central Acting Drugs*, Springer-Verlag, Berlin, pp. 369-89

Eysenck, H.J. (1980) *The Causes and Effects of Smoking*, Maurice Temple Smith, London

Feierabend, J.M. and Bättig, K. (1982) 'Effekte von Koffein auf physiologische Parameter während eines Lerntests', *Sozial- und Präventivmedizin*, 27, 240-241

Feinstein, M.B. (1963) 'Inhibition of caffeine rigor and radiocalcium movements by local anesthetics in frog sartorius muscle', *Journal of General Physiology*, 47, 151-73

File, S.E., Bond, A.J. and Lister, R.G. (1982) 'Interaction between effects of caffeine and lorazepam in performance tests and self-ratings', *Journal of Clinical Psychopharmacology*, 2, 102-6

Flexner, J.B. and Flexner, L.B. (1975) 'Puromycin's suppression of memory in mice as affected by caffeine', *Pharmacology Biochemistry and Behavior*, 3, 13-17

Flood, J.F., Bennett, E.L., Orme, A.E., Rosenzweig, M.R. and Jarvik, M.E. (1978) 'Memory: modification of anisomycin-induced amnesia by stimulants and depressants', *Science*, 199, 324-6

Flory, C.D. and Gilbert, J. (1943) 'The effects of benzedrine sulphate and caffeine citrate on the efficiency of college students', *Journal of Applied Psychology*, 27, 121-34

Fog, R. (1969) 'Stereotyped and non-stereotyped behaviour in rats induced by various stimulant drugs', *Psychopharmacologia*, 14, 299-304

Foltz, E., Ivy, A.C. and Barborka, C.J. (1942) 'The use of double work periods in the study of fatigue and the influence of caffeine on recovery', *American Journal of Physiology*, 136, 79-86

Foltz, E.E., Ivy, A.C. and Barborka, C.J. (1943) 'The influence of amphetamine (benzedrine) sulfate, d-desoxyephedrine hydrochloride (pervitin), and caffeine upon work output and recovery when rapidly exhausting work is done by trained subjects', *Journal of Laboratory and Clinical Medicine*, 28, 603-6

Foote, W.E., Holmes, P., Pritchard, A., Hatcher, C. and Mordes, J. (1978) 'Neurophysiological and pharmacodynamic studies on caffeine and on interactions between caffeine and nicotinic acid in the rat', *Neuropharmacology*, 17, 7-12

Forde, J.H. and Hirsh, K.R. (1976) 'Caffeine effects on reticular formation neurons in the decerebrate cat', *Neuroscience Abstracts*, 2, 867

Forrest, W.H., Jr., Bellville, J.W. and Brown, B.W., Jr. (1972) 'The interaction of caffeine with pentobarbital as a nighttime hypnotic', *Anesthesiology*, 36, 37-41

Foulks, J.G., Perry, F.A. and Sanders, H.D. (1971a) 'Effect of external potassium concentration on caffeine contractures in frog toe muscle', *Canadian Journal of Physiology and Pharmacology*, 49, 879-88

Foulks, J.G., Perry, F.A. and Sanders, H.D. (1971b) Augmentation of caffeine-contracture tension by twitch-potentiating agents in frog toe muscle', *Canadian Journal of Physiology and Pharmacology*, 49, 889-900

Freestone, S. and Ramsay, L.E. (1982) 'Effect of coffee and cigarette smoking on the blood pressure of untreated and diuretic-treated hypertensive patients', *American Journal of Medicine*, 73, 348-53

Fröberg, J., Karlsson, C.-G., Levi, L., Linde, L. and Seeman, K. (1969) 'Test performance and subjective feelings as modified by caffeine containing and caffeine-free coffee' in F. Heim and H.P.T. Ammon (eds.), *Coffein und andere Methylxanthine*, F.K. Schattauer Verlag, Stuttgart, New York, pp. 15-20

Fujii, S., Inada, S., Yoshida, S., Kusanagi, C., Mima, K. and Natsumo, Y. (1972) 'Pharmacological studies on doping drugs for race horses. II. Caffeine', *Japanese Journal of Veterinary Science*, 34, 141

Fülgraff, G. (1969) 'Xanthinderivate als Diuretica' in H. Herken (ed.), *Handbook der experimentellen Pharmakologie*, Vol. 24: *Diuretica*, Springer-Verlag, Berlin, Heidelberg,

New York, pp 594-640
Fuxe, K., Agnati, L.F., Corrodi, H., Everitt, J.B., Hökfelt, T., Löfström, A. and Ungerstedt, U. (1975) 'Action of dopamine receptor agonists in forebrain and hypothalamus: rotational behaviour, ovulation, and dopamine turnover' in D. Caline, T.N. Chase and A. Barbeau (eds.), *Advances in Neurobiology*, Vol. 9: *Dopaminergic Mechanisms*, Raven Press, New York, pp. 223-42
Fuxe, K. and Ungerstedt, U. (1974) 'Action of caffeine and theophyllamine on supersensitive dopamine receptors: considerable enhancement of receptor response to treatment with dopa and dopamine receptor agonists', *Medical Biology*, 52, 48-54
Gabrys, B.F., Nyhus, L.M., Van Meter, S.W. and Bombeck, C.T. (1973) 'The effect of aminophylline on pentagastrin-induced gastric secretion in the dog', *American Journal of Digestive Diseases*, 18, 563-6
Geller, I., Hartmann, R. and Blum, K. (1971) 'Effects of nicotine, nicotine monomethiodide, lobeline, chlordiazepoxide, meprobamate and caffeine on a discrimination task in laboratory rats', *Psychopharmacologia*, 20, 355-65
Gibbs, F.A. and Maltby, G.L. (1943) 'Effect on the electrical activity of the cortex of certain depressant and stimulant drugs — barbiturates, morphine, caffeine, benzedrine and adrenalin', *Journal of Pharmacology and Experimental Therapeutics*, 78, 1-10
Giddings, G. (1934) 'Normal sleep pattern for children: factors which derange such a pattern (physical factors)', *Journal of the American Medical Association*, 102, 525-9
Giese, H. and Schmidt, G. (1968) *Studenten-Sexualität; Verhalten und Einstellung, eine Umfrage an 12 westdeutschen Universitäten*, Rowohlt, Reinbek bei Hamburg
Gilbert, R.M. (1976) 'Caffeine as a drug of abuse' in R.J. Gibbins, Y. Israel, H. Kalant, R.E. Popham, W. Schmidt and R.G. Smart (eds.), *Research Advances in Alcohol and Drug Problems*, Vol. 3, J. Wiley and Sons, New York, pp. 49-176
Gilliland, A.R. and Nelson, D. (1939) 'The effects of coffee on certain mental and physiological functions', *Journal of General Psychology*, 21, 339-48
Goldberg, H., Horvath, T.B. and Meares, R.A. (1974) 'Visual evoked potentials as a measure of drug effects on arousal in the rabbit', *Clinical and Experimental Pharmacology and Physiology*, 1, 147-54
Goldstein, A. (1964) 'Wakefulness caused by caffeine', *Naunyn-Schmiedebergs Archiv für experimentelle Pathologie und Pharmakologie*, 248, 269-78
Goldstein, A. and Kaizer, S. (1969) 'Psychotropic effects of caffeine in man. III. A questionnaire survey of coffee drinking and its effects in a group of housewives', *Clinical Pharmacology and Therapeutics*, 10, 477-88
Goldstein, A., Kaizer, S. and Warren, R. (1965a) 'Psychotropic effects of caffeine in man. II. Alertness, psychomotor coordination, and mood', *Journal of Pharmacology and Experimental Therapeutics*, 150, 146-51
Goldstein, A., Warren, R. and Kaizer, S. (1965b) 'Psychotropic effects of caffeine in man. I. Individual differences in sensitivity to caffeine-induced wakefulness', *Journal of Pharmacology and Experimental Therapeutics*, 149, 156-9
Goldstein, H.R. (1982) 'No association between coffee and cancer of the pancreas', *New England Journal of Medicine*, 306, 997
Goldstein, L., Murphree, H.B. and Pfeiffer, C.C. (1963) 'Quantitative electroencephalography in man as a measure of CNS stimulation', *Annals of the New York Academy of Sciences*, 107, 1045-56
Goodman, R.R. and Snyder, S.H. (1982) 'Autoradiographic localization of adenosine receptors in rat brain using [^{3}H]cacloohexyladenosine', *Journal of Neuroscience*, 2, 1230-41
Gorodischer, R. and Karplus, M. (1982) 'Pharmacokinetic aspects of caffeine in premature infants with apnoea', *European Journal of Clinical Pharmacology*, 22, 47-52
Goyen, J.E. (1980) Food and drug administration news release No. P80-36, September 4, 1980
Greden, J.F. (1974) 'Anxiety or caffeinism: A diagnostic dilemma', *American Journal of Psychiatry*, 131, 1089-92
Greden, J.F., Fontaine, P., Lubetsky, M. and Chamberlin, K. (1978) 'Anxiety and depression associated with caffeinism among psychiatric inpatients', *American Journal of Psychiatry*, 135, 963-6

Gresham, S.C., Webb, W.B. and Williams, R.L. (1963) 'Alcohol and caffeine: effect on inferred visual dreaming', *Science*, *140*, 1226-7

Gualtierotti, T. (1955a) 'Contribution of spinal centers to the action of caffeine on frog's spinal reflexes', *Journal of Physiology*, *128*, 326-32

Gualtierotti, T. (1955b) 'Variations in the frog's spinal reflexes caused by the action on the brain of large doses of caffeine', *Journal of Physiology*, *128*, 320-5

Gubareff, T., de and Sleator, W., Jr. (1965) 'Effects of caffeine on mammalian atrial muscle, and its interaction with adenosine and calcium', *Journal of Pharmacology and Experimental Therapeutics*, *148*, 202-14

Gupta, B.D., Dandiya, P.C. and Gupta, M.L. (1971) 'A psychopharmacological analysis of behaviour in rats', *Japanese Journal of Pharmacology*, *21*, 293-8

Hach, B. and Heim, F. (1971) 'Vergleichende Untersuchung über die zentralerregende Wirkung von Coffein und Chlorogensäure an weissen Mäusen', *Arzneimittel-Forschung*, *21*, 23-5

Haldi, J., Bachman, G., Ensor, C. and Wynn, W. (1941) 'The effect of various amounts of caffeine on the gaseous exchange and the respiratory quotient in man', *Journal of Nutrition*, *21*, 307-20

Hall, M.J., Bartoshuk, L.M., Cain, W.S. and Stevens, J.C. (1975) 'PTC taste blindness and the taste of caffeine', *Nature*, *253*, 442-3

Hauty, G.T. and Payne, R.B. (1955) 'Mitigation of work decrement', *Journal of Experimental Psychology*, *49*, 60-7

Hearst, E. (1964) 'Drug effects on stimulus generalization gradients in the monkey', *Psychopharmacologia*, *6*, 57-70

Hennekens, C.H., Drolette, M.E., Jesse, M.J., Davies, J.E. and Hutchinson, G.B. (1976) 'Coffee drinking and death due to coronary heart disease', *New England Journal of Medicine*, *294*, 633-6

Heyden, S. (1980) 'Coffee and fibrocystic breast disease', *Surgery*, *88*, 741-2

Heyden, S. (1982) 'Fibrocystic breast disease and methylxanthine consumption', *Dixième Colloque Scientifique International sur le Café, Brazil, 1982*, Association Scientifique Internationale due Café, Paris pp. 333-8

Heyden, S., Tyroler, H.A., Heiss, G., Hames, C.G. and Bartel, A. (1978) 'Coffee consumption and mortality: total mortality, stroke mortality, and coronary heart disease mortality', *Archives of Internal Medicine*, *138*, 1472-5

Higgins, I., Stolley, P. and Wynder, E.L. (1981) 'Coffee and cancer of the pancreas', *New England Journal of Medicine*, *304*, 1605

Hirchie, H., Haralambie, G., Kunze, K., Langohr, H.D. and Lübbers, D.W. (1971) 'Probleme der skelettmuskeldurchblutung', *Arzneimittel-Forschung*, *21*, 366-76

Hire, J.N. (1978) 'Anxiety and caffeine', *Psychological Reports*, *42*, 833-4

Hirsh, K., Forde, J. and Pinzone, M. (1974) 'Caffeine effects on spontaneous activity of reticular formation neurons', Abstract prepared for the Society for Neuroscience, Fourth Annual Meeting, St. Louis, Missouri, 1974, p. 257

Hofmann, W.W. (1969) 'Caffeine effects on transmitter depletion and mobilisation at motor nerve terminals', *American Journal of Physiology*, *216*, 621-9

Hori, Y. and Yoshii, N. (1963) 'Cortical click-evoked potentials in the waking animal', *Medical Journal of Osaka University*, *13*, 395-412

Horning, M.G., Butler, C.M., Nowlin, J. and Hill, R.M. (1975) 'Drug metabolism in the human neonate', *Life Sciences*, *16*, 651-72

Horst, K., Buxton, R.E. and Robinson, W.D. (1934a) 'The effect of the habitual use of coffee or decaffeinated coffee upon blood pressure and certain motor reactions of normal young men', *Journal of Pharmacology and Experimental Therapeutics*, *52*, 322-37

Horst, K., Robinson, W.D., Jenkins, W.L. and Bao, D.L. (1934b) 'The effect of caffeine, coffee and decaffeinated coffee upon blood pressure, pulse rate and certain motor reactions of normal young men', *Journal of Pharmacology and Experimental Therapeutics*, *52*, 307-21

Hughes, R.N. and Greig, A.M. (1976) 'Effects of caffeine, methamphetamine and methylphenidate on reactions to novelty and activity in rats', *Neuropharmacology*, *15*, 673-6

Hull, C.L. (1935) 'The influence of caffeine and other factors on certain phenomena of rote

learning', *Journal of General Psychology, 13*, 249-74
Hume, K.M. and Rhys-Jones, E. (1961) 'The response to bronchodilators in intrinsic asthma', *Quarterly Journal of Medicine, 30*, 189-99
Imai, S. and Takeda, K. (1967) 'Effect of vasodilators upon the isolated *Taenia coli* of the guinea pig', *Journal of Pharmacology and Experimental Therapeutics, 156*, 557-64
Ishay, J.S. and Paniry, V.A. (1979) 'Effects of caffeine and various xanthines on hornets and bees', *Psychopharmacology, 65*, 299-309
Ito, Y., Osa, T. and Kuriyama, H. (1973) 'Topical differences of caffeine action on smooth muscle cells of the guinea pig alimentary canal', *Japanese Journal of Physiology, 24*, 217-32
Izquierdo, J.A., Costas, S.M., Justel, E.A. and Rabiller, G. (1979) 'Effect of caffeine on the memory of the mouse', *Psychopharmacology, 61*, 29-30
Jacob, J. and Michaud, G. (1961) 'Actions de divers agents pharmacologiques sur les temps d'épuisement et le comportement de souris nageant à 20°C', *Archives internationales de Pharmacodynamie et de Thérapie, 133*, 101-15
Jacobson, E.D. and Thompson, W.J. (1976) 'Cyclic AMP and gastric secretion: the illusive second messenger' in P. Greengard and G.A. Robison (eds.), *Advances in Cyclic Nucleotide Research*, Vol. 7, Raven Press, New York, pp. 199-224
Jick, H., Miettinen, O.S., Neff, R.K., Shapiro, S., Heinonen, O.P. and Slone, D. (1973) 'Coffee and myocardial infarction', *New England Journal of Medicine, 289*, 63-7
Johns, M.W. (1974) 'The sleep habits and lifestyle of cigarette smokers', *Medical Journal of Australia, 2*, 808-11
Jokela, S. and Vartiainen, A. (1959) 'Caffeine Poisoning', *Acta Pharmacologica et Toxicologica, 15*, 331-4
Jouvet, M., Benoit, O., Marsallon, A. and Courjon, J. (1957) 'Action de la caféine sur l'activité électrique cérébrale', *Comptes Rendus des Séances de la Socité de Biologie, 151*, 1542-5
Kallman, W.M. and Isaac, W. (1975) 'The effects of age and illumination on the dose-response curves for three stimulants', *Psychopharmacologia, 40*, 313-18
Kaplan, E., Holmes, J.H. and Sapeika, N. (1974) 'Caffeine content of tea and coffee', *South African Medical Journal, 48*, 510-11
Karacan, I., Thornby, J.I., Anch, A.N., Booth, G.H., Williams, R.L. and Salis, P.J. (1976) 'Dose-related sleep disturbances induced by coffee and caffeine', *Clinical Pharmacology and Therapeutics, 20*, 682-9
Katz, A.M., Repke, D.I. and Hasselbach, W. (1977) 'Dependence of ionophore- and caffeine-induced calcium release from sarcoplasmic reticulum vesicles on external and internal calcium ion concentrations', *Journal of Biological Chemistry, 252*, 1938-49
Keiner, F. and Weder, W. (1967) 'Funktionsdiagnostik bei der cholecystographie', *Röntgenblätter, 20*, 12
Keister, M.E. and McLaughlin, R.J. (1972) 'Vigilance performance related to extraversion-introversion and caffeine', *Journal of Experimental Research in Personality, 6*, 5-11
Khanna, N.N., Somani, S.M., Boyer, A., Miller, J., Chua, C. and Menke, J.A. (1982) 'Cross validation of serum to saliva relationships of caffeine, theophylline and total methylxanthines in neonates', *Developmental Pharmacology and Therapeutics, 4*, 18-27
Khera, K.S. (1981) 'Common fetal aberrations and their teratologic significance. A review', *Fundamental and Applied Toxicology, 1*, 13-18
Kihara, G. (1928) 'Toleration of the diuretic action of caffeine', *Proceedings of the Imperial Academy of Tokyo, 4*, 418-20
Klebelsberg, D. von and Mostbeck, A. (1963) 'Wirken sich mittlere Dosen von Kaffee und Koffein auf psychische Funktionen der Fahrtüchtigkeit aus?', *Psychologie und Praxis, 7*, 23-35
Knowles, J.B. (1963) 'Conditioning and the placebo effect: the effects of decaffeinated coffee on simple reaction time in habitual coffee drinkers', *Behavior Research and Therapy, 1*, 151-7
Knutti, R., Rothweiler, H. and Schlatter, C. (1982) 'The effect of pregnancy on the pharmacokinetics of caffeine', *Archives of Toxicology, Suppl. 5*, 187-92
Kotake, A.N., Schoeller, D.A., Lambert, G.H., Baker, A.L., Schaffer, D.D. and Josephs, H.

(1982) 'The caffeine CO_2 breath test: dose response and route of N-demethylation in smokers and nonsmokers', *Clinical Pharmacology and Therapeutics*, 32, 261-9
Kraeplin, E. (1892) *Ueber die Beeinflussung einfacher psychischer Vorgänge durch einige Arzneimittel*, Verlag von Gustav Fischer, Jena
Kretschmer, W. (1936) 'Ein Fall von Koffeinvergiftung', *Medizinische Welt*. 10, 232-3
Krönig, D. and Künkel, H. (1973) 'Simultane Vielkanal-EEG-Spectral-analyse in Echtzeit. Signalverarbeitung', Vorträge hrsg. v.W. Schüssler, *Nachrichtentechnische Gesellschaft im V.D.I.*, May 1973, pp. 185-93
Krueger, H., Zülch, J. and Gandorfer, M. (1979) 'Der Einfluss von Koffein auf die motorische Reaktions- und die visuell-mentale Verarbeitungszeit', *Zeitschrift für Ernährungswissenschaft*, 18, 51-61
Kuftinec, D.M. and Mayer, J. (1964) 'Extreme sensitivity of obese hyperglycemic mice to caffeine and coffee', *Metabolism, Clinical and Experimental*, 13, 1369-75
Kulkarni, A.S. (1972) 'Avoidance acquisition and CNS stimulants', *Naunyn-Schmiedebergs Archiv für Pharmakologie und experimentelle Pathologie*, 273, 394-400
Künkel, H. (1976) 'Vielkanal-EEG-Spektralanalyse der Coffein-Wirkung', *Zeitschrift für Ernährungswissenschaft*, 15, 71-9
Kusanagi, C., Fujii, S. and Inada, S. (1974) 'Evaluation of doping drugs by treadmill exercise in dogs. I. Caffeine', *Japanese Journal of Veterinary Science*, 36, 81-92
Lader, M. (1969) 'Comparison of amphetamine sulphate and caffeine citrate in Man', *Psychopharmacologia*, 14, 83-94
Lam, L.K.T., Sparnins, V.L. and Wattenberg, L.W. (1982) 'Isolation and identification of kahweol palmitate and cafestol palmitate as active consistuents of green coffee beans that enhance glutathione S-transferase activity in the mouse', *Cancer Research*, 42, 1193-8
Lawson, D.II., Jick, II. and Rothman, K.J. (1981) 'Coffee and tea consumption and breast disease', *Surgery*, 90, 801-3
Leathwood, P.D. and Pollet, P. (1983) 'Diet-induced mood changes in normal populations', *Journal of Psychiatric research*, 17, 147-54
Lehmann, H.E. and Csank, J. (1957) 'Differential screening of phrenotropic agents in man: psychophysiologic test data', *Journal of Clinical Psychopathology*, 18, 222-35
Lie, E. (1930) 'Caffeine and diuresis in man', *American Journal of Physiology*, 92, 619-24
Lienert, G.A. and Huber, H.P. (1966) 'Differential effects of coffee on speed and power tests', *Journal of Psychology*, 63, 269-74
Lin, W. and Bittar, E.E. (1974) 'Some observations on caffeine-induced contracture of barnacle muscle fibers', *Life Sciences*, 15, 1611-9
Linn, S., Schoenbaum, S.C., Monson, R.R., Rosner, B., Stubblefield, P.G. and Ryan, K.J. (1982) 'No association between coffee consumption and adverse outcomes of pregnancy', *New England Journal of Medicine*, 306, 141-5
Lovingood, B.W., Blyth, C.S., Peacock, W.H. and Lindsay, R.B. (1967) 'Effects of d-amphetamine sulfate, caffeine, and high temperature on human performance', *Research Quarterly. American Association for Health, Physical Education and Recreation*, 38, 64-71
MacMahon, B., Yen, S., Trichopoulos, D., Warren, K. and Nardi, G. (1981) 'Coffee and cancer of the pancreas', *New England Journal of Medicine*, 304, 630-3
Marbach, G. and Schwertz, M.T. (1964) 'Effets physiologiques de l'alcool et de la caféine au cours du sommeil chez l'homme', *Archives des Sciences Physiologiques*, 18, 163-210
Marcus, M.L., Skelton, C.L., Grauer, L.E. and Epstein, S.E. (1972) 'Effects of theophylline on myocardial mechanics', *American Journal of Physiology*, 222, 1361-5
Marks, V. and Kelly, J.F. (1973) 'Absorption of caffeine from tea, coffee, and Coca-cola', *Lancet*, 1, 827
Marriott, A.S. (1968) 'The effects of amphetamine, caffeine and methylphenidate on the locomotor activity of rats in an unfamiliar environment', *International Journal of Neuropharmacology*, 7, 487-91
Marshall, J., Graham, S. and Swanson, M. (1982) 'Caffeine consumption and benign breast disease: a case-control comparison', *American Journal of Public Health*, 72, 610-12
Mau, G. and Netter, P. (1974) 'Kaffee- und Alkoholkonsum — Risikofaktoren in der Schwangerschaft?', *Geburtshilfe und Frauenheilkunde*, 34, 1018-22
Means, J.H., Aub, J.C. and DuBois, E.F. (1917) 'The effect of caffeine on heat production',

Archives of Internal Medicine, 19, 832-9
Merton, P.A. (1954) 'Voluntary strength and fatigue', *Journal of Physiology, 123*, 553-64
Meyer, F. P., Walther, T. and Walther, H. (1983) 'Ueber den Einfluss von persönlichkeitsmerkmalen auf die Reaktionsleistungen freiwilliger Probanden nach einmaliger Applikation von Plazebo, Crotylbarbital und Coffein', *Pharmacopsychiatria, 16*, 13-18
Minton, J.P., Abou-Issa, H., Reiches, N. and Roseman, J.M. (1981) 'Clinical and biochemical studies on methylxanthine-related fibrocystic breast disease', *Surgery, 90*, 229-304
Mitchell, V.E., Ross, S. and Hurst, P.M. (1974) 'Drugs and placebos: effects of caffeine on cognitive performance', *Psychological Reports, 35*, 875-83
Mitznegg, P. (1976) 'Der Einfluss von methylxanthinen auf die Funktion glatter Muskulatur' in O. Eichler, *Kaffee und Coffein*, 2nd edn., Springer-Verlag, Berlin, pp. 125-31
Moyer, J.H., Tashnek, A.B., Miller, S.I., Snyder, H. and Bowman, R.O. (1952) 'The effect of theophylline with ethylenediamine (aminophylline) and caffeine on cerebral hemodynamics and cerebrospinal fluid pressure in patients with hypertensive headaches', *American Journal of the Medical Sciences, 224*, 377-85
Müller-Limmroth, W. (1973) 'Der Einfluss von coffeinhaltigem und coffeinfreiem Kaffee auf den Schlaf des Menschen', *Cinquième Colloque International sur la Chimie des Cafés, Lisbonne, 1971*, Association Scientifique Internationale du Café, Paris, pp. 375-82
Mullin, F.J., Kleitman, N. and Cooperman, N.R. (1933) 'Studies of the physiology of sleep. X. The effect of alcohol and caffein [sic] on motility and body temperature during sleep', *American Journal of Physiology, 106*, 478-87
Murray, S.S., Bjelke, E., Gibson, R.W. and Schuman, L.M. (1981) 'Coffee consumption and mortality from ischemic heart disease and other causes: results from the Lutheran Brotherhood Study, 1966-1978', *American Journal of Epidemiology, 113*, 661-7
Nash, H. (1962) *Alcohol and Caffeine: A Study of Their Psychological Effects*, Thomas, Springfield, Ill.
Nieschulz, O. (1968) 'Ueber Nachweismöglichkeiten zentraler Wirkungen des Coffein in Tierversuchen', *Troisième Colloque International sur la Chimie des Cafés Verts, Torréfiés et leurs Dérivés, Trieste, 1967*, Association Scientifique Internationale du Café, Paris, pp. 277-85
Nolen, G.A. (1982) 'A Reproduction/Teratology Study of Brewed and Instant Decaffeinated Coffees', *Journal of Toxicology and Environmental Health, 10*, 769-83
Novotny, I. and Vyskocil, F. (1966) 'Possible role of Ca ions in the resting metabolism of frog sartorius muscle during potassium depolarization', *Journal of Cellular and Comparative Physiology, 67*, 159-68
Oberman, Z., Harell, A., Herzberg, M., Hoerer, E., Jaskolka, H. and Laurian, L. (1975) 'Changes in plasma cortisol, glucose and free fatty acids after caffeine ingestion in obese women', *Israel Journal of Medical Sciences, 11*, 33-6
Ogilvie, R.I., Fernandez, P.G. and Winsberg, F. (1977) 'Cardiovascular response to increasing theophylline concentrations', *European Journal of Clinical Pharmacology, 12*, 409-14
Ohba, M. (1973) 'Effects of caffeine on tension development in dog papillary muscle under voltage clamp', *Japanese Journal of Physiology, 23*, 47-58
Ohiokpehai, O., Brumen, G. and Clifford, M.N. (1982) 'The chlorogenic acids content of some peculiar green coffee beans and the implications for beverage quality', *Dixième Colloque Scientifique International sur le Café, Brazil, 1982*, Association Scientifique Internationale du Café, Paris, pp. 177-86
Onodera, K. (1963) 'Effect of caffeine on the neuromuscular junction of the frog, and its relation to external calcium concentration', *Japanese Journal of Physiology, 23*, 587-97
Orentlicher, M. and Ornstein, R.S. (1971) 'Influence of external cations on caffeine-induced tension: calcium extrusion in crayfish muscle', *Journal of Membrane Biology, 5*, 319-33
Oswald, I. (1968) 'Drugs and sleep', *Pharmacological Reviews, 20*, 273-303
Pan-American Coffee Bureau (1972) *Annual Coffee Statistics*, No. 36, New York
Parsons, W.D. and Neims, A.H. (1981) 'Prolonged half-life of caffeine in healthy term newborn infants', *Journal of Paediatrics, 98*, 640-1
Patwardhan, R.V., Desmond, P.V., Johnson, R.F. and Schenker, S. (1980) 'Impaired

elimination of caffeine by oral contraceptive steroids', *Journal of Laboratory and Clinical Medicine*, 95, 603-8

Paul, O., Leper, M.H., Phelan, W.H., Dupertuis, G.W., MacMillan, A., McKean, H. and Park, H. (1963) 'A longitudinal study of coronary heart disease', *Circulation*, 28, 20-31

Paul, O., MacMillan, A., McKean, H. and Park, H. (1968) 'Sucrose intake and coronary heart disease', *Lancet*, 2, 1049-51

Perkins, R. and Williams, M.H. (1975) 'Effect of caffeine upon maximal muscular endurance of females', *Journal of Medicine and Science in Sports*, 7, 221-4

Perry, A. (1973a) 'The effect of heredity on attitudes toward alcohol, cigarettes, and coffee', *Journal of Applied Psychology*, 58, 275-7

Perry, A. (1973b) 'Heredity, personality traits, product attitude, and product consumption — an exploratory study', *Journal of Marketing Research*, 10, 376-9

Peters, J.M. (1967) 'Factors affecting caffeine toxicity: a review of the literature', *Journal of Clinical Pharmacology and the Journal of New Drugs*, 7, 131-41

Pettijohn, T.F. (1979) 'Effects of alcohol and caffeine on wheel running activity in the mongolian gerbil', *Pharmacology Biochemistry and Behavior*, 10, 339-41

Phillis, J.W. and Wu, P.H. (1981) 'The role of adenosine and its nucleotides in central synaptic transmission', *Progress in Neurobiology*, 16, 187-239

Portugal-Alvarez, J., Zamarrón, A., Yangüela, J., Perezagua, C. and Velasco, A. (1973) 'Lipolysis induced by coffee and tobacco: its modification by insulin', *Journal of Pharmacy and Pharmacology*, 25, 668-9

Rall, T.W. (1980) 'Central nervous system stimulants: The xanthines' in A.G. Gilman, L.S. Goodman and A. Gilman (eds.) *The Pharmacological Basis of Therapeutics*, 6th edn., MacMillan Publishing Co., Inc., New York, pp. 592-607

Rall, T.W., Sutherland, E.W. and Berthet, J. (1957) 'The relation of epinephrine and glucagon to liver phosphorylase. IV. Effect of epinephrine and glucagon on the reactivation of phosphorylase in liver homogenates', *Journal of Biological Chemistry*, 224, 463-75

Rapoport, J.L., Jensvold, M., Elkins, R., Buchsbaum, M.S., Weingartner, H., Ludlow, C., Zahn, T.P., Berg, C.J. and Neims, A.H. (1981) 'Behavioral and cognitive effects of caffeine in boys and adult males', *Journal of Nervous and Mental Diseases*, 169, 726-32

Regina, E.G., Smith, G.M., Keiper, C.G. and McKelvey, R.K. (1974) 'Effects of caffeine on alertness in stimulated automobile driving', *Journal of Applied Psychology*, 59, 483-9

Revelle, W., Humphreys, M.S., Simon, L. and Gilliland, K. (1980) 'The interactive effect of personality, time of day, and caffeine: a test of the arousal model', *Journal of Experimental Psychology: General*, 109, 1-31

Roberts, J.J. (1978) 'The repair of DNA modified by cytotoxic mutagenic and carcinogenic chemicals' in J.T. Lett and H. Adlers (eds.), *Advances in Radiation Biology*, Vol. 7, Academic Press, New York, pp. 212-436

Roberts, J.J. (in preparation) 'Mechanism of potentiation by caffeine of genotoxic damage induced by physical and chemical agents: possible relevance to carcinogenesis'

Robertson, D., Fröhlich, J.C., Carr, R.K., Watson, J.T., Hollifield, J.W., Shand, D.G. and Oates, J.A. (1978) 'Effects of caffeine on plasma renin activity, catecholamines and blood pressure', *New England Journal of Medicine*, 298, 181-6

Robertson, D., Wade, D., Workman, R., Woosley, R.L. and Oates, J.A. (1981) 'Tolerance to the humoral and hemodynamic effects of caffeine in man', *Journal of Clinical Investigation*, 76, 1111-7

Roggenkamp, K.F. (1982) 'Kaffee-Statistik anschaulich gemacht', *Kaffee- und Tee-Markt*, 32, 12-14

Rosenberg, E., Martin, J.R., Oettinger, R. and Bättig, K. (1983) 'Locomotion by rats through a tunnel maze: effects of caffeine on habituation modified by configuration', *Experientia*, 39, 639-40

Rosenberg, L., Mitchell, A.A., Shapiro, S. and Slone, D. (1982) 'Selected birth defects in relation to caffeine-containing beverages', *Journal of the American Medical Association*, 247, 1429-32

Roth, J.A. and Ivy, A.C. (1946) 'Caffeine und "Peptic" Ulcer', *Gastroenterology*, 7, 576-82

Sant'Ambrogio, G., Frazier, D.T. and Boyarsky, L.L. (1962) 'Effect of caffeine on spinal reflexes', *Proceedings of the Society for Experimental Biology and Medicine*, 109, 273-6

Sattin, A. and Rall, T.W. (1970) 'The effect of adenosine and adenine nucleotides on the adenosine 3',5'-phosphate content of guinea pig cerebral cortex slices', *Molecular Pharmacology, 6*, 13-23

Schallek, W. and Kuehn, A. (1959) 'Effect of drugs on spontaneous and activated EEG of cat', *Archives internationales de Pharmacodynamie et de Thérapie, 120*, 319-33

Schwertz, M.T. and Marbach, G. (1965) 'Effets physiologiques de la caféine et du meprobamate au cours du sommeil chez l'homme', *Archives des Sciences Physiologiques, 19*, 425-79

Seashore, R.H. and Ivy, A.C. (1953) 'The effects of analeptic drugs in relieving fatigue', *Psychological Monographs, 67*, 1-13

Severson, R.K., Davis, S. and Polissar, L. (1982) 'Smoking, coffee and cancer of the pancreas', *British Medical Journal, 285*, 214

Shibata, S. and Hollander, P.B. (1967) 'Effects of caffeine on the contractility and membrane potentials of rat atrium', *Experientia, 23*, 559

Siegmund, P. and Wolf, M. (1952) 'Eine einfache Methode der Motilitätsmessung an Mäusen', *Naunyn-Schmiedebergs Archiv für Pharmakologie und experimentelle Pathologie, 216*, 323-6

Skinner, B.F. and Heron, W.T. (1937) 'Effects of caffeine and benzedrine upon conditioning and extinction', *Psychological Record, 1*, 340-6

Smith, B.D., Wilson, R.J. and Jones, B.E. (1983) 'Extraversion and multiple levels of caffeine-induced arousal: effects on overhabituation and dishabituation', *Psychophysiology, 20*, 29-34

Smith, D.L., Tong, J.E. and Leigh, G. (1977) 'Combined effects of tobacco and caffeine on the components of choice reaction-time, heart rate, and hand steadiness', *Perceptual and Motor Skills, 45*, 635-9

Smith, J.R. and Jensen, J. (1946) 'Observations on the effect of theophylline aminoisobutanol in experimental heart failure', *Journal of Laboratory and Clinical Medicine, 31*, 850-6

Snyder, S.H., Katims, J.J., Annau, Z., Bruns, R.F. and Daly, J.W. (1981) 'Adenosine receptors and behavioral actions of methylxanthines', *Proceedings of the National Academy of Sciences of the U.S.A., 78*, 3260-4

Soldatos, C.R., Kales, J.D., Scharf, M.B., Bixler, E.O. and Kales, A. (1980) 'Cigarette smoking associated with sleep difficulty', *Science, 207*, 551-3

Solyom, L., Enesco, H.E. and Beaulieu, C. (1968) 'The effect of RNA, uric acid and caffeine on conditioning and activity in rats', *Journal of Psychiatric Research, 6*, 175-83

Somers, G., Devis, G., Van Obberghen, E. and Malaisse, W.J. (1976) 'Calcium antagonists and islet function. II. Interaction of theophylline and verapamil', *Endocrinology, 99*, 114-24

Somlyo, A.V. and Somlyo, A.P. (1967) 'Electromechanical and pharmacological coupling in vascular smooth muscle', *Journal of Pharmacology and Experimental Therapeutics, 159*, 129-45

Spengler, J. (1957) 'Untersuchungen über die Wirksamkeit leistungssteigernder Pharmaka', *Schweizerische Zeitschrift für Sportmedizin, 5*, 97-124

Spilker, B. and Callaway, E. (1969) 'Effects of drugs on "augmenting/reducing" in averaged visual evoked responses in man', *Psychopharmacologia, 15*, 116-24

Starr, I., Gamble, C.J., Margolies, A., Donal, J.S., Jr., Joseph, N. and Eagle, E. (1937) 'A clinical study of the action of 10 commonly used drugs on cardiac output, work and size; on respiration, on metabolic rate and on the electrocardiogram', *Journal of Clinical Investigation, 16*, 799-823

Statland, B.E. and Demas, T.J. (1980) 'Serum caffeine half-lives: healthy subjects vs. patients having alcoholic hepatic disease', *American Journal of Clinical Pathology, 73*, 390-3

Stradomsky, N. (1970) 'Untersuchungen über Schlafbewegungen nach coffeinhaltigem und coffeinfreiem Bohnenkaffee', *Medizinische Klinik, 65*, 1372-6

Strasser, H. and Müller-Limmroth, W. (1971) 'Vergleichende Untersuchungen über die Wirkung von Koffein und Chlorogensäure auf die Psychomotorik des Menschen', *Aerztliche Forschung, 25*, 209-17

Studlar, M. (1973) 'Ueber den Einfluss von Coffein auf den Fett- und Kohlenhydrat-Stoffwechsel des Menschen', *Zeitschrift für Ernährungswissenschaft, 12*, 109-20

Sullivan, F.M. and McElhatton, P.R. (1983) 'The teratogenic and other toxic effects of drugs on reproduction' in P.F. D'Arcy and J.P. Griffin (eds.), *Iatrogenic Diseases Annual Update 1983*, 2nd ed., Oxford University Press, Oxford

Sunano, S. and Miyazaki, E. (1973) 'Effects of caffeine on electrical and mechanical activities of guinea pig *Taenia coli*', *American Journal of Physiology*, 225, 335-9

Sutherland, E.W. and Rall, T.W. (1958) 'Fractionation and characterization of a cyclic adenine ribonucleotide formed by tissue particles', *Journal of Biological Chemistry*, 232, 1077-91

Sutherland, E.W., Robison, A. and Butcher, R.W. (1968) 'Some aspects of the biological role of adenosine 3',5'-monophosphate (cyclic AMP)', *Circulation*, 37, 279-306

Takagi, K., Saito, H., Lee, C.-H. and Hayashi, T. (1972) 'Pharmacological studies on fatigue I', *Japanese Journal of Pharmacology*, 22, 17-26

Tang-Liu, D.D.-S., Williams, R.L. and Riegelman, S. (1983) 'Disposition of caffeine and its metabolites in man', *Journal of Pharmacology and Experimental Therapeutics*, 224, 180-5

Thomas, C.B. (1973) 'The relationship of smoking and habits of nervous tension' in W.L. Dunn, Jr. (ed.), *Smoking Behavior: Motives and Incentives*, J. Wiley and Sons, New York, pp. 157-70

Thorpe, W.R. and Seeman, P. (1971) 'The site of action of caffeine and procaine in skeletal muscle', *Journal of Pharmacology and Experimental Therapeutics*, 179, 324-30

Tibblin, G., Wilhelmsen, L. and Werko, L. (1975) 'Risk factors for myocardial infarction and death due to ischemic heart disease and other causes', *American Journal of Cardiology*, 35, 514-22

Timson, J. (1977) 'Caffeine', *Mutation Research*, 47, 1-52

Ulrich, R. (1968) 'Die Kaffeeverträglichkeit des psycholabilen Patienten', *Troisième Colloque International sur la Chimie des Cafés Verts, Torréffiés et leurs Dérivés, Trieste, 1967*, Association Scientifique International sur le Café, Paris, pp. 325-7

Valdes, J.J., McGuire, P.S. and Annau, Z. (1982) 'Xanthines alter behavior maintained by intracranial electrical stimulation and an operant schedule', *Psychopharmacology*, 76, 325-8

Vener, A.M. and Krupka, L.R. (1982) 'Caffeine use and young adult women', *Journal of Drug Education*, 12, 273-83

Vinegar, R., Truax, J.F., Selph, J.L., Welch, R.M. and White, H.L. (1976) 'Potentiation of the anti-inflammatory and analgesic activity of aspirin by caffeine in the rat', *Proceedings of the Society for Experimental Biology and Medicine*, 151, 556-60

Vitiello, M.V. and Woods, S.C. (1977) 'Evidence for withdrawal from caffeine by rats', *Pharmacology Biochemistry and Behavior*, 6, 553-5

Von Voigtlander, P.F. and Moore, K.E. (1973) 'Turning behavior of mice with unilateral 6-hydroxydopamine lesions in the striatum: effects of apomorphine, L-DOPA, amantadine, amphetamine and other psychomotor stimulants', *Neuropharmacology*, 12, 451-62

Waldeck, B. (1971) 'Some effects of caffeine and aminophylline on the turnover of catecholamines in the brain', *Journal of Pharmacy and Pharmacology*, 23, 824-30

Wanner, H.U. and Bättig, K. (1965) 'Pharmakologische Wirkungen auf die Laufleistung der Ratte bei verschiedener Leistungsbelohnung und verschiedener Leistungsanforderung', *Psychopharmacologia*, 7, 182-202

Warszawski, D. and Gorodischer, R. (1981) 'Tissue distribution of caffeine in premature infants and in newborn and adult dogs', *Pediatric Pharmacology*, 1, 341-6

Watanabe, H., Watanabe, K., Hagino, K. and Ikeda, M. (1978) 'Effects of dopaminergic stimulating agents, caffeine and antipsychotic drugs and rotational behaviour in mice with unilateral striatal 6-hydroxydopamine lesions', *Journal of the Pharmaceutical Society of Japan*, 98, 1613-18

Weathersbee, P.S., Olsen, L.K. and Lodge, J.R. (1977) 'Caffeine and pregnancy: a retrospective study', *Postgraduate Medicine*, 62, 64-9

Webb, D. and Levine, T.E. (1978) 'Effects of caffeine on DRL performance in the mouse', *Pharmacology Biochemistry and Behavior*, 9, 7-10

Weiss, B. and Laties, V.G. (1962) 'Enhancement of human performance by caffeine and the amphetamines', *Pharmacological Reviews*, 14, 1-36

Wenzel, D.G. and Rutledge, C.O. (1962) 'Effects of centrally-acting drugs on human motor and psychomotor performance', *Journal of Pharmaceutical Sciences, 51*, 631-44

White, B.C., Lincoln, C.A., Pearce, N.W., Reeb, R. and Vaida, C. (1980) 'Anxiety and muscle tension as consequences of caffeine withdrawal', *Science, 209*, 1547-8

White, R.P. (1963) 'Relationship between cholinergic drugs and EEG activation', *Archives internationales de Pharmacodynamie et de Thérapie, 145*, 1-17

Wilson, J.G. and Scott, W.J., Jr. (in preparation) 'The teratogenic potential of caffeine in laboratory animals'

Winstead, D.K. (1976) 'Coffee consumption among psychiatric inpatients', *American Journal of Psychiatry, 133*, 1447-50

Winter, L., Jr, Appleby, F., Ciccone, P.E. and Pigeon, J.G. (1983) 'A double-blind, comparative evaluation of acetaminophen, caffeine, and the combination of acetaminophen and caffeine in outpatients with post-operative oral surgery pain', *Current Therapeutic Research, 33*, 115-22

Zett, L. (1966) 'Membranpotential und Ionengehalt des isolierten Skelettmuskels bei Koffein-, Säure- und Kaliumkontractur', *Acta Biologica et Medica Germanica, 17*, 603-13

GLOSSARY

ABA	Abscisic acid, a plant growth regulator that antagonises the effect of growth-promoting hormones, and promotes senescence, leaf fall, dormancy and other effects.
Acarina	Mites.
Acetaminophen	Therapeutic used to reduce pain.
Acetylcholine	A reversible acetic acid ester of choline; it is released at the end of stimulated 'cholinergic' neurons (primarily peripheral parasympathetic neurons, motor neurons and central neurons) and plays a crucial role in the transmission of nerve impulses across the synpatic junction.
Action potential	The brief electrical impulse resulting from brief changes in membrane permeability to sodium and potassium ions which provides the basis for conduction of information along nerve or muscle fibres.
Actogram	The graphic tracing made by the actometer, an instrument for measuring activity in the horizontal plane.
Adenosine receptors	Specific cell membrane proteins binding with adenosine.
Adrenergic synapses	Nerve endings at which epinephrine, or substances with similar activity, are liberated when a nerve impulse passes through the nerve fibre.
Agar gel electrophoresis	An electrophoretic system for separating proteins by net charge.
Airveying	A system of transferring low density solid materials, e.g. coffee beans, from one location to another, by use of fast moving air in pipes.
Albumins	A class of proteins which are water-soluble at pH 7.
Albuminuria	Presence of protein in the urine, usually resulting from disease.
Aldol condensation	A reaction between two aldehydes (or two molecules of the same aldehyde) one of which must contain an active methylene group. The product is a hydroxycarbonyl, which if it also contains an active methylene group, may react again in a similar manner.

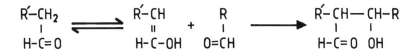

Allele	One of the two or more forms of a locus on a particular chromosome.
Allelozyme or electromorph	Isoenzymes arising from multiple alleles at the same locus.

Glossary

Allogamous	Cross-pollinating.
Allopatric	Populations or species which are mutually exclusive, but usually inhabit adjacent geographical regions.
Allotetraploid	An organism with four sets of chromosomes in the somatic cells, made up of the entire somatic complement of two species.
Amorphous	Having no visible differentiation in structure.
Amyloplast	Leucoplast of colourless starch-forming granule in plants.
Angiotensin	A substance, produced by the action of renin on serum alpha-2 globulin, which causes constriction of blood vessels resulting in a rise in blood pressure.
Anhydrosugar	A compound formed from a sugar or sugar residue in a polysaccharide by the loss of water resulting in the production of an ether linkage between two carbons. The carbons may be in the same residue (intra-residue anhydro) or in adjacent residues (inter-residue anhydro).
Anisotropic	Doubly refracting in polarised light.
Anthesis	The opening of a flower bud and the dehiscence of pollen by the anthers.
Anthribidae	A family of beetles, some of which are pests of stored products.
Anthracnose	Dark, sunken, necrotic disease lesion.
Arthropods	Animals with a hard external skeleton and jointed limbs. Includes insects, mites, spiders, millipedes, etc.
Aspirin	Therapeutic used to reduce pain, fever, and inflammation.
Atropine	An alkaloid which functions as an anticholinergic to relax smooth muscles, increase heart rate by blocking the vagus nerve and to dilate the pupil of the eye.
Autogamous	Self-pollinating.
Autotetraploid	An organism with four sets of chromosomes in the somatic cells derived from the same species.
Auxins	A group of plant growth regulators; natural auxin is indole-acetic acid, which is produced in growing points and transported to co-ordinate development.
Axil	The angle between a leaf and the stem which bears it.
Beck depression scale	Beck's Inventory for Measuring Depression, was developed from observations in ambulant psychoanalytic therapy of depressed patients and contains 91 items distributed over 21 symptom-attitude categories.
Beneficial insects	Insects that benefit man, especially those that attack pests, but also includes pollinating insects and those that feed on weeds etc.
Beta agonist	Substance which inhibits or blocks beta-adrenergic receptors, thus in-

	hibiting the response of an effector organ (heart, bronchia, lung) to autonomic nerve impulses; used therapeutically to reduce cardiac output and hypertension.
Beta-elimination reaction	The elimination, often of a water molecule, particularly when carbohydrates are heated, but other similar eliminations are known. In this case the hydroxyl β to the carbonyl group is protonated and eliminated as water. An unsaturated compound is produced.

$$\begin{array}{c} \text{H-C=O} \\ \alpha \;\; \text{H-C-OH} \\ \beta \;\; \text{H-C-OH} \end{array} \rightleftharpoons \begin{array}{c} \text{H-C-OH} \\ \| \\ \text{C-OH} \\ | \\ \text{H-C-OH} \end{array} \xrightarrow{-H_2O} \begin{array}{c} \text{H-C=O} \\ | \\ \text{C-OH} \\ \| \\ \text{H-C} \end{array}$$

Beta receptors	Alpha$_1$, alpha$_2$, beta$_1$, beta$_2$ are the receptor subtypes in postsynaptic fibres for which catecholamines have affinity.
Biotope	Particular habitat where a community of different kinds of plants and animals are living together.
Birefringence or double refraction	Phenomenon observed when a ray of light enters an anisotropic crystal. This ray is in general divided into two rays, which travel through the crystal with different velocities, and in different directions.
Bradycardia	Excessively slow heart rate, usually less than 60 beats per minute.
C$_3$ plants	Plants that have a form of photosynthesis in which atmospheric CO_2 is fixed directly through the reductive pentose phosphate pathway (3-carbon compounds) as opposed to the C$_4$-dicarboxylic acid pathway.
Capping	Cutting the top off an orthotropic stem.
Catecholamines	Group of substances including adrenaline and noradrenaline which function in the transmission of nerve impulses at certain junctions and which have a sympathomimetic action.
Cerambycidae	A family of beetles with wood-boring larvae.
Cerebellum	Part of the metencephalon located behind the brain stem which controls muscular coordination.
Cerebrovascular resistance	Constriction of the arteries of the brain.
Chlorosis	Yellowing of foliage.
Cholecystokinetic	Causing or promoting contraction of the gallbladder.
Cholinergic agonists	Drugs whose action resembles that of the neurotransmitter acetylcholine or which inhibit the enzyme which breaks down acetylcholine thereby prolonging its action.
Cimetidine	A reversible, competitive antagonist of the actions of histamine exerted on H$_2$ receptors; it inhibits histamine-evoked gastric acid secretion.

Cisternae	Closed spaces which serve as reservoirs for body fluids.
Coccidae	A family of scale insects belonging to the order Hemiptera.
Cognitive	Internal processes involving knowing, anticipating and perceiving.
Coleoptera	Beetles. An order of insects with biting mouth-parts and hard or leathery forewings.
Collar region	Where roots change to stem; at soil level.
Corticosterone	A steroid secreted in the adrenal cortex which is dependent upon the secretion of adrenocorticotrophic hormone by the pituitary gland.
Cova	A group of trees planted at one location, so that they form effectively one multiple-stem tree.
Cutworms	The ground living larvae of some of the Noctuidae — a family of moths.
Cytokinins	A group of plant growth regulators produced by roots and other organs; natural cytokinins include kinetin and zeatin, which promote cell division and regulate development.
Cytoplasm	Protoplasm of cell body other than the nucleus.
Decussate	Having pairs of opposite leaves with succeeding pairs at right angles.
Dentate gyrus	Serrated strip of grey matter located under the border of the hippocampus and in its depths.
Diastolic blood pressure	The minimum blood pressure which occurs late in the period of ventricular diastole or relaxation.
Diploid	An organism with two chromosomes of each kind.
Diptera	Two-winged flies. One of the major insect orders.
Dopamine	A precursor of epinephrine and norepinephrine, it functions as a central neurotransmitter.
E_a	Net assimilation rate.
Ectrodactyly	Congenital absence of all or part of a digit.
Endoparasite	In this context, animals that live and feed inside the plant.
Endosperm	Nutritive tissue of most seeds, residue of female prothallus surrounding an embryo.
Enteral	Method of drug administration involving the alimentary canal.
Epinephrine	Hormone secreted by the adrenal medulla and liberated at the ends of sympathetic nerve fibres; it is a powerful vasopressor and also increases glycogenolysis and glucose release.
Ergograph	Instrument which records the work done in muscular exertion.
Evapotranspiration	The loss of water by a plant through the stomata.

Exoparasite	In this context, animals that inhibit the soil and attack the plant externally.
Field Capacity	The maximum quantity of water which can be held in the soil. Excess water over this amount is drained away; if it cannot drain away the soil waterlogs.
Flurazepam	Therapeutic used as a sedative-hypnotic.
French Roast	A qualitative term sometimes used, especially in the USA, to describe a degree of roasting applied to green coffee, which provides a darkly roasted appearance (i.e. dark brown in colour, with the natural oil of the coffee brought to the surface).
Gametophytic	Expression of genes at the sexual phase (e.g. S-alleles of incompatibility system in the pollen).
Genome	The basic chromosome set of an organism in Eukaryotes, consisting of a species-specific number; hence the sum total of its genes.
Gibberellins	A group of plant growth regulators, which promote cell elongation and have other developmental effects. First discovered in the fungus *Gibberella fujikuroi*.
Globulins	A class of proteins which are not soluble in water at pH 7 but which are soluble in neutral salt solution.
Glutelins	A class of proteins which are insoluble in water at pH 7, insoluble in alcohol, but which are soluble in dilute acid or dilute alkali.
Haematomata	Plural haematoma – localised collection of blood, usually clotted, in an organ, space or tissue, resulting from a break in a blood vessel wall.
Haploid	Organism having a single set of chromosomes.
Hemiptera	An order of insects with piercing mouth-parts and usually two pairs of wings, including capsids and aphids.
Heteroglycan	A polysaccharide consisting of significant quantities of more than one monomeric unit (sugar).
Heterosis	Hybrid vigour such that an F_1 hybrid falls outside the range of the parents with respect to characters like yield, growth, size, etc.
Hippocampus	A forebrain structure of the temporal lobe which is an important part of the limbic system and of Papez's circuit and is involved in the regulation of emotionality, in spatial concept formation and possibly also in short-term memory.
Histamine	An amine which is released from tissues during conditions of stress, inflammation and allergy; it functions to decrease blood pressure, constrict the bronchial smooth muscle of the lungs and to activate gastric secretion.
Homoglycan	A polysaccharide consisting predominantly of one type of monomeric unit (sugar).
Hymenoptera	An order of insects that usually have two pairs of membraneous wings, comprising the bees, wasps, ants, and allied groups.

Glossary 445

Hypertension	High blood pressure.
Hypotension	Low blood pressure.
Ischaemic heart disease	A condition caused by the deficiency of blood supply to the heart due to obstruction or constriction of the coronary arteries.
Isoptera	An order of social insects, commonly called termites.
Isotropic	Single refracting in polarised light.
Italian Roast	As for French roast, but where the appearance of the roasted coffee is even darker (i.e. burnt or black in colour). Both French and Italian roasts are often referred to as 'Continental'.
Karyotype	Particular chromosome set of an individual or a related group of individuals, as defined both by the number and morphology of the chromosomes usually in mitotic metaphase.
L	Leaf area index. The total leaf area divided by the ground area below it.
Lamiidae	A family of beetles with wood-boring larvae.
Latosol	A highly weathered, friable, tropical soil consisting largely of kaolinitic clay with small amounts of smectite, illite or vermiculite-type clay. These soils usually have a red colour; the depth of the colour increases with the amount of iron present.
Lepidoptera	Butterflies and moths. In this order, insects have wings and bodies that are usually covered in small scales.
Lung	One mature orthotropic stem carrying secondaries which will flower the following year; this stem is left on the tree when all other stems are removed during the operation of pruning by stumping.
Maragogipes	Green coffee beans of very large size of a *Coffee arabica* variety, resulting from false botanical development of the two seeds, also known as 'elephant beans'. First noted in some coffees from Bahia, Brazil.
Medial thalamic nuclei system	One of the main groups of nuclei of the thalamus in which information is transformed on the way to the cerebral cortex.
Melolonthidae	A family of Coleoptera. Commonly known as Chafers.
Membrane	Thin layer consisting of phospholipids and proteins, limiting the cytoplasm, nucleus and organelles of a cell.
Meprobamate	Therapeutic used as a minor tranquilliser.
Miridae	A family of the Hemiptera.
Mitochondria	Double-membraned cytoplasmic organelles containing enzymes involved in respiration.
Mocca (Mokka or Mocha)	A type of green coffee exported from the Yemen, (strictly speaking grown only within Arabia), formerly through the port of Mocca, now generally through Aden or Hodeida, has special brew or liquoring characteristics when roasted, with quality connotations. Longberry Mocha, now referred to as Longberry Harer is an Ethiopan coffee,

	though maybe also exported through Aden.
Monophyletic	Individuals derived in the course of evolution from a single interbreeding population.
Monosynaptic reflexes	Stimulus response reaction involving just one sensory and one motor neuron.
Myocardial infarction	Heart attack; it occurs as a result of arteriosclerosis of the coronary arteries; the lack of a sufficient supply of oxygen leads to the death of a segment of heart muscle.
Necrotic	Of dead or dying cells.
Nematode	A non-segmented round worm; plant parasitic nematodes are microscopic in size and usually soil borne.
Neonate	Newly born animal or human.
Net assimilation rate	Rate of increase in plant or crop dry weight per unit of leaf area or weight. Also called Unit Leaf Rate.
Neuronal firing	The release of an action potential, the brief electrical impulse which is the basis of information conduction in a nerve cell; often presented as firing rate.
Nociceptive stimulation	Stimulation of a neuron receptive for pain.
Non-enzymic browning	An extensive and complex series of reactions which classically involve amino compounds and carbonyl compounds, particularly reducing sugars. This complex of reactions typically leads to the formation of a brownish colour and a cooked or roast flavour, their precise nature depending upon the precise nature of the reactants and the precise reaction conditions.
Norepinephrine	A neurotransmitter with powerful vasoconstrictor action.
Orthoptera	The order of insects that includes grasshoppers and crickets. Hind legs are usually modified for jumping.
Orthotropic	Tending to grow vertically upwards (stems), or downwards (roots).
Osmiophilic	Staining readily with osmic acid as, for example, unsaturated lipids in tissues.
Oxisols	Lateritic or ferritic soils, i.e. containing iron and aluminium hydroxides.
Palynology	Study of living or fossil plant spores and pollen.
Parenchyma	A ground tissue composed of relatively undifferentiated living cells which may differ in size, shape, structure and function (nutrition, storage).
Parenteral	Method of drug presentation involving channels outside the alimentary canal, such as subcutaneous or intramuscular injection.
Pathogen	A parasitic organism causing a disease.

Glossary

Peaberries (caracoli)	Small coffee beans (usually of *Coffea arabica*) of nearly ovaloid form resulting from the development of a single seed in the fruit.
Pedology	The scientific study of soils.
Pentagastrin	A synthetic penta-peptide which stimulates secretion of gastric acid and pepsin.
Pentosan	A polysaccharide consisting predominantly of pentose sugar units.
Pericarp	The fruit wall which develops from the ovary wall.
Peripheral vascular disease	Diseases of the arteries, veins and capillaries, such as arteriosclerosis.
Periplasm or perimembraneous space	Space located between the cell wall and the plasmalemma.
Perisperm	A nutrient tissue of the seed, similar to the endosperm, but of nucellar origin.
Phenological	The timing of periodic phenomena in plants, such as budburst or flowering, in relation to seasonal or annual changes in the climate.
Phenylbutazone	Therapeutic used to reduce inflammation.
Phloem	The soft part of the conducting tissue of plants.
Photo-system II	The stage of photosynthesis in which cytochrome f is reduced to produce O_2 from H_2O.
Phylloplane	Leaf surface.
Physiologic sink	Part of a plant such as fruits or growing points to which nutrients and other substances flow.
Pink Disease	A disease of plant stems or branches caused by the fungus *Corticium salmonicola* and characterised by pale pink encrustations or eruptions.
Plagiotropic	Tending to grow horizontally or obliquely, like branches and lateral roots.
Plasmalemma	Double membrane covering the outer surface of the cytoplasm and adjacent to the cell wall.
Plasmodesmata	Cytoplasmic threads penetrating cell wall and forming intercellular bridges.
Polyacrylamide gel electrophoresis (PAGE)	An electrophoretic system for separating proteins by size.
Polycross	An isolated group of plants or clones arranged in some way to facilitate random mating.
Polysynaptic reflexes	Stimulus response reactions involving multiple sensory and motor fibres.
Predispose	To make susceptible to, e.g. a disease.

Glossary

Propagule	Part of a plant such as a spore, sclerotia, or part of the vegetative structure from which new individuals are propagated.
Pseudococcidae	Mealy bugs. A family of the order Hemiptera.
Psychotropic	Capable of modifying mental activity.
Renin	A hormone secreted by the kidneys which breaks down protein and produces a rise in blood pressure.
Retroaldolisation	The degradation of a hydroxy carbonyl compound, such as a sugar, to yield two new carbonyl compounds or two molecules of the same carbonyl compound. The reverse of the aldol condensation q.v.
Rhizomorph	Root-like mass of fungal hyphae which can grow through the soil.
Root stock	A cultivar which is planted as the base of a tree. It provides the root system and the lower section of the main stem.
Sarcoplasmic reticulum	The internal membrane system of a striated muscle cell which controls the level of calcium ions in the cytoplasm.
Scion	A cultivar which is used to provide the frame and canopy of a tree. It therefore carries the flowers and fruit.
Sclerotia	Hard, perennating organs of some fungi, often resembling seeds.
Second messenger	Intermediate metabolic substance which is essential for the initiation of further metabolic changes.
Seed-at-stake	Planting by placing one, or more, seeds directly in the field at the final tree location.
Smectic phase	In liquid crystals, the molecules are parallel and in layers with their long axes usually normal to the plane of the layers.
Spielberger questionnaire	See State-trait anxiety inventory.
Sporophores	Organs of a fungus from which spores are dispersed, e.g. the fruit bodies of mushrooms.
Spodosol	Soil with a strongly bleached eluvial horizon (Podzol).
State-trait anxiety inventory	Testing instrument developed by Spielberger to measure anxiety through self-report; it consists of two scales: Anxiety-State (transitory emotional state) and Anxiety-Trait (frequency of anxiety states over time).
Stereotypy	The repetition of senseless movements or persistence in maintaining a bodily attitude.
Strecker-active amino acids	Those amino acids capable of participating in the Strecker degradation, a reaction between the amino acid and a dicarbonyl compound, which results in the formation of a new odiferous aldehyde by decarboxylation and deamination of the amino acid. The amino group is transferred to the original dicarbonyl compound, e.g.

$$\underset{\underset{NH_2}{|}}{R-CH_2-COOH} + \underset{\underset{O\ \ O}{||\ \ ||}}{R'-C-C-R''} \longrightarrow R-CHO + CO_2 + \underset{\underset{O\ \ NH_2}{||\ \ |}}{R'-C-CH-R''}$$

Synchronous Generations	When most individuals of an animal population are of similar age and at the same stage in the life history.
Systolic blood pressure	The maximum blood pressure which occurs near the end of the stroke output (contraction) of the left ventricle of the heart.
Tachycardia	Excessively rapid heart rate, usually above 100 beats per minute.
Tachypnoea	Excessively rapid respiration.
Teleologically	Referring to adaptive plant characteristics or phenomena as if they were evolved with a preconceived purpose.
Thysanoptera	Thrips. An order of small insects that usually have two pairs of feather-like wings.
Tracheomycosis	Disease caused by growth of the fungus *Gibberella xylarioides* in vascular tissue of the root and stem.
Trypetidae	A family of Diptera, commonly called fruit-flies.
µE	Micro-Einsteins, a quantity of photoenergy.
Unossified sternebrae	Segments of the sternum (longitudinal plate of bone in the wall of the thorax) in early life which have not yet developed into bone.
Urticating	Stinging.
Vacuole	A small, usually spherical space within a cell cytoplasm bounded by a membrane and containing fluid, solid matter, or both.
vpm	Volumes per million.
Water activity	The ratio of the partial pressure of water in a food to the vapour pressure of pure water at the same temperature. The lower the water activity for a given water content the more strongly is the water bound to structural elements in the food. Low water activities inhibit reactions requiring aqueous conditions but may favour oxidative reactions.
Web blight	Spreading necrotic disease of plants in which the affected area is covered with web-like mycelial strands.
Wilting-Point	The amount of water in the soil, at which plants cannot absorb water. If no additional water reaches the soil at this point, the plants will wilt.
Xylem	The hard, lignified part of the conducting tissue of plants.

INDEX

Abyssinicae 16, 42
Acarina 209, 211
Acidity 57, 238, 341, 352-3, 355-6
Acids
   aliphatic 237, 341-2
   humic 345-6
   quinic 328, 331, 340
   see also amino acids, chlorogenic acids
Agobio (Abobiado) 173, 179
Aluminium 146
Amines 311-12
Amino acids
   free 311-12. 347-8, 357
   in protein 312
Anagyrus kivuensis 214
Analysis
   carbohydrates 317-18, 320, 322, 359
   chlorogenic acids 334-5, 358
   foliar 139, 149-50
   lipids 325
   moisture 305-7
   soil 103-5, 139
Antestia 211, 212, 213-14, 215, 219
Antestiopsis *see* Antestia
Anthores leuconotus 210
Anthribididae 212
Ants 210, 216
Aphids 210, 216
Arabica
   biometric studies 56
   Colletotrichum coffeanum
   resistance 67-70
   composition *see* individual components
   cultural characteristics 51, 97-8, 100-1, 104, 164
   Hemileia vastatrix resistance 65-7
   physical characteristics 57, 305-6, 342, 379
   see also Coffee arabica
Arabinogalactan 317, 318, 324
Arabusta 50, 81, 234, 309, 325, 327, 336-7, 349
Araecerus 216
Araecerus fasciculatus 212
Argocoffea 16
Armillaria mellea 227
Ascotis 214, 216
Ascotis selenaria reciprocaria 211, 213
Ash 236, 307-8, 359
Association Scientifique Internationale du Café (ASIC) 11-12
Asterocalanium coffeae 214
Astringency 238, 338, 354, 355, 357

Atractyloside 328-9
Aufschluss polysaccharid 317, 318

Bacillus thuringensis 215-16
Bacterial blight (*Pseudomonas syringae*) 228
Beans
   composition *see* individual components
   defectives 223, 234, 237, 238, 240, 244, 246-8, 267, 272, 338, 379; *see also* peaberries
   pests 211-12, 216, 246
   physical characteristics 57, 305-6, 379
   processing *see* processing
   quality *see* quality, green bean structure 230; *see also* microscopy
   yield *see* yield
Berry
   diseases 219, 226-7
   pests 211-12
   see also cherry, *Colletotrichum coffeanum*
Berry blotch (*Cercospora coffeicola*) 58, 226, 227
Berry borer (*Hypothenemus hampeii*) 211-12, 214
Biennial bearing
   diseases 219-20
   pests 212
   pruning 172
Birds 209
Bitterness 238-9, 309, 328, 353, 355
Black rot (*Rosellinia* spp.) 227
Blending 376-7
Boring beetles 209
Boron 146
Botrytis cinerea 227
Branch borer (*Xyleborus morstatti*) 58
British standards 307, 342, 381, 382
Brown blight 219
Brown eyespot (*Cercospora coffeicola*) 58, 226, 227
Brown scale (*Saissettia coffeae*) 214
Budding 161, 162

Cafamarine 328-9
Café en parche *see* parchment coffee
Cafestol 242, 326-8, 342, 398
Caffeine
   analysis 308-9
   beans 289-92, 302, 308-10, 334, 354, 358
   beverage 309-10, 353, 356, 394-5

cherry processing 239
decaffeination 309, 325, 338, 358, 377-8
germplasm and hybridisation 30, 52, 58, 81-2
see also physiological effects
Calcium 142, 290, 303
Candelabra 73, 177, 179
Capping 173-4, 189
Carbohydrates
  analysis 317-18, 320, 322, 359
  green bean 287, 289, 296, 300, 315-20
  roasted products 320-4, 346
  see also polysaccharides, sugars
Catimor 66, 77, 225
Cation exchange capacity (CEC) 103
Caturra 51, 66, 76-7, 82, 121, 149, 157-8, 168-9
Celluloses 301, 318
Cera 28, 51
Cerambycidae 210
Ceratitis capitata 211
Ceratocystis spp. 221, 228
Cercospora coffeicola 58, 226, 227
Cherry
  pests 211-12
  processing 232-43
Chicory 316, 317, 332, 334
Chlorine 146-7
Chlorogenic acids
  analysis 334-5, 358
  biosynthesis 332
  caffeine complex 302, 308, 354
  cherry processing 238-9
  diCQA 332, 333, 338, 343, 354-5, 356-7
  FQA 332, 333, 338, 348, 356-7
  nomenclature 330-2
  pigments 242, 342-3
  total 289-302 *passim*, 328, 330-41, 356-8
  transformation (degradation) 338-41, 346, 347, 349, 358
  see also physiological effects, scopolin
Chlorosis 120, 220, 221
Chromosome number 26
Cicadidae 210
Climate
  diseases 48, 67
  geographical aspects 34, 59, 104, 106, 110-11
  meteorological aspects 97-102, 106, 171
  see also rainfall
Coccidae (Scales) 210, 211, 214, 216
Cock chafers (Melolonthidae) 210

Coffea arabica
  breeding populations 52-3
  chromosome number & polyploids 26
  Coffea arabica var. *arabica* (*typica*)
  (*Nyasa*) 50, 65, 157; *C.arabica* cv.
  Amfillo 53; *C.arabica* cv. BA series
  53, 65; *C.arabica* cv. *Barbuk Sudan*
  53; *C.arabica* cv. *Blue Mountain*
  53, 69; *C.arabica* cv. *Boma Plateau*
  53; *C.arabica* var. *bourbon* 50, 65, 157; *C.arabica* cv. *Catuai* 286;
  C.arabica cv. *Colombia* 225;
  C.arabica cv. *Dalle* 53; *C.arabica* cv.
  Dilla 53; *C.arabica* cv. *French*
  Mission 50; *C.arabica* cv. *Geisha*
  53; *C.arabica* cv. *Gimma Mbuni* 53;
  C.arabica cv. *Harar* 53; *C.arabica*
  cv. *Iarana* 66; *C.arabica* cv. *Kent*
  and K series 52-3, 65, 69, 157;
  C.arabica cv. *Mundo Novo* 51, 82,
  136; *C.arabica* cv. *Padang* 76-7;
  C.arabica cv. *Ruma Sudan* 53,
  69, 76; *C.arabica* cv. S series 53, 65;
  C.arabica cv. SL series 71-2, 76-7,
  83, 169, 226
  deoxyribonucleic acid 42
  enzyme polymorphism 41, 315
  living collections 24-6, 53
  natural habitats 19-20
  origin 39-41
  pollination 28, 51
  reproductive systems 28
  wild populations 30
  see also arabica
Coffea buxifolia 318
Coffea canephora
  breeding populations 53, 55
  chromosome number 26
  Coffea canephora var. *canephora* 158
  C.canephora Congolese group 55
  C.canephora Guinean group 55
  C.canephora cv. *kouilou* (*quillou*) 55, 66, 158, 195
  C.canephora cv. *nganda* 55, 158
  C.canephora cv. *petit indenié* 55
  C.canephora cv. *robusta* 55, 286
  deoxyribonucleic acid 42
  enzyme polymorphism 41, 315
  interspecific hybridisation 36-8, 82
  living collections 24
  nana population 30-31, 79
  natural habitat 20
  pollination 29
  reproductive systems 29
  wild populations 29
  see also robusta
Coffea spp.
  species, varieties and hybrids *see* Tables 2.3, 2.4, 2.6, 2.7, 2.8, 2.9, 2.10, 3.3, 3.4, 3.5, 3.6, 3.7, 3.8 and Figures 2.5, 2.7, 2.8, 3.7

Coffea congensis 22, 24, 34-43 *passim*, 48, 65, 79, 82
Coffea dewevrei 22, 36, 38, 41-2
Coffea eugenioides 20, 24, 32-42 *passim*, 82
Coffea excelsa 48, 65, 158
Coffea fadenii 20
Coffea humilis 20, 34, 41
Coffea kianjavatensis 38
Coffea liberica
  deoxyribonucleic acid 42
  enzyme polymorphism 41
  interspecific hybridisation 36-8, 48, 82
  multispecific population 34
  natural habitat 20
  reproductive systems 29
  wild populations 29
Coffea pseudozanguebariae 31
Coffea racemosa 32, 42, 65
Coffea resinosa 22
Coffea rhamnifolia 41
Coffea richardii 22
Coffea salvatrix 42, 334
Coffea stenophylla 31, 36, 63
Coffea vianneyi 328
Coffea zanguebariae 31-2
Coffee
  anecdotes 1-8
  consumption 395-6
  contracts 267, 272-4, 275
  dealers 261-3, 267, 272
  development of domestic market 10-11
  origin of word 2
  spread across world 2-3, 48-50
Coffee extenders 316-17, 320, 332, 334, 359
Coffee gene pool
  Colletotrichum coffeanum resistance 69, 224
  evolution of 34-42
  Hemileia vastatrix resistance 224
  see also individual genes
Coffee houses 4-5, 10
Coffee berry disease (CBD) *see Colletotrichum coffeanum*
Coffee leaf rust *see Hemileia vastatrix*
Coffee selection and breeding
  breeding populations 50-6
  caffeine 58
  disease resistance 58, 224
  drought resistance 59
  mechanical harvesting 59
  pest resistance 58, 214
  quality 59, 224
  yield 58-9, 224
Coffee wilt (*Ceratocystis fimbriata*) 221, 228

Colchicine 26, 81-2
Coleoptera 209
Collecting missions 23-4
Colletotrichum coffeanum
  control 67-70, 226-7
  leaves 226
  progeny testing 63-4. 68-9
  resistance 53, 58, 67-70, 195
Congusta 29, 79
Consumption 395-6
Contracts 267, 272-4, 275
Copper
  disease resistance 307
  fungicides 7, 145, 222, 225-8
  nutrient 145
Corticium spp. 220, 221, 228
Cova 169, 173
Cover crops 192, 198-9, 201
Cultural practice
  choice of cultivar 157-8
  cover crops 198-9
  disease 221-2
  field planting 170-2
  fruit abscission 200-1
  harvesting 203-4
  intercropping 166-7, 199-200
  land clearance 164
  mulch 197-8
  nurse crops 166-7, 199, 201
  pruning 172-94
  shade 164-7, 171-2, 198-9
  weed control 201-3
  windbreaks 166
  see also propagation
Curing *see* processing, hulling
Cuttings 161-2
Cytotaxonomy 26

Dealer trading 261-3, 267, 272
Decaffeination 309, 325, 338, 358, 377-8
Decorticage *see* hulling
Deoxyribonucleic acid (DNA) 42, 421
Dewaxing 309, 325-6, 327, 338, 358, 378
Die back 118, 127, 137, 141, 165, 172, 188
  diseases 219-20, 227-8
Diptera 209
Dirphya nigricornis 210
Disease
  control 187-8, 221-5, 225-8 *passim*, 334
  yield 55, 67-8
  see also individual diseases
Diterpene glycosides 328-9
Drainage 102, 164

Embryo 286, 300-1
Enzymes 292, 314-15
  pectin degrading 236-7

polymorphism 41
polyphenol oxidase 312-15, 338, 343-3, 358
Epidermis 297-300
Erecta 51, 52
Erythrocoffea 16, 34-7
Ethylene 71, 121
Eucoffea 16
Eucosoma spp. 211, 216
European Economic Community Standards 307, 309
Evaporation, open-pan $-E_0$ 97, 195-6
  see also evapotranspiration
Evapotranspiration $-E_t$ 97-8, 101, 102, 196
  see also evaporation, open-pan
Exports 258-61, 264, 266-7, 268-9, 278, 280
Extraction 307, 383-6

Fertilisers *see* nutrients
Field capacity 195
Floral biology 14-15, 59-60
Flower bud 122-6, 211
Flowers 15, 28-9, 212
  rainfall 34, 59, 97, 98-100, 108, 110
Fruit abscission 200-1, 219
Fruit growth 125-7
Fumigants 216
Fungicides
  application 222-3; operating hazards 223; residues 223
  Colletotrichum coffeanum 67-8, 70, 72
  copper 145, 222, 225, 226, 228
  Hemileia vastatrix 72
  pesticides 213, 221-3, 225
  phytotonic effects 70-1, 120-1, 145, 163
  rainfall 223
  resistance 222, 226-7
  yield 72-3
Fusarium spp. 58, 221, 227-8
Feve puante *see* Stinkers

Gascardia brevicauda 214
Germination 159-60
Giant looper (Ascotis) 211, 213
Gibberella xylarioides 55, 58
Grading 57, 76-9, 83, 243-8
Grafting 161, 162
Grey rust (*Hemileia coffeicola*) 58, 225-6
Grinding 381-3

Harvesting 59, 203-4, 231-2
Hemicellulose 301-2, 317-18
Hemileia coffeicola 58, 225-6
Hemileia vastatrix 72, 225-6
  climate 48
  resistance 52-3, 55, 58, 195

resistance genes 65-7
Hemiptera 210-11
Herbaria 18-19
  see also living collections
Herbicides 163, 202-3
  see also weed control
Heteroptera 210
Hibrido de timor 53, 65, 77
Historical anecdotes 1-8
Holocellulose 317, 319-20
Honey fungus (*Armillaria mellea*) 227
Hormones 116, 161, 200-1
Hot and cold disease 101-2, 221
Hulling 233-4, 242-3
Hybridisation 26, 60-2, 81-2
Hybrids 34-40, 65-7, 79-83, 84-9, 224, 336-7
  propagation 158
  vigour 75, 79, 81
  see also individual hybrids
Hymenoptera 209
Hypothenemus hampeii 58, 211-12, 214

Icatu 66-7
Imports 258, 263, 265-7, 268-9
Insecticides *see* pesticides
Insolation 118, 194-5
  see also chlorosis
Intercropping 166-7, 199-200
International Coffee Agreement 251-8
International standards (ISO) 244, 307, 342
Iron 145
Irrigation 194-7
  see also rainfall
Isoptera 209

Jasminum arabicanum 14

Kahweol 242, 326-7, 342, 356, 358, 398
Koleroga (*Corticium* spp.) 220, 221, 228

Lace bugs (*Tingidae*) 211
Lamiidae (*Anthores leuconotus*) 212
Laurina 51-2
Leaf area index 120, 121
Leaf area ratio 111
Leaf miner (Leucoptera) 58, 211, 212
Leaf skeletoniser (*Leucophema doheryti*) 214
Leaves
  diseases 220, 225-6
  pests 209, 211, 212, 214, 215
  see also chlorosis, fungicides, *Hemileia vastatrix* phytonic effect
Lepidoptera 209, 211
Leucophema doheryti 214
Leucoptera, 58, 211, 213-16
Liberica 145, 158, 173, 281

Libericae 16
Lignin 289, 294
Limacodidae 211
Lipids
   analysis 325
   crude (total) 287-8, 325-6
   free fatty acids 287
   triglycerides 287, 288, 290, 297, 300, 302, 326, 357
   unsaponifiables: diterpene glycosides 328-9; diterpenes 326-8, 398; hydrocarbons 326, 328; hydroxytryptamides (C-5-HT, serotonin) 297, 298, 301, 325-6, 342, 358; phospholipids 326; sterols 328, 358; tocopherols 328; waxes 297-8, 299, 300, 301, 325-6
Living collections 24-6, 53
   see also herbaria
Lung 179, 186, 191, 192
Lyonetiidae (Leucoptera) 211

Magnesium 142-3
Mammals 209
Mannan 317, 319-20, 324, 357
Manganese 145-6
Manures 147-8, 160
Maragogipe 51, 157, 246, 282
Mascarocoffea
   chromosome number 26
   collecting missions 24
   interspecific hybridisation 36, 82
   living collections 24-6
   pollination 29
   reproductive systems 29
   taxonomy 16-18, 328, 334, 336
   wild populations 32
Mascaroside 328-9
Mbuni 204
Mealy bugs (Pseudococcidae) 208, 210-11, 214, 216
Mediterranean fruit fly (*Ceratitis capitata*) 211
Melanocoffea 16, 37
Meloidogyne spp. 210
Melolonthidae 210
Microscopy
   light 284-303 *passim*, 323, 359
   electron 285-303 *passim*
Milling *see* hulling
Millipedes 209
Minerals *see individual minerals*, nutrients
Miridae 213, 214
Mites (Acarina) 209, 211
Moisture content
   analysis 305-7, 358
   cherry 233, 234, 240
   coffee storage 248

parchment coffee 240
   trading 273
   see also soil
Mocca (mokka) 51-2, 157
Molybdenum 146
Mozambicoffea 16, 37, 42
Mucilage 300, 302
Mulches 147-8, 160, 172, 197-8, 200, 201-2, 214, 221
Multiples verticales 173
Mycena citricola 226
Myriapods 209

Nanocoffea 16, 37
Natural habitats 19-22, 27, 100-1, 128-9
Nematodes 58, 161, 163, 209, 210, 216, 227, 228
Nematospora spp. 211, 219
Net assimiliation rate — E_a 120, 122, 123, 127
Niacin 310
Nitrogen
   fixation 166-7, 198, 199
   non-protein 308-12
   nutrition 116, 128, 139-40, 163, 199
   total (Kjeldahl) 312-13
Nurse crops 166-7, 199, 201
Nurseries
   diseases 228
   maintenance 162-3
   pests 209, 216, 228
   rotation 163
   seedling 160-1
Nutrients
   availability 136-7, 140-6 *passim*; *see also* soil, pH value
   cover crops 198
   crop quality 148-9, 307, 334
   deficiency symptoms 116, 141-6 *passim*, 151-2
   foliar analysis 149
   interactions 140-6 *passim*
   intercropping 199
   requirements 127, 135-9, 140-6 *passim*
   toxicity 140, 141, 144-7 *passim*, 151-2
   see also individual nutrients

Orthoptera 209
Overbearing 116-18, 127, 128, 179, 185, 212, 220

Pachycoffea 16, 17, 24, 37-8, 41
Packaging 248, 251, 389-91
Paracoffea 16, 18
Paracoffea ebracteolata 42
Parchment coffee 234, 239-43
Parenchyma 286-97, 301-2

Parras 173
Peaberries 76, 82, 127, 243, 244, 376
Pergamino *see* parchment coffee
Pectic substances 236, 293, 294, 296, 317, 318-19, 324
Pest control
  biological control 214
  cultural aspects 214
  integrated pest management 209, 213, 216-17
  see also pesticides
Pesticides 160, 163, 172, 315
  application 213, 214-16; operating hazards 215-16, 228
  rainfall 216
  residues 217
  resistance 217
  see also pest control
Pests
  bean 212, 216, 246
  biennial bearing 212
  cherry 211
  crop losses 208-9, 212
  infestation assessment 212-13
  leaves 212
  nurseries 209, 216, 228
  roots 210, 216
  seedlings 209, 216
  shoots 209
  woody tissue 210
  see also individual pests
Phoma spp. 226, 228
Photosynthesis 110, 111, 118-23, 220
Physiological effects (consumer)
  caffeine: absorption 396; elimination 397; metabolism 396-7
  cardiovascular system 405-6
  cellular action 399-401
  central nervous system 406-8
  chlorogenic acids 398
  diterpenes 398
  intersubject variations 418-20
  learning performance 413-14
  mood 417-18
  motor performance 409-11
  psychomotor performance 411-13
  sleep 414-17
  spontaneous activity 408-9
  toxicity 420-4; acute 420; behavioural disorders 424; behavioural teratology 423-4; fibrocystic breast disease 422; gastro-intestinal disorders 422; methodological shortcomings 421-2; mutagenicity 421-2; myocardial infarction 420-1; teratogenic effects 422-3
  vegetative effects: diuresis 403-4; endocrine secretions 403; exocrine secretions 402; metabolic rate 404; smooth muscle 401-2
Physiology (plant)
  adaptation to natural habitat 128-9
  bean yield 111-18
  flower bud development 123-4
  fruits growth 125-7
  net assimilation rate (E_a) 120, 122, 123, 127
  net carbon fixation 118-22
  respiratory rate 127
  seasonal growth cycle 108-12
Pigments
  allagochrome 342
  carotenoids 342
  chlorophyll 242, 342
  flavonoid, taxonomy 42, 342
  humic acids 345-6
  roasted products 344-6
Pink disease (*Corticum*) 220, 221, 228
Planococcus citri 210
Planococcus kenyae 208, 214
Planting density 51, 56, 75, 76, 114-15, 121, 167-70, 196-7, 221
  see also shade
Planting out 160-1
Pollen 17-18, 41, 59-63
Pollen mother cells (PMC) 36-7
Pollination 28-9, 50-1, 100, 116, 117
Polysaccharides *see individual polysaccharides*
  component sugars 317-19, 324
Potassium 128, 141-2, 303, 307, 359
Pratylencus coffeae 210
Prices 253-5, 275, 278-9
Processing
  blending 376-7
  cleaning 377
  decaffeination 377-8
  dewaxing 378
  domestic brewing 382
  dry 232-4
  extract drying 386-9; agglomeration 388; aromatisation 388; freeze drying 388-9; preconcentration 387; spray drying 388
  extraction 383-6
  grading and sorting 243-8; air classification 246; colour sorting 247; manual sorting 246-7; size grading 243-6; UV excitation 247-8
  grinding 381-3; granulising 382; normalising 382
  packaging 248; roast coffee 389-91; soluble powder 391
  polishing 248
  roasting 378-81
  storage and transportation 248

wet 243-243; berry disease 219; chlorogenic acids 334; classification 235; drying 239-42; fermentation 235-9; hulling 233-4, 242-3; pulping 235
Production 253-8, 260-1, 264, 266-7, 268-71
Propagation
seed 63-4, 84-9, 158-61
vegetative 64-5, 84-9, 158, 161-2
Protein
amino acid composition 312
crude 236, 287, 292-5 *passim*, 299, 302, 312-13, 358
electrophoresis 312, 314
humic acids 345-6
see also enzymes
Pruning
agobio (agobiado) 173, 179
biennial bearing 172
candelabra 173, 177, 179
choice of system 188-9, 191
die back 118, 165, 172
disease 172, 187, 221
flower formation 113-14
irrigation 194-7
leaf/fruit ratio 122, 128
multiple stem systems 174, 175, 177, 179-94
multiples verticales 173
nutrient loss 135-6
overbearing 128
parras 173
pest control 214
planting density 173
rejuvenation 192-3
root pruning 163
shade 192
single stem systems 173-4, 175, 176, 177, 178
temperature 172, 173, 177
yield 167-8, 169-70, 172, 173, 191
Pseudococcid (Planococcus) 208, 210-11, 214, 216
Pseudomonas syringae 228
Psilanthopsis kapakata 36, 42
Psilanthus spp. 14-15, 18, 26, 28, 42
Pulp composition 235-7
Purpurescens 28, 51

Quality
assurance 355-9
breeding 57
cherry processing 232
cup quality 57, 74-9, 235, 237, 238, 239; *see also* sensory aspects
definitions 57, 355
green bean 57, 74-9, 148, 240-1
heritability 74-9
Quenching 380, 382

Radolphus similis 210
Rainfall
disease and pest control 216, 223-4
flowering and pollination 34, 59, 97, 98-100, 108, 110
fruit growth 126
natural habitat 20
statistics 98-9
water requirement 51, 98, 168
yield 98
Red blister disease (*Cercospora coffeicola*) 58, 226, 227
Reserve mannan 317, 319-20, 324, 357
Reproductive systems 26, 28-9, 59-64
see also propagation
Respiratory rate 127
Roasting 378-81
Robusta
Colletotrichum coffeanum resistance 67-70
composition *see individual components*
cultural characteristics 98, 101, 104
Hemileia vastatrix resistance 55, 65-7
physical characteristics 57, 305-6, 342, 379
Tracheomycosis 55
see also Coffea canephora
Robustae 16, 42
Root systems
absorptive ability 103; *see also* nutrients, availability
diseases 220, 221, 227
growth 100, 102, 171, 179, 189, 195, 197
pests 210, 216
pruning and transplanting 163, 171
Rosellinia spp. 227
Rotylenchus spp. 210

Saissetia coffeae 214
San ramon 51-2, 157
Sao bernardo 51-2, 157
Scale insects (Coccidae) 210, 211, 214, 216
Scopolin 338
Seed-at-stake 170
Seedlings 160-1, 209, 216
Sensory aspects
aroma 346-50, 351, 355, 357
aromatisation 326, 388
body 57, 354-5
colour 344-5
staling 350-2, 355, 359, 390-1
taste 309, 328, 352-5, 356; *see also individual taste sensations*
physiological effects 398-9
Serological affinities 42
Serotonin 297, 298, 301, 325-6, 342, 358
Shade
cultural practice 164-6, 171-2, 198-9

disease 221
nutrition 137
pest control 214
pruning 192
temporary 166-7
see also planting density
Shoot miners 209
Shoots
growth 127-8; *see also* dieback
pests 209, 210
structure 113-14
see also pruning
Shot hole borer (*Xylotrechus quadripes*) 210
Silverskins 234, 243, 248
Slugs 209
Snails 209
Sodium 143-4
Soil
chemical characteristics: analysis 103-5, 139; cation exchange capacity (CEC) 103; pH value 103-5, 139-40, 142, 145, 164
fertility 103
physical characteristics: drainage 102, 164; water holding capacity 100; water logging 100, 159, 164
Soluble powder
development of 10-11
exports 264, 278
imports 265
packaging 391
physical properties 306
production: extract drying 386-9; extraction 383-6; grinding 382-3
see also individual components
Sophronica ventralis 212
Sourness 352-3
South American leaf spot (*Mycena citricola*) 226
Staling 350-2, 355, 359, 390-1
Starch 292, 293, 302, 319
Star scale (*Asterocalanium coffeae*) 214
Stephanoderes coffeae see Hypothenemus hampeii
Stephanoderes hampeii see Hypothenemus hampeii
Stinkers 238, 247
Stomatal index 122
Storage
green bean 248, 274; biochemical changes 306, 311, 315, 326, 338, 357; moulds 338; pests 208-9, 212, 216
pollen 60-3
roasted products 306, 350-2
seed 63-4
Stumping 171, 174, 175, 179, 185-93
Sugars
cherry processing 236, 239
chicory 316, 317
green bean 315-17, 357
roasted products 321-3, 348
Sulphur 144

Taxonomy 14-18, 41-2
Terminal markets 275-7, 280-2
Thrips (Thysanoptera) 209, 214
Tingidae 211
Tip-boring caterpillar (Eucosoma) 211, 216
Toxoptera aurantii 211
Tracheomycosis (*Gibberella xylarioides*) 55, 58
Transport 248, 274
Triaging 246
Trigonelline 238, 310, 347, 353, 358
Trylenchus coffeae 161

Virus diseases 226
Volatiles
green beans 338, 343, 358
roasted products 310, 312, 340-1, 346-52, 355, 358-9

Warty berry (*Botrytis cinerea*) 227
Water stress 70, 97, 194-5, 221
Web blight (*Corticium*) 220, 221, 228
Weed control 160, 167, 192, 198-9, 201-3
see also herbicides
White borer (*Anthores leuconotus*) 210
White waxy scale (*Gascardia*) 214
Wild populations 29-34
Wilting 102, 195, 196, 220, 227-8
Wind 102
Windbreaks 102, 165, 166, 171
Woody tissue 210, 216, 220, 228

Xyleborus morstatti 58
Xylotrechus quadripes 210

Yellow-headed borer (*Dirphya nigricornis*) 210
Yield
climatic aspects 98
cultural practice: planting density 56, 114-15, 168-70; pruning 167-8, 169-70, 172, 173, 191
determinants of 113-18
disease 55, 67-8
fungicides 72-3
genetic factors 56, 72, 74-9
mathematical models 57, 114
physiological aspects 113-17
solubles extraction 384

Zinc 144-5

Printed in Poland
by Amazon Fulfillment
Poland Sp. z o.o., Wrocław